Water Resources Research in Northwest China

Yaning Chen
Editor

Water Resources Research in Northwest China

Springer

Editor
Yaning Chen
Xinjiang Institute of Ecology
and Geography
Chinese Academy of Sciences
Xinjiang
People's Republic of China

ISBN 978-94-017-8016-2 ISBN 978-94-017-8017-9 (eBook)
DOI 10.1007/978-94-017-8017-9
Springer Dordrecht Heidelberg New York London

Library of Congress Control Number: 2014930889

Printed on acid-free paper

Springer is part of Springer Science+Business Media (www.springer.com)

Preface

With the increasing concern of global environmental and ecological degradation, there has been an urgent need to investigate the related water cycle changes. Designed for both academic and business sectors, this book examines the major issues in water resources research in Northwest China, approximately one fourth of the nation's entire land area and one of the world's largest arid regions. The arid region of Northwest China is characterized by its extremely vulnerable water resources and associated ecological environment. The large alpine snow and ice cover has contributed to the development of numerous inland streams, forming a unique landscape characterized by mountain-oasis-desert ecosystems. Water, largely originated from the mountain areas, has been a most critical factor to drive the energy and mass circulation in this region, which responds sensitively to the global climate change. This book focuses on some possible impacts of climate change on hydrology and water resources in the arid region of Northwest China. The contributing authors for this book include Yaning Chen, Weihong Li, Zongxue Xu, Zhongqin Li, Jianhua Xu, Xianwei Wang, Yanjun Shen, Zhi Li and Huaijun Wang, all of whom are active researchers in water resources research in arid and cold environments.

This book comprises 11 chapters discussing various aspects in water resources research. Specifically, the book begins with an introductory chapter (Chap. 1) discussing the physical geography and socioeconomic aspects in Northwest China. Chaps. 2-7 discuss the climate system and hydrologic system changes, some implications of these changes in relation to potential evapotranspiration, the hydrological cycle, and the spatiotemporal variations of the snow cover and glaciers through remote sensing, geographic information systems, and statistical analysis. Chaps. 8 and 9 focus on the model description and experimental design to interpret the hydro-climatic process, emphasizing the integration of water, climate, and land ecosystems through field observations and computer-based simulations. Chaps. 10 examines some extreme hydrological events and presents a study using the historical trend method to investigate the spatial and temporal variability of changing temperature and precipitation extremes in the hyper-arid region of Northwest China. And the Chap. 11 of this book discusses some possible strategies for sustainable watershed management. We believe that the lessons from this study area can be useful for other arid areas in the world.

The research reported in this book has been supported by the National Basic Research Program of China (973 Program: 2010CB951003). The Editor is grateful to all of those who contributed papers and revised their papers one or more times and those who reviewed papers according to our requests and timelines. The book project would not have been completed without the help and assistance from several staff members at Springer, especially Margaret Deignan and Takeesha Moerland-Torpey. Acknowledges are due to Dr. Xiaojun Yang for his help in many ways.

Urumqi, Xingjiang, China
15 July 2013

Yaning Chen

Contents

Contributors

Yaning Chen State Key Laboratory of Desert and Oasis Ecology, Xinjiang Institute of Ecology and Geography, Chinese Academy of Sciences, Urumqi, Xinjiang, China

Zhongsheng Chen State Key Laboratory of Desert and Oasis Ecology, Xinjiang Institute of Ecology and Geography, Chinese Academy of Sciences, Urumqi, Xinjiang, China

Yuting Fan State Key Laboratory of Desert and Oasis Ecology, Xinjiang Institute of Ecology and Geography, Chinese Academy of Sciences, Urumqi, Xinjiang, China

Baofu Li State Key Laboratory of Desert and Oasis Ecology, Xinjiang Institute of Ecology and Geography, Chinese Academy of Sciences, Urumqi, Xinjiang, China

Qihu Li College of Geomatics, Xi'an University of Science and Technology, Xi'an, China

Weihong Li State Key Laboratory of Desert and Oasis Ecology, Xinjiang Institute of Ecology and Geography, Chinese Academy of Sciences, Urumqi, Xinjiang, China

Zhi Li State Key Laboratory of Desert and Oasis Ecology, Xinjiang Institute of Ecology and Geography, Chinese Academy of Sciences, Urumqi, China

Zhongqin Li State Key Laboratory of Cryospheric Sciences/Tianshan Glaciological Station, Cold and Arid Regions Environmental and Engineering Research Institute, Chinese Academy of Sciences, Lanzhou, China

Tiangang Liang College of Pastoral Agriculture Science and Technology, Lanzhou University, Lanzhou, China

Zhaofei Liu Institute of Geographic Sciences and Natural Resources Research, Chinese Academy of Sciences, Beijing, China

Key Laboratory of Water and Sediment Sciences, Ministry of Education, College of Water Sciences, Beijing Normal University, Beijing, China

Yanjun Shen Key Laboratory of Agricultural Water Resources, Center for Agricultural Resources Research, Chinese Academy of Sciences, Shijiazhuang, China

Guili Sun College of Forestry and Horticulture, Xinjiang Agricultural University, Urumqi, China

Meiping Sun College of Geography and Environment Sciences, Northwest Normal University, Lanzhou, China

Huaijun Wang State Key Laboratory of Desert and Oasis Ecology, Xinjiang Institute of Ecology and Geography, Chinese Academy of Sciences, Urumqi, China

Puyu Wang State Key Laboratory of Cryospheric Sciences/Tianshan Glaciological Station, Cold and Arid Regions Environmental and Engineering Research Institute, Chinese Academy of Sciences, Lanzhou, China

Xianwei Wang Center of Integrated Geographic Information Analysis, School of Geography and Planning, and Guangdong Key Laboratory for Urbanization and Geo-simulation, Sun Yat-sen University, Guangzhou, China

Hongjie Xie Laboratory for Remote Sensing and Geoinformatics, Department of Geological Science, University of Texas at San Antonio, San Antonio, TX, USA

Changchun Xu Key Laboratory of Oasis Ecology, School of Resources and Environmental Science, Xinjiang University, Urumqi, China

Jianhua Xu The Research Center for East-West Cooperation in China, The Key Lab of GIScience of the Education Ministry PRC, East China Normal University, Shanghai, China

Zongxue Xu Key Laboratory of Water and Sediment Sciences, Ministry of Education, College of Water Sciences, Beijing Normal University, Beijing, China

List of Figures

List of Tables

Chapter 1
Exordium

Yaning Chen

Abstract The arid region of Northwest China is characterized by its extremely vulnerable water resources and associated ecological environment. Water is a critical factor to drive the energy and mass circulation in this region, which responds sensitively to the global climate change. The contradictions between ecology and industrial, agricultural production are very conspicuous in arid region. This chapter discusses the physical geography, climatic characteristics and socioeconomic aspects in Northwest China, the main aspects are as follows: (1) By a waving geomorphic land surface with inter-spacing mountains and depressed basins, the typical landscape comprised vertically with mountain—oasis—desert ecosystems. (2) Situated far from the sea, the relatively secluded Northwest region has a typical continental climate marked by scarce precipitation, high evaporation, wide temperature fluctuations and strong winds. (3) The multicultural population is mainly distributed in the oases, where oasis agriculture is the mainstay. The ecosystems are extremely fragile due mainly to the internal factors of ecological processes and the interference of human activities.

Keywords Northwest China · Landform features · Socio-economic characteristics · Population composition · Ecological degradation

1.1 Natural Geography

The arid region of Northwest China is located in the hinterland of the Eurasian continent, spanning an area from 35° to 50° N and 73° to 106° E. It lies west of the Helan and Wushao Mountains, north of the Kunlun Mountains, and includes the Tarim River Basin, the Qaidam Basin, the Badan Jaran Desert and the Tengger Desert. Geopolitically, the area incorporates the Xinjiang Uyghur Autonomous

Y. Chen (✉)
State Key Laboratory of Desert and Oasis Ecology,
Xinjiang Institute of Ecology and Geography,
Chinese Academy of Sciences, No. 818 South Beijing Road,
Urumqi 830011 Xinjiang, China
e-mail: chenyn@ms.xjb.ac.cn

Y. Chen (ed.), *Water Resources Research in Northwest China*,
DOI 10.1007/978-94-017-8017-9_1,

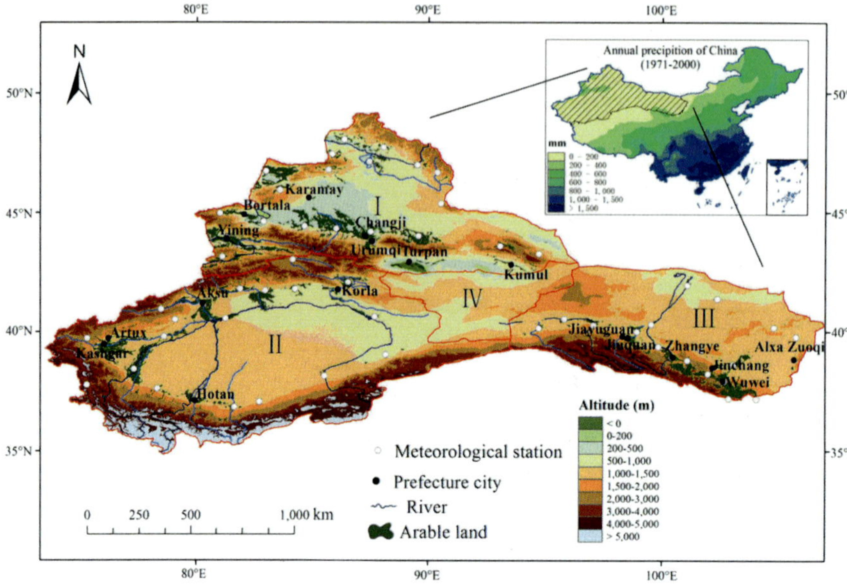

Fig. 1.1 The geomorphologic setting and zoning of the arid region in Northwestern China

Region (Xinjiang for short), the Hexi Corridor in Gansu Province, and the western portion of the Helan Mountains in Inner Mongolia. The land mass comprises approximately 2.5 million km^2 and accounts for one quarter of the total land area of China (Fig. 1.1). It is a major component of the arid region of Central Asia and also the typical region of desert in the world.

The Northwest arid area is bounded to the south by the Kunlun Mountains and and Qilian Mountains, north by the Altai Mountains (Fig. 1.1). These ranges, together with the Tibetan Plateau, effectively block the Indian Ocean moisture from reaching the Northwest. Likewise, the Altai Mountain to the north prevents moisture from the Arctic Ocean from reaching the area.

Xinjiang is divided by the Tianshan Mountains into two parts: north and south. Conventionally, the area south of the Tianshan Mountains is called South Xinjiang and the area to the north is called North Xinjiang. To the east, the Helan Mountains block the moisture from the eastern monsoons. The area north of Qilian Mountain and west of the Helan Mountains are known as the Hexi Corridor.

Northwest China arid zone has diverse and extreme terrain. Its typical topography and geomorphology patterns include towering glaciers, widespread deserts (e.g., the Gobi), vast grasslands, sporadic oases, and numerous mountains and basins. The Tarim River Basin, which is China's largest, spans an area of approximately 530,000 m^2. Additionally, China's largest (and the world's second-largest) mobile desert, the Taklimakan Desert, is in the central portion of the basin. The Turpan Basin, in eastern Xinjiang, has an elevation of − 145 m, making it the lowest natural point in the country. The Tarim River and the Heihe River are, respectively, China's longest and second longest inland rivers.

As well as being diverse and made up of extremes, the Northwest arid area is a secluded inland region featuring a continental climate. Being very dry, the region tends to evaporate all precipitation it does receive, the average annual precipitation being only 130 mm (Chen et al. 2009). Because precipitation in the plains area does not produce surface runoff, this is one of the most severely arid areas in the world. The ecological environment is fragile and unstable, and species are extremely poor. Moreover, as the area is situated far from the sea, moisture is unable to reach it (Li 2012). Instead, precipitation mainly occurs in the Yili River Valley (an area strongly influenced by westerlies), north of the Altai Mountains (which is influenced by Arctic Ocean water vapor), or at the edge of the Hexi Corridor region. The highest precipitation levels (up to 800 mm) occur in the eastern mountains of the Yili Valley (Park et al. 2010; Wu et al. 2010; Ye et al. 2000; Yuan et al. 2003; Zhong et al. 2007, 2010).

In light of the area's large geographical span and unique landscape pattern, precipitation and water resources are unevenly distributed (Chen et al. 2009). Some oases with agricultural production activities have annual precipitation of less than 50 mm, with potential evaporation of up to 2,500 or 3,000 mm, which is 50–60 times the average precipitation. Clearly, the arid region of Northwest China can easily be described as one of the world's most arid regions.

The snow and glaciers coating the Northwest mountains give birth to more than 600 rivers. Other than for the Irtysh River in northern Xinjiang (which is part of the Arctic Ocean water system), all are inland rivers. Due to scant river runoff and closed terrain conditions, no tributaries pour into the main stream and the waters flow mainly through deserts until finally evaporating. The major rivers in this area are the Tarim River, the Yili River, the Irtysh River, the Heihe River, and the Shiyang River.

The Yili River originates in the Tianshan Mountains and is fed by high levels of precipitation due to the effect of the prevailing westerlies. The annual runoff of the Yili is 17 billion m^3, making it the largest in the region (Ye et al. 1996). The Tarim River Basin is composed of the Aksu River, the Hotan River, the Yarkand River and the Kaidu-Kongque River. Measuring 2,179 km in length, the Tarim River is China's longest inland river, and flows from west to east through the northern basin. All of these rivers, along with numerous streams, are products of glacier snowmelt water. They wind through the arid region's oases and serve as the main source of irrigation water for local farmland.

The Northwest features a large diurnal temperature range with adequate heat and light sources, making it suitable for thermophilic plants and crops. However, because of the large geographic span from north to south and the uneven spatial distribution of heat resources, there are significant differences in the cropping systems and crop varieties. In southern Xinjiang, and where heat resources permit, cotton is grown and there are both summer and winter wheat crops. In contrast, the cooler climes of northern Xinjiang and the Hexi Corridor yield only one annual growing season, the main crops being wheat and spring corn. The uneven spatial distribution of heat resources directly impacts inter-regional differences in crop growth periods.

Currently, Northwest China is the main gathering point for ethnic minorities. Despite harboring a vast and rich supply of minerals, oil, gas and other natural resources, the area remains underdeveloped but is slated to become an important resource replacement for China in the twenty-first century. Indeed, with the focus of national economic construction moving ever westward, the arid zone has become the fulcrum for China's economic growth. However, the resources presently in place to support the development of the ecological environment are inadequate, and the conflict of water utilization is very conspicuous. How to achieve sustainable use of resources in the Northwest, along with ecologically sustainable management and sustainable economic and social development, has become the focus of attention of all levels of government as well as the affected communities.

1.1.1 Landform Features

As outlined above, the arid area of Northwest China is a vast territory of complex topography that includes mountains, plateaus, deserts and basins, with mountain runoff supplying most of the water resources. From a macroeconomic viewpoint, the region is mainly composed of large basins (the Tarim, Junggar and Turpan-Hami Basins and the Hexi Corridor) and continuous mountain ranges (the Tianshan, Karakoram, Kunlun, Qilian and Altai Mountains).

1.1.1.1 Terrain

Generally speaking, the terrain in the Northwest arid zone is high in the west and low in the east. In the mountainous western area, which features the Qilian, Kunlun, Tianshan and Altai Mountains, some peaks reach as high as 7,000 m. The foot of the Qilian Mountain is connected to the Alashan Plateau and the Hexi Corridor, the latter which forms part of the depression zone at the edge of the Qilian Mountains and is about 1,000 km long. The piedmont sloping plains, composed of numerous alluvial fans of the Qilian Mountains. These plains, situated among the Heli Hill and the Longshou and Qilian Mountains, are between 1,000 and 1,500 m above sea level.

Being located west of the Yellow River, the Hexi Corridor terrain tilts from southeast to northwest and is an important road leading to the Western Regions from eastern China. Historically, the famous ancient "Silk Road" passed through there and it is still the main road connecting eastern China with Xinjiang. Although the annual precipitation is less than 200 mm, the abundant snowmelt from the Qilian Mountains enables irrigation, thereby creating the mainstay agricultural base of the Northwest territories (Zhao et al. 1995).

The Alashan region is high and flat. It belongs to the temperate dry desert region, which is 1,000–1,500 m above sea level. In winter, this area is strongly influenced by high pressure centered on Mongolia, and so the climate is more arid than in the Junggar Basin. However, in summer, the southeast is impacted by Pacific monsoons, bringing sudden and thunderous showers. Annual precipitation in the region hovers around 200 mm, while the lower reaches of the Heihe River receive only about 50 mm. The main desert vegetation is made up of extremely sparse shrubs and semi-shrub desert (Zhao et al. 1995).

1.1.1.2 Mountains

The Northwest has an abundance of mountains, including the Kunlun Mountains, the Tianshan Mountains, the Altai Mountains, the Altun Mountains, the Qilian Mountain and Helan Mountain. The Karakorum and Kunlun Mountains form part of the "roof of the world", with multiple peaks higher than 6,000 m. Chogori, at 8,611 m, is ranked as the second highest peak in the world.

The Tianshan Mountains are the largest mountain range in Central Asia, spanning a total length of 2,500 km from Kyrgyzstan in the west to Hami in Xinjiang in the east. After forming as a Mesozoic and Cenozoic denudation plain, the Tianshan Mountains emerged as the late Paleozoic fold from a strong fault block uplift in the Himalayan movement, followed by a more recent period of tectonic movement. The Xinjiang portion of the Tianshan Mountains is about 1,700 km long, with a width of 250–300 km and an average elevation of 5,000 m. This portion of the Tianshan Mountains includes the highest peaks in the range—Tuomuer (7,435 m) and Khan Tengri (6,995 m)—and is composed of a complex series of mountains and basins. According to Yamagata, the Tianshan Mountains divide into the North Tianshan Mountains, South Tianshan mountains and valleys. The Barkol basin, the Yiwu Basins, the Yule Meadows Basin, the Yanqi Basin, the Turpan-Hami Basin, and the transit Shanxi end of the famous Yili Valley are all located within the confines of this mountain range.

As well as spanning countries and regions, the Tianshan Mountains range across numerous biological and climatic zones. In the north is the desert climate; in the south, is warmer desert climate while to the west, the Yili Valley belongs to the temperate desert steppe. Due to the different zones of hydrothermal conditions reflected in the vertical belt system, different aspects of soil and vegetation vertical structure are likewise significantly different.

The Altai Mountains, which form the northern boundary of Xinjiang, share a 200-kilometer-long border with Mongolia and Russia. The northwest/southeast-oriented mountains have an average elevation of 3,000 m (the highest peak, friendship Peak, tops out at 4,374 m) and a width of 200–350 km. The climate is arid desert and semi-arid desert, the natural landscape is mainly forest and grassland, and the precipitation ranges from 200 to 600 mm, depending on altitude. The runoff of Altai Mountains, which develops into the Irtysh and Ulungur Rivers, roughly follows a south or south-westerly direction into the hills and plains.

The Kunlun Mountains, located to the east of the Pamir Plateau, are part of the Qinghai-Tibet Plateau. The mountains in this range cover a length of 2,500 km, are 100–250 km wide, and have an average altitude of 6,000 m, although some mountains are higher than 7,000 m. For instance, the elevation of the Gongel and Qiongmuzitag peaks is 7,719 and 7,546 m, respectively. These mountains are the most arid in central Asia, with vegetation less rich than in a desert landscape.

The Altun and Qilian Mountains, located on the Gansu and Qinghai border, are each around 4,000 m high. Altun is located in the eastern section of the Aer Mountains, in the northern margin of the northern Tibetan Plateau, which forms the north-south boundary between the Qaidam Basin and Tarim River Basin, east-west.

The Qilian Mountains and the Kunlun Mountains are connected to the two mountain ranges. From east to west, the mountains span 730 km, while their width, north to south, ranges from 60 to 100 km. Compared to the Kunlun Mountains, the Altun experiences more droughts due to lower precipitation levels. Additionally, the mountain's glaciers are thin and its rivers shallow. Indeed, the whole edifice is almost completely enveloped by a desert climate and is unofficially the world's most arid mountain. Few plants grow below the 3,400 m altitude mark, only sparse vegetation and poor quality mountain pasture suitable for grazing goats, shag sheep and other livestock (Zhao et al. 1995).

The Qilian Mountains are located between the Hexi Corridor and the Qaidam Basin, running west from the Dangjin mouth and connecting to the Altun phase, east of the Gansu provincial boundaries. With a length of about 850 km and widths ranging from 200 to 300 km, the Qilians are a fold-block formation and consist of a series of NWW-ESE-aligned mountains hovering over a valley basin. The Qilians are high in the northwest and low in the southeast, with the vast majority of the mountains being 3,500–5,000 m above sea level. The highest peak is Unity, with an elevation of 5,826.8 m.

The Qilian Mountains are glacial landforms, rich in grassland. Their forest zonal distribution gave birth to the three mighty inland river systems of Heihe River, Shule River and Shiyang River. The main basins are the Sahara Lake Basin and the Qinghai Lake Basin, Heihe, while the main rivers are the Mustang River, the Shule River, the Dang River, and the Harteng River. Due to their alpine climate, the Qilian barely reach an average annual temperature of 0 °C. Precipitation, mainly from the southeast monsoons, is high in the southeast, lowest in the northwest, and increases with altitude. This region is dominated by the *S. krylovii breviflora*, a cold Artemisia mountain grassland-based dominant species, and cold temperate coniferous forests blanket the eastern elevations from 2,500 to 3,300 m (Zhao et al. 1995).

The Helan Mountains are at the junction of the Ningxia Hui Autonomous Region and the Inner Mongolia Autonomous Region. With an average altitude of 2,000–3,000 m, the range also boasts a 3,556 m peak (Aobaoyida). Vegetation in this mountain range grows along vertical bands such as alpine shrub meadow, deciduous broad-leaved forest, mixed forest, the Qinghai spruce forest, the oil pine forest, mountain steppe type, and so on. The Helans form the dividing line between the temperate desert and temperate desert steppe and are the watershed of the Northwest area of influx and outflow regions. Towering terrain and good vegetation play an important role in the affluent Yinchuan Plain's ecological environment protection.

1.1.1.3 Basins

As the above overview attests, Northwestern China is vast in territory and wide-ranging in geological scope, and mountains, basins, rivers, plains and oases alternate in distribution. Below, three major basins (the Tarim, the Junggar, and the Qaidam) will be described and their importance delineated.

The Tarim River Basin in southern Xinjiang is China's largest inland basin. It is located between the Tianshan, Kunlun, and Altun mountains and has a length of 1,400 km from east to west, a north-south width of 550 km, and an area of 53×10^4 km^2. The Tarim River Basin is flanked to the south by the Kunlun, Karakorum, and Altun Mountains, to the east by the Kitayama hills and Kumtag, to the north by the Tianshan Mountains, and to the west by the Pamirs and Kunlun Mountains. Due to the depths of the hinterland and the barriers formed by the Tianshans and Kunluns blocking the entry of moist Indian Ocean and Pacific air, the climate is extremely arid. There is little precipitation and only sparse vegetation, which is not surprising that China's largest desert, the Taklimakan (which is also the world's second largest mobile desert), also finds its home base in this region. The Taklimakan Desert features rolling and shifting sand dunes measuring 70–80 to 250 m in height and little to no vegetation.

In contrast, ringing the edges of the basin and nurtured by mountain snowmelt water, are numerous fertile oases known collectively as the "Silk Road". These oases serve as the main point of contact for the region, both internally and externally, and as the focal point of the area's economic and social development. For instance, there is the southern Xinjiang and Hetian oasis, the oasis of Yarkand, the Kashgar oasis, the Aksu oasis, and the Korla oasis.

The Junggar Basin, located between the Tianshan and Altai Mountains, is China's second-largest basin, with a length of 850 km from east to west, a widest point of 380 km from north to south, and an area of 38×10^4 km^2. The contour of the basin is roughly an equilateral triangle, and the elevation is high in the east and low in the west. The Altai gives birth to the Urumqi River and the Tianshan feeds the Manas and Kuitun Rivers, all of which flow into the Junggar Basin. On the western margin, water pools into Aibi Lake, Manas Lake, and so on. However, due to the large number of irrigation projects and decreasing water resources, Manas Lake has recently dried up and Ebinur Lake area has been significantly reduced. The central part of the basin features China's second largest desert: the Gurbantungate Desert. The southern margin of the basin is irrigated by the Tianshans' snowmelt water and consequently developed oasis-based agriculture and urbanization.

China's third-largest basin, the Qaidam Basin, is located 2,600–3,100 m above sea level. Its structure belongs to the Qaidam Depression in the East Kunlun fold system around the basin, surrounded by the Kunlun, Altun, and Qilian Mountains The basin is about 700 km long, running in an east-west direction, and its north-south width is 300 km, giving a total area of about 22×10^4 km^2. The Qaidam Basin has a cold dry climate (with underground permafrost) and scarce rainfall, but it also has plenty of sunshine and thus abundant solar energy resources. Additionally, aeolian landforms are widely developed there, and water is delivered via short rivers and mountain snowmelt. As befits such an environment, vegetation is mainly super-xerophytes shrubs and semi-shrub, suitable for camel grazing. However, there are many lakes and salt marshes and abundantly rich deposits of ferrous and non-ferrous metals, rare metals, and oil. Despite the arid and cold climate, the basin's river- and lakesides provide fertile pasture land, so animal husbandry plays an important role in the local economy.

The Hexi Corridor, located between the Qilian Mountains and Beishan, is a narrow mountain basin. About 1,000 km and 50–60 km wide, the Corridor's main features are the Shule River, the Heihe River Basin, and the Shiyang River Basin.

From a landform perspective, Northwest China includes a wide array of different types and sub-types, such as lacustrine landforms, fluvial landforms, aeolian landforms, dry landforms, loess landforms, periglacial landforms, glacial landforms, volcanic landforms and karst landforms. Fluvial, aeolian and dry landforms comprise 24.65, 26.10 and 31.34 % of the landscape, respectively (Chen 2010). In the mountains, the exterior forces and surface composition of the material form a unique topography of glacial, periglacial, sand, and loess (Zhao et al. 1995). In the plains area, due to strong wind erosion, the formation of a number of different types of aeolian landforms can be divided into aeolian landforms and wind erosion. The former is carried by wind, and sandy material accumulates in the desert as a typical representative (such as the Taklimakan Desert, the Gurbantunggut Desert, Kumtag Badan Jaran Desert, Tengger Desert, and Ulan Buh).

1.1.2 Climatic Characteristics

Situated far from the sea, the relatively secluded Northwest region has a typical continental climate marked by scarce precipitation, high evaporation, wide temperature fluctuations and strong winds. Other than for the high mountainous areas of Yili Tacheng in the west, annual precipitation is less than 250 mm. Southeasterly, southwesterly, westerly, and Arctic air flows are effectively barred from entering the Northwest by the mountains. What little precipitation and moisture does make it past these natural barriers is soon evaporated. Generally, precipitation decreases from west to east, with the junction of Gansu and Xinjiang receiving the least. In the Turpan Basin, the average rainfall is only 6.3 mm, with 1 year recording only 0.5 mm.

In addition to precipitation amounts, there is also a seasonal distribution element in different regions within the arid zone. The Yili and Tacheng mainly receive spring precipitation; the Hexi Corridor and eastern Xinjiang get 60 % of their annual rainfall in summer; while in the Junggar and Tarim River Basins to the west, summer precipitation accounts for only about 40 % of the annual rainfall.

Temperature differences in the zone are also significant, both in annual range and diurnally. The annual average temperature in the north is lower than that in the south, and the mountainous regions are colder than the plains. The hottest temperatures can be found in the Turpan Basin, where summertime daytime maximums can peak at 50 °C.

Cloudy days are a rarity in the Northwest, which boasts China's longest sunshine hours (2,550–3,500 h annually). Solar radiation heat is 54.4×10^8–71.2×10^8 J/($m^3 \cdot a$), making it a rich resource. This is not only beneficial for crop growth and development, but also provides favorable conditions for the extensive use of solar energy. In addition, because the ground is dry, less solar energy is expended for evaporation (Zhao et al. 1995).

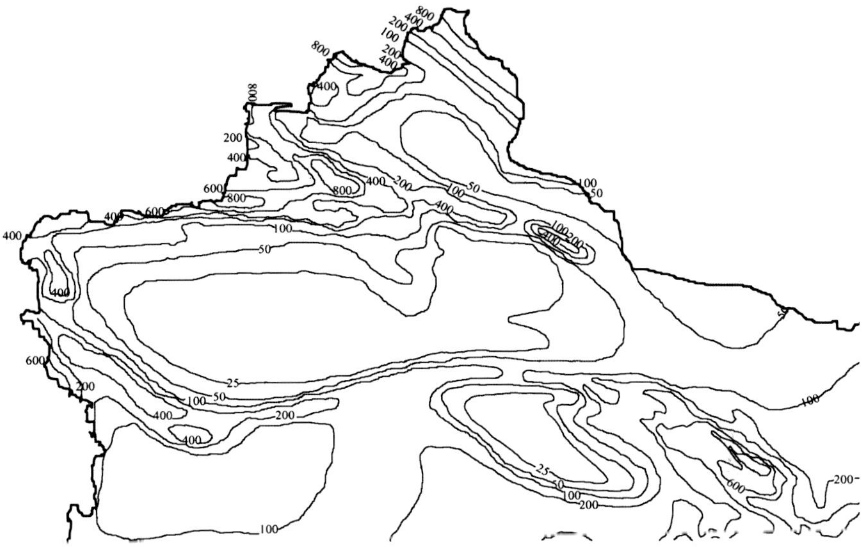

Fig. 1.2 Annual precipitation contour map for the arid Northwest region

1.1.2.1 Precipitation Characteristics

Generally speaking, China's precipitation decreases from the sea-fringed southeast to the inland northwest. Hence, the Northwest arid area suffers a serious water shortage, with annual precipitation often below 400 mm. In the region demarcated by Sunitezuoqi, Bailingmiao, Etuokeqi and Yanchi, annual precipitation ranges between 300 and 400 mm. This is considered a semi-arid area that can be farmed but production is very unstable and the potential for severe desertification is high. Between this line and the Helan Mountains, annual rainfall reaches only 200–300 mm. Natural vegetation growths are mainly desert grassland and agriculture must be irrigated. In the broad desert region to the west of the Helans, annual rainfall is less than 200 mm, with inland basins such as the Qaidam and the Tarim receiving less than 50 mm.

The mountains receive the highest levels of precipitation. Southeast of the northwest region, the East and South Asian monsoons strongly influence precipitation, with annual rainfall around 600 mm. In the western portion of the arid zone, Xinjiang feels the affects of the prevailing westerlies, with annual rainfall averaging between 400 and 600 mm in the Tianshan Mountains, and up to 800 mm in the Pamirs (Fig. 1.2). However, in low-lying areas like the Gobi Desert, the southern Xinjiang Basin, the Qaidam Basin, the western end of the Hexi Corridor, and the Badan Jaran Desert in western Inner Mongolia, annual precipitation hovers in the 50–200 mm range, leading to the formation of extreme droughts (Fig. 1.2). Most of the water resources of Northwest China come either from precipitation or mountain snowmelt.

Arid areas in China have a vertical distribution of annual precipitation. In general, precipitation increases with altitude, but usually just up to the tree line of mountains. In the western mountains of the Altai Mountains and Junggar Basin, most of the annual precipitation falls as snow. From November to March, the snowfall in these mountains averages 270 mm, accounting for 46 % of the annual precipitation, while the hilly areas receive 50 mm and the basin 30 mm, or 35 and 23 % of their annual allotment, respectively. Then, come spring, when temperatures rise, the snowmelts and recharged rivers bring the spring floods.

Xinjiang is located inland, far from sea. The surrounding mountains prevent the maritime air stream from entering the inland basin, thus creating an extremely dry continental climate. Water vapor moving in from the west is therefore the main source of precipitation in Xinjiang, whereas water vapor from the north accounts for only about a quarter of the westerly flow, and its influence is confined to the Altai Mountains vicinity. According to Xinjiang Meteorological Bureau estimates, water throughput is $11{,}540 \times 10^8\ m^3$ in Xinjiang and the average precipitation from vapor is $2{,}400 \times 10^8\ m^3$, accounting for 20.8 % of the total water vapor transportation. This is equivalent to an average annual rainfall of 150 mm (Xinjiang hydrological Station 1985). In terms of regional distribution, the Tarim River Basin accounts for 51–56 % of the total water vapor, the Junggar Basin accounts for 36–43 %, and the mountains only for 6.5–8 %. Regarding seasonal distribution, the summer receives the most precipitation, accounting for 42.7 % of the annual rainfall, the winter gets the least (12.7 %), and spring and autumn precipitation accounts for 24.4 and 20.2 % of the annual amounts, respectively.

Along with being scarce, precipitation in Xinjiang is very unevenly distributed. For instance, more than half of the regions receive less than 100 mm annually, and only one-sixth of the Xinjiang area (mainly the mountains) gets more than 300 mm. Annual precipitation is most abundant (900–1,000 mm) in the forest farm region upstream of the Yili River Valley and least (only 7.3 mm) in Tuokexun County in the Turpan Basin.

Despite being uneven, the geographical distribution of precipitation in Xinjiang reveals a clear pattern. Specifically, northern Xinjiang receives more precipitation than the southern regions, the west more than the east, and there is a general reduction from the northwest to the southeast. The Tianshan mountain ridge forms the boundary: North Xinjiang gets 255 mm annually, and the southern border 106 mm. In the plains, the average annual precipitation of the northern Yili River Valley and Tacheng Basin is 250–350 mm, the fringes of the Junggar Basin receive 150–200 mm, the abdomen of the basin get 100–150 mm, the northern and western edge of the Tarim River Basin in southern Xinjiang get 50–70 mm, the eastern and southern area receive more than 50 mm, and the Qiemo and Ruoqiang to the southeast area get only about 25 mm.

Here, again, a pattern emerges, in that precipitation is higher in the mountains than the plains, with the leeward slopes getting more than the windward sides, and the mountains overall receiving 2–3 times more rain and snowfall than the adjacent plain areas. Average annual precipitation in the mountains is above 400 mm. The maximum precipitation (more than 600 mm) occurs in the Altais; west of the Junggar Basin and northwest of Baerluke have annual precipitation of around 500 mm; and precipitation in the Tianshans measures about $1{,}000 \times 10^8$ t, accounting for more than 40 % of the total amounts in Xinjiang. Differences in annual precipitation even within a single region are exemplified in the following: the Tianshans upstream of the Yili River receive 600–800 mm; the middle of the mountains receives

500–600 mm; the eastern portion of the mountains gets 400–500 mm; and the southern slopes receive between 300 and 500 mm.

Precipitation in Xinjiang also exhibits significant seasonal differences, with most precipitation occurring during the summer months. The northern mountains receive 40–50 % of their annual precipitation in the summer; the higher altitudes receive 60 %; and the southern border mountains get 50–70 %. Mountain precipitation is much less in winter, with precipitation accounting for only about 5–15 % of annual amounts.

In contrast, the plain areas have more uniform precipitation distribution throughout all four seasons, with spring or summer receiving slightly more. However, there are some regional disparities. For instance, precipitation in the Junggar Basin is mostly concentrated in the summer (50 % of the annual rain and snowfall), with winter receiving only about 10 %. A similar pattern occurs in the southern Xinjiang Turpan Basin, where summer rainfall is 50–70 % of the annual allotment and winter brings less than 10 %.

The main moisture source for the Hexi Corridor is moist air flowing in from the Pacific and Indian Oceans. Hexi Corridor's distribution of annual precipitation is contrary to Xinjiang's in that it decreases from east to west. During the height of the summer, masses of warm moist air drift in from the Indian Ocean to blanket the Qinghai-Tibet Plateau east of the Hexi Corridor. Circulation then occurs due to thermal low pressure. In the Shule region, high pressure brings little precipitation, whereas in the Zhangye Qilian Mountains, the low pressure northern airflow clashes with the Qinghai-Tibet Plateau, leading to higher levels of rainfall in Shiyang and the Heihe River Basin at the junction and the Mongolian Gazelle and Zamu Rivers. In the western areas of this region, westerly circulation of Atlantic water vapor results in Shule River Basin precipitation during the year in the Hexi Corridor, although the eastern areas fare significantly differently. To the west, the precipitation is concentrated from May to August, while to the east, rainfall is heaviest from June to September.

Annual precipitation within the mountainous region of the Hexi Corridor has three high-value centers: the Mongolian Gazelle Lenglongling River, the Heihe Daejeo Ma River, and the Lai River. In the eastern Shiyang River Basin, precipitation can top out at 600 mm or more, while the western portions receive only about 200 mm. In the piedmont, annual precipitation is 80–160 mm, decreasing from south to north and from east to west, Wuwei gets about 162 mm, Zhangye about 128 mm, Jiuquan about 86 mm, and Dunhuang around 45 mm.

Annual precipitation in the Hexi Corridor occurs mostly in the warm seasons. The western portions, in spring, receive between 10–20 mm, while the central and eastern parts get around 30 mm. In summer, the northwest gets generally less than 50 mm, while the central and eastern portions receive between 50 and 150 mm. In autumn, the western Hexi Corridor gets between 3–15 mm and the central and eastern areas between 15–40 mm.

The Qaidam Basin gains precipitation from the Bay of Bengal's tropical southwest monsoon moisture, which flows northward from east to west, thereby significantly decreasing the precipitation trend from east to west. The precipitation engendered by the moist air flow also has a great impact on river runoff recharge and river hydrological regimes. Nonetheless, the eastern part of the basin (e.g., Zelin ditch, Xiangride) features an annual rainfall of only around 200 mm or more, and the northwest less than 20 mm.

1.1.2.2 Temperature Characteristics

Temperatures in the Northwest arid area are characterized by hot summers, cold winters, and large diurnal differences. In winter, the main factor affecting the climate in Xinjiang is cold air streaming down from Siberia. Urumqi averages about 20° below zero in winter, although it has been known to dip into the −40 °C range or even lower. For instance, on twenty-first February, 1960, Fuyun county reached −51.5 °C, which is the coldest temperature recorded in the region.

In summer, the southern regions of Xinjiang are hotter than the northern areas. The Turpan Basin generally records the highest temperatures, averaging about 40 °C in July °C, with extremes occasionally nearing 50 °C. In many parts of Xinjiang, the diurnal temperature range is between 20 and 25 °C. In the arid desert portions of Turpan, the maximum diurnal temperature range is 50 °C. However, in Ruoqiang county along the south of the Taklimakan Desert, the annual average daily temperature is poor to 16.2 °C, and the maximum diurnal temperature range is around 27.8 °C. However, such stark changes between daytime and nighttime temperatures are conducive to the accumulation of carbohydrates of fruits.

The Hexi Corridor has a warm arid continental climate. In January, the average temperature, from the northwest to the southeast, is − 10 to −2 °C, respectively, and the average temperature in July is 23 °C. Regarding seasonal temperature variations, Mongolia in the winter experiences cold high pressure control, leading to temperatures of − 11 to −7 °C, with extreme minimums dipping to −32 to −27 °C. In spring, due to variations in solar radiation, the cold and warm air masses boost the daily temperature readings in the Hexi Corridor to between 6 and 12 °C. In summer, daily temperatures fluctuate between hot days and cool nights, while in autumn, as the solar radiation gradually weakens, the cold air forces rapid cooling. In the Hexi Corridor in fall, temperatures generally range from 5 to 9 °C.

The Qaidam Basin experiences a typical alpine continental desert climate: cold, dry, plenty of sunshine, intense solar radiation, and gusty wind. The average temperature in the Basin in January is −9.8–13.9 °C, and in July 13.6–19.2 °C. The annual temperature range in Basin is 25.2–30 °C.

1.1.2.3 Light, Heat and Wind

The Northwest arid area is rich in solar energy resources. In Xinjiang, the annual average number of sunshine hours is 2,500–3,550, which is the highest in China, and the eastern portion of the arid zone receives more sunlight than the western portion. For instance, the annual average sunshine time in the east is more than 3,100 h, and this decreases as one moves from the basins to the mountains. However, not all mountains receive less sunshine than the plains. The southwest slope of the Altai

Mountains, along with the southern and northern slopes of the Tianshan Mountains and the western Junggar Basin, decrease sunshine hours with increasing altitude, but in the Pamir Mountains, the annual average sunshine hours actually increases the higher one goes.

In the growing season (April–September), in the plains of the northern and eastern Xinjiang oases, there are 1,700 h of sunshine, while in southern Xinjiang there is less. The highest tally of sunshine hours occurs in July, with more than 300 h, and the fewest (82–204 h) happen in December in the northeast of South Xinjiang and North Xinjiang. Meanwhile, in the southwest of South Xinjiang and East Xinjiang in January, there are about 174–212 sunshine hours.

Xinjiang not only has plenty of sunshine hours, it also receives large amounts of solar radiation, ranking it second in the country after the Qinghai-Tibet Plateau. According to recent estimates, the solar radiation beamed onto Xinjiang for 1 year is 100 times the total energy generated by the world in 1980. However, different parts of Xinjiang receive different amounts of solar radiation. Overall, the total amount of radiation in Xinjiang reduces as the latitude increases. The highest total radiation is in Hami, and the lowest occurs in Wusu, Jinghe and Karamay in North Xinjiang. This is attributed to the sandy soil of the desert. In the southwestern corner of the Tarim River Basin, due to constant dust storms, the global solar radiation is less than it could be otherwise.

Seasonally, solar radiation varies according to numerous factors. In winter (December–February), the minimum amount of total solar radiation is 520–890 MJ/($m^2 \cdot$ quarter), accounting for 10–16 %. In spring (March–May), solar radiation is 1,490–1,930 MJ/($m^2 \cdot$ quarter), accounting for 28–33 %. In summer (June–August), the largest solar radiation is 1,800–2,200 MJ/($m^2 \cdot$ quarter), accounting for 32–39.6 %, while in autumn (September to November), the solar radiation ranges from 900 to 1,450 MJ/($m^2 \cdot$ quarter), accounting for 18.0–23.5 % of the total annual amount of radiation. Solar radiation can be absorbed by plants, which is called photosynthetically active radiation. Currently, there is no observed data for photosynthetically active radiation in meteorological stations in Xinjiang. The calculated active radiation measures from 2,400 to 3,100 MJ/($m^2 \cdot$ a); in southern and eastern Xinjiang, from 2,800 to 3,100 MJ/($m^2 \cdot$ a); and in North Xinjiang, from 2,400 to 2,600 MJ/($m^2 \cdot$ a). In the Hexi Corridor, the total annual solar radiation is from 5,800t to 6,400 MJ/($m^2 \cdot$ a) and 5,800–6,200 MJ/($m^2 \cdot$ a), with annual sunshine hours numbering 2,200–3,030 h, giving a frost-free period of around 85–165 days.

The annual average number of sunshine hours in the Qaidam Basin is generally 3,100, although in the western portion it can reach 3,200–3,500 h (the highest number ever recorded being 3,532.3 h). The percentage of sunshine in the basins is more than 70 %, resulting in solar radiation and sunshine amounts that are second only to Tibet.

Due to low levels of precipitation, large diurnal temperature fluctuations and sparse vegetation, the Northwest arid zone often experiences extremely windy weather, especially in spring. In northwestern Turpan, along the Lan-Xin Railway and 13 Rooms in Hami, strong winds blow up to 100 days a year, while in Karamay, the wind gusts on average for 75 days. The wind can be so extreme at times that it disrupts railway transportation and oil production. However, if the strength of such winds could be converted to heat and electrical energy, it would be of great benefit.

Based on the current world level of technology, the seven windiest areas in Xinjiang could generate 7.1 million W of power, which is equivalent to four times the generating capacity of Xinjiang in 1978.

Strong winds also dominate the Hexi Corridor region in the summer monsoon season. Due to the uplift of the Qinghai-Tibet Plateau, a downdraft airflow is formed in the Corridor, where strong winds near the cold front are exacerbated by the narrow and flat terrain. The annual maximum wind speed is generally 14–28 m/s, although in the Corridor's northwest, it is 25–28 m/s.

As wind energy is virtually inexhaustible and has almost no environmental pollution, it is worthwhile to develop and utilize. However, the potential for dust storms that wind farms might create should also be taken into consideration.

1.1.2.4 Evaporation Characteristics

The mountains in arid region of the Northwest have high precipitation levels and little evaporation capacity, while in the plains there is low precipitation and high evaporation. Using this perspective, the evaporation of the land surface has a vertical zonal pattern.

Evaporation is generally low in northern Xinjiang and high in southern Xinjiang. Likewise, it is generally low in the west and high in the east; low in the mountains and high in the basin; and low at the center of the basins and higher at the basins' edge. The annual evaporation in northern Xinjiang measures between 1,300 and 2,300 mm; in the Yili River Valley, 1,200 and 1,600 mm; in Karamay, Bole and Alashankou, more than 3,400 mm; in south Xinjiang, 2,000–3,000 mm; and in Turpan and Hami, 3,000–4,000 mm. Qijiaojing and Balikun in Hami experience the largest evaporation in the Xinjiang area, with an annual evaporation of between 4,000 and 4,500 mm. In the mountains, evaporation usually measures below 1,500 mm in the north and below 2,000 mm in the south. In spring and summer, evaporation rates are highest from April to August, accounting for more than 70 % of annual evaporation.

As the temperature increases sharply in spring, drought results when high evaporation rates do not kept pace with irrigation. In winter, however, the low temperatures, low solar radiation, low wind speeds and high relative humidity cause the least amount of evaporation. In terms of months, the lowest levels of evaporation occur in January and the maximum evaporation happens in July (or in June in windy areas).

1.1.3 Rivers and Lakes

Water is important not only for maintaining the ecological balance of the natural environment in the Northwest, but also for enabling socio-economic development. As outlined in previous sections, China's arid Northwest zone has unique natural and geographical conditions that render it susceptible to extreme dryness and even drought. Although there are various types of rivers and lakes, their water amounts are small, and they are mostly seasonal bodies of water supplied primarily by ice- and snowmelts from the mountains and from forest precipitation recharge. Precipitation is generally so low in the plains areas that there is little to no runoff.

Moreover, most of the rivers originate in the surrounding mountains and converge inside the basins, constituting a radial river system. These rivers rely mainly on glacial meltwater for their existence and usually have a short flow length. According to recent statistics, the number of inland rivers in Northwest China is 676 in total, which breaks down to 570 rivers in Xinjiang, 55 in the Hexi Corridor, and 51 in the Qaidam Basin. Due to snow and ice-melt water regulation, interannual river runoff in the Northwest changes little, but river water is mainly concentrated in the summer. Hence, summer accounts for more than 60 % of the full year's runoff. Glaciers and snow cover provide a unique form of water resources in the arid region, covering an area of approximately 3.24×10^4 km^2 (Zhang 2002; Zhao et al. 1995).

In addition to seasonal rivers, the region has many seasonal inland lakes. However, due to generally low runoff and little groundwater, the lakes are subject to strong evaporation and often devolve to lagoons or salt lakes.

Water resources in the Northwest mainly come from mountain snowmelt water and mountain forest precipitation. The Altai Mountains, Kunlun Mountains, Tianshan Mountains and Qilian Mountains, along with other high mountains in the area, are blessed with abundant rain and snow. This results in mountain glaciers and permanent snow cover, which plays a regulatory role in surface runoff and the formation of inland rivers, of which there are about 676 (Qin 2002). The largest of these are the Irtysh River, the Manas River, the Kuitun River, and the Urumqi River in northern Xinjiang; the Yarkand River, the Aksu River, the Hotan River, the Kaidu-Kongque River, and the Tarim River (four sources and one mainstream) in southern Xinjiang (Chen 2010). The Hexi Corridor also has sizeable rivers such as the Heihe River, the Shule River and the Shiyang River. The mountains and inland basins in these regions have a topographic distribution pattern, which makes the rivers originate in the mountainous areas and flow together to form a larger river at the heart of the basin. This feature aids the survival and development of oases. However, the surface water and groundwater relationship is complex. In this particular arid zone's water cycle, the seepage of the surface water into subterranean chambers accounts for 60–87 % of the groundwater recharge (Tang 1995). The cycle of water resource determines the characteristics of the inland river basins with regards to the surface runoff and groundwater resources. These are likewise closely linked not only to ecological balance but economic development.

1.1.3.1 The Tarim River

The Tarim River contains the Hotan, Yarkand, Aksu, Kaidu-Kongque, Kashgar, Cherchen, Dina, Weigan, Kuche, and Keriya Rivers. These are the nine major river systems of the 144 rivers in the area. Currently, only the Aksu, Hotan, Yarkand and Kaidu-Kongquehe Rivers have surface water contacts with the Tarim River (Chen 2010). From the Yarkand Laskaimu River to Taitema Lake, the total length of the Tarim River is 2,437 km. The confluence of the three major tributaries—Xiaojake—marks the beginning of the river's main stream.

The Tarim is a typical braided river, forming a north-south alluvial plain with a width of more than 100 km. By river erosion and stacking interactions, the riverbed is very unstable and the river swing strong. In the 1950s, Taitema Lake was the Tarim's tail lake. However, when the Daxihaizi reservoir was built in 1972, it replaced the Taitema Lake as the end of the Tarim River, leaving 266 km of the Yingsu dry. In order to protect and improve the ecological environment and the cross-strait green corridor of the lower reaches of the Tarim River, the Daxihaizi Reservoir began to deliver water downstream of the Tarim River in 2000.

The Aksu River includes two major tributaries. The first of these is the Kumalake River (also known as the Kunmalik River), which originates in the Kyrgyzstan Tianshan glacier. The other tributary, the Tuoshigan River, has its source at the border of China and Kyrgyzstan in the southern Tianshan Mountains. These two rivers flow together in Wensu County, forming the Aksu. From this point of confluence in Wensu to the Tarim River, the Aksu measures 130 km in length and is the source of the Tarim River. During the summer flood season, the Hotan and Yarkand Rivers also flow into the Tarim River.

The Hotan River, formerly known as the Khotan, runs in the southwestern Tarim River Basin. It is the largest river in the northern slope of the Kunlun Mountains. The headstream, Yurungkax, comes from the northern foot of the Kunlun Mountains, while the smaller headstream, the Karakax River, descends from the Karakoram Mountains. Where the Yurungkax and the Karakax converge, the Hotan River begins. The Hotan is divided into two tributaries. The eastern branch of the old Hotan River downstream has completely dried up, but the western branch of the new Hotan River runs for 806 km through the Taklimakan Desert, in the vicinity of the upstream reservoir and the Aksu and Yarkand River confluence, where the Tarim River begins.

The Yarkand River, which, at 996 km, is the Kashi region's longest river, originates in the Karakoram Pass of the Karakoram mountains in northern Kashmir. It forms where the Laskaimu and Arcatagai Rivers flow together at the Helbalongke of Karakoram Pass. The Yarkand flows from the southwest to the northeast, running through the Kashi region, the Kirgiz Autonomous Prefecture, and the Hotan and Aksu Prefectures. The drainage area is 108,000 km^2, and the irrigation areas are the Taxkorgan, Yecheng, Poskam, Yarkand, Markit and Bachu counties as well as ten mission fields in the Agricultural Division, giving a total land mass of 2,883.78 million m^2. Water from Chogori feeds the Yarkland, along with snowmelt water, spring river floodwaters, and rain.

The Kaidu River is the third largest river in Tianshan and has its source in the Albion mountain in Tianshan Mountains. Nearly 560 km in length, the Kaidu River flows through the Yurdusi Basin at the upstream section of the river and ends at Bosten Lake. Bosten Lake is the source of the Peacock River in the southwest area of the Yanqi Basin. The Peacock then flows into the Tarim River Basin along Tiemenguan injection.

1.1.3.2 Urumqi River

The Urumqi River, with a total length of 214.3 km, has its source in the northern slopes of the Tianshan Kalawucheng Mountains. After meandering through the city of Urumqi, it then disappears in the Northwest Territories of Miquan County. The Urumqi is a typical transverse river, running downstream along the northern slope of the Tianshans and being fed by runoff whose multi-year average volume is small (230 million m^3). Nonetheless, more than 2 million people in Urumqi, the capital of Xinjiang, rely on the water from the Urumqi River for their domestic and industrial needs. In recent years, with the gradual increase in water consumption, issues surrounding water supply and demand in the region have become increasingly prominent.

1.1.3.3 Heihe River

The Heihe River is Northwest China's second largest inland river. Starting from Shandan in the east portion of the arid zone, the Heihe River flows to the west of Jiuquan, south of the Qilian Mountains from Qilian County, and north of Gashunnoel and Suoguonoel of Ejinaqi. The river is 820 km long and has a basin area of 1.3 million km^2. The Heihe River proper is fed by the Shandan, Liyuan, Fengle, Daduma, Balang and Taolai River tributaries.

The Heihe River has its source in the Qilian Mountains. From there, it flows to the valley south of the Qilian Mountain community, bordering the Republic of Mongolia to the north, then eastward adjacent to the Shiyang River Basin, westward with the Shule River, and finally north again to the Inner Mongolia Autonomous Region within the Juyanhai in Ejinaqi and the Mongolian People's Republic border. The Heihe River Basin has 35 small tributaries and three separate sub-river systems. The western sub-water system includes the Taolai River, covering an area of 21,000 km^2; the central sub-river systems includes the Maying and Fengle Rivers and ends in the Gaotaiyanchi-Minghua Basin, with an area of 0.6 million km^2; and the eastern sub-water system includes the Hei and Liyuanhe Rivers and covers an area of 116,000 km^2.

The upper portion of the Heihe River is 303 km long, covering an area of 10,000 km^2. There, the river banks are formed by high mountains and deep valleys, making for a steep riverbed. This area serves as the Heihe River Basin runoff zone. The climate is damp and cold, with an annual average temperature of less than 2 °C and annual rainfall of 350 mm.

The 185-kilometer span of the Hei River from Yingluoxia to Zhengyixia form the middle reaches of the river, covering an area of 25,600 km^2. The land flanking this portion of the Hei river is flat with plenty of light and heat resources but also severe drought. Annual precipitation measure a scant 140 mm, the average temperature is 6–8 °C, and yearly sunshine hours are in the 3,000–4,000 range. The evaporation capacity on an annual basis is 1,410 mm, and there is extreme land salinization.

The third sub-river system is the Justice Gap, the 333-km long downstream portion of the River Hei. It covers an area of 80,400 km^2 and includes smaller rivers and deltas as well as most of the Gobi desert. The annual precipitation averages only 47 mm, and the annual average temperature is 8–10 °C. The extreme minimum temperature is −30 °C and the extreme maximum temperature exceeds 40 °C. Annual sunshine hours are in the 3,446 range, and annual evaporation capacity can reach as high as 2,250 mm. Not surprisingly, given these conditions, the climate is very dry. The drought index is 47.5, categorizing the area as extremely and consistently drought-ridden, and the sandstorms are monumental. Indeed, this area is the main source of dust storms for all of northern China.

1.1.3.4 The Shule River

The Shule River, with a total length of 580 km, is located in the western part of the Hexi Corridor and is mainly composed of the Petroleum, Changma, Poplar, Tashi, Dang and Annanba Rivers. The Shule is the second largest river in the Hexi Corridor in Gansu Province. It originates in the Qilian Mountains' Unity Peak and flows through the alpine meadows of Subei County, across the Snowy Mountains—Canyons in Tuolainan Mountains and then through the Changma Basin, and ends in Lop Nur. The area before Changma Gap is the upstream of the Shule River, the water current is swift; the midstream is between the Changma Gap and plain, and then the river is divided into East and West sections, the west branch is the mainstream; the downstream is below the Anxi Twin Towers reservoir.

1.1.3.5 The Shiyang River

The Shiyang River is located in the eastern part of the Hexi Corridor. Originating in North Lenglongling in the Eastern Qilian Mountains, the Shiyang is formed by the confluence of numerous tributaries that flow through Wuwei, Gulang, Yongchang, and across the plains. The river then flows across Minqin Basin to Qingtu Lake, for a total length of more than 300 km.

From east to west, the Shiyang River is mainly composed of the Dajing, Gulang, Huangyang, Zamu, Jinta, Xiying, East and West Rivers. All of these rivers flow into the middle of the river near the city of Wuwei, where they merge to create the Shiyang River's main stream. Afterwards, the combined waters flow into the Hongyashan Reservoir and then into the Minqin Basin. The western rivers and part of the East River to the north of the city of Yongchang merge into Jinchuan, and finally end into Jinchang basin.

The Shiyang River Basin, which lies lower in the south than in the north, can be divided into four major geomorphic units: the southern Qilian Mountains, the Central Corridor plain area, the northern hilly area, and the desert areas. In the southern Qilian Mountains, the higher elevations are about 2,000–5,000 m, with the mountains ranging in a roughly northwest-southeast direction. The Central Corridor plain area is divided into northern and southern basins. The South Basin includes the Dajing, Wuwei, and Yongchang Basins, and has an altitude of 1,400–2,000 m. The North, Minqin, and Jinchuan-Cangning Basins are 1,300–1,400 m above sea level. The northern hilly areas, with an altitude of less than 2,000 m, are relatively low for hills and sprawl into a desert formation.

1.1.3.6 Bosten Lake

Bosten Lake is located in the southeast portion of the Yanqi Basin, south of the Tianshan Mountains. With an area of approximately 990 km^2, Bosten is the largest lake in Xinjiang and also China's largest inland freshwater lake. Bosten Lake is not only the Kaidu River's tail lake but also the source of the Kongque River; it has the function of flood irrigation and protection of Korla and Yuli (the two oases of production and live water) and plays an important role in the lower reaches of the Tarim River's ecosystem and the environment (Tan et al. 2004). Bosten Lake is 14 km from Bohu City and 24 km from Yanqi county, at an elevation of 1,048 m. The lake is 55 km long, running in an east-west direction, and is 25 km wide from north to south, showing a slightly triangular form. The water storage capacity of Bosten Lake is 9.9 billion m^3. There are about 10 rivers flow into the Bosten Lake, Kaidu River, Huangshuigou River, and Qingshui River are the main rivers, runoff that from Kaidu River accounts for about 87.4 % of the water entering the Bosten Lake. The water flows out towards the southwestern through the Kongque River, and ends in Lop Nur. When the water level of Bosten Lake is 1,045 m, with the area of 886.5 km^2, the capacity of the lake is 5.37 billion m^3; while the water level of Bosten Lake is 1,049 m, with the area of 1,262.3 km^2, and the capacity of the lake is 1,262.3 km^2. The highest water level recorded was 1,049.39 m in 2002, while the lowest level recorded was 1,044.81 m in 1986.

Bosten Lake is surrounded by high mountains and experiences low precipitation, high evaporation, and an average sunshine rate of 67–68 %, which is characteristic of a temperate arid desert climate. The spring climate is variable, the summer is hot and dry, the autumn cools off quickly, and winter is frigid with high evaporation. Overall, the sun shines throughout the year, giving an annual average temperature of 7.9 °C. This is broken down to an average temperature in January of − 12.7 °C (extreme low − 30.2 °C) and an average temperature in July of 22.8 °C, with extreme temperatures up to 38 °C.

1.1.3.7 Aibi Lake

Aibi Lake, located 35 km north of the city of Jinghe County, is the largest lake of the Junggar Basin and is also the largest saltwater lake in Xinjiang. The Boertala, Jinghe, and Kuitun Rivers flow into Aibi Lake from the west, south, and east, respectively, and are the lake's main water sources. Estimated at around 189 m above sea level, Aibi Lake is oval-shaped and has an area of 650 km^2 and an average depth of 2–3 m.

Aibi Lake is fed by 11 large rivers, which gives it an annual supply of 600 million m^3 of surface water. Earlier in the past century, the lake area comprised 1,200 km^2, but that has now shrunk to 500 km^2 due to desertification resulting from falling water tables. Indeed, desertification has greatly accelerated in the surrounding area of the basin, and is now estimated at 38 km^2 per year. The main reason for the deterioration of the ecological environment is the total water resources available due to the contradiction between supply and demand. This arose mainly from the human-based activities of the 1950s to the 1970s, due to population growth and large-scale land and water development. As Aibi Lake has an average depth of only 1.4 m, the lake inflow is very sensitive to increases or decreases in water. Hosting an abundance of wildlife and plant species, the lake is listed as a Xinjiang Uygur Autonomous Region Wetland Nature Reserve.

1.1.4 Vegetation Characteristics

Northwest China arid zone has low vegetation coverage and thus experiences continuous soil erosion. The arid climate and barren salt soil severely limit plant growth, development, and dissemination. The plant species that do survive in this zone are mainly small trees, shrubs and semi-shrubs that have adapted to harsh desert habitat conditions through a variety of physiological and ecological mechanisms. Primarily, most plants have reduced or degraded foliage instead of green shoots to activate photosynthesis. Being sparse, the forest coverage rate is less than 1.6 %, which means that the amount of residual litter falling into the soil surface is scant. In hot and dry climatic conditions, soil organic matter mineralizes rapidly, so soil organic matter content is very low (e.g., below 0.5 % or 0.3 %, and rarely more than 1 %).

1.1.4.1 Relatively Poor Species

Although the Northwest arid zone is vast, the severity of the environment makes plant species relatively poor. According to statistics, there are only 1,704 kinds of seed plants in the region. These belong to 82 subjects and 484 genera, accounting for only 6.31 % of similar species in the whole country, 24.34 % of the subjects, 15.53 % of the genera. Among these, Gymnosperms have only 3 subjects (Ephedra, Panacea and Cupressaceous), 4 genera, and 17 species; Dicots have 63 subjects, 384 genera, and 1,349 species; and Monocots have 16 subjects, 96 genera, and 338 species.

Plants in the arid zone include a variety of early and super-early shrubs, small shrubs and semi-woody plants. Specifically, Chenopodiaceae, Asteraceae Tamaricaceae, Tribulus terrestris Branch, ephedra and Polygonaceae play a leading role in the zone's ecosystems. The extremely harsh ecological conditions and special topography make a more ancient type remain within the region, along with a large number of endemic species and genera, such as the sand holly and dwarf sand holly. These are early relic species of the Paleogene subtropical evergreen broad-leaved forest, and also the only evergreen shrub in the desert area. As well, there are cotton thorns, Prunus mongolica, and bare fruits, which are classified as isolated relic plants. The most well-known endemic genera are core Brassica and lily Artemisia. All together, there are 84 endemic species, accounting for 4.93 % of the region's total number of species. Compared with other regions, species endemism in the arid zone is higher than in Inner Mongolia but less than in the northeast, north, central and southwestern regions (Wang 1997).

The geographical distribution of arid zone plants is extremely uneven. Overall, Alashan has the most abundant seed plant species, with a total of 59 subjects, 309 genera and 606 kinds, followed by the Hexi Corridor region, with 54 subjects, 156 genera and 326 kinds. The Qaidam Basin's seed plants have 41 subjects, 130 genera and 255 species; the Junggar Basin's seed plants have 30 subjects, 121 genera and 245 kinds; the massive (50,104 km^2) Tarim River Basin area has only 35 subjects of seed plants, 105 genera and 165 species; and the eastern Xinjiang gashun Gobi is extremely poor in seed plants, with a total of only 34 species (Pan et al. 2010).

The ecosystem of arid areas is generally fragile, and Northwest China is no exception. Over the years, human activities have exacerbated the deterioration of the environment, leading to even poorer plant diversity. According to statistics, at present there are 36 rare and endangered plants. (1) Agricultural activities have severely damaged shrub, grassland, desert forest, and local vegetation. For instance, in 1949, the total area of irrigated oases was 1.3×10^4 km^2. Then, by the late 1980s, this area had extended to 5.87 km^2, resulting in a sharp decline in oasis peripheral vegetation. (2) Predatory firewood collection and digging have also caused a wide range of vegetation recession and a reduction in species diversity. In the Junggar Basin desert regions, the average household burns 2t H. ammodendron firewood each year, cutting 6–7 hm^2 of the natural Haloxylon forest, creating dunes. The Tarim River Basin's original Euphrates poplar forests, which once measured at 53×104 hm^2, was decreased by more than half over the course of only one decade. Xinjiang's original tamarisk shrub, estimated at 400×10^4 hm^2, has also been cut more than a half. Destructive excavation of herbs and precious medicinal herbs, such as licorice, ephedra and cynomorium, likewise caused severe damage. In addition, abusive animal husbandry methods and extreme overgrazing have also led to serious degradation of grassland resources. (3) The lack of water, construction of reservoirs, and flood-control intercept a large number of water resources, resulting in rivers downstream of the water supply suffering a water shortage or even drying up completely. Since the 1960s, China's infamous dry lakes include Lop Nur Lake, Manas Lake, Taitema Lakes, Aydin Lake, and East and West Juyanhai. Lakes that have been drastically reduced include Aibi Lake, Wulungu Lake, Ancient Lake and Bosten Lake. The reduction of a downstream water conveyance leads to the shrinkage or drying up of lakes. As a consequence, the presence of lake plants will either sharply drop or disappear altogether, which is what occurred at Lop Nur, the Manas River and Knoll Lake. Aibi Lake has seen a severe reduction in reeds, while Bosten Lake and its downstream Lake District saw its reed production decreased by one-third over the past 30 years.

Moreover, due to development interventions, river portions downstream of the water supply are reduced, resulting in a large area of dead or reduced natural or planted stands. This occurred in the Tarim River, where numerous agricultural irrigation water projects upstream made the downstream flow of the river sharply decline, resulting in a reduction in Populus euphratica, from Yingsu to the Kuergan, from 4.5×10^4 hm^2 to 1.6×10^4 ha. Similarly, agricultural irrigation in the middle region of the Hexi Corridor's Heihe River, blocked a huge amount of water, thereby reducing the downstream flow from 12×10^8 m^3 in the 1960s to less than 5×10^8 m^3 in the 1980s. This led to a reduction in *Populus euphratica* downstream, and the Tamarisk forest was reduced from 15×10^4 to 10×10^4 hm^2. The Shiyang River's downstream portion suffered similar problems, reducing the Minqin oasis plantation, the natural Populus, and the Tamarisk forest by two-thirds.

1.1.4.2 Significant Vertical Zone Effects

Temperature generally declines with mountain altitude, a natural phenomenon that is called vertical zoning. The basic condition to form a vertical band is the uplift of the mountain, while the direct cause is the moisture and heat change that occurs rapidly with increasing height. Temperatures usually drop 6.5 °C for each kilometer of altitude. The basic outline of the Northwest arid area is formed by a series of mountains, basins, hills and plateaus. Kunlun and Karakoram, whose ridge lines' average height is 6,000 m, are located in the south and southwestern margins of the arid areas, respectively, and combine with the Qinghai-Tibet Plateau into one feature. The western margin of the zone has the Pamirs, Junggar, and Altai Mountains. To the east are the Helan and Yinshan Mountains, while Wolf Mountain demarcates the start of the extreme arid region. From here, the Mongolian Plateau is situated to the northeast, and the Tianshan and Altun-Qilian Mountains cross the central portion. These intersect with the Altai and Kunlun Mountains, with the eastern Tianshan internal rift forming the Turpan-Hami Basin 154 m below sea level. The Tianshan bifurcation forms the Ili Valley. To the west, this portion of the arid region is divided into the Junggar, Tarim and Qaidam Basins and the Alashan Plateau. Each zone features a mountain in the Hong alluvial fan Basin Desert (Gobi) terrain system, which is also referred to as a mountain-basin system (Zhang 2001).

In a mountain-basin system, vegetation distribution has clearly defined zonal properties. In Xinjiang, for example, the northern slope of the Tianshan Mountains, the southern slope of Tianshan, and the Kunlun Mountains are all mainly populated by Tianshan spruce. In the lower portions of the mountains and valleys, there is also some Poplar and Birch distribution. In the Altai forest in Xinjiang, larch predominates, followed by Xinjiang spruce. In the Haba River and the Burqin forest, which is northwest of Altai, we find Xinjiang firs and Xinjiang pines. The western Junggar, Saul and Baerluke Mountains mainly feature valley spruce, Siberian larch forest, and Tianshan spruce, while willows flourish in the low reaches of the valley. In contrast, the northern slope of the Kunlun Mountains and Pamil forest have an arid climate, and vegetation is sparse there.

On the plains, broadleaf forests are the predominant growth types. Among these, Desert Populus is liberally distributed in the Tarim and Junggar Basins on both sides of the river terraces and floodplain. The main species are Populus and Gray Poplar, in the valley's secondary forest, in the Ili and Irtysh Rivers' riverbed shallow zones, the main tree species are E-river Poplar and Miye Poplar, Bitter Poplar, Silver Poplar, Populus nigra, lobular, and white willow. The desert shrub is widely distributed in Xinjiang's vast desert region, especially in the alluvial fan and fan-edge areas in the southern and northern rivers, the floodplain, the riparian terraces, around the river basin, on the fixed or semi-fixed dunes, and on the dune slacks. Desert shrubs generally have strong fecundity and are important vegetation for establishing sand dunes. As well, they maintain the desert's ecological balance and protect the ecological security of peripheral oases. Shrubs that are resistant to drought, salinity and wind include the Tamarix and Haloxylon (Yuan et al. 2008).

Approximately 58% of the grassland in Xinjiang is located in the mountains. South of the Kunlun Mountains, in central Tianshan and north of the Altai Mountains, rising ladder fault-block mountains occur. Thus, enormous amounts of sediment are formed at the top of the mountains. Additionally, there are sedimentary deep loess in the valleys and basins between the mountains. These provide an excellent soil base for prairie plant development. In the alpine and subalpine regions of the Kunlun, Pamir, Tianshan and Altai Mountains, alpine steppe and alpine desert grassland flourish in the harsh environment. The vegetation mainly occurs in cold dry bunchgrass, resistant to cold, drought, and wind. Artemisia and pad-like semi-shrub occur, but the main dominant species are Stipa, Xinjiang silver spike grass, hard-leaf sedge, alpine Artemisia silk, Ceratoides, cushion-like Ceratoides, Tibet Chrysanthemum, and short flower Stipa. In the north of the Tianshan Mountains watershed subalpine, temperate meadow steppes and mountain meadows abound that feature vegetation such as xeric and wide xeric perennial grasses. In the middle mountain zone of South Xinjiang and the low mountain zone of North Xinjiang, the temperate steppe vegetation is mainly perennial xeromorphic bunchgrass plants. Here, the dominant species are fescue, ditch fescue, Stipa, Northwest Stipa, Kunlun Stipa, and wheatgrass, followed by Artemisia shrubs and shrubs, which are mainly cold Artemisia, Xinjiang Chrysanthemum, alpine Artemisia silk, and golden pheasant. In the northern piedmont plains, low mountains, southern middle mountains and subalpine, vegetation is of the temperate desert steppe variety, composed of small perennial grasses, desert Artemisia half-shrubs, salt firewood class, semi-shrubs and drought shrubs. In the plain areas, which feature lake basins, low mountain valleys, and river basins, the valleys mainly have flat meadow and temperate desert grassland, and the vegetation is raw, dry, wet grass, legume grass and miscellaneous grasses and shrubs. These are mainly Haloxylon, Haloxylon persicum, Calligonum, Tamarix, Ephedra przewalskii, King, Ceratoides, Reaumuria, Anabasis, Artemisia, reeds, Splendens, licorice, bitter beans, A. sparsifolia, Iris lactea, big-leaf white linen and caspica (Yuan et al. 2008).

1.1.4.3 Significant Levels of Differentiation

Regarding the spatial distribution of vegetation, the southeast Tarim and Qaidam Basins and the western Hexi Corridor are arid in their central portions though become wetter towards their fringe areas. Higher plant-life is rare in the desert areas of the Gobi and the salt flats. Similarly, in the north and east, the vegetation transitions to types commonly found in deserts and desert steppes, connecting with the deserts in Central Asia to the west. To the south and southwest, the transition is to an alpine desert, desert steppe and alpine grasslands. Here, the differences in latitudes reflect the effects of temperature and rainfall. From north to south, there is a temperate steppe desert (Altai Mountains, north of East Alashan), a temperate desert (Junggar Basin,

Alashan, and most of the high plains), the temperate plateau desert (Qaidam Basin) and warm temperate desert (Tarim River Basin). Specifically, desertification holds sway in this arid zone, and the constructive species are the super-xerophytic trees, half-shrubs, small shrubs and bushes. The desert occupies almost all sandy alluvial plains, wide terraces, alluvial fans, former hills and low mountains in the basins, in addition to the Altai Mountains south and east in northern Alashan. It even sprawls into the Kunlun Mountains, up the northern slope of the middle zone and south of the valley alpine zone that demarcates the Kunluns' main plateau ridge. Grassland, shrub land, forest and meadow vegetation are mainly distributed in the rainy mountain runoff water accumulating in the desert areas on both sides of the river, where the fan margin of the water overflows the low-lying land around the lake (Ni et al. 2005).

1.2 Socio-Economic Characteristics

The main human activities of China's arid Northwest are agriculture and animal husbandry. The multicultural population is mainly distributed in areas of sufficient moisture, such as the oases, where oasis agriculture is the mainstay. However, productivity is not as high as it could be, even given the generally arid conditions of the natural environment.

1.2.1 Oasis Economy Characteristics

From a global perspective, an oasis economy has three main characteristics. The first of these is dispersion, meaning the economy is distributed along the water and disperses in the desert (e.g., the Gobi). Such dispersion is generally a few kilometers or even hundreds of kilometers between oases. The size of the oasis and the degree of concentration depends on the density of the river size. The second characteristic of an oasis economy is its enclosed property. Oases are usually small and surrounded by desert. This quality means they are likely also prone to poor communication networks, long transport lines, high shipping costs, and sporadic interaction with the outside world. The third main characteristic of an oasis economy is self-sufficiency. Such a market is usually in the transition phase to a commodity economy.

The evolutionary trajectory of the oasis economy in Northwest China arid area is as follows: hunting (including hunting and fishing) → nomadic → livestock and agricultural or different emphases, the rise of gardening → the development of agriculture and animal husbandry and handicrafts, commerce → agricultural development mainly nomadic, the rise of industrial → modern oasis economy with completely categories. An oasis economy performs as a significant characteristic of the arid zone and the main body of an agricultural economy. It features small-scale production, low levels of concentration, and has transformed from a semi-closed to an open, export-oriented system of trade.

The main body of oasis agriculture has the following three aspects:

1) An oasis economy is an agricultural industry supplemented by the development of non-agricultural industries. Statistics show that the total income in rural areas of the arid oasis region includes a non-agricultural income of 46.2 %. In Xinjiang's rural regions, per capita net income accounts for only 7.45 %.
2) The agricultural output structure is based on farming and pastoral structures, but the share of agriculture predominates. In 2011, the proportion of agricultural output value accounted for 73.54 %, whereas the proportion of animal husbandry output value was only 21.22 %.
3) Food crops in oasis areas are mainly corn and wheat, and the main cash crops are sugar and cotton. In 2011, in the Xinjiang oasis, the proportion of corn produced was 43.11 % and the proportion of wheat was 48.02 %. Economically, the proportion of crop yields was 59.27 % for sugar and 33.10 % for cotton.

1.2.2 Oasis Distribution

Oasis distribution is determined by the availability of water resources in the arid zone. Hence, oases in Northwest China are mainly distributed south of the Tianshan Mountains and north of the Kunlun and Qilian Mountains and the Qaidam Basin. Generally, oases develop along a river whose geomorphological structure has likely been affected by human activity. The following is a brief introduction of the distribution and characteristics of three typical oases in Northwest China.

1.2.2.1 Xinjiang Oasis Region

Xinjiang has the most widely distributed and largest oases in China, mainly in the South Tianshans, north Kunlun-Altai Mountains, the Yili Valley and the Irtysh River Basin. The Xinjiang oases includes the Junggar oasis with 12 countries, the Tarim oasis with 42 countries, the Turpan oasis with 3 county, and the Hami oasis with 1 county.

(1) Oases along rivers
Water distribution is a key factor in determining the layout of an oasis. In Xinjiang, the distribution of water resources has obvious regularity. Specifically, there are more water resources in the east and fewer in the west, more in the mountains and fewer in the plains, and more on the windward slopes and fewer on the leeward sides of the mountains. A clear line can be drawn in a northeast-southwest direction from Qinghe in the Altai Mountains, past Changji in the Tianshan Mountains, and on through to Hotan in the north Kunlun mountains. This area is affected by the prevailing westerly winds and steep mountains which cause higher levels of precipitation, form large glaciers and swell river waters, all of which results in large oases.
However, southeast of this line, one finds the opposite conditions. For example, from west to east along the Tianshan Mountains, the average annual precipitation is 300–400 mm in the Yili River valley, 200–300 mm in Changji, and

50 mm in Hami. In contrast, the western portion is wet, the central part is mid-wet, and the eastern part is dry. The western oases areas of the north Tianshan Mountains and the western part of the south Tianshan Mountains are basically zonal contiguous distribution areas, while the eastern part of the north Tianshan Mountain oases feature small block distribution. In the more arid south, small oases consisting of only a dozen or even a few acres of land dot the landscape. These are irrigated by waters from the Karez in the Turpan-Hami Basin. Kunlun and Altun have less rainfall and lack rivers and glaciers, but their northern reaches support a large oasis space.

(2) Oases around basins

Oases are generally located in mountain passes along the rivers, plains, lakes, and river deltas. However, wherever there is an adequate and reliable water supply, an oasis may form. Oases need a special combination of soil, water, light and heat to develop. For example, a few large oases in the south of the Tianshan mountains are located at the alluvial plain of the middle and lower reaches of the Yarkand, Aksu, Hotan, Weigan, Kashgar, and Kongque Rivers, whose annual runoff is more than 10×10^8 m^3. Other large oases can be found in the north of the Tianshan Mountains and alluvial-diluvial plains of rivers such as the Manas, Kuitun, Hutubi, Toutun, Urumqi, Boertala, and Jinghe. In the alluvial-diluvial plains of these rivers or the alluvial fan in the lower part of the water, the soil is deep and suitable for farming. An oasis develops with soil and water; as channels extend and improve the basin irrigation area, spring wells can be drilled to expand land reclamation, and the oasis will continue to spread. Oases are generously distributed around the Tarim and Junggar Basins. Especially in the Tarim River Basin, artificial oases within the natural oasis ring basin is in the form of a discontinuous distribution pattern of the diamond or circle. This is closely related with the topography, water distribution and the formation of the alluvial plain. The southern Tarim River Basin oasis stretches 1,100 km from east to west, and its width ranges from a few to more than 100 km.

(3) Dotted distribution

Different geomorphological parts form different types of oases. For instance, alluvial fans are more likely to produce oases, as shown in Changji, Manas, and Shawan of the Tianshan Mountains. The presence of terraces on both sides of a river valley may also lead to the formation of an oasis, such as occurred at the Yili and Wushi oases. Additionally, downstream of a river where the water diffuses and disappears may give rise to a dry delta oasis, such as the Jiashi or Bachu oases. A flood alluvial plain at the middle and lower reaches of a river may also develop into a large oasis, such as the middle and lower reaches of the Tarim River oasis, the middle and lower reaches of the Yarkand River oasis, and oases generated at of the sides of plains lakes such as the Yanqi oasis at the Bosten Lake West Bank and the Jinghe oasis at Aibi Lake's south bank. Oasis distribution is different in the two basins. The Junggar Basin, north of the Tianshan Mountains, is an open semi-enclosed basin that features larger amounts of precipitation and hence larger oases, whereas Tianshan south of the Tarim River Basin is a fully-enclosed basin where there is little precipitation. Hence, this oasis area is generally small, with beaded and blotchy distribution.

1.2.2.2 Oases in the Hexi Corridor

The Hexi Corridor measures about 1,000 km in length from east to west and 10 km in width from north to south. The Corridor includes 17 cities and counties. Gansu, Yumen, Anxi, and the Dunhuang plain belong to the Shule River system; most of Zhangye, Gaotai, and the Jiuquan plain belong to the Heihe River system (although a small proportion belongs to the North River system); and Wuwei and the Minqin plain belong to the Shiyang River system. The Hei, Kuantai, and Dahuangshan Mountains separate the Corridor into the Shiyang, Hei, and Shule interior-drainage rivers. These three rivers originate in the Qilian Mountains and are supplied by snowmelt water and rainfall. The river infiltrates the Gobi desert to form a subsurface flow, or is used for oasis irrigation. In the Hexi Corridor, the accumulation of material carried down by the mountain rivers form the adjoining piedmont sloping plains. Downstream in the larger rivers, one finds the formation of alluvial plains, where soil is either a fine, sandy loam, or loam-based composition. This is the main distribution area of the Hexi Corridor's arable land, registering a thickness of 1 m. Under irrigation and with healthy organic matter content, soil fertility is generally good. This feature, combined with of the long hours of sunshine throughout the year (up to 3,500 h annually), yields rich growing conditions particularly suitable for the accumulation of fruits and sugar.

1.2.2.3 Oases in the Qaidam Basin

Qaidam is a fall basin with an elevation of 2,600–3,000 m. Precipitation in the basin region is low, with a general decreasing trend from east to west. Chaka, which is located at the eastern margin of the basin, receives about 200 mm of precipitation annually, but the western part of the basin gets only 20 mm. the aridity index is 2.0–9.0 in the east, and 20 in the west. Due to the extreme dryness, there is neither agriculture nor irrigation in the basin, but there are about 40 inland rivers that originate in the Kunlun and Qilian Mountains. These rivers are short, with annual runoff of surface water measuring 44.1×10^8 m^3 (Ren 2004). After 1949, the construction of agricultural oases along rivers such as the Chahanwusu, Xiangride, Golmud and Bayan, created the mohair agricultural areas of Golmud, Nuomuhong xiangride, Chahanwusu, Chachaxiangka and Delhi (Bai 1997).

Overall, oases regions in China's arid Northwest cover an expanse of more than 7×10^4 km^2, accounting for 4.5 % of the total land area. These oases are not only the product of long-term development and operation, but also the carriers and supporters of human civilization. In China's arid zone, the vast majority of the population, economic activity and social wealth are concentrated in oases. Hence, the Northwest's economy is essentially an oasis economy, especially in Xinjiang.

1.2.3 Population and Ethnic Structure

The socio-economic activities of Northwest China arid zone are intrinsically connected to the water resources and ecosystems. Indeed, population distribution and

Table 1.1 Northwest arid area's population composition in 2011

Region	Population/10^4	National population		Rural population		Agricultural population	
		Minority/10^4	Proportion/%	/10^4	/%	/10^4	/%
Hexi Corridor	342.41	28.39	9.2	319.57	93	251.68	74.0
Qaidam	48.93	30.38	62.08	18.27	37.3	21.19	43.3
Xinjiang	2,208.71	1,306.72	59.16	1,073.72	48.6	1,318.51	59.7
Nationwide	134,735	11,132.48	8.26	65,656	48.7	88,521	66.7

ethnic composition are directly impacted by the water resources' location and utilization and also have an effect on ecological environment construction and protection as well as the content and quantity of socio-economic activities.

1.2.3.1 Population Distribution

Natural conditions form the basis for the survival of mankind. These provide basic human living space and in natural conditions and regional differences that affect population distribution. Northwest China arid area is a vast and complex region with diverse types of terrain and climate. Within this framework, the natural and socio-economic conditions vary greatly.

The ethnic distribution of the different regions of the arid Northwest also has unique characteristics. Ethnic minorities in Xinjiang include Uygurs, Kazaks and others. The Uygur and Kirgiz peoples are concentrated in the south; the Kazakhs have settled in the north; the Xibe live in the Yili; and the Hui are more concentrated in Ningxia and Qinghai Province. Meanwhile, the living habits of different ethnic groups also play an important role in where they choose to settle. The Han, Uygur, and Hui are engaged in agriculture, industry and commerce, especially handicrafts. These peoples live mainly in the basin plains. The Kirgiz, Kazak and Mongols, with their traditional form of animal husbandry, are located in the middle zone, and the Tajiks are perched in the alpine zone.

1.2.3.2 Ethnic Composition

The ethnic composition of the Northwest is as complex as its land variations. At present, Xinjiang has 46 ethnic minorities, including Uygur, Kazak, Hui, Mongolian, Kirgiz, Xibe, Tajik, Uzbek, Daur, Tatar, and Russian. In the Hexi Corridor oasis region, there are 40 ethnic minorities, the main ones being Hui, Tibetan, Dongxiang, Tu, Man, Yugu, Baoan, Mongolia, Sala, and Kazakh. The Qaidam oasis also has dozens of ethnic minorities from Tibet, Mongolia, Kazakhstan, and so on. In these three areas—Xinjiang and the Hexi Corridor and Qaidam oases—the minority population outnumbers the national average at 59.16, 9.2 and 62.08%, respectively, meaning that minorities form the main population base of Northwest China (Table 1.1).

1.2.3.3 High Proportion of Agricultural Population

Because the arid area's oasis economy is agricultural-based, the rural population and the proportion of agricultural population are high. The proportion of rural population in Wuwei City is 72.5 %, much higher than the national average of 48.73 %. Similarly, the proportion of agricultural population is relatively high. In Xinjiang and Qaidam Basin, the numbers are slightly lower than the national average of 66.7 %, with the Hexi Corridor agricultural population accounting for 74.0 %.

1.2.3.4 Oasis Population Density

Currently, the population density of the Northwest arid oases (12.33/km^2) has exceeded the population density of arid zone limits (12/km^2). This is especially problematic in Xinjiang, where more than 95 % of the population live in less than 6×10^4 km^2 of the oasis area, giving an oasis population density close to 300 people/km^2, which is close to the average population density of the coastal provinces in eastern China. This causes a great deal of pressure on the ecological environment, especially on the oasis environment. Meanwhile, with the increase of the population in Xinjiang, the total population distribution has greatly expanded (Table 1.2). After 1994, there was a general movement in population from the mountain alluvial fan edge zone to the center of the basin or downstream of the river. Many new oases have been built in the natural desert and continue to extend to the desert. The handover part, which transitions from artificial oasis to desert, is often the area that is prone to reversal in ecology and environment. This distribution and expansion of population greatly exacerbates the difficulties inherent in ecological and environmental protection, and also exacerbates the possibility of ecological and environmental degradation.

1.3 Production and Ecological Problems

1.3.1 Productivity Development Lags Behind

The Northwest arid area is a typical continental arid and semi-arid zone. The region's average annual precipitation is 186.8 mm, which is 28.8 % of the corresponding national average annual precipitation (648.4 mm). Surface water and groundwater is 1 710.8 × 10^8 m^3, accounting for only 7 % of the total water resources. This is vastly disproportionate to the land size (one-third of the country's total area), which means there is a scarcity of water resources, and the spatial and temporal distribution is extremely uneven. The land area in southeastern Xinjiang is almost the same size as the northwest, but the region's water resources account for only 7 % of the total area. Similarly, the Hexi Corridor makes up 68.2 % of the land area in Gansu province, but its water resources account for only 27.3 %. The water resources per

Table 1.2 The main ethnic groups in Xinjiang population growth from 1991 to 2011 (Unit: 10^4). (Source: 2012 China Statistical Yearbook, Statistical Yearbook of Gansu and Xinjiang Statistical Yearbook)

Ethno-	1991	1993	1995	1997	1999	2001	2003	2005	2007	2009	2011	Average annual growth rate /%
Uighur	736.3	758.9	780.0	802.0	825.0	860.6	882.4	923.5	965.1	1002	1037	2.04
Han	584.2	603.7	631.8	660.1	687.2	742.2	771.1	795.7	823.9	741.7	844.4	2.23
Kazakh	116.1	119.6	123.8	127.1	130.5	131.9	135.2	141.4	148.4	151.2	154.3	1.64
Hui	70.37	73.23	74.76	77.06	79.26	84.42	86.67	89.35	94.30	98.04	100.3	2.13
Kirgiz	14.69	15.42	15.78	16.27	16.64	17.01	17.37	17.15	18.19	18.93	19.4	1.60
Mongolian	14.47	14.92	15.28	15.75	16.13	16.19	16.69	17.17	17.71	17.96	17.89	1.18
Xibe	3.51	3.68	3.82	3.90	4.00	4.13	4.03	4.15	4.24	4.28	4.27	1.08
Russia	0.80	0.86	0.90	0.92	0.94	1.08	1.11	1.12	1.16	1.17	1.16	2.25
Tajik	3.45	3.61	3.82	3.90	4.01	4.12	4.09	4.40	4.48	4.72	4.73	1.86
Uzbek	1.17	1.28	1.33	1.35	1.39	1.37	1.46	1.51	1.61	1.67	1.74	2.44
Tatar	0.44	0.44	0.47	0.46	0.48	0.47	0.49	0.47	0.47	0.49	0.49	0.57
Man	1.75	1.89	1.99	2.06	2.12	2.30	2.39	2.46	2.56	2.62	2.64	2.54
Daur	0.55	0.58	0.62	0.63	0.65	0.67	0.67	0.65	0.67	0.69	0.69	1.27
Other	6.75	7.09	7.00	6.56	6.75	9.75	10.32	11.37	12.42	12.74	13.23	4.80
Total population	1,554	1,605	1,661	1,718	1,775	1,876	1,933	2,010	2,095	2,058	2,209	2.11

capita in Gansu, Ningxia, and Inner Mongolia are also far below the national average, especially in the densely populated and economically developed oasis areas, which continuously experience serious water shortages.

On the other hand, the precipitation in the region has space differences. Gansu, Ningxia and Qinghai, west of Lanzhou, have an annual precipitation of less than 300 mm; 100×10^4 km^2 of the desert and Gobi region have a low annual precipitation of 50 mm; the Ruoqiang area (southeastern Tarim River Basin) has less than 10 mm of precipitation, with the annual number of precipitation days clocking in at around 13, making it the country's most arid region. Seasonal distribution of precipitation in the region shows more obvious differences. In the Hexi Corridor (in the eastern part of Xinjiang), rainfall in July and August accounts for 60 % of the annual precipitation. Conversely, in northern Xinjiang and the western Tarim River Basin, the summer rains comprise only 40 % of annual precipitation, while Yili-Tacheng has more precipitation in spring.

The shortage of water resources and fragile ecological environment are important factors constraining economic and social development in the Northwest. Large-scale, high-intensity activities change the water/space allocation, destabilizing the relative balance. Moreover, with rising economic development, ecological problems become increasingly prominent. Over the past 50 years, the oases area in the Xinjiang region has increased by nearly three-fold. At the same time, desertification has expanded to encompass 8,600 km^2; grassland degradation involves an area of 670×10^4 hm^2; desert vegetation has declined 50–70 %; lake water has reduced nearly 50 %; and soil salinization has increased to more than 30 % of the existing arable land. Additionally, the integrity of the mountain ecosystem has been severely damaged, water conservation efforts have declined, and flood and drought disasters have compounded.

1.3.1.1 The Expansion of Desertification Areas

The most important ecological problem of the Northwest is desertification. In Xinjiang, the natural desert formed after the Quaternary Pleistocene, especially during the late Pleistocene to Holocene. It should be noted that this period of development was a natural process. In contrast, in the historical period (especially in the recent past), humans have either changed the original oases into desert or contributed to the deterioration of the original desert environment.

In the arid area, desertification takes three main forms: (1) blowing sand and dune movement encroaching on the original non-desert land; (2) irrational human economic activities, such as overgrazing and firewood scavenging, which contributes to the drying up of rivers and lakes, the changing of surface conditions, and the exacerbation of aeolian erosion, forming new quicksand; and (3) fixed and semi-fixed desert destruction of vegetation desert activation. In the past, people paid more attention to the first form while ignoring the other two forms. Today, modern desertification is called the "ecological processes of desertification", accurately showing the real causes of desertification.

Table 1.3 The Hexi Corridor land desertification: degree and distribution characteristics (10^4 hm^2)

Area	Jiuquan	Wuwei	Zhangye	Jiayuguan	Jinchang	Total
Desertification area	340.1	154.14	33.46	0.45	8.37	536.5
Mild desertification area	107.24	41.03	4.78	0.01	0.34	153.4
Moderate desertification area	67.04	15.95	13.86	0.16	7.96	104.97
Serious desertification area	165.8	97.16	14.82	0.28	0.07	278.13
Proportion of serious desertification area	48.75	63.03	44.29	62.22	0.84	51.84

Historical desertification of Xinjiang's land area is approximately 3.05×10^4 km^2, accounting for 17.27% of China's desert land, of which 30% occurred in the past century. Take as an example the edge of the Taklimakan Desert and the desert on both sides of the river. Over the past 2,000 years, the formation of desertification land area measured 2.83×10^4 km^2, while in just this past 100 years 0.86 km^2 of desertification was added. These statistics do not take the dunes into account.

Desert activation is the main form of desertification in Xinjiang and is a direct threat to adjacent oases. In the Gurbantungate Desert, for example, the original quicksand area was only about 3% of the land area but has now grown to encompass 15–21%, increasing to 5856~8784 km^2. This increase occurred over a 40-year period, between the 1950s and the late 1990s. At present, desertification exists in 80 counties and cities and 90 mission fields in Xinjiang, and 1200×10^4 of the population is impacted. More specifically, an average of 66.67×10^4 hm^2 farmland is impacted, as well as 0.08×10^8 km^2 of grassland. This loss through rapidly encroaching desertification results in annual economic losses of up to 25×10^8RMB. Desserts are expanding at an annual rate of about 400 km^2, accounting for about 15% of China's overall desertification. Considering the rate of desertification expansion, Xinjiang is facing a grave crisis. Its desert environment (including the Gobi Desert) comprises about half of its natural environment. When oasis expansion space is limited, and salinity, drought and other factors reach a certain level, desertification's net growth will rapidly increase and quickly affect the existing oases.

The Hexi Corridor in Gansu Province is also one of China's worst land desertification hazards. As we can see from the status of land desertification and the extent of its distribution characteristics (Table 1.3), except Jinchang, almost all of the desert land accounts for more than 44.3% of the total land area of the Hexi Corridor. Of this area, Wuwei and Jiayuguan account for more than 62%, showing the extent of desertification in the region.

According to 1996 land desertification census results from the Forestry Survey of Design Institute of Gansu Province, the development of land desertification in the western part of the Hexi Corridor had progressed rapidly over the previous 50 years. By the mid-1980s, the drive for economic development in the area had spurred desertification to an annual growth rate of 2.15 %, and a similar degree of desertification was occurring in the Jinta area at the edge of the Badan Jaran Desert. Prior to the 1980s, the evolution of land desertification was substantially less, at 0.37 %. In the Zhangye Prefecture, land desertification was also occurring, but to a lesser extent. Since the 1990s, however, land desertification in the Hexi Corridor area has significantly slowed.

Overall, then, the desertification in the Corridor region was most extreme during the 1970s and 1980s, after which it slowed and even in some sections experienced a partial reversal. Moreover, desertification in the western part of the Corridor is significantly stronger than in the eastern portion. In the middle reaches of the Heihe River Basin, the process of desertification block phenomenon is most obvious (Wang and Cheng 1999). Since the 1980s, the artificial oases area in the middle reaches of the basin has increased by an average annual rate of 2.1 %, while desertification increased by only 0.21 %. This is related to the large-scale Sand Control efforst in the basin's middle reaches.

Despite attempts to intervene in the desertification process or even to undo the damage done previously, in most parts of the Heihe River Basin the formation and development of land desertification is controlled by the river water changes. Reduced surface runoff is the most important driving factor in land desertification of downstream areas (Zhu and Chen 1994). The Ejinaqi, situated in the lower reaches of the Heihe River, has a land area of 70,712 km^2. The natural oasis area there was reduced from 3,655 to 3,328 km^2 in the mid-1980s, a decrease of 8.95 %. Since that time, desertification has increased 31.75 %, from 25,834 to 34,038 km^2.

Salinization of the land is another major form of ecological degradation. The Hexi Corridor region's arable land salinization covers 11.24×10^4 ha, of which 7.09×10^4 ha is secondary salinization. In recent years, the land salinization process between the middle and lower reaches of the basin showed opposite changes. In the upper and middle reaches of the Shiyang in the Heihe River Basin zone, salinization of the land area has been gradually reduced by falling water tables, while the downstream region, such as northern Minqin Basin, shows an increasing trend. Salinized arable land grew by 1.05×10^4 ha in 1958; 1.23×10^4 ha in 1963, 2.21×10^4 ha in 1978, and 4.08×10^4 ha in the mid-1990s. By 1995, Ejinaqi's salinized land measured 66.41×10^4 hm^2, accounting for 49.5 % of the area's degraded land.

1.3.1.2 Shrinking Rivers and Lakes

In tangent with desertification, rivers and lakes in China's arid Northwest region are shrinking and even completely drying up. There are 570 small rivers in Xinjiang; the vast majority of these are genus rivers with a small amount of water and a relatively short run. Due to the continuous development of artificial canals, the water levels in many of the rivers downstream have sharply dropped or dried up. For instance, the Tarim River, which is China's largest inland river, saw its downstream water reduced from 13.5×10^8 m^3 in the 1950s to the current 2.84×10^8 m^3. Similarly, the Daxihaizi reservoir is almost dried up, and the northern slopes of the Tianshan water system have been changed. Major rivers that once formed reservoirs have been reduced so that they no longer flow into the desert. Even the famous Manas River has been completely replaced by an artificial canal.

Likewise, lakes have been adversely affected by human activity. Xinjiang is a multi-lake area with more than 139 lakes greater than 1 km^2. Overall, there is a total lake-water area of 9,700 km^2, accounting for 7.3 % of China's total lake water and ranking the area fourth in the country. However, by the late 1970s, the total lake water dropped to 4,748 km^2, which was a loss of nearly 5,000 km^2. According to a recent survey made of 16 representative lakes, nearly 3,000 km^2 of Lop Nur Lake and 550 km^2 of Manas Lake, Aydin Lake, and the Taiwan Lakes have shrunk by 1,200 km^2. Aibi Lake has been reduced to 500 km^2.

Unfortunately, as lake areas are reduced, lake salinity increases and water deteriorates. Until the end of the 1950s, the salinity of the lower reaches of the Tarim River did not exceed 1 g/L. Now, however, even in the wet season of 1994, Daxihaizi reservoir salinity is still as high as 3.04 g/L; in the dry season, the river has a high salinity of more than 5 g/L, resulting in reservoirs along the river having a salinity of 1 g/L or more. The highest is up to 8 g/L (Mirzali reservoir, the Dakarquga reservoir, June 1997). As river water quality deteriorates, so does the quality of the lake water. From the 1960s to the 1980s, the salinity of China's largest inland freshwater lake—Bosten Lake—increased by 0.4–1.8 g/L. After 2000, lake salinity began to decrease, but since 2006, it started a rebound trend, rising to 1.2–1.5 g/L in 2012. Changes in the lake environment pose a serious threat to drinking water in Korla. River water quality deterioration is a direct threat to agriculture and aquaculture as well, further exacerbating the water crisis.

The Tarim River and its tributaries, along with Bosten and Aibi Lakes, can change the overall ecological impact of Xinjiang. In this, natural ecosystems play a crucial role. For instance, Bosten Lake is important for the regulation of the river, the protection of agricultural production, aquaculture development, and the development of new tourist venues. Aibi Lake is a saltwater lake that contributes to the salt industry development and Artemia production. Additionally, the lake serves as a habitat and breeding ground for millions of migratory birds. Today, however, the lake has atrophied, resulting in a large area of exposed lake bottom and the formation of new sources of sand. Furthermore, the Tarim River's outlet is a huge threat to the economic belt on the northern slope of the Tianshan Mountains and the New Eurasian Continental Bridge. As a result, there is an urgent need in Xinjiang to implement protection of the ecological environment of key waters in order to curb further deterioration of the environment.

The Heihe River inflow from Inner Mongolia into the Ejinaqi River plays a decisive role in the fertility and growth of the Ejina Oasis. Irrigation and water consumption have increased year by year, drying out the river. At the end of the Ejina River is Juyanhai, whose catchment area prior to industrial development in the region was Multi-lake Lakes. In the 1940s, the water depth of western of Juyanhai is 2.9 m, with the area of 190 km^2; while in the eastern of Juyanhai, the water depth is 4.1 m, with the area of 35.5 km^2. Until the 1960s, the western of Juyanhai has dried up, the eastern of Juyanhai has changed a lake. In the early 1980s, the lake shrank to 23.65 km^2 and a depth of only 1.8 m; by 1999, it was completely dried up. Since then, through the implementation of ecological water transport to the lower reaches of the Heihe River, the downstream ecological landscape has changed significantly, with varying degrees of ecological recovery reproducing the landscape of the past.

1.3.1.3 Decrease in Natural Forest Area

The Xinjiang forest coverage rate increased from 1.03 % at the end of the 1970s to the current level of 1.68 %, a gain of 0.65 % points, which is equivalent to an area of 107×10^4 ha. This is mainly due to a substantial increases in plantation area. However, there were also measurement standard changes, including canopy density standards and forest types. Only part of the natural forests, including montage forest, valley forest and desert riparian forest, have recovered. The area experienced a serious reduction (more than one-third) in mountain forest resources, where only about 40 % of the forests remain. Of this, growth is poor, and the young forest does not compare with the original natural forest in terms of ecological benefits (e.g., virgin forest water conservation water is 1 % compared to 0.1 % of new-growth forests). The valley forest suffered more severe damage. Although there is some degree of recovery in the desert Populous, the desert shrub area was reduced by up to 50 %, and due to deterioration in water conditions and the lack of natural regeneration conditions, closing measures alone are inadequate to restore the area to its former state. Therefore, in addition to water resources control, it is also necessary to deal with the implementation of appropriate protection measures in key areas.

Ejina likewise faces serious degradation of its forest ecosystems. From 1958 to 1982, in the downstream delta area along the river, Populus, Elaeagnus, and tamarisk willows were reduced by 5.73×10^4 hm^2, with an average annual reduction of 0.26×10^4 hm^2. Since the 1980s, Populus angustifolia, Tamarix ramosissima and shrub areas decreased 19.24×10^4 hm^2, with an annual decline of 1.374×10^4 ha. In the Minqin Basin in the 1950s, the natural Chai Wan covered nearly 13.33×10^4 hm^2. However, in a survey in 1981, only 7.242×10^4 ha remained, and in the 1990s, nearly 3.6×10^4 hm^2 of the stand were close to death.

1.3.1.4 Grassland Degradation

Like the rivers and lakes, Xinjiang's natural grassland area is shrinking rapidly. Grassland degradation has reached $2{,}133 \times 10^4$ hm^2, with the severely degraded area measuring 670×10^4 ha and the available grass area reduced to 240×10^4 hm^2. The identified grass degradation rate has likely reached 29×10^4 ha, with some researchers setting the rate even higher, at 66.7×10^4 ha. The grassland degradation means a rapid decline in productivity. In yield per unit, area grass in the 1960s decreased by 30–50 %, and by as much as 60–80 % in some places. Such a drastic decline in grassland has also produced landslides, making the development of grassland animal husbandry difficult and underscoring the urgent need to establish suitable grassland protection practices.

The continuing deterioration of the ecological environment in Xinjiang due to the maintenance of the status quo is a prominent issue. It indicates that the complexity of the task of environmental protection and construction is arduous and long-term, and that there is a present dearth of leadership in ecological matters. Society, nature, people and the environment form a continuum; only if they live together in harmony will they be able to make progress together and create sustainable development.

After nearly 50 years of deleterious human activity, the Hexi Corridor grassland region has sustained reduced forage yields and grassland desertification and has significantly reduced its carrying capacity. The degradation of the Heihe River Basin's upstream grassland area has reached 52 % of the available grassland, and the overall degradation of the basin-wide grass area is at 46.72 %, resulting in the desertification of grassland in an area covering 28.101×10^4 ha. In addition, sand has further reduced the grass area by 29.35×10^4 ha. Meanwhile, the Shiyang River watershed is suffering degradation of around 45.07 % of its total grassland area, with a grass overload rate of 111.2 %. The Shule River, although not yet in an overloaded condition, has a grassland degradation of 27.01 %. Over the past 50 years, the Hexi Corridor area's grass overloading rate reached 69.1 %, with a degradation rate of 46.86 %. At the same time, due to the limited moisture factor and the continuous reduction of the amount of water basin downstream of plant species, the density, coverage and yield consistently dropped. Since the early 1980s, the vegetation in Ejinaqi has had a coverage of more than 70%. Specifically, shrub meadow with the coverage of 70 % has been reduced 78.26 %; while the shrub meadow with the coverage of 30–70 % has been reduced 40.45 %, and the grassland reduced by nearly 50.5 %; on the contrary, the Gobi desert with the coverage of 10–30 % and the desert with the coverage less than 10% expanded 67.63 %.

1.3.2 *Serious Ecological Degradation*

Due to limitations in natural conditions, ethnic characteristics, weak infrastructure and many other factors, the economy in the region has been at a low level and its productivity development has lagged behind other regions. The main reasons for this state of affairs are as follows:

1.3.2.1 Weak Economic Advantage

China's arid Northwest covers an enormous amount of territory. The terrain of this region features huge mountain ranges and vast basins and valleys. The oases are mostly widely dispersed and relatively small, and are surrounded and isolated by the Gobi desert and desert grasslands. The vast majority of the arid zone's social and economic activities occurs in the oasis regions, while there is little or no human activity in the periphery areas. Moreover, there is little economic activity with regards to transportation lines because the oases are so far apart.

If we view the oases as a discrete system, we see that this system's surface region is highly dispersed. Each node has direct contact only with a small number of other nodes, mainly through one or two communication lines, resulting in less contact with internal and external discrete systems. This rudimentary level of contact and communication renders the flow rate too slow, detrimentally impacting both the economic concentration and the diffusion effect. These properties cause the oasis economy to be closed, limited and unbalanced, slowing the arid region's development and giving it a distinct disadvantage in the national and international marketplace.

Northwest China's arid zone accounts for about one-quarter of the country's total area. Although rich in solar, mineral and petroleum resources, it features sparse vegetation, soil desertification and salinization. The ecological environment is very fragile and forms the weakest link in the population-resources- environment chain. Since 1949, due to ever-changing economic policies, the oases areas have undergone sporadic expansion and development, improving the area for human habitation but destroying many aspects of the natural ecological environment. Human activities in the Northwest have led to drying lakes, worsening desertification, expanding deserts, loss of biodiversity, and narrowing transition zone between oases and deserts. Worst of all, especially in light of future development plans, the vulnerability of the arid zone's ecological environment causes a low bearing capacity, which can have a serious impact on economic development.

1.3.2.2 Irrational Industrial Structure

Although agriculture has been the mainstay of the oasis economy in Northwestern China, the irrational industrial structure restricts further economic development in the region. This structure is mainly manifested by the disproportionate role that agriculture plays in the economy. For example, in Aksu, Xinjiang, in 2011, agricultural output accounted for about 53.43 %, industrial output for about 27.67 %, and tertiary industry 18.90 %. We can see from this division that the proportion of agricultural is too large. Moreover, within the structure of agriculture, farming accounted for about 88.37 %, animal husbandry for 5.49 %, forestry for about 1.07 %, and other activities for about 4.65 %. The unreasonable and dominant proportion agriculture constrains further development of the oasis economy.

China's oasis economy is based on agriculture but also includes animal husbandry and the primary processing of materials. At present, the Sinotrans supplies in Xinjiang are: crude oil, oil products, coal, salt, Glauber's salt, cotton, cotton, wool, leather, fruit, sugar and other raw materials and primary processed products. In the Hexi Corridor, the supplies are: commodity grain, cotton, sugar beets, oil, fruits and vegetables, oil, coal and nickel and other raw materials, and primary processed products. The seed industry also plays an important role in Hexi Corridor's oasis economy.

With its focus on agriculture, the region's economic development depends on high water consumption. Generally speaking, the economy in the arid zone is different from the eastern part of China, where precipitation is abundant. The oasis economy grew because natural rainfall was insufficient to nourish crops in the arid region; hence, agricultural production and human economic activities are closely related to the development and utilization of water resources. The main general characteristics of water resources in China's arid zone are low water resources and uneven geographical distribution. Specifically, the area is more mountainous than flat, has more windward slopes than leeward ones, and has few perennial rivers. Most water sources are highly seasonal and include intermittent streams, temporary surface runoff, and short-run rivers fed by mountain snowmelt water and glaciers. The annual runoff of surface water volume is 9.46×108 m^3, accounting for only 3.34 % of the water in the entire country.

Not surprisingly, the water consumption in the arid region is significantly higher than the national average. For instance, water consumption in the Hexi Corridor is 1.3 times higher than the national average and Xinjiang's consumption is nearly 6 times higher. Furthermore, the water-use efficiency of the gross national product in developed countries is 10.3 \$/m^3, whereas in China it is 10.7 yuan/m^3, which only accounts for 1/8 and 1/25, respectively. In Xinjiang, the water-use efficiency of the gross national product is 27 yuan/m^3, which is only 1/4 of the average value in the whole country, and far behind the developed countries.

Overall, there is a high water diversion rate and a low water utilization rate in the arid region. Although the water diversion rate of rivers is around 60–80 %, the water is not yet fully utilized in irrigation areas. Other than for the Tianshan Mountains and the Shiyang River in the Hexi Corridor, the gross irrigation quota ranges between 1,800 and 27,000 m^3/hm^2, while in the Heihe River it ranges between 7,500 and 12,000 m^3/hm^2. The canal system utilization factor is 0.35–0.45 in most parts, whereas in the Tianshan Mountains and eastern Hexi Corridor it reaches 0.5–0.6. In the high-level irrigation area of the Manas River and the Hexi Corridor, 1 m^3 of water can produce 0–1.0 kg grain (the water consumption of grain being 1.0–1.6 m^3/k). Likewise, 1 m^3 of water can produce 0.16 kg cotton (the water consumption of cotton being 6.3 m^3/kg). In the low-level irrigation of the Tarim River Basin, Qaidam Basin and Shule River in the Hexi Corridor, 1 m^3 of water can produce 0.15–0.3 kg grain (the water consumption of grain is 3.0–6.6 m^3/kg) and 1 m^3 of water can produce 0.03 kg cotton (the water consumption of cotton is 8–28 m^3/kg). Industrial water consumption accounts for only 0.6 % of the total amount of water used. For instance, the water consumption of 10,000 yuan industrial output value is 141–223 m^3, which is close to the national average. From the above, it is can be seen that the agricultural economy in the oasis arid region is supported by water resources. Despite this dependency, water-use efficiency is low, and the investment in water resources and output in agricultural are asymmetric. Economic development in this region is based on high water consumption.

1.3.2.3 Low Level of Productivity

Water resources in the arid Northwest region give birth to the oases and the extent of the water resources effectively determines the size and productivity of the oases. A small oasis area per capita translates to a low GDP of per capita and hence to a low economic productivity level. As outlined in previous sections, Xinjiang is a typical example of an oasis economy. Studies (Wu et al. 2011; Abaydulla et al. 2006) have shown that the distribution and scale of an oasis has a close relationship with the water diversion rate of surface and groundwater. Taking the water consumption in Xinjiang in 1995 as an example, 1×10^8 m^3 of water can create an artificial oasis measuring 1.58×10^4 hm^2 (equivalent to 23.74×10^4 acres). Per acre, an oasis in South Xinjiang uses about 500 m^3 of water, while in North Xinjiang, an acre uses 343 m^3 of water, and in Turpan it uses 780 m^3 of water. Moreover, the absolute land area in Xinjiang is large but the available area is small. Although the land area totals 166×10^4 km^2, land suitable for agricultural production comprises only 5 % of that total. As of 2011, the cultivated land area in Xinjiang measured 412.6×10^4 hm^2, accounting for only 2.48 % of the total area. The desert, barren lands, and other unused land areas totaled around 1.02×10^8 hm^2, giving a land utilization rate of 38.64 %.

As can be seen from the above, the oases in Xinjiang account for only a small amount (5 %) of the total land area. However, under the dual constraints of water shortages and landscape structure, oasis development is limited. Despite these restrictions, the population in Xinjiang has grown rapidly, from 4,333,400 in 1949 to 22,087,100 in 2011. This represents a growth rate of 409.70 % over the past 62 years, which is far higher than the national average of 148.74 %. In the Xinjiang oases, this means that 95 % of the population lives in only 5 % of the total area (Committee of Natural Resources Series 1995). Such an imbalance clearly inhibits the region's potential social and economic development and is also serious threat to the environmental carrying capacity.

At the same time, and not surprisingly, the economic benefits of capital construction investment in the oasis area are relatively low. Since 1994, the delivery rate of fixed assets of infrastructure has been much lower than the national average, as have been funds interest rate and labor productivity. During the "Sixth Five-Year Plan" period, the rate of capital profit tax was only 50 % of the country's; however, by the "Ninth Five-Year Plan" period, most of the enterprises began to gain significant earnings.

Numerous factors can cause the phenomenon of low economic efficiency. In addition to the afore-mentioned water constraints, such factors include a weak economic foundation, the small scale of enterprises, outdated production equipment, and long transport lines. Agriculture and animal husbandry have formed the mainstay of the traditional oasis economy for years, but productivity is low and economic accumulation weak. Some progress has been made since 1949, but small-scale production, even in an industrialized setting, rarely garners larger economic benefits. The tools of agricultural production in the oasis region are traditional, which, while quaint, causes low productivity. Further compounding the situation are long transport lines and high transportation costs, which reduce the region's ability to compete with the east. The limited capacity of transportation also leads to a backlog of supplies, resulting in a further weakening of competitiveness in the market-based economy. For example, due to transport issues, the abundant coal resources are confined to west of Lanzhou.

1.3.3 Causes of Ecological Degradation

Damage caused to an ecosystem is generally associated with that system's own fragile ecological environment. A normal ecosystem can maintain a dynamic balance between the biological and the natural environments, but if the structure and function of the ecosystem are changed, the equilibrium of the original ecosystem will be disrupted, and the structure and function of the system will change or be obstructed. The ecosystems in the arid region of Northwest China are extremely fragile due mainly to the following two factors: (1) large-scale regional environmental factors, that is, the internal factors of ecological processes; and (2) the interference of human activities, or so-called external factors.

1.3.3.1 Natural Factors

(1) Special Conditions of Landform Terrain The arid region of Northwest China is located in the hinterland of the Eurasian continent, far from any ocean. The terrain of this region is characterized by huge mountain ranges and vast basins or valleys; the oases are widely-dispersed, relatively small, and surrounded by the Gobi desert and desert grassland. The Junggar and Tarim River Basins are closed inland basins, the Tarim being fully enclosed. At the edge of the Basin is gravel, then sheet oasis, and finally the wide expanse of the Taklimakan Desert. Irrigation drains into the basin, so the soil has a high salt content, which then flows into the farmland with the surface water. Due to strong evaporation, the land is prone to secondary salinization. In addition, the Heihe River also ends in the Ejina Desert due to the topography.

(2) Fragile Ecological Environment The climate in the arid region is drought-prone and the harsh natural conditions create extreme sand and soil salinization. The long-term dry weather is a key factor affecting the development of plant groups, which are exposed not only to dangerously low levels of precipitation and irrigation but also wind erosion and desertification (Huang 1993). Although there are azonal hydrothermal conditions in the arid region, alien species find it difficult to survive here, so plant species in this region are simple. For example, the forest coverage rate in Xinjiang is only 1.68 %, indicating a vulnerable ecosystem. Moreover, this vegetation community has a great deal of irreversibility; once destroyed, it is difficult to restore, which further exacerbates land degradation.

1.3.3.2 Human Factors

Recent human activities have greatly changed the distribution pattern of the original natural ecological environment in Xinjiang. An artificial ecosystem is slowly replacing the natural ecosystems. Consequently, we see farmland and planned forests instead of natural forests and grassland, artificial channels instead of natural rivers, artificial reservoirs instead of natural lakes, and so on. These human activities have seriously impacted the stability of the desert ecosystems.

(1) Population Growth The excessive growth of the region's population has already put extreme pressure on the fragile ecological environment in Xinjiang. Coupled with the discordant development of the basin regional (basin), the consumption of water resources and desertification process has been significantly exacerbated. More than 95 % of the population lives in oases that account for only 8.89 % of the land area in Xinjiang. In 1950, Xinjiang's population was just 4,100,000. This increased to 16,613,600 by 1995 (representing an oasis population density of 111 people per km^2) and to 22,087,100 by 2011. Such rapid growth in population requires an appropriately expanded scale of production and construction. Hence, Xinjiang's arable land area grew from 121.0×10^4 hm^2 in 1950 to 310.0×10^4 hm^2 in 1990. By 2011, the population had increased to 412.46×10^4 hm^2, placing a tremendous strain on the environment.

The land area in the Hexi Corridor is about 27.4×10^4 km^2, comprising 60 % of the total land area in Gansu Province. However, the arable land only accounts for 18.2 % of the total area of cultivated land in Gansu Province, and this 18.2 % of land provides 35 % of the food and 90 % of the cotton output in Gansu Province. The oasis area has expanded threefold over the past 50 years, and the population has increased exponentially, leading to a population density of around 15/km^2, far more than the international standard of 7/km^2 for arid regions. According to the statistics released in 1998 (Wang 2002), the population density of the Hexi Oasis is around 200 people per km^2, and in the Wuwei Oasis, it is a staggering 570 people per km^2.

The pressure on the environment is increasing with the rising population. In order to meet the needs of the rapidly expanding population, development strategies are being proposed to reclaim wasteland and expand the scale of agricultural production. However, desert vegetation is invariably destroyed in the reclamation process. According to recent statistics, in 1950, arable land area in Xinjiang measured 121.0×10^4 hm^2; this increased to 412.46×10^4 hm^2 by 2011. Thus, arable land in the region increased by about nearly four-fold over a 61-year period. More recently, in just the past 10 years, irrigated arable land in the middle of the Heihe River increased 721.01 km^2. The Hexi Corridor provides 70 % of the commodity grain, 90 % the cotton and sugar beets, and about 40 % of the oil, melons, vegetables and other agricultural products of the entire Gansu Province. In this process, water resource utilization exceeds 65 %.

Overall, the processes of land reclamation and expanding farmland did not bring long-term economic benefits but instead produced a series of ecological and environmental problems. First of all, the rivers were generally shortened during the large-scale reclamation of wasteland, unplanned irrigation, and dam building. This led to a decrease in the middle and lower reaches of Tarim River waters, resulting in 321 km of the river's downstream portion drying up. Furthermore, the coastal green corridor is in danger of disappearing as millions of acres of grassland that relied on the river's downstream waters become desertified. As well, the floodplain forest is over-cultivated in the Ebinur Lake, damaging about 8×10^4 hm^2 of Populus euphratica and 4.7×10^4 hm^2 of Phragmites communis. This, in turn, has caused the original fixed and semi-fixed sand dunes to turn into liquidity dunes. In addition, the Ejina and Juyanhai Rivers downstream of the Heihe River are drying up, and the Ejina Oasis is atrophying.

Lakes are also suffering serious shrinkage. In the early 1950s, there were more than 100 lakes in Xinjiang, with a lake area of 97.0×10^4 km^2. However, by the end of the 1970s, the lake area was reduced to 47.5×10^4 km^2, a drop of nearly 50 %. Many of these lakes dried up completely or had the quality of their water significantly compromised.

Some land reclamation projects caused soil and water imbalances which led to further deterioration of the ecological environment as well as failed projects. According to statistics, in Hetian, the accumulated land reclamation was 2.33×10^4 hm^2 from 1988–1996, but the arable land area was only 15.96×10^4 hm^2 in 1996, an increase of a mere 8×10^4 hm^2 compared with 1988. Despite such dismal showings, the scale of reclamation was further expanded in the early twenty-first century, with most of the new wasteland (which in the meantime had become "man-made sand") being abandoned due to water shortages.

Almost all of the desert vegetation in the arid region is associated with rivers and lakes. However, as most the rivers have been rerouted into artificial irrigation channels, their courses and lengths have been substantially shortened. This has led to significant declines in the groundwater table as well as the degradation of riparian vegetation. It has also expanded the desertification of the lower reaches of rivers.

Because rivers have either been rerouted or their lower reaches have disappeared altogether, the lakes they once fed have either shrunk or dried up. Manas Lake, for example, has succumbed to drought brought on by a rerouted river, leading to the death of Phragmites communis and the extinction of numerous aquatic species. Sincere but misguided attempts to prevent this from happening elsewhere include 8.7×10^8 m^3 of water being injected into the Wulungu Lake in northwest edge of the Gurban desert since the 1950s. In recent 10 years, due to water scarcity in the Tarim River's downstream portion, desert vegetation around the riverside has withered and the desertified land area expanded from 106 to 140 km^2. The original Populus in the green corridor in Korla and Ruoqiang has decreased from 6.1×10^4 to 1.6×10^4 hm^2, and approximately 0.7×10^4 hm^2 of the *Populus euphratica* and *Haloxylon* are now dead.

Groundwater exploration has likewise increased to meet the needs of the expanding arable land areas. This has caused a significant decline in groundwater levels at the edge of the oases and in newly reclaimed land. For example, after more than 2,000 new wells were drilled in the Tarim River, the groundwater level declined 30–50 m. Over the past 10 years, the desert vegetation that depended on groundwater has died, biodiversity has been damaged, and the vegetation coverage reduced.

(2) Extensive Management Water usage of irrigated land in Xinjiang is 930 m^3 per acre, indicating relatively low efficiency. The utilization coefficient of the channel system is about 0.41, and the irrigation quota is around 13,500–16,500 m^3/hm^2. There is also extreme seepage of water used in farmland, and water conservation facilities cannot keep pace with land development. This leads to declines in groundwater levels and increased soil salinity of arable land, causing secondary salinization. Currently, the total area of secondary salinization in the Tarim River is about 3.6×10^4 hm^2, which accounts for 43–60 % of the arable land. The worst of the secondary salinization has occurred in the middle of the Tarim River. Depleted groundwater and rising river salinity are the two most important causes of soil salinization in irrigation areas.

Overgrazing also destroys ecological balance. In Xinjiang, the degradation of grassland caused by overgrazing is widespread and prominent, as the expansion of livestock-related industries has exceeded the capacity of grass to sustain the herds. Between 1949 and 1996, the number of livestock in Xinjiang nearly tripled or in some cases more than quadrupled. At the same time, the available area of pasture decreased by 266.7×10^4 hm^2, reducing the average portion of grass from 5.39 to 1.77 hm^2. Consequently, about $2{,}000 \times 10^4$ livestock barely survive the hunger and cold of winter and spring, economic losses are in the 10×10^8 yuan range, and 40×10^8 kg of grassland is being degraded. To make up for this loss, human production must generate 20×10^8 yuan, but this does not take into account the impact of such production on the ecological environment. At present, the grassland in spring and autumn is overloaded by more than 70 %; in winter, the overload is around 50 %, and only summer sees a surplus. Grasslands overloads occur to different degrees in different regions. In Urumqi, the grassland overloads about 2.7 times in spring and autumn, and 1.25 times in winter, which causes a decrease in grass production of 50–70 %, while in Changji, the grassland overloads about 4.1 times in spring and autumn.

There is also severe overgrazing of grassland in the upstream and middle reaches of the Heihe River Basin, at an average rate of 21.68 %. The overload rate of the local Sunan pasture has reached 70 %, and the available area of pasture in winter and spring only takes 39 % of the total grassland area. Although the theoretical stocking rate is 98.76×10^4 of sheep, the actual stocking rate is 260.4×10^4, so the overloading rate is as high as 163.67 %. The overgrazing phenomenon of grassland also happens in summer in the regions of Minle, Gaotai, Zhangye and Linze and averages $2{,}264 \times 10^4$. The overgrazing of grassland has likewise become an important factor in grassland degradation in the upper and middle reaches of the Heihe River, where there is an over-concentration of livestock grazing the shrub meadow in the river's downstream portion.

Moreover, excessive felling results in the demise of large swathes of desert vegetation. Desert vegetation plays an irreplaceable role in fixing moving sands, reducing wind erosion, and improving the environment, and many desert herbs and small shrubs are nutritious forage and also have medicinal value. Due to the seemingly endless drive for economic development, excessive hunting, deforestation and other destructive human activities damage the area's natural fixing plants such as Tamarisk and Haloxylon. Consequently, the desertification process expands rapidly and biodiversity is sharply reduced. As a sad example, the desert Xinjiang tiger and bullhead are now extinct.

In the 1950s, there were 193×10^4 hm^2 of licorice resources in Xinjiang, but now there are only 67×10^4 hm^2 and licorice storage has been reduced from 400×10^4 to 100×10^4 t. The annual excavation of licorice exceeds 3×10^4 t, but digging up 1 km of licorice will damage 5 km^2 of grassland. In total, 1.5×10^4 hm^2 of grassland is damaged in this way per year. Xinjiang has around eight enterprises to product ephedrine, which has a production capacity of about 30–40 t. This number translates to a rate of 1/8 of total storage per year. So far, the digging of wild herbs has destroyed more than 33×10^4 hm^2 of the grassland area. In addition, the damage to grassland caused by mining and gold digging is also extreme. The desert shrub in the Junggar Basin and Populus euphratica in the Tarim River Basin decreased 68.4 and 84.0 %, respectively, over the past 50 years.

(3) Weak Sense of Environmental Consciousness The ecological environmental degradation is the combined result of natural factors, socio-economic factors, technological innovation, and policies and regulations. Degradation in the arid region is closely related to a low level of consciousness of ecological and environmental protection and outright ignorance of the feedback effect on the ecological environment. Therefore, publicity and education should be strengthened to improve the ecological and environmental protection consciousness and establish a positive concept of, and concern for, the ecological environment.

The current laws, regulations and policies in China are inadequate to address land desertification issues. For example, there are no criminal laws specifically addressing land desertification caused by the unreasonable use of land. Because of this lack of existing laws, regulations and policies, the misuse of resources is not only permitted but repeated again and again, which is the main cause of human-induced land desertification.

The degradation of desert ecosystem is caused by the synergism of nature and humans. The unique natural conditions of the arid region create a fragile ecological environment, but dense population, oasis reclamation and overgrazing causes degradation and destruction. However, the people still live in poverty, a condition that drives them to exploit the natural resources more intensely, leading to a vicious cycle. Due to poverty and rudimentary production processes, people lack the necessary financial resources, material resources and technology to cope with the impact of natural disasters in a fragile ecological environment. At the same time, due to the vulnerability of the environment, people lack the opportunity to choose not to excessively utilize their natural resources, thereby putting pressure on the environment and eventually creating imbalances across the entire ecological spectrum.

1.3.4 Changes in Ecology and Environment

Global warming has become an irreversible present and future trend. Give this inescapable condition, the evolution of desert ecosystems in Northwest China arid region will be determined by the following aspects:

1.3.4.1 Water Resources

Global warming has accelerated the melting of the glaciers, which will result in an even further reduction of already dwindling water resources in the near future. Since the 1970s, global warming has hit the high latitudes of the Northern hemisphere especially hard, including arid Northwest China. There, temperature and precipitation have significantly changed. Research shows that temperatures have increased substantially over the past 50 years, with the most significant warming trends happening in North Xinjiang and the Qaidam Basin. Meanwhile, precipitation not only in the mountains and hills but also in the basin and desert has continued an increasing trend, especially in South Xinjiang, and evaporation has likewise increased. However, precipitaiton of some meteorological stations in North Xinjiang have shown decrease trend (Ren and Yang 2008). Climate change in most of Xinjiang means rising temperatures, increased precipitation, melting glaciers and increased runoff.

1.3.4.2 Changes in Glaciers

There are 46,377 glaciers in China, covering a combined area of 59,425 km^2. More than 80 % of the glaciers are located in the Qinghai-Tibet Plateau and surrounding mountains. Nowadays, glaciers in China are decreasing at an average rate of 131 km^2 per year due to the increase in CO_2 emissions. Based on this estimate, about one-third of China's glaciers will disappear by 2050. However, the trend towards increasing precipitation in the mountains will offset the melting trend of glaciers to a certain extent. It is predicted that if temperature rises 1 °C, precipitation will increase 10 %, and the glacial runoff in northern Xinjiang will increase 7.8–12 %. At the same time, glaciers will shrink 4.3 % to 6%. Therefore, even if temperatures and precipitation both increase, the melting trend of glaciers cannot be completely offset.

These changes in glaciers have a major impact on water resources. The runoff recharge proportions from glaciers will increase significantly in the short-term, which will have a positive effect on water resources and the ecological environment. However, the accelerating melting rate will reduce the water resources over a longer period, which will then bring severe challenges to the Northwest.

Although climate change accelerates the melting trend of glaciers, it will have some positive effects on the mountains' ecological system. The gradual recovery of forest and vegetation will enhance the ecological completeness and improve soil and water conservation.

1.3.4.3 Increased Frequency of Floods

Floods often occur in the Tarim River Basin area, especially in the Tianshan and Kunlun Mountains. Being at high elevations, the mountains retain their winter snows due to the effect of the Siberian cold air up until the summer, when the snowline retreats from subtropical highs. In addition, there is large gap in elevation between the mountains and basins, which leads to flooding of low altitude areas. Analysis of climate change in the Tarim River Basin over the past 50 years shows that the temperature and precipitation of the basin experienced a jump-point in the mid-1980s. Since then, temperature and precipitation have substantially increased mainly in the autumn and winter, while increasing precipitation mainly occurs in the summer. The mountains are experiencing increased warming, whereas humidity is rising in the plain region. Warming in the mountains will increase the probability of snow floods and increased precipitation leads to increased flood frequency of rain-fed rivers. According to statistics, flood frequency generally increased from the 1950s to 2000 but has become especially prominent since 1987. Likewise, disaster-related losses in Xinjiang have increased nearly ten-fold over the past few decades.

The accelerated melting of snow and permafrost due to rising temperature causes soil erosion. It is predicted that about 80–90 % of the permafrost in the Qinghai-Tibet Plateau will disappear by 2050 and that the surface area of permafrost will be reduced by 10–15 %, while the permafrost lower bound will uplift 150–250 m (Qin 2002). In summary, increases in precipitation and loss of permafrost, coupled with increased flood frequency, will increase the probability of geological disasters such as mudslides and landslides now and into the future.

1.3.4.4 Expanded Areas of Artificial Oases

With increases in population, the pressure on land is rising. As a result, artificial oases increase and natural oases decrease. Even though the protected forest area within artificial oases is expanding, the natural oases and transitional zones between oases and deserts is atrophying. Hence, the natural barrier of ecological benefits at the peripheral oasis zone is reduced, as the desert forest and desert grassland area are decreasing. Due to rising temperatures, surface runoff has increased. This ensures irrigation, meaning that oasis ecosystem productivity will significantly improve and the carrying capacity of the oasis will be enhanced. However, the productivity of natural organisms in the desert area of the peripheral oasis will decrease due to the limited water supply, as the ecosystem tends to be fragile and unstable in these natural areas.

Rising temperatures will shift plant-growth boundaries in the mountains, and grasslands could expand their domain up to 400 m (Qin 2002). At the same time, it is predicted that the area of desertification is likely to increase by nearly 43 % by 2050 due to increasing evaporation in the plains and desert areas. Of course, this figure depends on many factors, including management investments in anti-desertification projects as well as changes in precipitation forecasts. The plains region in Xinjiang has become warmer and wetter in the past five decades, and such changes may well reverse many present instances of desertification. However, even as desertification moderates in some regions, in other areas rising temperatures will worsen the desertification process.

1.3.4.5 Influence on Biodiversity

Increases in temperature and precipitation will change the phenophase and habitat conditions of original species, especially affecting plants and insects. Species are more likely move up (vertical band) to find a suitable living environment, meaning that the diversity of the desert plains will be reduced and the mountain region's biological diversity will increase. In addition, climate change will increase C_4 plants and decrease C_3 plants of some herbs and shrubs (Epstein et al. 2002). The study shows that the concentration of CO_2 in the air will cause thousands or even tens of thousands of biodiversity hot spot regions to disappear, and that even the most vulnerable species will grow in the mountains. Climate change also affects the migration of ecosystem zones because plant or vegetation zones are determined by a combination of water and heat conditions. In light of current warming trends in the arid region, vegetation zones may move upward, resulting in an extension of the plain desert area and atrophying the steppe and forest ecosystems.

1.3.4.6 Change of the Utilization of Water Resources

The expansion of artificial oases and arable land as well as increases in domestic water usage will strain water resources, as much more water is used for artificial oases than natural ones. As well, soil salinization remains a serious problem, and climate change will aggravate salinization in the plains. Rising temperatures will exacerbate land surface evaporation, which will inevitably result in salt accumulation in the shallower areas of groundwater depth. The effect of this salt accumulation on the oases regions will be especially significant.

The fragile ecological environment and limited available water supply in the desert areas heighten the conflict between ecological utilization of water resources and the economy. Nonetheless, in this battle, the ecological use of water should be of greater concern, as sustainability of water resources is the key element in social and economic development. How to maintain the stability of desert ecosystems in arid areas is a major ongoing challenge. The past and present method of robbing natu ral rivers to build artificial irrigation channels has resulted in a drastic drop in the groundwater table, resulting in the decline in shallow root plants essential for fixed and semi-fixed dunes and expediting desertification. Increased precipitation alone cannot undo this process. What is needed are strict laws that prevent further irrigation projects that rob natural resources and a general education program that heightens people's awareness of the importance of preserving their natural environment.

References

Abaydulla D, Yimit H, Wang ZM (2006) Relationship of oasis distribution and water in the Tarim River Basin. J Xinjiang University 23(2):216–218

Bai XL (1997) Water resources of agriculture in oasis in the Qaidam Basin. Agric For Qinghai (2):32–33
Chen YN (2010) Analysis of ecohydrology in Tarim river basin. Beijing, Science Press
Chen YN, Xu CC, Yang YH, Hao XM, Shen YP (2009) Hydrology and water resources variation and its responses to regional climate change in Xinjiang. Acta Geograph Sin 64(11):1331–1341
Compilation Committee Natural Resources Series in China (1995) China natural resources series (comprehensive volume). Beijing, China Environmental Science Press
Epstein HE, Gill RA, Paruelo JM, Lauenroth WK (2002) The relative abundance of three plant functional types in temperate grasslands and shrublands of North and South America: effects of projected climate change. J Biogeogr 29(7):875–888
Huang PY (1993) Ecology in arid region. Xinjiang University Press, Urumqi
Li QH (2012) Climate change and its impact on hydrological processes in arid region of Northwest China. University of Chinese Academy of Sciences, Beijing
Ministry of Agriculture, Xinjiang agricultural development research team (2008) Researches of Xinjiang agricultural development (II). China Agric Resour Reg Plan 2:8–12
Ni J, Guo K, Liu HJ (2005) Ecological regionalization in arid region of Northwest China. Acta Phytoecol Sinica 2:175–184
Pan XL, Dang RL, Wu GH (2010b) Geography and resource utilization of flora in desert in the arid region of Northwest China. Beijing, Science Press
Park JH, Duan L, Kim B, Mitchell MJ, Shibata H (2010) Potential effects of climate change and variability on watershed biogeochemical processes and water quality in Northeast Asia. Environ Int 36(2):212–225
Qin DH (2002) Assessment report of environmental evolution in Western China. Beijing, Science Press, pp 56–74
Ren ME (2004) Outline of physical geography of China. Beijing, Commercial Press
Ren ZX, Yang DY (2008) Climate change in arid region during the past 50 years. Quat Sci 26(2):299–300
Tan W, Wang YJ, Ning JZ (2004) Analysis on the change of the ecological environment in the Bosten Lake area, Xinjiang. Arid Zone Res 21(1):7–12
Tang QC (1995) Oasis development and the rational use of water resources. J Arid Land Resour Environ 9(3):107–112
Wang HS (1997) Geography of flora. Beijing, Science Press
Wang LM (2002) Discussion on the problems of the water resources shortage in Hexi Corridor oasis. J Arid Land Resour Environ 16(4):48–52
Wang GX, Cheng GD (1999) Land desertification status and developing trend in the Heihe River Basin. J Desert Res 19(4):368–374
Wu Z, Zhang H, Krause CM et al (2010) Climate change and human activities: a case study in Xinjiang, China. Climatic Change 99(3):457–472
Ye BS, Lai ZM, Shi YF (1996) The effect of climate change on runoff in the Yili river in the Tianshan Mountains. J Glaciol Geocryol 1:31–38
Ye W, Dong G, Yuan Y et al (2000) Climate instability in the Yili region, Xinjiang during the last glaciation. Chinese Sci Bull 45(17):1604–1609
Yuan ZY (2008) Plant resources in Xinjiang—the value and resource utilization of ecosystem services. Xinjiang University, Xinjiang
Yuan YG, He Q, Zhang JB, Mu G (2003) The primary research of climate change features and their causes for the recent 40 years in Xinjiang. Ecosyst Dyn Ecosyst-Soc Interact Remote Sens Appl Semi-Arid Arid Land Pts 1 and 2 4890:373–379
Zhang SX (2001) Ecological restoration and sustainable agricultural paradigm of Mountain-Oasis-Ecotone-Desert system in the North of the Tianshan Mountains. Acta Bot Sin 43(12):1294–1299
Zhang TP (2002) Characteristics and utilization of water resources in the arid region of Northwest China. Groundwater 24(3):167–168
Zhang ZW, Yang DG, Zhang XL, Chen HJ, Zhang YQ (2011) Correlation between comprehensive urbanization scale and water resource in urumqi during 1995-2007. J Desert Res 31(2):536–542

Zhao J, Chen YW, Han YF et al (1995) Physical geography of China. Higher Education Press, Beijing, pp 288–290

Zhong W, Xue JB, Shu Q, Wang LG (2007) Climatic change during the last 4000 years in the southern Tarim River Basin, Xinjiang, Northwest China. J Quat Sci 22(7):659–665

Zhong W, Xue JB, Li XD, XU HJ, Ouyang J (2010) A Holocene climatic record denoted by geochemical indicators from Barkol Lake in the Northeastern Xinjiang, NW China. Geochem Int 48(8):792–800

Zhu ZD, Chen GT (1994) Desertification of land in China. Science Press, pp 115–200

Chapter 2
Climate System in Northwest China

Yaning Chen, Baofu Li and Changchun Xu

Abstract A comparative analysis for detecting temperature and precipitation changes in the period 1960–2010 was conducted using data from 74 meteorological stations in China's arid northwest. Significant increasing trends ($P<0.01$) were revealed through the investigation, showing rates of 0.343 °C/10a and 6.07 mm/10a, which is higher than the global average. The winter temperature rate increased 0.486 °C/10a, while the summer temperature changed at a rate of 0.232 °C/10a. However, although precipitation increased 2.51 mm/10a in summer, it slowed in winter to a rate of 1.16 mm/10a. Regionally, the highest increasing rates of temperature and precipitation occurred in northern Xinjiang, and the lowest rates were recorded in southern Xinjiang and the Hexi Corridor. We found that, among the four seasons, winter saw the greatest temperature changes. We also found that the winter temperature in this region had a strong association with the Siberian High (correlation coefficient: $R=-0.715$) and greenhouse gas emissions ($R=0.51$).

Keywords Increased temperature change · Dynamic precipitation change · Extreme temperature and precipitation · Past climate change · Future forecast

Over the past century, carbon dioxide emissions into the atmosphere caused by human activities increased from 9.4 billion tons in 1960 to 33.5 billion tons in 2010. Accordingly, concentrations of carbon dioxide increased from 315 ppm in 1960 to 389 ppm in 2010, respectively. Meanwhile, other greenhouse gas emissions are also on the rise. Significant changes in atmospheric composition and radiative forcing

Y. Chen (✉) · B. Li
State Key Laboratory of Desert and Oasis Ecology,
Xinjiang Institute of Ecology and Geography,
Chinese Academy of Sciences, No. 818 South Beijing Road,
830011 Urumqi, Xinjiang, China
e-mail: chenyn@ms.xjb.ac.cn

B. Li
e-mail: libf@ms.xjb.ac.cn

C. Xu
Key Laboratory of Oasis Ecology, School of Resources
and Environmental Science, Xinjiang University,
No.14 Shengli Road, 830046 Urumqi, China
e-mail: xcc0110@163.com

Y. Chen (ed.), *Water Resources Research in Northwest China*,
DOI 10.1007/978-94-017-8017-9_2, © Springer Science+Business Media Dordrecht 2014

result in a severe damage to climate systems with regards to the natural energy balance, which then causes varying degrees of changes in the evolutionary processes of the natural environment (Guo 2010). The fourth assessment report of IPCC (2007) points out that the linear growth rate of the global average temperature was 0.74 °C from 1906 to 2005 and that it exceeded the rate from 1901 to 2000. On average, temperature increases were more prevalent in high latitude areas of the Northern Hemisphere, and warming was more prevalent in continents than in oceans.

Overall, climate change has exerted extensive and profound influences on the natural world. In response to these challenges, scholars have stepped up their research efforts into climate change variability, effects and forecasting. However, two disadvantages exist in climate change analysis and simulation on a global scale: the first is that simulated results are not always the same and may even vary for different simulation methods; the second is that simulated results obtained on a global scale are usually not applicable to a regional scale, which has negative impacts on the suitability of comprehensive policy-making. For these reasons alone, it is important to carry out research on regional climate change and its effects on natural and man-made environments (Tebaldi et al. 2005; Pierce et al. 2009; Mariotti et al. 2011; Ashrafi et al. 2012).

The climate of China's arid northwest region is characterized mainly by dryness. In earlier studies in this area, researchers focused on the formation and evolution of arid climacteric conditions (Zhang and Jiang 1992; Zhang et al. 2000), along with the characteristics of climate change (Chen et al. 1984; Li et al. 1995) and the projection of future climate (Shi and Zhang 1995). However, the scope and depth of research was limited. In 2002, Shi et al. (2002) pointed out that, over the past 100 years since the end of the Little Ice Age in the nineteenth century, Northwest China was in a fluctuating warm-dry period. Specifically, and using the area to the west of the Tianshan Mountains as being representative, a warm-dry to warm-wet climate emerged after 1987. Accompanying it was increasing precipitation, ice-snowmelt water and surface runoff, which resulted in increases in lake levels, flood disasters, vegetation improvements and decreasing dust storms.

Meanwhile, increasing precipitation and runoff were also found in the middle and west of the Qilian Mountains and other areas of Xinjiang. Projections by existing regional climate modes confirmed the transformation into a warm-wet climate, but the rate and magnitude of the transformation still needs further discussion. Subsequent studies confirmed the transformation (Li et al. 2003, 2011; Zhang et al. 2010a), although questions were also raised about the regional consistency of the change. Zhang et al. (2000) pointed out that Northwest China showed a warm-dry climate as a whole, except in some areas where warm-wet signals appeared. In view of the recent noticeable climate change, precipitation varied greatly, increasing in some places and decreasing in others. The evaporation potential also increased due to increasing temperatures. Overall, the arid index rose, accompanied by more frequent drought disasters. As can be seen from this brief summary, the climate change of Northwest China is complex, and viewing it from a short-term and isolated perspective cannot accurately reveal the process and development of precisely how the climate is changing.

2.1 Temperature Change

In recent years, the issues and challenges of climate change have attracted widespread social attention, and numerous scholars have researched the impacts and predictions of changing climate patterns (Abouabdillah et al. 2010; Stahlschmidt et al. 2011). Recently, regional effects have been detected to make strategies to address climate change (Moriondo et al. 2011; Segev 2010).

With recent socioeconomic development, China's northwest has become an important energy base due to its vast area and abundant resources. At the same time, its ecological system is becoming increasingly fragile and sensitive to regional climate effects. Therefore, it is important to know how changes in climate will affect the region not only from a weather aspect but also from environmental, economic, and human aspects.

To study the impact of climate change on the natural environment and human activities, it is necessary to know how the regional climate has changed in recent years. Previous scholars' research on climate changes in China's northwest reveals that the climate has generally trended warmer (Shi and Zhang 1995; Shi et al. 2007). However, analyses of temperature and precipitation changes in this region have been limited to the late 20th century or earlier (Shi et al. 2003; Xu et al. 2003; Zhang and Li 1982; Zhang et al. 2010).

In contrast, in our study, we use the latest (1960–2010) meteorological data from meteorological stations to explore historical and predicted temperature variations from different perspectives (i.e., years, seasons, decades, and so on). We also analyze relationships among temperature, precipitation, latitude, longitude, elevation and atmospheric circulations. Our intention here is to identify the direct causes of temperature change, hoping that results will provide a scientific basis for predicting future climate trends, assessing the impact of climate change, and formulating strategies to address its impact.

2.2 Characteristics of Temperature Change

2.2.1 Methods

(1) Mann–Kendall Non-Parametric Statistical Test The non-parametric Mann–Kendall method was used to detect change trends in precipitation and temperature which do not require data to comply with a certain statistical distribution. The method is used to detect possible trends in meteorological elements and is widely used in trend analysis. In the test, the null hypothesis (H0) states that the data (x_1, x_2,..., x_n) comprise a sample of n independent and identically distributed random variables. The alternative hypothesis (H1) of a two-sided test is that the distribution of x_k and x_i is not identical for all k and i. The test statistic (Kendall's S) is calculated as follows:

$$S = \sum_{i=1}^{n-1} \sum_{k=i+1}^{n} \operatorname{sgn}(x_k - x_i) \tag{2.1}$$

Time series x_i are ranked from $i = 1, 2\ldots\ldots n-1$ and x_j from $j = i+1,\ldots, n$. Each data point x_i is used as a reference point and is compared with all other data points x_j, such that:

$$\operatorname{sgn}(\theta) = \begin{cases} 1, & \theta > 0 \\ 0, & \theta = 0 \\ -1, & \theta < 0 \end{cases} \tag{2.2}$$

If the data set is identically and independently distributed, then the mean of S is zero and the variance of S is:

$$\operatorname{var}[S] = \frac{\left[n(n-1)(2n+5) - \sum_t t(t-1)(2t+5) \right]}{18} \tag{2.3}$$

in which n is the length of the data set, and t is the extent of any given tie and denotes the summation overall ties. For larger sample sizes, statistic S is converted to Z_c (a statistic that is approximated by a normal distribution) and is calculated as follows:

$$Z_c = \begin{cases} \dfrac{S-1}{\sqrt{\operatorname{var}(S)}}, & S > 0 \\ 0, & S = 0 \\ \dfrac{S+1}{\sqrt{\operatorname{var}(S)}}, & S < 0 \end{cases} \tag{2.4}$$

The magnitude of the trend is given as

$$\beta = Median\left(\frac{x_i - x_j}{i - j}\right), \quad \forall j < i \tag{2.5}$$

in which $1 < j < i < n$. A positive value of β indicates an 'upward trend' and a negative value of β indicates a 'downward trend'.

With the null hypothesis (H0) of $\beta = 0$, where β is the slope of trend, the Mann–Kendall test rejects H0 if $|Z_c| > Z_{1-\alpha/2}$, in which $+Z_{1-\alpha/2}$ and $-Z_{1-\alpha/2}$ are the standard normal deviates and α is the significance level for the test. In this study, we use significance levels of $\alpha = 0.1$ and 0.05, with the corresponding standard normal deviates of 1.28 and 1.96, respectively.

(2) Abrupt Change Point Analysis The MK non-parametric test is widely applied to determine the occurrence of abrupt change points of meteorological and hydrologic series. Advantages of the method include simplicity of calculation, confirmation of the starting time of abrupt changes, and clear identification of the area of abrupt changes.

Let $x_1,\ldots, x_n$ be the data points. For each element x_i, the numbers r_i of elements x_j preceding it ($j<i$), such that $x_j<x_i$ are computed. Under the null hypothesis (no abrupt change point), the normally distributed statistic S_k can be calculated via the formula:

$$S_k = \sum_{i=1}^{k} r_i \quad (2 \le k \le n) \tag{2.6}$$

Mean and variance of the normally distributed statistic Sk can be given by:

$$\overline{S}_k = \frac{E(S_k) = k(k-1)}{4} \tag{2.7}$$

$$\mathrm{var}(S_k') = \frac{k(k-1)(2k+5)}{72} \tag{2.8}$$

The normalized variable statistic UF_k is estimated as:

$$UF_k = \frac{(S_k - \overline{S}_k)}{\sqrt{\mathrm{var}(S_k)}} \tag{2.9}$$

The normalized variable statistic UF_k is the forward sequence, and the backward sequence UB_k is calculated using the same equation but with a reversed series of data. When the null hypothesis is rejected (i.e., if any of the points in the forward sequence are outside the confidence interval), the detection of an increasing ($UF_k>0$) or a decreasing ($UF_k<0$) trend is indicated. The sequential version of the test used here enables detection of the approximate time of occurrence of the trend by locating the intersection of the forward and backward curves of the test statistic. If the intersection occurs within the confidence interval, it indicates an abrupt change point.

(3) Correlation Analysis Method The correlation analysis method is applied to detect the relationship between or among two or more variables. We used this method to detect correlation degrees and properties among variables. The correlation coefficient (r) is usually calculated as follows:

$$\mathrm{r} = \frac{\sum (x_i - \overline{x})(y_i - \overline{y})}{\sqrt{\sum (x_i - \overline{x})^2 \cdot \sum (y_i - \overline{y})^2}} \tag{2.10}$$

The distribution range of the correlation coefficient r is between $-1 \le r \le +1$. When the correlation coefficient is positive, the correlation between variables indicates a positive correlation; when the correlation coefficient is negative, it indicates a negative correlation. The greater the absolute value of the correlation coefficient $|r|$, the closer the degree of correlation between two variables. In addition, to determine if the correlation coefficient is meaningful or not, it is necessary to carry out a significance test.

(4) R/S Analytic Method R/S analysis is also called rescaled range analysis. It is usually applied to analyze long-term records of runoff time series (Rehman 2009; Rehman and Siddiqi 2009). Its principle is as follows. In considering the time series of annual runoff in a certain river, X(t), for any positive integer $\tau \ge 1$, the mean value series is defined as:

$$\langle X \rangle_\tau = \frac{1}{\tau}\sum_{t=1}^{\tau} X(t) \quad \tau = 1, 2, \ldots \tag{2.11}$$

The accumulative deviation is

$$X(t,\tau) = \sum_{t=1}^{t} (X(u) - \langle X \rangle_\tau) \quad 1 \le t \le \tau \tag{2.12}$$

The extreme deviation is

$$R(\tau) = \max_{1 \le t \le \tau} X(t,\tau) - \min_{1 \le t \le \tau} X(t,\tau) \quad \tau = 1, 2, \ldots \tag{2.13}$$

The standard deviation is

$$S(\tau) = \left[\frac{1}{\tau}\sum_{t=1}^{\tau}\left(X(t) - \langle X \rangle_\tau\right)\right]^{\frac{1}{2}} \quad \tau = 1, 2, \ldots \tag{2.14}$$

When analyzing the statistic rule of $R(\tau)/S/(\tau)R/S$, Hurst discovered a relational expression

$$R/S \propto \left(\frac{\tau}{2}\right)^H \tag{2.15}$$

which can be used to identify the Hurst phenomenon in the time series and where H is called the Hurst exponent. Apparently, H is given by the slope coefficient of R/S versus $\tau/2$. According to (τ, R/S), H can be obtained by the least squares method (LSM) in a log-log grid.

In 1965, Hurst proved that, in an independently random series with limited variance, the exponent H = 0.5 and H ($0 < H < 1$) is dependent on an incidence function C(t):

$$C(t) = 2^{2H-1} - 1 \tag{2.16}$$

When H>0.5, C(t)>0. This signifies that the process has a long-enduring characteristic and that the future trend of the time series will likely be consistent with the past. In other words, if the past showed an increasing trend, the future will also show an increasing trend. When H<0.5, C(t)<0, which indicates that the process has an anti-persistence characteristic and that the future trend of the time series will likely be opposite from the past. In other words, if the past showed an increasing trend, the future will assume a reducing trend. When H=0.5, C(t)=0, meaning that the process is stochastic.

(5) Kriging Method Kriging is an advanced, computationally intensive, geostatistical estimation method that generates an estimated surface from a scattered set of points with z-values. Consider a continuous field variable z(x) defined in a domain Ω. The domain is represented by a set of properly scattered nodes x_I, I=1, 2, …, N, where N is the total number of nodes in the whole domain. Given N field values, $z(x_1)$, …, $z(x_N)$, the problem is to obtain an estimate value of u at a point $x_0 \in \Omega$. Let us consider a set of nodes x_i, i=1, 2, …, n, surrounding point x_0 inside a sub-domain $\Omega x_0 \subseteq \Omega$. Here, a small letter index is used in place of a capital latter to emphasize that the numbering is referred to the sub-domain Ωx_0. The Kriging estimated value $Z^*_v(x)$ is a linear combination of $Z_v(xi)(i = 1,2,...,n)$, i.e.,

$$Z^*_v(x) = \sum_{i=1}^{n} \lambda_i Z(xi) \tag{2.17}$$

where λi are the (Kriging) weights and n is the number of nodes inside Ωx_0. Kriging weights λi can be obtained by solving Kriging equations and are determined by requiring that the estimator $Z^*_v(x)$ be unbiased and optimal (minimum squared error of estimation), i.e.,

$$E\left[Z^*_v(x) - Z_v(x)\right] = 0 \tag{2.18}$$

$$Var\left[Z^*_v(x) - Z_v(x)\right] = E\left[Z^*_v(x) - Z_v(x)\right]^2 \rightarrow \min \tag{2.19}$$

Using the Lagrange method for constraint optimization problems, the requirements of minimum variance and unbiased estimator lead to the following Kriging equation system.

Ordinary Kriging is a linear geostatistical method. The Kriging approach uses the semivariogram to express spatial continuity (autocorrelation) and requires two separate steps. The first step is to calculate and model/fit the semivariogram for the whole area; the second step is a Kriging estimation of unmeasured points in the area. The semivariogram measures the strength of the statistical correlation as a function of distance. Typically, an authorized variogram model (e.g., exponential or spherical) is fitted to the experimental semivariogram values, calculated from data, for given angular and distance classes. In bounded models (e.g., spherical and exponential), semivariogram functions increase with distance until they reach a maximum, named sill, at an approximate distance known as the range. The range is the distance at which the spatial correlation vanishes, and the sill corresponds to the maximum variability in the absence of spatial dependence.

$$E\left[Z(x)\right] = m \tag{2.20}$$

$$c(h) = E\left[Z(x)Z(x+h)\right] - m^2 \tag{2.21}$$

$$Zv(x_0) = \frac{1}{V}\int_{v} Z(x)dx \tag{2.22}$$

Ordinary Kriging is based on the assumption that the set of unknown values is a set of spatially dependent random variables; hence, each measurement z(xi) is a particular realization of the random variable Z(xi). In (2.23), the optimal Kriging weights λi are determined by solving the Kriging equations that result from minimizing the estimation variance while ensuring unbiased estimation of Z (x0) by Z*(x0). In Ordinary Kriging, the Kriging weight decreases as the datum location gets farther away from the location being estimated, and so negative Ordinary Kriging weights can occur. Typically, negative weights arise when the influence of a specific datum is screened by that of a closer one.

$$Z^{*}v = \sum_{i=1}^{n} \lambda_i Z(xi) \tag{2.23}$$

(6) Inverse Distance Weighted (IDW) Interpolation The inverse distance weighted (IDW) method was used to interpolate climate change trends. It is a simple and widely used spatial interpolation method that estimates the value at an unsampled location using the values of nearby sampled points, which are weighted inversely by their distance to the location (Ruelland et al., 2008). With a set of points, their coordinates and values are (X_i, Y_i) *and* Z_i (i=1, 2, …, n), the Z value at a location (X, Y) is calculated by the following equation:

$$Z = \frac{\left[\sum_{i=1}^{n-1} \frac{z_i}{d_i^2}\right]}{\left[\sum_{i=1}^{n} \frac{1}{d_i^2}\right]} \tag{2.24}$$

Where Z is the interpolated value for the location, n is the number of nearby points (a value of 12 was used in the study), and d_i is the distance from the ith nearby point to the location, which is calculated by the following equation:

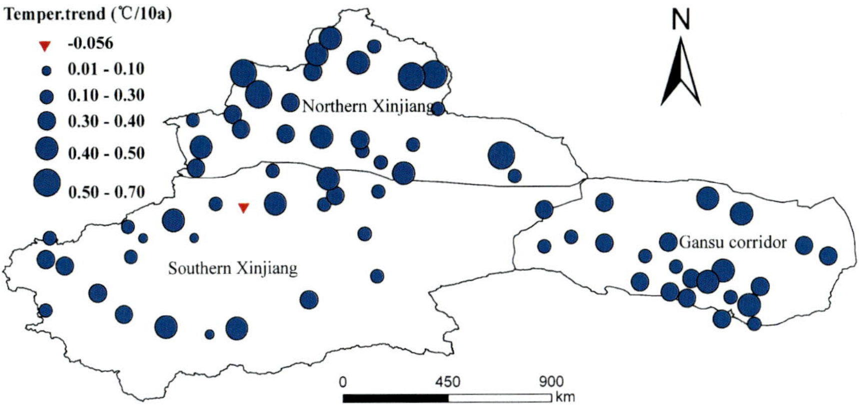

Fig. 2.1 Trends in temperature during 1960–2010

$$d_i^2 = (X - X_i)^2 + (Y - Y_i)^2 \tag{2.25}$$

2.2.1.1 Inter-Annual Variability

(1) Temperature Change in Northwest China The linear tendencies of temperature over Northwest China for the period 1960–2010 (Fig. 2.1) shows that the air temperature trended toward higher values at 73 (98.65 %) of the 74 sites analyzed in the study area. All of the positive trends are statistically significant ($P<0.05$, under the Mann–Kendall test). The one site that does not show a significant trend has a decreasing temperature value. The increasing rate in mean annual air temperature is 0.343 °C/10a in China's northwest (Fig. 2.2), which is noticeably higher than the global rate of 0.13 °C/10a (IPCC 2007).

Regionally, temperature increases show stark differences. The temperature in northern Xinjiang increased the fastest, at a rate of 0.386 °C/10a, followed by the Hexi Corridor at a rate of 0.352 °C/10a. Southern Xinjiang showed the slowest increasing rate (0.283 °C/10a).

(2) Temperature Change in the Mountain-Oasis-Desert Areas

Trend of mean annual temperature The Mann–Kendall test found that the mean temperatures of all three landscape types showed significant increasing trends during the period 1960–2010 ($P<0.01$), with Fig. 2.3 charting the linear tendencies of temperature at 51 stations. To some extent, this map presents the spatial distribution of temperature variation in China's arid northwest for the period under study. Of the 51 stations, only one has a slightly decreasing trend, showing a slow rate of 0.056 °C per decade; all of the others have increasing trends. The mean temperature readings of all 51 stations reveal a slight increasing trend at a rate of 0.341 °C per decade.

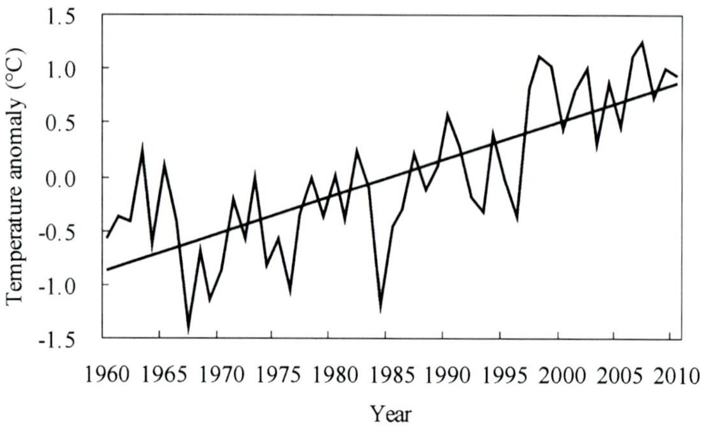

Fig. 2.2 Time series of temperature and its linear trend in ANC from 1960 to 2010

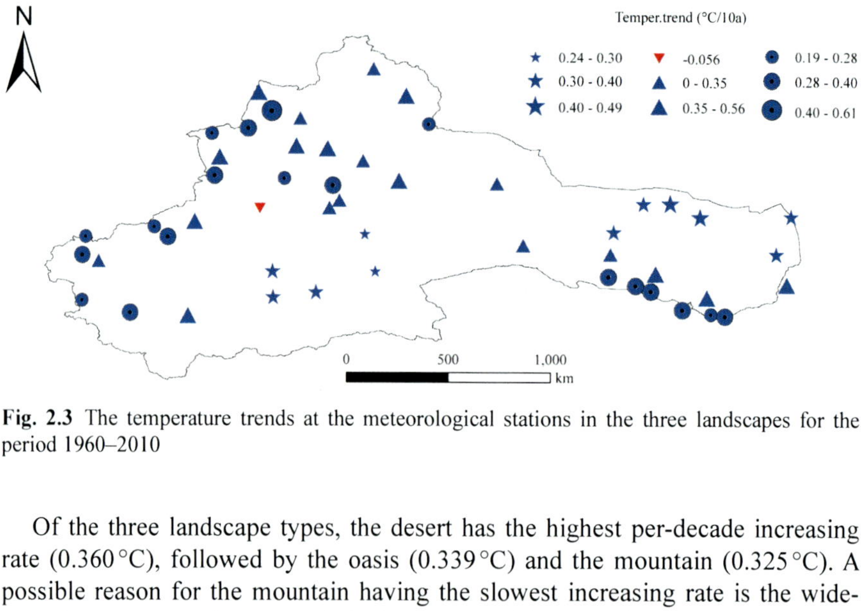

Fig. 2.3 The temperature trends at the meteorological stations in the three landscapes for the period 1960–2010

Of the three landscape types, the desert has the highest per-decade increasing rate (0.360 °C), followed by the oasis (0.339 °C) and the mountain (0.325 °C). A possible reason for the mountain having the slowest increasing rate is the widespread occurrence of snow and glaciers, along with vegetation diversity and high ecosystem stability. Mountain areas impose a certain buffer action on global climate change, while the desert region does the opposite.

To further clarify the temperature change of each area in different periods, the time series were divided into two periods: prior to the 1990s, and after the 1990s.

Prior to the 1990s, temperature trends among the 51 stations were inconsistent. For instance, seven stations (all located in the mountain and oasis areas) had decreasing trends (Fig. 2.4a). Overall, however, the trend in the 51 stations was a weak rising one, with a mean rate of 0.151 °C per decade. Sixteen of the 19 mountain

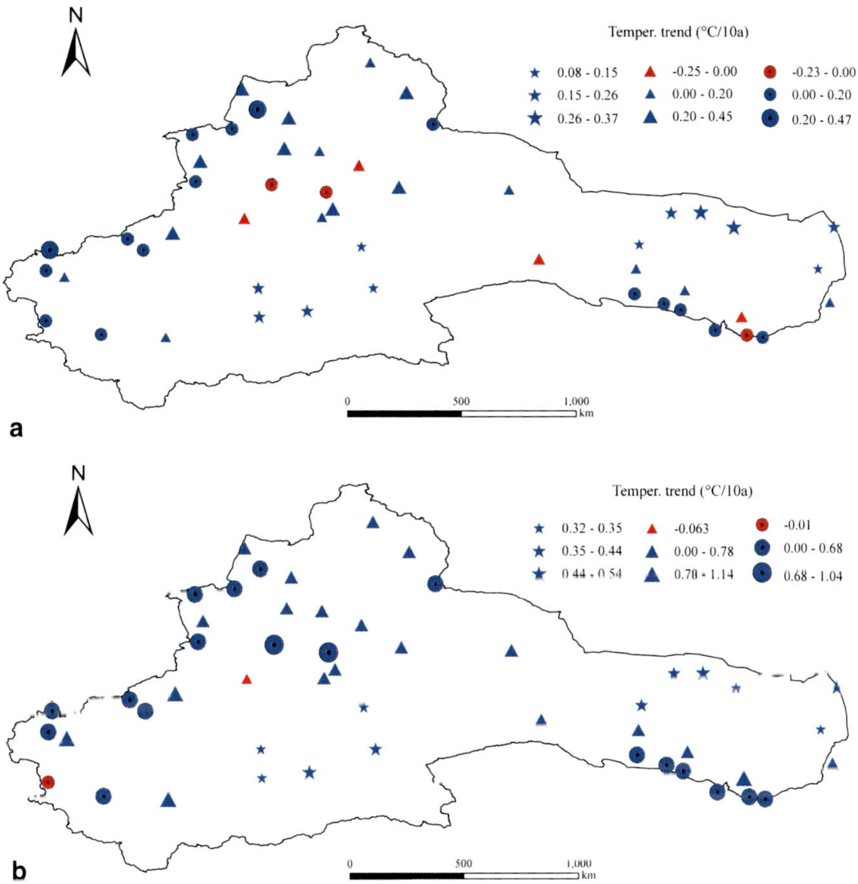

Fig. 2.4 The temperature trends in different landscapes for period **a** 1960–1989 and **b** 1990–2010

stations had increasing trends, with the mean increasing rate of the mountain landscape at 0.103 °C per decade. Seventeen of the 21 oasis stations had increasing trends, with the mean increasing rate of this landscape being 0.138 °C per decade. All stations in the desert areas had increasing trends during this period, with a mean rate of 0.214 °C per decade. One reason for these variations among landscape types is that the ecosystem in the mountain area has high stability, whereas the desert ecosystem has low stability. Another reason is that the arid area of the northwest had a small population density during this time period and thus there was little human interference with the natural order of the oasis and mountain areas.

After the 1990s, the entire area experienced rapid temperature increases (Fig. 2.4b). The mean of the increasing rates at the 51 stations was 0.517 °C per decade. The oasis areas had the highest rising rate (0.60 °C per decade), followed by the mountain area (0.542 °C per decade). Both oasis and mountain areas had

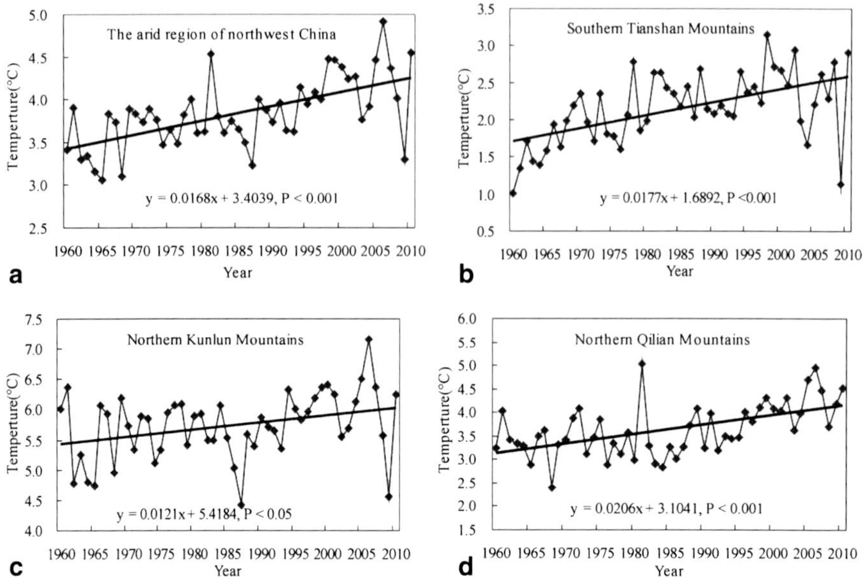

Fig. 2.5 Time series of temperature of snowmelt period and its linear trend for **a** entire northwest China, **b** southern Tianshan Mountains, **c** northern Kunlun Mountains, and **d** northern Qilian Mountains from 1960 to 2010

95 % of their stations showing increasing trends, and while all stations in the desert areas had increasing trends, their mean rate was only 0.402 °C per decade, the lowest among the three landscapes. We suspect that human activity was a major factor contributing to the formation of this pattern. Since 1990, China's arid northwest has experienced rapid population growth along with rapid expansion of industrialization, tourism, and urbanization. Greenhouse emission and the "urban heat island" effect may contribute to variations among different landscapes over the background of global warming and regional climate change.

Trend of Snowmelt Period Temperature in the Mountain Area During the 1960–2010 timeframe, the mean air temperature of the snowmelt period in the northwest mountainous areas of China exhibited an upward trend, with an increasing rate of 0.168 °C/10a. This increasing trend carries a $P<0.001$ level of significance (Fig. 2.5a). Over the past 51 years, the air temperature in the snowmelt period increased by 0.857 °C, which is higher than the world average (Brohan et al. 2006; IPCC 2007) for the same period.

Figs. 2.5b, c and d show the temporal variations of linear tendencies of the temperature of snowmelt periods over different areas from 1960 to 2010. From this figure, it can be seen that all parts of the northwest mountainous areas of China have a positive temperature tendency during their snowmelt periods. The northern Qilian Mountains have a relatively high positive tendency, whereas low positive tendencies are observed in the southern Tianshans and northern Kunlun Mountains.

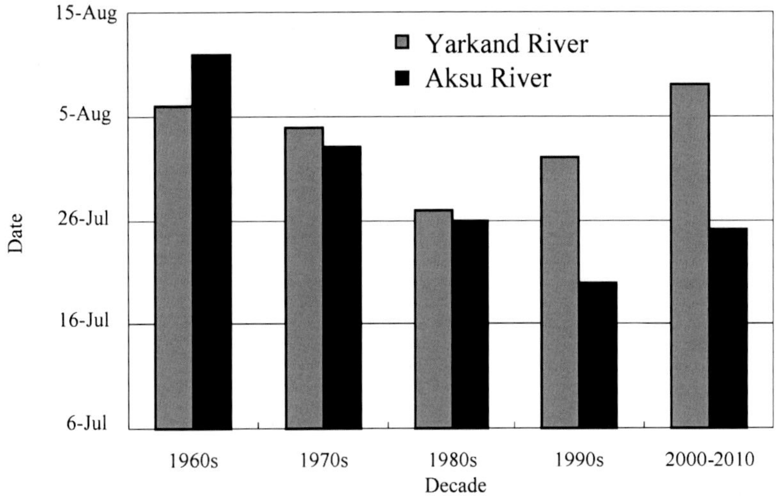

Fig. 2.6 The date of maxima daily temperature (9 day moving mean) in every decade over Aksu River and Yarkand River

However, all of these trends are statistically significant ($P<0.05$, under the Mann–Kendall test).

Regionally, temperatures during the snowmelt period in the northern Qilian Mountains increased the fastest (1.05 °C), followed by those in the southern Tianshan Mountains (0.903 °C). However, temperatures in the northern Kunlun Mountains increased the slowest, by only 0.617 °C.

Next, we take the Aksu and Yarkand Rivers as objects for further analysis on the date change of the maximum daily mean temperature in different decades (a 9-day moving average). For the period 1960–2000, the date of the maximum daily temperature in the Aksu River moved ahead 22 days (Fig. 2.6) compared with that in 1990s, while the date was delayed 5 days for the period 2000–2010 (which is basically consistent with that in the 1980s). The maximum daily temperature in the Yarkand River occurred 10 days ahead in 1960s–1980s, while the date was delayed in 1990–2010. However, the date for the period 2000–2010 was basically consistent with that in the 1960s.

Overall, the dates for maximum daily temperature in the mountains appear earlier in the 1970s and 1980s, while the date has been delayed in the first decade of the twenty-first century.

(3) Temperature Changes in Typical Rivers In this section, 11 rivers of five typical river areas in three typical areas in China's arid northwest zone form the research topic (Table 2.1). The three typical areas are northern Xinjiang (the south slope of the Altai Mountains and the north slope of the Tianshan Mountains), southern Xinjiang (the north slope of the Kunlun Mountains and the south slope of the Tianshan Mountains) and the Hexi Corridor (the north slope of the Qilian Mountains), which basically includes all of the main inland rivers in China's arid region

Table 2.1 The rivers and weather stations information in the arid region of Northwest China

Areas	Typical river areas	Rivers	Weather stations (Information sessions)
Hexi Corridor	The north slope of Qilian Mountains	Heihe	Qilian (1957–2010)
		Shiyang River	Wushaoling (1957–2010)
		Shule River	Tuole (1957–2010)
Northern Xinjiang	The south slope of Altai Mountains	Kelan River	Altai (1957–2005)
	The north slope of Tian-shan Mountains	Urumqi River	Daxigou, Yingxiongqiao (1957–2006)
		Manas River	Kenswatt (1957–2006)
		Kuitun River	Jialeguola (1957–2006)
Southern Xinjiang	The south slope of Tian-shan Mountains	Aksu River	Toergate, Aheqi (1957–2010)
		Kaidu River	Bayinbuluke (1957–2010)
	The north slope of Kunlun Mountains	Hotan River	Hotan (1957–2010)
		Yarkand River	Taxkorgan (1957–2010)

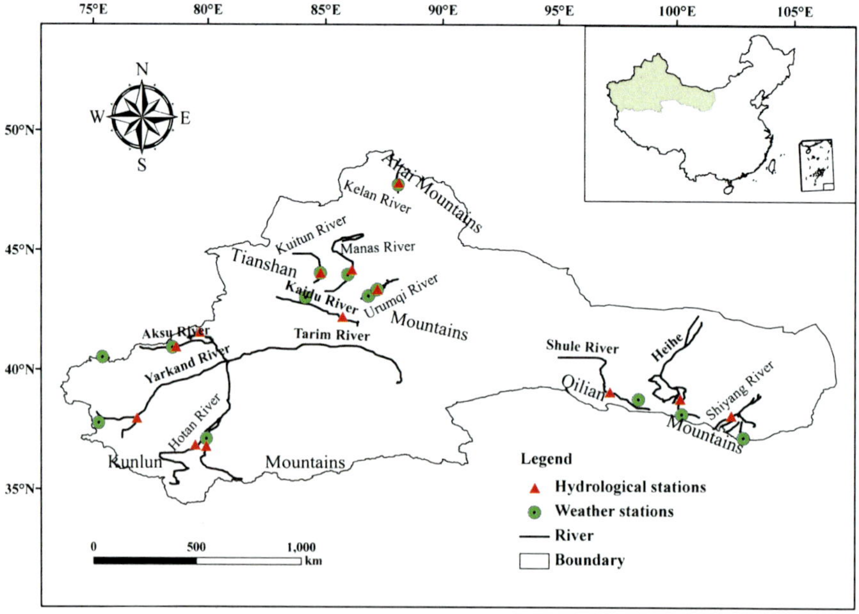

Fig. 2.7 The weather and hydrological stations of rivers in the arid region of northwest China

(Fig. 2.7) (Li et al. 2012a). To minimize the impact of human activities and make the regional climate changes more clear, data from the selected weather stations represent climate changes for each mountain area. We use the mean value of the data to represent temperature and precipitation.

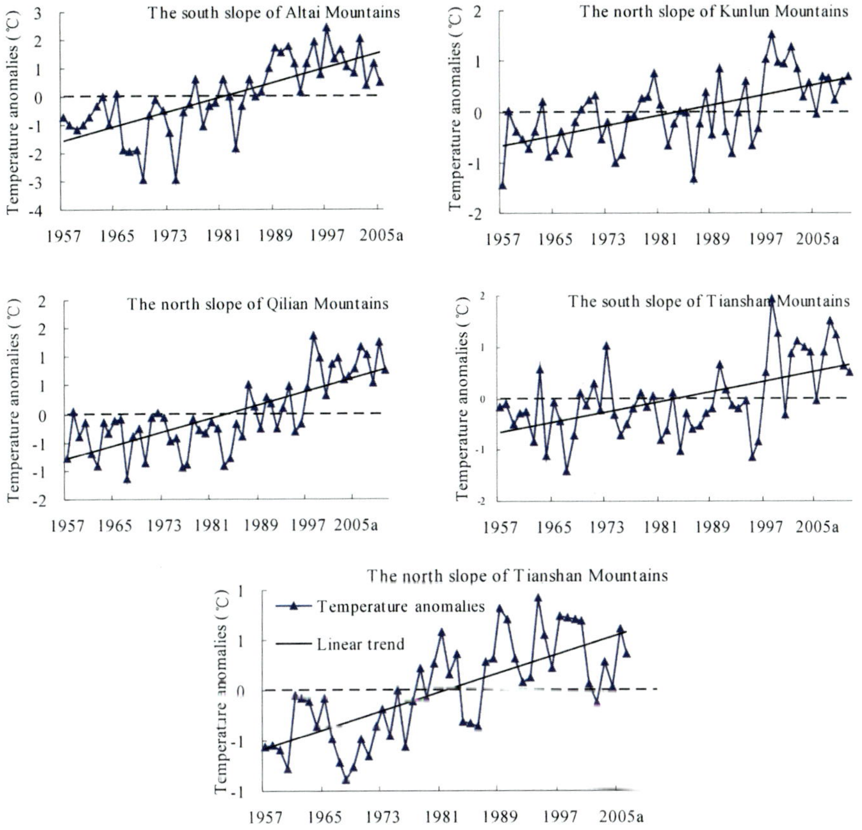

Fig. 2.8 Temperature anomalies trends in typical river areas in the arid region of Northwest China

Over the past five decades, the temperature in each river area (the south slope of the Altai Mountains, the north slope of the Kunlun Mountains, the north slope of the Qilian Mountains, and the south and north slopes of the Tianshan Mountains) has increased, fluctuating at an average rate of 0.32 °C per decade (Fig. 2.8) (Li et al. 2012). The temperature on the south slope of the Altai Mountains rose the fastest, increasing by 0.64 °C/10a; the second fastest increase (0.29 °C/10a) occurred on the north slope of the Qilian Mountains, while the slowest increases occurred on the south (0.24 °C/10a) and north (0.24 °C/10a) slopes of the Tianshan Mountains and on the north slope (0.25 °C/10a) of the Kunlun Mountains. In addition, the Mann–Kendall test showed that the increasing temperature trend in each river area was statistically significant, at $P < 0.01$ (Table 2.2).

From the perspective of coefficients of variation, temperature variations were the largest on the north slope of the Qilian Mountains, indicating strong variability. This was followed by temperature variations on the southern slopes of the Altai and Tianshan Mountains, suggesting moderate variability, while temperature variations

Table 2.2 Temperature trends and tests in different typical river areas in the arid region of northwest China

Areas	Typical river areas	Cv	Zc	Step change points	Periods (a)
Hexi Corridor	The north slope of Qilian Mountains	1.252	4.68**	1994**	19**
Northern Xinjiang	The south slope of Altai Mountains	0.327	5.53**	1984**	19**
	The north slope of Tianshan Mountains	0.087	5.14**	1978**	17**
Southern Xinjiang	The south slope of Tianshan Mountains	0.148	2.65**	None	20*
	The north slope of Kunlun Mountains	0.083	3.76**	(1980,1987)**	19**

*$P<0.05$; **$P<0.01$

on the north slopes of the Tianshan and Kunlun Mountains were the smallest, pointing to weak variability (Table 2.2).

2.2.1.2 Seasonal Variability

From 1960 to 2010, spring air temperatures trended to larger values at 73 (98.65 %) of the 74 sites (Fig. 2.9), declining at only 1 site. Most of the upward trends were statistically significant ($P<0.05$, under the Mann–Kendall test). At sites registering no significant trends, temperature values increased at 8 stations (10.96 % of all sites), and declined only at 1. The increasing rate of mean air temperature was 0.268 °C/10a across the entire arid region. Specifically, the increasing rate of temperature in northern Xinjiang was the highest (0.286 °C/10a), followed by the rates in the Hexi Corridor (0.270 °C/10a) and southern Xinjiang (0.248 °C/10a) (Table 2.3).

In summer during the period under study, the air temperature generally increased at 67 (90.54 %) sites and decreased at the remaining 7, and the majority of the positive trends were significant ($P<0.05$, under the Mann–Kendall test). Of the stations showing no significant trends, 6 sites (8.96 %) showed rising temperature trends and 3 showed declining temperature trends. The rising rate of mean summer air temperature was 0.232 °C/10a in the study area, which is lower than that in spring. The rising rate of temperature was the highest in the Hexi Corridor (0.292 °C/10a), followed by northern Xinjiang (0.231 °C/10a) and southern Xinjiang (0.173 °C/10a) (Table 2.3).

In autumn for the same time period (1960–2010), the air temperature increased at 72 sites (97.30 % of total) and decreased at the other 2 sites. The trends were significant ($P<0.05$) at all sites except for the 2 (2.78 % of total). The rising rate of mean temperature was 0.360 °C/10a, which is notably higher than the rates in spring and summer. Northern Xinjiang had the highest rate of temperature increase

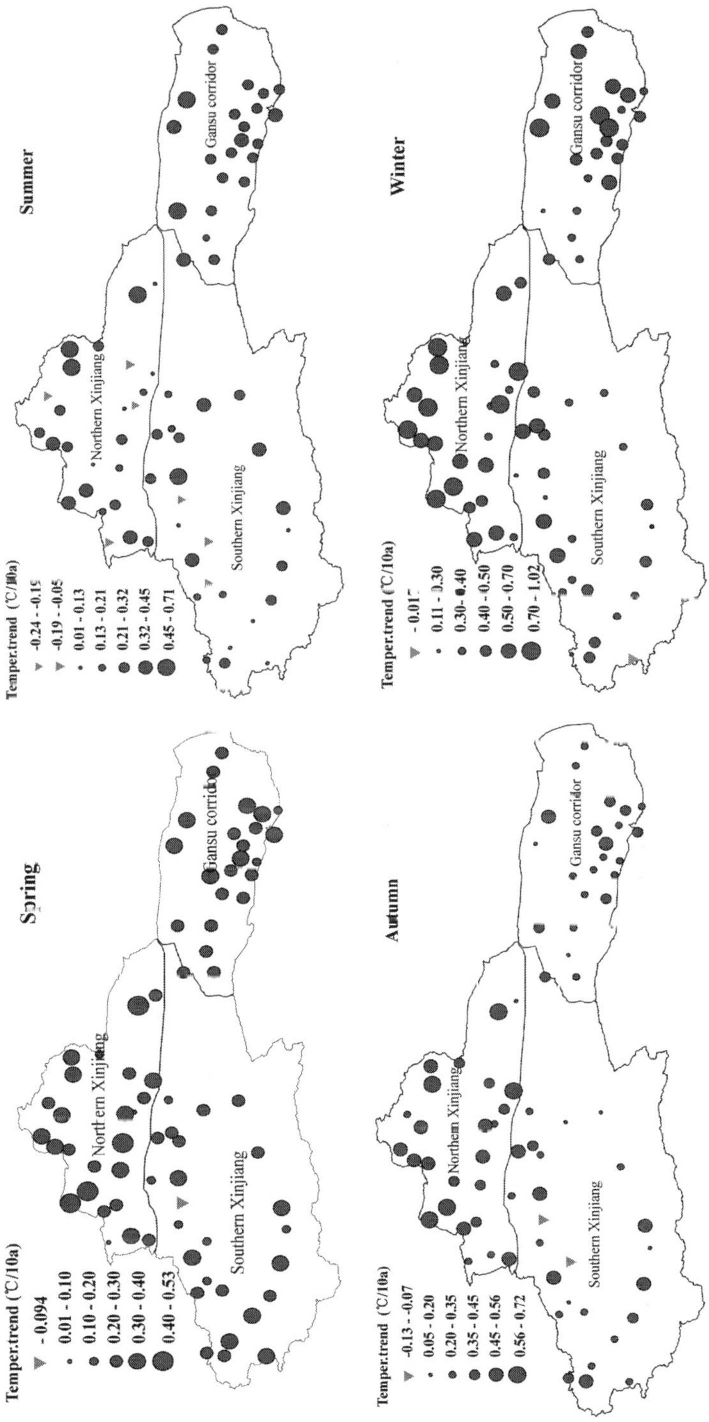

Fig. 2.9 Seasonal temperature trends in the arid region of Northwest China

Table 2.3 The annual and seasonal trend in temperature over different regions

Region	Spring (°C/10a)	Summer (°C/10a)	Autumn (°C/10a)	Winter (°C/10a)	Annual (°C/10a)
ANC	0.268**	0.232**	0.360**	0.486**	0.343**
Northern Xinjiang	0.286*	0.231**	0.451**	0.587**	0.386**
Southern Xinjiang	0.248*	0.173**	0.303**	0.385**	0.283**
Hexi Corridor	0.270**	0.292**	0.326**	0.486**	0.352**

*$P<0.05$; **$P<0.01$

(0.451 °C/10a), followed by the Hexi Corridor (0.326 °C/10a) and southern Xinjiang (0.303 °C/10a).

In winter over the same 50-year time frame, the air temperature trended toward higher values at 73 (98.65 %) of the 74 stations (Fig. 2.9) and toward a lower value at the 1 remaining site. The trends were all statistically significant ($P<0.05$) except at 5 sites (6.85 % of total) that showed rising temperatures and 1 site that showed declining temperature. The increasing rate of mean air temperature was 0.486 °C/10a, which is higher than that in spring, summer and autumn. The increasing rate of temperature in northern Xinjiang was the highest (0.587 °C/10a), followed by the Hexi Corridor (0.486 °C/10a) and southern Xinjiang (0.385 °C/10a) (Table 2.3).

2.2.1.3 Decadal Variability

In the 1960s, the mean temperature in China's arid northwest was 6.453 °C; by the 1970s, it had increased only by 0.045 °C. Overall, however, the temperature had risen at 55.41 % of the meteorological stations. The sites with temperature reductions were located mainly in the southern Tianshan Mountains and in eastern Xinjiang. Temperature increases occurred in southern Xinjiang (0.11 °C), followed by the Hexi Corridor (0.029 °C) and northern Xinjiang (0.016 °C) (Fig. 2.10).

In the 1980s, the mean temperature in the study area rose by 0.297 °C, which was a more substantial increase than during the 1970s. Overall, the temperature at 83.78 % of the meteorological stations had increased. The sites with temperature declines were located mainly in the Qilian Mountains. Temperature increased the most in northern Xinjiang (0.466 °C), followed by the Hexi Corridor (0.249 °C) and southern Xinjiang (0.153 °C).

In the 1990s, the mean temperature increased by 0.507 °C, which was a faster rate of increase than the 1960s, 1970s, or the 1980s. Overall, the temperature increased at 87.84 % of the meteorological stations. Temperatures in the Hexi Corridor and in northern Xinjiang increased by 0.597 °C and 0.564 °C, respectively, while the temperature in southern Xinjiang increased only slightly (0.356 °C).

During the period 2000–2010, the mean temperature showed a continuous increasing trend. However, the increasing rate was 0.487 °C, which is less than in the 1990s, and 9.5 % of the stations showed declining temperatures. Moreover, there

Fig. 2.10 Temperature trends by decade in the arid region of northwest China

Table 2.4 Results of Mann–Kendall test for annual and seasonal precipitation and temperature

Parameter	Annual	Spring	Summer	Autumn	Winter
Precipitation (mm/decade)	8.01[b] (172.38)	1.78 (43.93)	2.55[a] (78.48)	1.72[a] (35.50)	1.73[b] (13.80)
Temperature (°C/decade)	0.35[b] (7.20)	0.24[a] (9.18)	0.26[b] (17.75)	0.41[b] (7.40)	0.39[b] (–8.83)

The data in the parenthesis are the mean of the study area
[a, b] delineate the significance at 5 %
[b] delineate the significance at 1 % significance level

were only slight differences in the extent of temperature increases in the Hexi Corridor, northern Xinjiang, and southern Xinjiang, with increases of 0.483 °C, 0.415 °C, and 0.571 °C, respectively.

2.2.1.4 Spatio-Temporal Distribution of Monotonic Trends

In Table 2.4, the regional MK trends for annual and seasonal air temperatures are investigated. For air temperature, significantly increasing trends occurred across the entire region for all seasons. The most significant change occurred in autumn, with a regional trend magnitude of 0.41 °C/decade.

The spatial distribution pattern of temporal trends for air temperature at both individual stations and at basins (Fig. 2.11 and Fig. 2.12) shows that, for annual air temperature, about 97 % of the stations had a significant trend ($P<0.05$), and that all of the basins showed statistically significant increases ($P<0.01$), especially the Cherchen River, the Bayi Basin and the Heihe River (Fig. 2.12a). About 61 % and 73 % of the stations had statistically significant increasing trends in spring and winter, respectively (Figs. 2.11b, e). Moreover, the spatial distribution of trends in spring also had the same pattern as those in winter, with stations in southern Xinjiang and the Hexi Corridor showing larger trend magnitudes (Figs. 2.12b, e). Regarding summer-time air temperature, 80 % of the stations exhibited statistically significant trends (Fig. 2.11c), and all of the basins (except for the rivers in the eastern section of the north slope of the Tianshan Mountains [the NSET Rivers], the Weigan River and the Yarkand River) showed significant changes at a 0.05 significance level (Fig. 2.12c). Similarly, in autumn, approximately 95 % of the stations demonstrated statistically increasing trends (Fig. 2.12d), with the Hotan and Aksu River basins exhibiting the most significant changes compared to other basins (Fig. 2.12d). It can therefore be concluded that the air temperature increased significantly, resulting in a warmer climate in the arid region of northwestern China.

2.2.1.5 Spatial Distribution Characteristics of Temperature Extremes

Minimum temperature The spatial distribution of minimum temperature in the arid region of northwest China is inconsistent, ranging from 2.4 to 4.7 °C

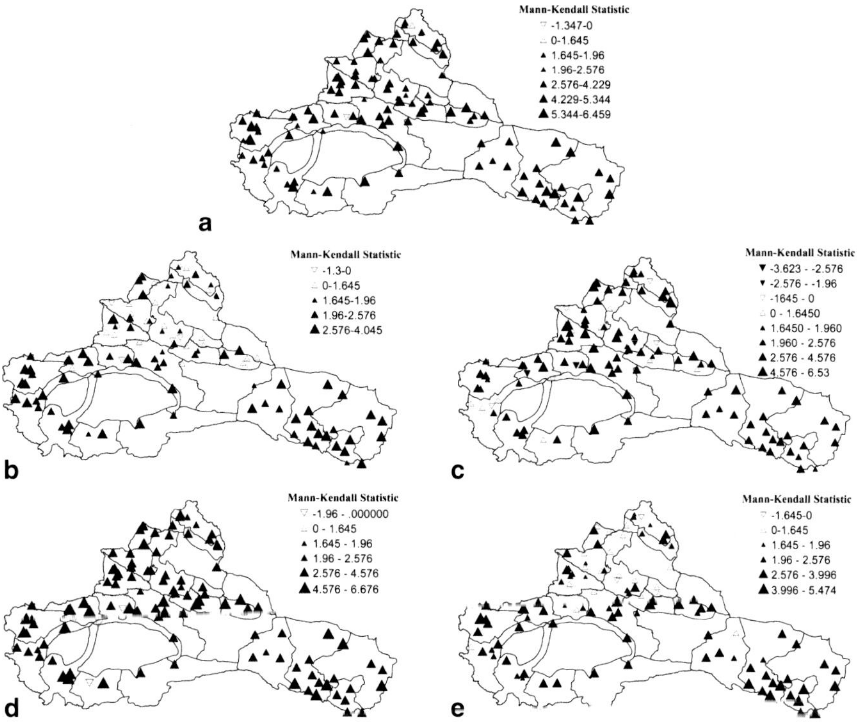

Fig. 2.11 Distribution of Mann–Kendall trends of air temperature: **a** annual, **b** spring, **c** summer, **d** autumn, **e** winter

(Fig. 2.13a). The highest minimum temperature region is located in the eastern portion of the northern slope of the Kunlun Mountains, at a range of 3–4.7 °C. In the outer ring of this area, the minimum temperature reduces to 2–3 °C. In the vast area of the Tianshan Mountains and part of eastern Xinjiang, the minimum temperature ranges from 0 to 2 °C. The minimum temperature in the northern Tianshan Mountains, including the central Hexi Corridor, is the lowest in the study area, with an average annual minimum temperature reaching −2.4 °C. Thus, we can see that the minimum temperature distribution is quite similar to the mean temperature distribution, which, to a certain extent, suggests that the minimum temperature has a larger influence on the average annual temperature.

The trend magnitude (Z value) for minimum temperature (Fig. 2.13b) shows a clear rising trend. Even the lowest Z value can reach up to 4.0 ($P < 0.01$). In most of the region under study, Z values range from 5 to 6, except in three sites, where the Z value is greater than 6. In southern Xinjiang and the northern Qilian Mountains, the Z value reaches 4.5. The uneven distribution of the Z value indicates that the minimum temperature has increased sharply, far more than the variation of mean temperature, though the scopes of minimum temperature are different.

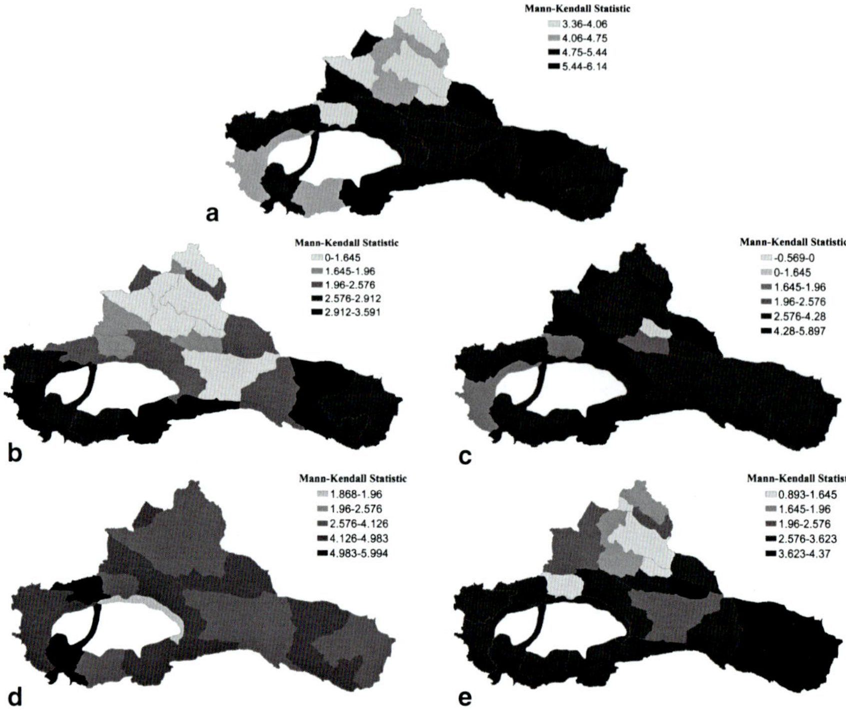

Fig. 2.12 Regional trends of air temperature at basin scale: **a** annual, **b** spring, **c** summer, **d** autumn, **e** winter

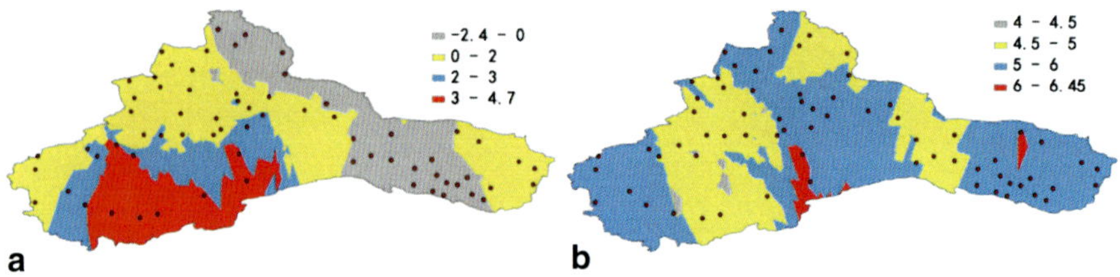

Fig. 2.13 **a** Minimum temperature distribution and **b** trend variation Z in the arid region of Northwest China

Maximum Temperature The distribution change for maximum temperature is similar to that of average and minimum temperature (Fig. 2.14a), showing a latitude-orientation characteristic. In other words, the higher the latitude, the lower the maximum temperature. Maximum temperature ranges from 10 to 19 °C, with great spatial variation. In the northern slope of the Kunlun Mountains, maximum temperature ranges from 17 to 19 °C. In the Altai, Alashan and the northern slope of the Qilian Mountains, maximum temperature ranges from 11 to 14 °C. In the rest of study area, it ranges from 14 to17 °C.

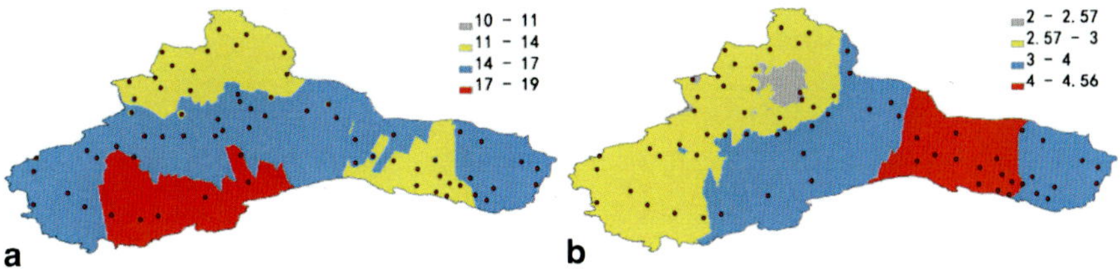

Fig. 2.14 **a** Maximum temperature distribution and **b** trend variation Z in the arid region of Northwest China

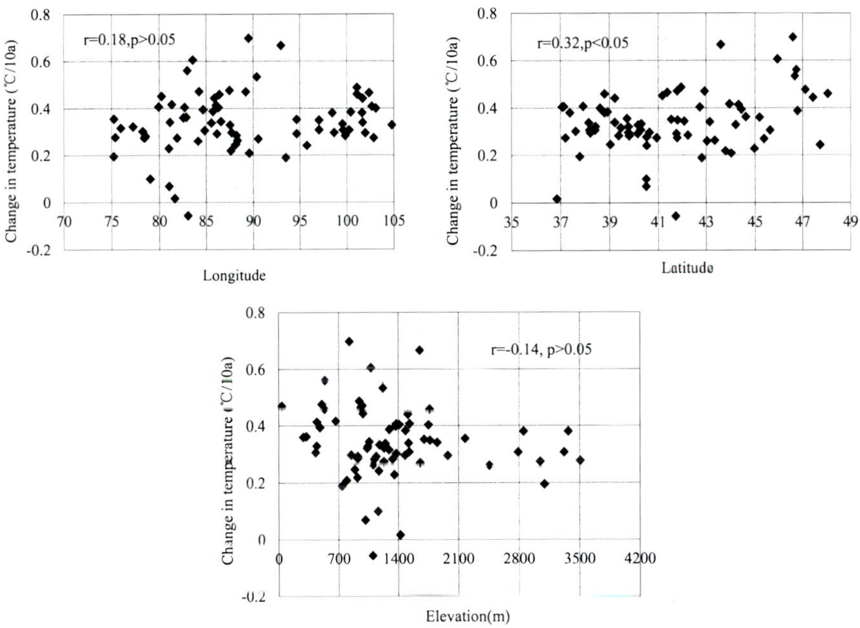

Fig. 2.15 Relationship between the trends in temperature and elevation, latitude and longitude

The Z value distribution for maximum temperature shows that it increases gradually from west to east, and then reduces slightly beyond the central Hexi Corridor, with a Z value from 3 to 4 (Fig. 2.14b). Except for a small area on the northern slope of the Tianshan Mountains, the Z values for the rest of the area meet the significance level of 0.01. Overall, the Hexi Corridor sees the greatest maximum temperature increases.

The influences of elevation, longitude and latitude on seasonal temperature trends are examined by correlation analysis (Fig. 2.15, Table 2.5). The results show that there is no significant association between the spring temperature trend and any other factor. However, the correlations between summer temperature trends and longitude are significant (at the $P<0.05$ level of significance) and the impact

Table 2.5 The relationship between seasonal temperature trends and longitude, latitude and elevation

Item	Spring (°C/10a)	Summer (°C/10a)	Autumn (°C/10a)	Winter (°C/10a)	Annual (°C/10a)
longitude	0.02	0.31*	0.05	0.16	0.18
latitude	0.14	0.06	0.38*	0.42**	0.32*
elevation	0.18	0.13	0.06	0.36*	−0.14

$*P<0.05$; $**P<0.01$

of latitude on autumnal temperature trends is also significant ($P<0.05$), as are the effects of latitude and elevation on winter temperature trends ($P<0.05$).

Although our results show only a weak correlation between trends in temperature and elevation and temperature and latitude/longitude, some researchers find that climate changes are related to elevation (Beniston and Rebetez 1996). Nonetheless, our results are consistent with other recent studies (Vuille and Bradley 2003). Although the relationship is not simple or linear, reductions in latitude generally coincide with decreases in the rate of temperature change, and some increases in latitude coincide with increases in the rate of temperature change (Fig. 2.15), which is consistent with previous research (He et al. 2005; Lu 2009).

2.2.2 *Step Change Detection for Temperature*

The Mann–Kendall method was used to detect step changes in temperature in the arid region of Northwest China. Results show that step changes probably occurred in 1988 ($P<0.01$) for regional annual air temperature (Fig. 2.16), but no similar results can be found at a seasonal scale. Step changes in 1998, 1996, 1987 and 1986 ($P<0.05$) were detected for spring, summer, autumn and winter for the whole region, respectively. Regarding basin scales, step changes in annual air temperature occurred in 1988 for 17 basins and in 1996 for 10 basins. Step changes at a seasonal scale were found in 1998 for spring (15 basins), in 1996 for summer (21 basins), in 1986 for autumn (15 basins), and in 1985 for winter (19 basins). This shows that the step-change point years on a basin scale are in accordance with the results for the entire arid region. Thus, for example, step changes for annual air temperature were observed in 1988.

Step change detection was done for temperature in the mountain, oasis and desert areas (Fig. 2.17). It was found that step changes occurred first in the desert area in 1987, followed by changes in the oasis area in 1990 and in the mountain area in 1997 ($P<0.05$). For the five typical river areas, step changes ($P<0.01$) were found in the northern slope of the Qilian Mountains in 1994, the southern slope of the Altai Mountains in 1984, the northern slope of the Tianshan Mountains in 1978, and in the northern slope of the Kunlun Mountains in 1980 and 1987. However, the southern slope of the Tianshan Mountains, a step change was not found (Table 2.2).

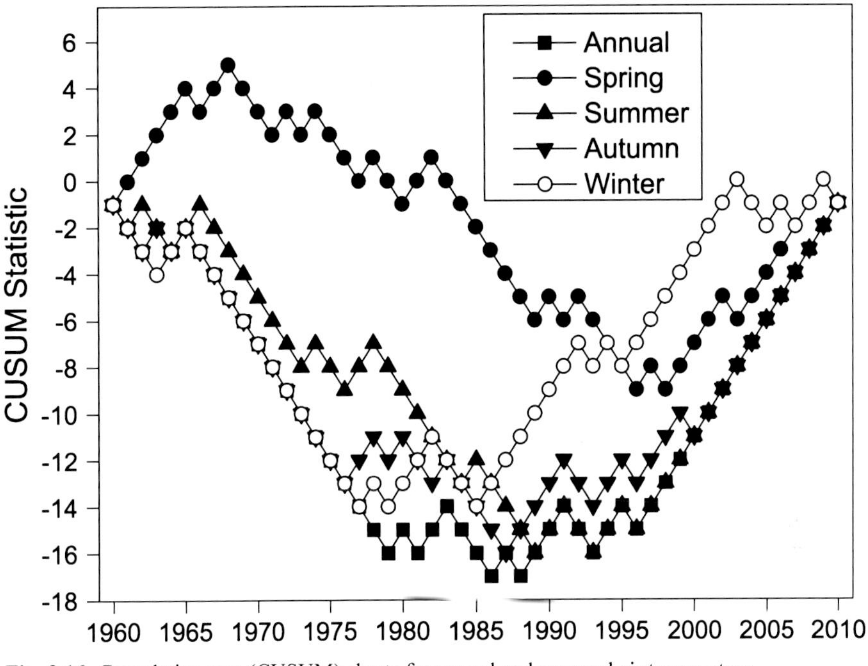

Fig. 2.16 Cumulative sum (CUSUM) charts for annual and seasonal air temperature

2.2.3 *Potential Causes of Winter Temperature Change*

2.2.3.1 Increasing Temperature

The Mann–Kendall (MK) statistical test revealed a significant rising trend in air temperature in the arid region of Northwest China from 1960 to 2010 ($P<0.001$), at a rate of 0.343 °C/decade. These values are consistent with results from previous research (Shi and Zhang 1995; Wang et al. 2008; Zhang et al. 2010). A rising trend was found in both oasis areas and mountainous areas, with a lower changing rate in the mountainous areas (0.325 °C/decade) and a higher rate in the oasis areas (0.35 °C/decade). Even the lower rate in the mountainous areas, however, is much higher than the average for China (0.25 °C/decade) (Ren et al. 2005) or the average of the entire globe (0.13 °C/decade) (Brohan et al. 2006; IPCC 2007). Due to the relatively low population density in China's arid region, we do not consider human activities as a major factor driving the overall regional temperature change to deviate considerably from the national and global trends. The finding that the temperature of China's northwest is quickly on the rise is consistent with results from previous research (Wang et al. 2008; Zhang et al. 2010; Sun et al. 2010) and with the statement in the Fourth Assessment Report of Intergovernmental Panel on Climate Change (IPCC) that the mid-high latitude regions of the northern hemisphere are experiencing a high rate of temperature increase.

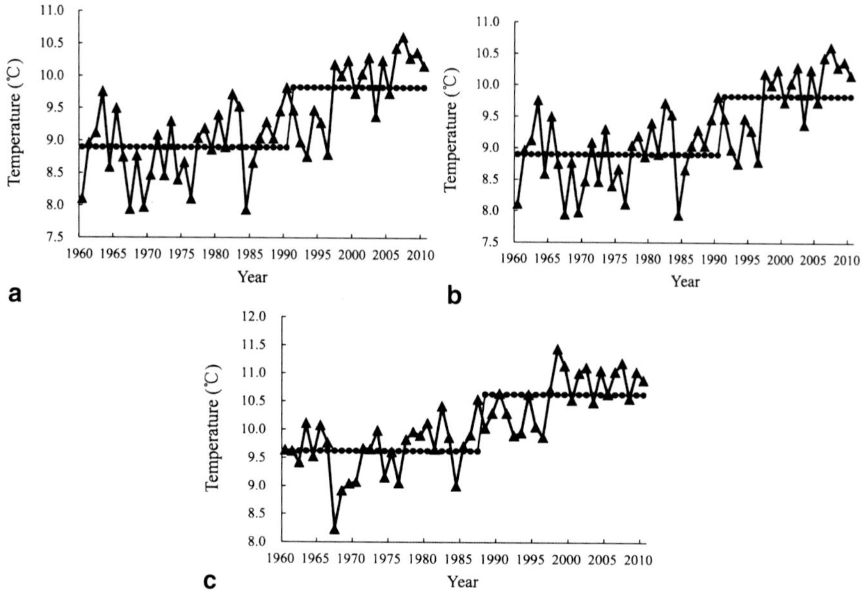

Fig. 2.17 Temperature step changes in **a** mountains, **b** oasis, **c** desert landscapes

The importance of the change of one season in an annual change is quantified as its proportion in the total change of the four seasons. This proportion was calculated for the four seasons of each year from 1981 to 2010, a process that can be represented as follows:

$$I_{s,i} = \frac{|T_{s,i} - \overline{T}_{s,1960-1980}|}{\sum |T_{s,i} - \overline{T}_{s,1960-1980}|} \times 100 \qquad (2.26)$$

$$s \in \{spring, summer, autumn, winter\}$$

where $I_{\text{winter},\ i}$ is the importance of the winter temperature variation of year i in the overall temperature variation of the same year, and in this study i = 1981, 1982, …2010; $T_{\text{winter},\ i}$ is the winter temperature of year i; $\overline{T}_{\text{winter},1960-1980}$ is the average winter temperature from 1960 to 1980; and $T_{s,\ i}$ is the seasonal temperature of year i.

Using Eq. (2.26), we calculated the seasonal importance in the yearly change for each year between 1981 and 2010. Over this period, the mean importance values for spring, summer, autumn, and winter were 19.3 %, 13.7 %, 23.6 %, and 43.4 %, respectively (Fig. 2.18). For the period 1984–1995, during which SHI was the weakest in the past 50 years, the importance of winter change was 57.01 %. These results suggest that winter temperature change is the most important factor affecting the rising annual air temperature in China's arid northwest.

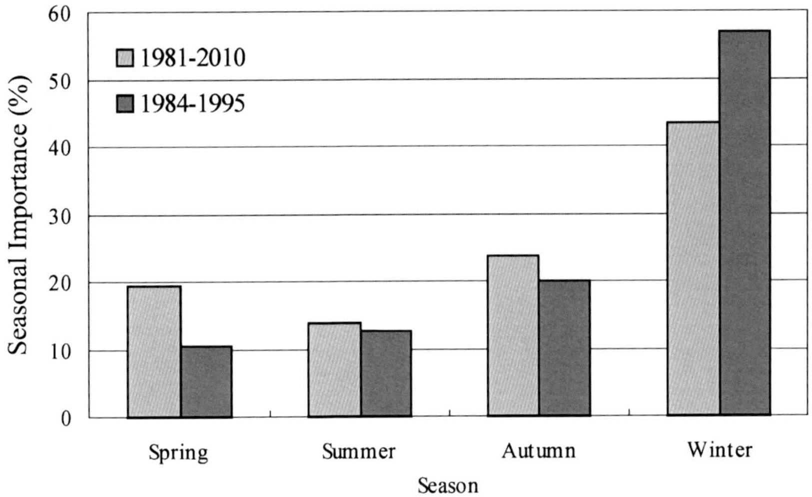

Fig. 2.18 The seasonal importance of temperature changes in different periods (1981–2010, 1984–1995)

Table 2.6 The correlation coefficients between the winter temperature in the entire region (the Arid Region of Northwest China), mountains, oases and certain factors that may affect the temperature

Factor	SOI	AAOI	NAOI	AOI	PNAI	WCI	SHI	CDE
Entire region	−0.256	0.128	0.329*	0.371*	0.023	0.377*	−0.715**	0.510**
Mountains	−0.238	0.078	0.220*	0.276	0.021	0.236	−0.629**	0.498**
Oases	−0.255	0.177	0.350*	0.388*	0.033	0.402*	−0.723**	0.504**

$^*P<0.05$; $^{**}P<0.01$

2.2.3.2 Potential Causes of Winter Temperature Change

The Pearson's correlation coefficient values (Table 2.6) show that the winter temperatures in Northwest China have a strong and significant correlation with the Siberian High Index (SHI, $R=-0.715$, $P<0.001$) and with China's carbon dioxide emissions (CDE, $R=0.51$, $P<0.001$). For all of the other tested atmospheric circulations, the correlations are much weaker and less significant. Another feature worth mentioning is that the correlations in oases are higher than in those in mountains, which means that the impact of atmospheric circulation on temperatures in oases are stronger than in mountains. In general, the Siberian High and the winter temperatures in the region have an almost perfect "mirroring" relationship during the 1960–2010 period, suggesting a direct impact of the former on the latter (Fig. 2.19). Particularly, the Siberian High intensity in the region had a deep "valley" between the late 1980s and the mid-1990s, which confirms the findings of D'Arrigo et al. (2005), Panagiotopoulos et al. (2005), and Gong and Ho (2002), who contend that the 1980s and 1990s was the weakest period for the Siberian High in the past 100 years. The aforementioned 'valley' of the Siberian High intensity coincides with

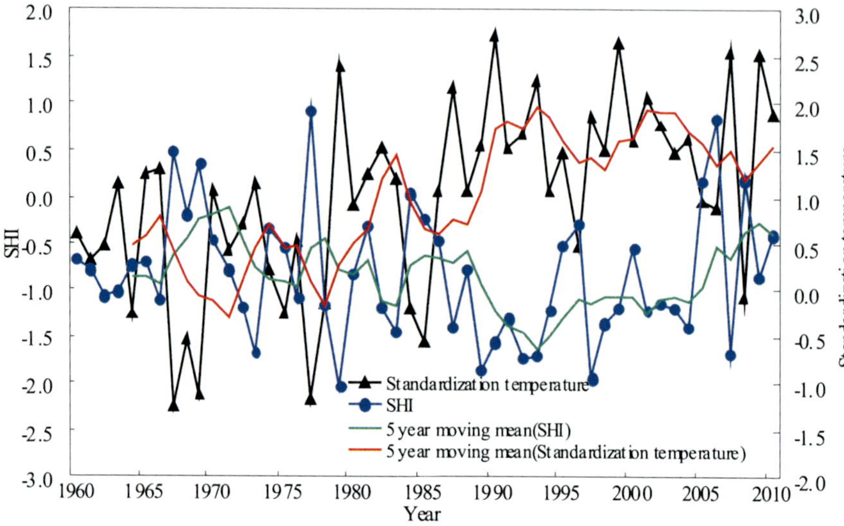

Fig. 2.19 The Siberian High Intensity and the winter temperature in the arid region of northwest China

the most obvious peak in winter temperatures in the region. Recently (2005–2010), however, the winter air temperature has been falling, along with the Siberian High intensity's recovery.

Excessive greenhouse gas emissions are generally regarded as the main cause of global warming (Crowley 2000; Mahlstein and Knutti 2010; IPCC 2007). However, as shown by Fig. 2.20, while the carbon dioxide emissions in the past 20 years maintained a strongly increasing trend, the winter temperature did not follow suit. It can be inferred from this that the rising winter temperature during the 1960–2005 period seems to be more associated with the Siberian High than with China's carbon dioxide emissions.

The R value calculated by Gong and Ho (2002) for the Siberian High intensity and air temperature in mid-to-high latitude Asia (30N–70N, 30E–40E) was −0.58, lower than that of China's arid region (−0.715). This suggests that the Siberian High had a stronger influence on air temperature in China's northwest than in other regions.

2.2.3.3 Contribution of the Siberian High to Increasing Winter Temperatures

The relationship between the change rate of the Siberian high in the 1961–2010 period and the winter temperature change rate of the same period was investigated (Fig. 2.21a). Not only did the fitting equation pass the 0.01significance level test, but in addition, the slope of the fitting equation (0.0466) was the elasticity coefficient of winter temperature change to the Siberian high change, meaning that if the

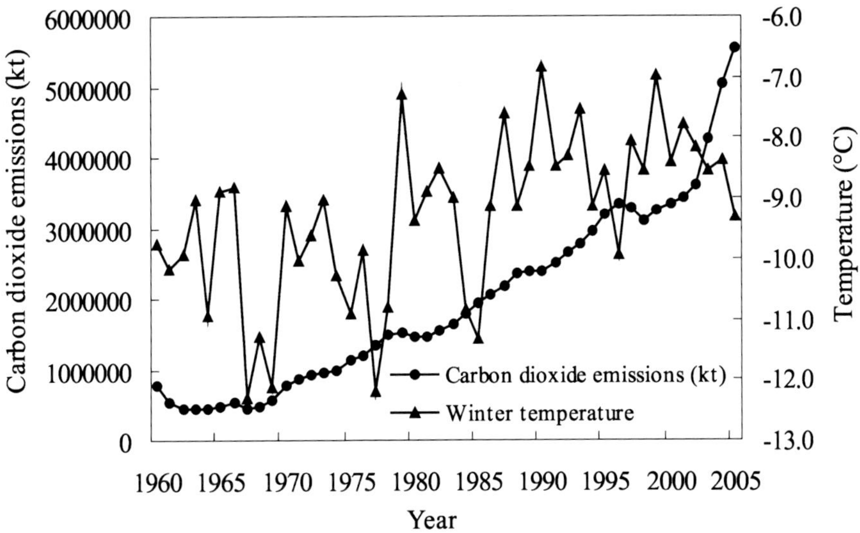

Fig. 2.20 The winter temperature in the arid region of northwest China and the yearly carbon dioxide emission in China

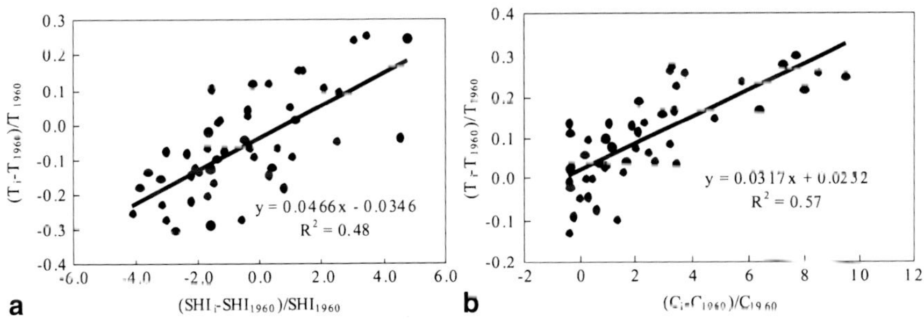

Fig. 2.21 Relationships between proportional changes of **a** annual Siberian High Index (SHI), CDE and winter temperature for the arid region of northwest China, **b** relationships between proportional changes of CDE and annual temperature for the arid region of northwest China

Siberian high index changed by 1 %, the temperature of China's arid region would change by 0.0466 %.

As the temperature in Northwest China dramatically changed in 1987, the 1960–1986 time frame is defined as a normal temperature fluctuation period, while 1987–2010 is defined as a climate change period. Furthermore, the contribution of the Siberian High to winter temperature change after 1987 (relative to that prior to 1987) is analyzed quantitatively.

Calculations based on a contribution rate formula and Table 2.7 indicates that the Siberian High contributed to winter temperature changes at a rate of 29.2 % for the period 1987–2010.

Table 2.7 The statistical characteristics of Siberian High Index, CDE, annual temperature and winter temperature for different period

Items	1960–1986	1987–2010	Variation rate
Siberian High Index (SHI)	0.318	−0.007	102.3 %
Winter temperature (°C)	−9.98	−8.35	1.63
Average annual temperature (°C)	6.25	7.20	0.94
Carbon dioxide emission (kt)	1,047,662	4,067,658	288.3 %

According to the above analysis, in 1987–2010, 40.9 % of the average annual temperature change was caused by winter temperature change, and 29.2 % of the winter temperature change was caused by changes in the Siberian High. Hence, we can suggest preliminarily that 11.9 % of the average annual temperature change was caused by changes in the Siberian High in China's northwest arid region.

2.2.4 *Future Temperature Trends*

2.2.4.1 Method

Wavelet Analysis As a wavelet transform can expand a time series into time frequency space, it can find localized intermittent periodicities. Wavelet analysis can also reveal localized time and frequency information without requiring the time series to be stationary. A continuous wavelet function (CWT) is described from a single function φ by translations and dilations:

$$\varphi_{a,b}(t) = |a|^{-\frac{1}{2}} \varphi\left(\frac{t-a}{a}\right), a, b \in R \tag{2.27}$$

where a is the scale parameter, b is the position parameter, and t denotes time. The CWT of the signal $f(t)$with the analyzing wavelet φ is the convolution of $f(t)$ with a set of scaled and translated wavelets:

$$W_f(a,b) = \langle f(t), \varphi_{a,b} \rangle = |a|^{-\frac{1}{2}} \int_R f(t)\varphi^*\left(\frac{t-a}{a}\right)dt \tag{2.28}$$

where * indicates the complex conjugate and $W_f(a, b)$ denotes the wavelet coefficient. Thus, the concept of frequency is replaced by that of scale, which can characterize the variation in the signal, $f(t)$, at a given time scale. The choice of wavelet φ depends on the signal to be analyzed. In this case, we select the Morlet wavelet as φ. The Morlet wavelet is defined by

$$\varphi(t) = \pi^{-1/4} e^{i\omega_0 t} e^{\left(-t^2/2\right)} \tag{2.29}$$

where ω_0 is the non-dimensional frequency (usually taken to be 6 to satisfy the admissibility). The relation between scale a and period T of the Morlet wavelet is given as: $T = \frac{4\pi a}{\omega_0 + \sqrt{2+\omega_0^2}} \approx 1.033a$ Hence, the Morlet wavelet (with ω_0=6) and the Fourier period (T) are almost equal to the scale (a).

The wavelet variance used to detect the periods present as the power density at different time scales, a, is calculated as:

$$E_a = \frac{1}{N}\sum_{b=1}^{N}\left|W_f(a,b)\right|^2 \tag{2.30}$$

where N is the length of data. The cone of influence (COI) is the region of the wavelet spectrum in which edge effects become important. It is defined here as the e-folding time for the autocorrelation of the wavelet power at each scale. Here, COI is used to ignore the edge effects. The significance of the global wavelet spectrum is using a red noise model to compare it with the theoretical global wavelet power spectrum.

2.2.4.2 Period Analysis

The temperature change period is an important factor that affects the accuracy of temperature simulations and future projections. In our study, we performed a time-series analysis of variance extrapolation to analyze temperature change periods of river areas in the northwest arid zone. The results show that the northern slope of the Qilian Mountain, the south slope of the Altai Mountains and the northern slope of the Kunlun Mountains have the same temperature change period of 19 years, while the temperature change period in the northern slope of the Tianshan Mountains is the shortest (17 years) and that in the southern slope of the Tianshan Mountains is the longest (20 years). Overall, there is little difference among the temperature change periods for each river area.

Further, the wavelet analysis method was used to analyze temperature periodicity in the four headstreams of the Tarim River Basin (Fig. 2.22). The results show that temperature in the Aksu River Basin had two weak periods of 3 and 11 years and two strong periods of 18 and 22 years. Because the data length is only 50 years in total, 18 years was taken as a reliable period. Meanwhile, the wavelet time-frequency diagram of temperature shows that the wavelet coefficients real part of the 18-year period within the 1960–1974 and 1986–1997 time frames is in the negative phase, which indicates that, in these years, temperatures were low. Conversely, the periods of time exclusive of these years were in a positive phase, which indicates that the temperatures were high.

Temperature in the Yarkand River basin had two weak periods of 4 and 11 years and two strong periods of 17 and 22 years. In view of the data length (as mentioned

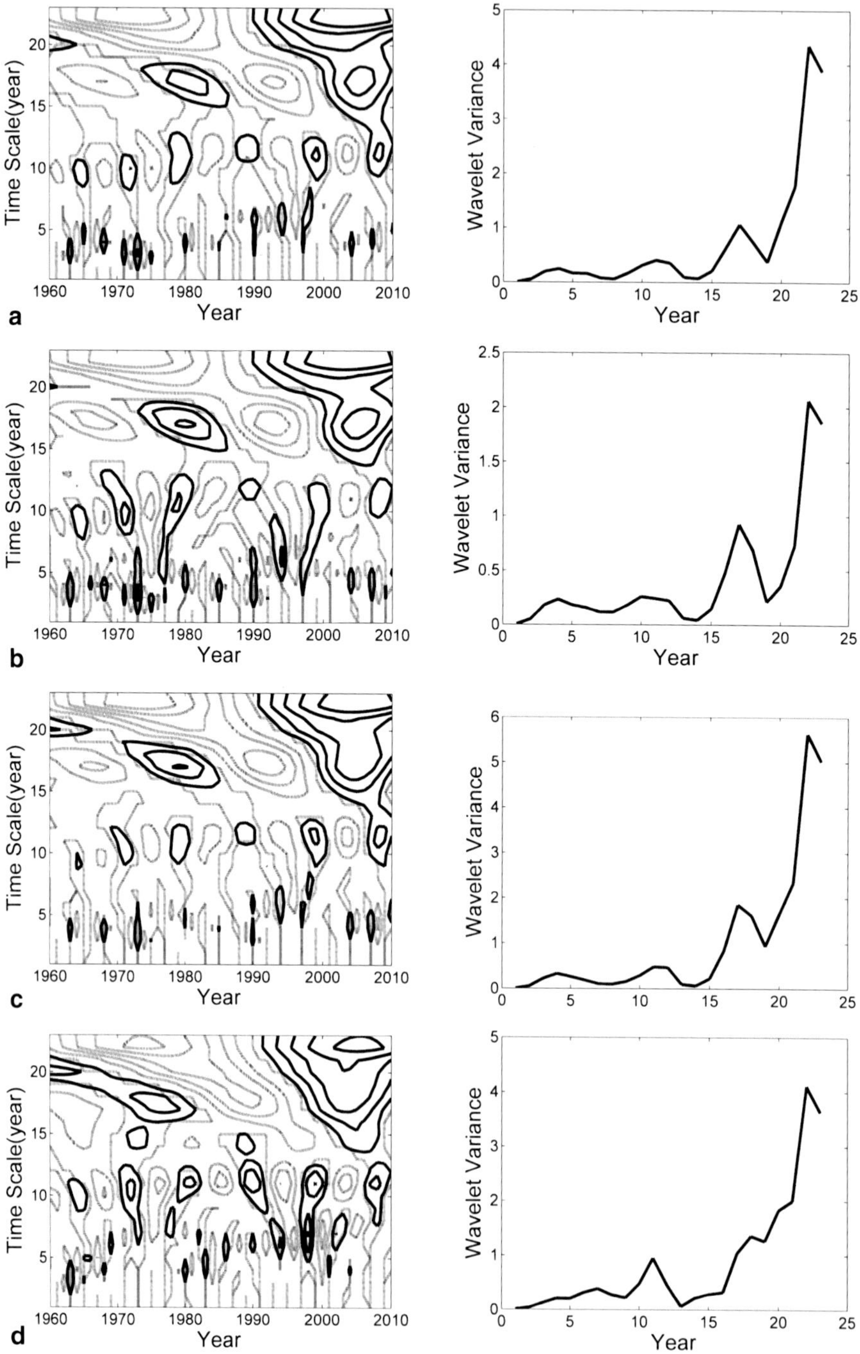

Fig. 2.22 Wavelet time-frequency distribution (*left*) and variance (*right*) of temperature in the four headstreams of Tarim river basin **a** Aksu River, **b** Yarkant River, **c** Hotan River, **d** Kaidu River

above), we considered the 17-year period as being reliable. The time-frequency distribution diagram of temperature indicated that the real part of the wavelet coefficients for the 17-year period was in a positive phase from 1973 to 1983 and again from 1997 to 2007, when temperatures were high.

Temperatures in the Hotan River basin had two weak periods of 4 and 11 years and two strong periods of 17 and 22 years, from which the 17-year period was assumed to be reliable, given the overall 50-year data collection timeframe. The real part of the wavelet coefficients of the 17-year period for the 1970–1980 and 2000–2007 stretches were in a positive phase, indicating that temperatures during these years were high.

Temperatures in the Kaidu River basin showed three weak periods of 7, 16 and 18 years and two strong periods of 11 and 18 years. In consideration of the data consistency, an 18-year cycle was taken as being reliable. The wavelet coefficient time-frequency distribution indicated that temperatures during the period 1964–1974 and after 1994 were high.

2.2.4.3 Future Temperature Forecast

To further clarify temperature changes for each river area in both historical periods and in the future, a superimposed trend prediction model was used (Table 2.2) to predict the average temperature for the next full cycle. This was then compared with the average temperatures for the period 1990–2010, the historical time period 1957–1989, and the next 10 years (2011–2020). For the 2006–2010 timeframe, simulated temperature values were used to replace temperatures where no measured data existed.

The linear regression coefficient test (Fig. 2.23) shows that the superimposed trend prediction model is highly accurate. The temperature simulation achieved $P<0.05$ level of significance in the southern slope of the Tianshan Mountains, while other river areas reached $P<0.01$, indicating that the model is suitable for temperature forecast in the study area.

In comparing the average temperatures in the periods 1990–2010 and 1957–1989, temperature on the southern slope of the Altai Mountains showed the greatest increase trend (Fig. 2.24), followed by temperature on the northern slope of the Qilian Mountains, the southern slope of the Tianshan Mountains, and the northern slope of the Kunlun Mountains. The temperature increments for these areas were 1.91, 0.94, 0.81 and 0.75 °C, respectively. The average temperature on the northern slope of the Tianshan Mountain increased least, with an increment of only 0.63 °C. Overall, the average temperature increased the most in northern Xinjiang, followed by the Hexi Corridor and southern Xinjiang.

A comparison of average temperatures between 2011–2020 and 1990–2010 (Fig. 2.24) shows increasing trends in the southern slope of the Altai Mountains, the northern slope of the Kunlun Mountain and the northern slope of the Tianshan Mountains, with increments of 0.76, 0.19 and 0.01 °C, respectively. However, temperatures on the southern slope of the Tianshan Mountains and the northern slope of

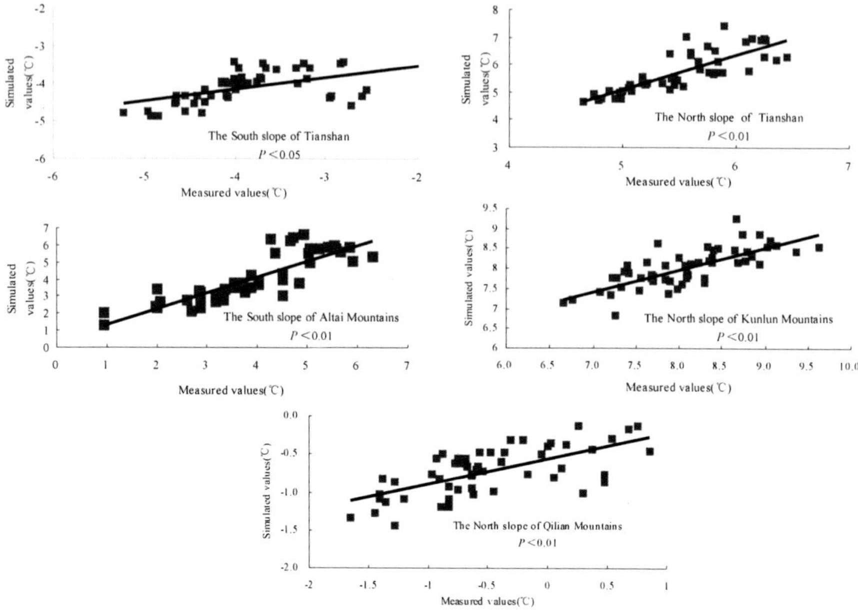

Fig. 2.23 Relations between simulated and measured values of temperature

Fig. 2.24 Temperature variations in each river area in different periods

the Qilian Mountains show a decreasing trend, with decrement of 0.52 and 0.39 °C, respectively. The decreasing temperatures may be attributed to anticipated fluctuant changes in temperature during the 2011–2020 cycle's low period. Therefore, while the average temperature in some river areas may decrease over the next few years, it does not represent a long-term climate trend in the area.

2.3 Precipitation Change

Of all the natural resources required for economic development, water is the most important, particularly in arid and semi-arid regions. It is widely recognized that anthropogenic activities will affect natural ecosystem and available water resources. The most obvious activity is climate change, which has resulted in modified precipitation patterns. In recent years, the occurrence of extreme events such as droughts and floods has been on the rise worldwide. The elasticity of runoff and water resources is particularly higher for drier climates, with a higher percentage change in runoff resulting from even small changes in precipitation in arid and semi-arid regions. It is therefore very important for water resources managers to know about and prepare to deal with the effects of precipitation change on hydrological cycles and streamflow regimes (Sullivan 2001). A good place to start is to know the characteristics of precipitation changes in recent years.

In our investigation, we used the latest (1960–2010) meteorological data from meteorological stations to explore historical and predicted precipitation variations in typical river and mountain-oasis-desert areas. In addition, we analyzed relationships between and among temperature, precipitation, latitude, longitude, elevation and atmospheric circulations, and detected step changes for precipitation. The results will provide a scientific basis for predicting future climate trends and for assessing the impact of climate change on water resources.

2.3.1 Characteristics of Precipitation Change

2.3.1.1 Inter-Annual Precipitation Change

(1) Precipitation Changes in Northwest China

Over the past 50 years, mean precipitation has measured approximately 153.37 mm in China's arid northwest region. Of the 74 sites analyzed in the study area, 69 (93.24%) (Fig. 2.25) showed precipitation trending toward larger values during the 1960–2010 period, while decreases were noted at only five sites. Many of these upward trends reached the level of statistical significance ($P<0.05$, under the Mann–Kendall test); however, of the sites with trends that did not pass this level, the precipitation values increased at 30 sites (43.48%) and decreased at 4 (80%). Thus, the overall increasing rate of mean annual precipitation was 6.07 mm/10a. This trend is significant ($P<0.01$, under the Mann–Kendall test), which indicates that the climate has become wetter in the arid region.

Our study revealed varying degrees of regional differences in precipitation increases. For example, northern Xinjiang had the highest rising rate (9.20 mm/10a), followed by southern Xinjiang (5.35 mm/10a) and the Hexi Corridor of (3.96 mm/10a).

(2) Precipitation Changes in the Mountain-Oasis-Desert Area

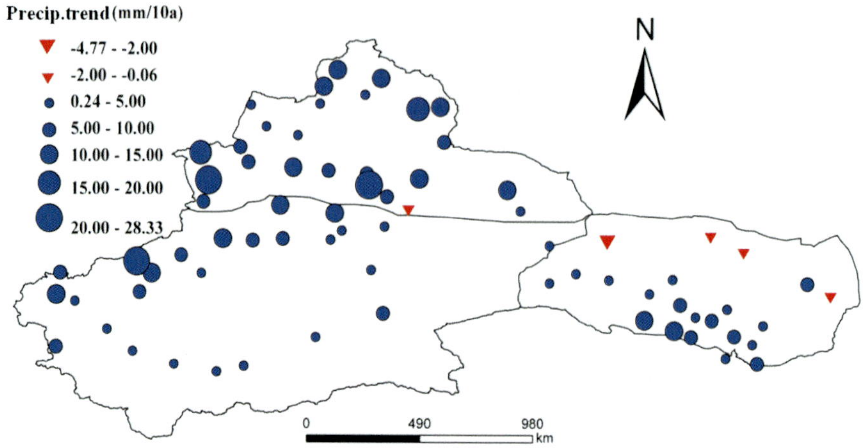

Fig. 2.25 Trends in precipitation during 1960–2010

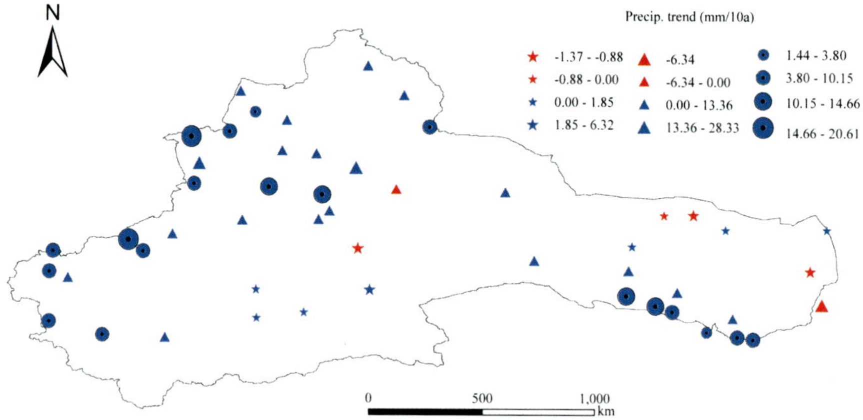

Fig. 2.26 The precipitation trend of each meteorological station during the period 1960 to 2010

Trends of Annual Precipitation The linear tendencies in precipitation measured at 51 stations over an approximate 50-year period are examined (Fig. 2.26). Among these stations, 45 had increasing trends for the period, which indicates that the overall trend is one of increased precipitation, with the mean rate at all 51 stations measuring 5.77 mm per decade.

The mountain landscape had an increasing rate of 10.15 mm per decade, the fastest among the three landscape types. This was followed by the oasis landscape at 6.29 mm per decade and the desert landscape, with only 0.87 mm per decade. The Mann–Kendall test shows that the increasing trends of the mountain and oasis landscapes are statistically significant ($P<0.01$), while the trend for the desert

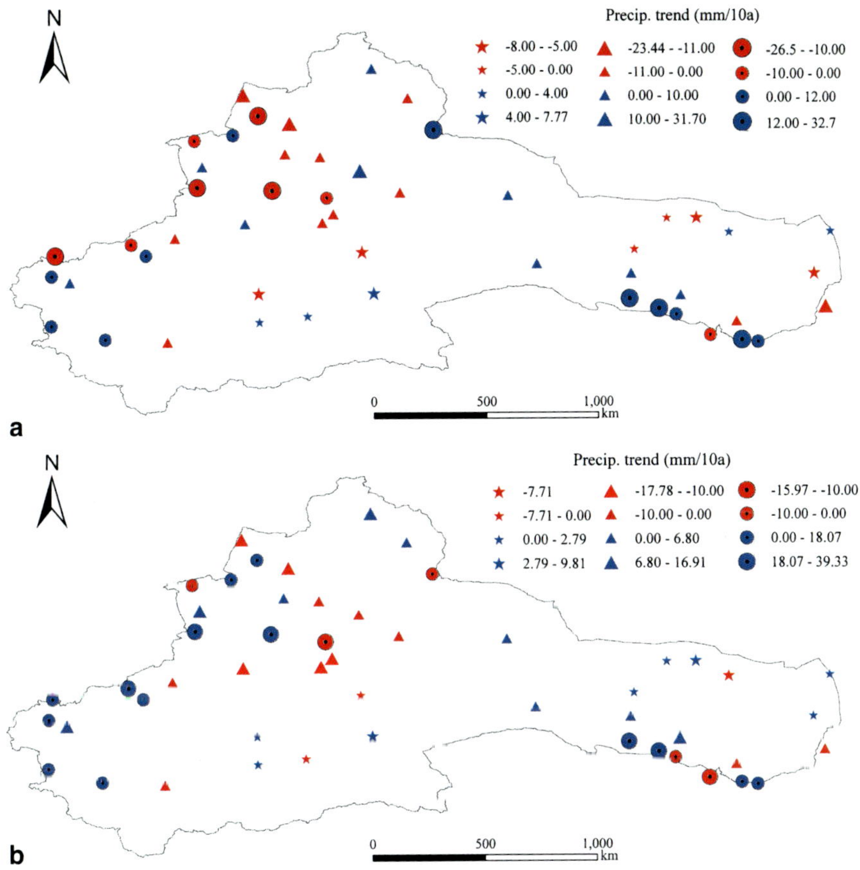

Fig. 2.27 The precipitation trend of each the meteorological station during the periods **a** 1960–1986, and **b** 1987–2010

landscape is not. There are four main reasons that explain the increasing precipitation (Yang et al. 2009): (1) local climate characteristics like high water vapor content in the air and favorable weather conditions; (2) temperature rises leading to increased amounts of glacial and snowmelt water and further promoting the formation of mountain precipitation conditions; (3) rising temperatures strengthening local circulation between the plain areas and the mountains and improving air humidity in the oasis area, leading to increased precipitation in the latter; and (4) the process of global warming-driven water circulation speeding up. Bengtsson (1997) estimated that the doubling of greenhouse gases might cause temperature rises that could increase atmospheric water vapor content by 15 % and precipitation by 8 %.

To further clarify the precipitation change of each area over different periods, the time series has been divided into two periods: 1960–1986 and 1987–2010.

During the 1960–1986 period, about a half of the 51 stations experienced increasing trends and the other half decreasing trends (Fig. 2.27a). The mean changing

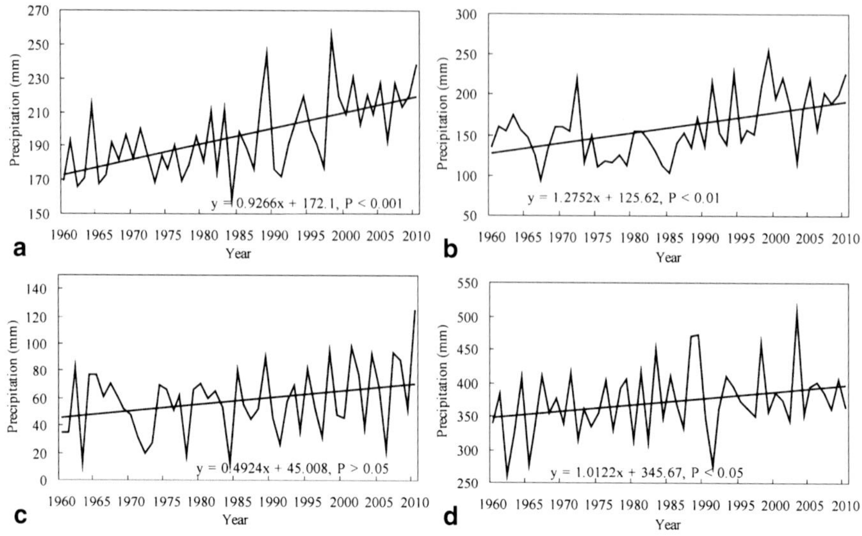

Fig. 2.28 Time series of precipitation of the snowmelt period and its linear trend for **a** entire northwest China, **b** southern Tianshan Mountains, **c** northern Kunlun Mountains, **d** northern Qilian Mountains

rate was −4.35 mm/10a. The rate of mountain landscape was 0.98 mm/10a, with 11 of 19 stations showing increasing trends. The oasis landscape had a rate of −1.17 mm/10a, with 12 of 21 stations showing decreasing trends. The desert landscape had a similar rate of −1.12 mm/10a, with 6 of 11 stations showing decreasing trends. Overall, precipitation changes across all types of landscape area were relatively insignificant during the 1960–1986 timeframe.

From 1987 to 2010, more stations (31 stations) showed increasing trends (Fig. 2.27b), with the mean rate of the 51 stations measuring 3.59 mm/10a, but precipitation changes were inconsistent across the different landscapes. The mountain areas had a rising trend with a mean rate of 12.20 mm/10a, and 14 of 19 stations had increasing trends, which was considerably more than during the 1960–1986 period. The precipitation in the oasis area jumped in 1987, and then experienced large variances while maintaining a relatively high mean (23.02 mm, considerably higher than the period of 1960–1986). Precipitation in the desert area increased slightly, at only 1.31 mm/10a. Overall, increases in precipitation during the 1987–2010 period occurred mostly in the mountain and oasis areas

Trend of Snowmelt Period Precipitation in the Mountain Area Since 1960, the mean precipitation of snowmelt shows an increasing trend in the study area, with a linear trend of 9.27 mm/10a at $P<0.001$ level in the significance test (Fig. 2.28a). The rates of increase in precipitation in southern Tianshans and the northern Qilian Mountains are higher, measuring 12.75 mm/10a and 10.12 mm/10a, respectively, and reaching a significance level of $P<0.05$. However, in the northern Kunlun

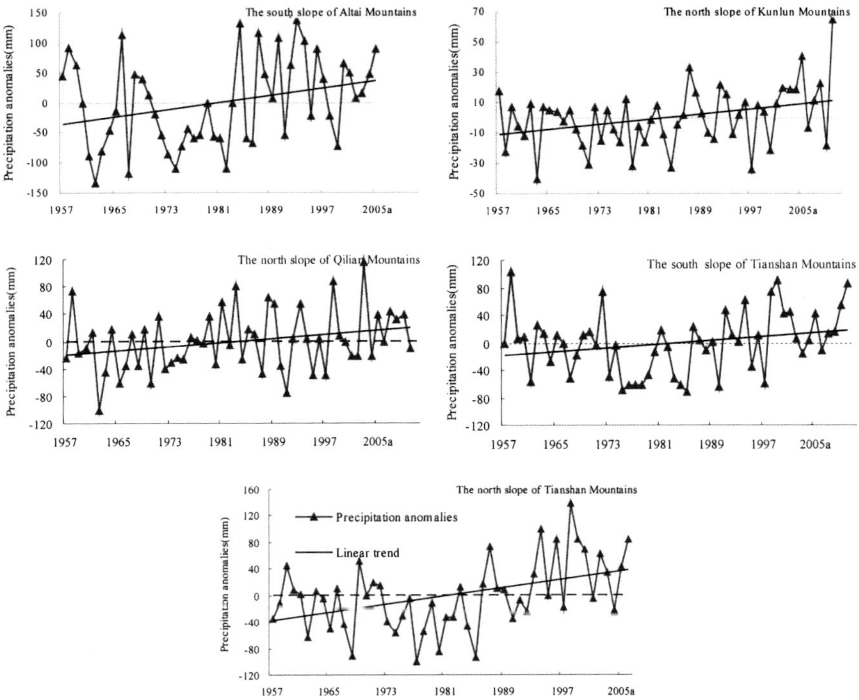

Fig. 2.29 Precipitation anomalies trends in typical river areas of the arid region of Northwest China

Mountains, the rate of increase is lower (4.92 mm/10) and does not reach significance level (Figs. 2.28b, c, d).

(3) Precipitation Changes in Typical Rivers

Precipitation Trend Precipitation in each river area shows clear increasing trends (Fig. 2.29) (Li et al. 2012). On the north slope of the Tianshan Mountain and the south slope of the Altai Mountains, the increasing trend of precipitation is statistically significant at $P<0.01$ and $P<0.05$, respectively, while trends on the north slope of the Qilian Mountains, the north slope of the Kunlun Mountains and the south slope of the Tianshan Mountains are not significant.

Since 1960, the average increasing rate of precipitation in each river area was 9.12 mm per decade. The rates of precipitation on the north slope of the Tianshan Mountains and the south slope of the Altai Mountains increased the fastest, at 15.48 and 15.30 mm/10a, respectively, followed by those on the north slope of the Qilian Mountains and the south slope of the Tianshan Mountains, at 7.64 and 6.75 mm/10a, respectively, while the north slope of the Kunlun Mountains showed increase rates of only 4.25 mm/10a.

The variation coefficient of precipitation in each river ranged from 0.12 to 0.34, which indicates moderate variability (Table 2.8). The variation coefficient on the

Table 2.8 Precipitation trends and tests in typical river areas of the arid region of Northwest China

Typical areas	Typical river areas	Cv	Zc	Step change points	Periods (a)
Hexi Corridor	The north slope of Qilian Mountains	0.12	1.25	None	22**
Northern Xinjiang	The south slope of Altai Mountains	0.26	1.97*	1987*	18**
	The north slope of Tianshan Mountains	0.17	2.64**	1992*	17*
Southern Xinjiang	The south slope of Tianshan Mountains	0.17	0.75	None	12*
	The north slope of Kunlun Mountains	0.34	1.72	1999*	20**

*$P<0.05$; **$P<0.01$

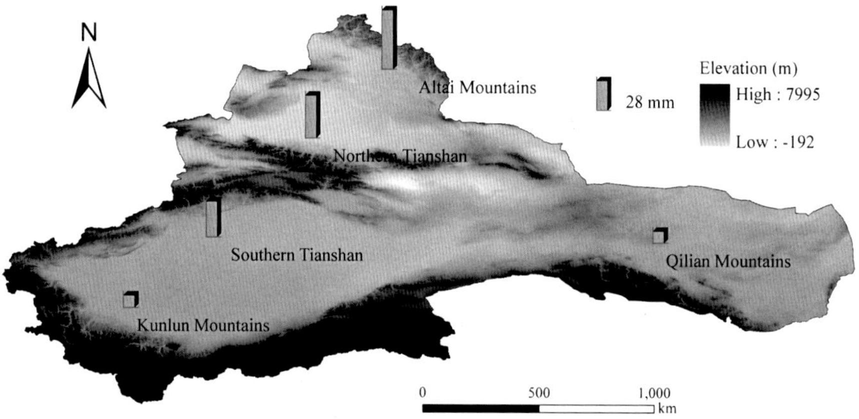

Fig. 2.30 Precipitation changes of each river in different periods in the arid region of China

northern slope of the Qilian Mountains was the smallest (0.12). while that on the northern slope of the Kunlun Mountains was the largest (0.34).

Precipitation Changes in Different Periods Compared to 1957–1989, precipitation during the period 1990–2010 on the south slope of the Altai Mountains showed the largest increase (55.41 mm) (Fig. 2.30), followed by the north and south slopes of the Tianshan Mountains, which increased by 39.56 mm and 32.91 mm, respectively. Precipitation increases on the northern slopes of the Kunlun and Qilian Mountains were the smallest, increasing by 11.33 mm and 10.77 mm, respectively. In general, northern Xinjiang registered the largest increases, followed by southern Xinjiang and the Hexi Corridor.

2.3.1.2 Seasonal Variability of Precipitation

In the 1960–2010 timeframe, spring precipitation exhibited a significant rising trend at 58 (78.38 %) of the 74 sites (Fig. 2.31) while decreasing at the remaining 16 sites. Most of these upward trends are not statistically significant; however, the precipitation increases at 5 sites (8.62 %) did reach significance level ($P<0.05$, under the Mann–Kendall test). Overall, the mean increasing rate in precipitation was 1.29 mm/10a for the entire northwest region. Furthermore, precipitation in northern Xinjiang showed the fastest rate of increase (2.41 mm/10a) while southern Xinjiang exhibited the slowest rate (0.55 mm/10a) (Table 2.9).

Summer precipitation increases were recorded at 54 (72.97 %) of the 74 sites during the period 1960–2010 (Fig. 2.31). Precipitation values increased at 18 sites and decreased at 1, reaching significance level ($P<0.05$, under the Mann–Kendall test). The mean increasing rate in precipitation was 2.51 mm/10a in the study area, which was higher than in the other seasons. The precipitation rate increased the fastest (3.80 mm/10a) in southern Xinjiang, followed by northern Xinjiang (2.67 mm/10a) and the Hexi Corridor (0.99 mm/ 10a) (Table 2.9).

In autumn, precipitation generally trended toward larger values at 64 (86.49 %) of the sites (Fig. 2.31). Although most of these trends were not at the level of statistical significance, 11 sites did achieve this level ($P<0.05$, under the Mann–Kendall test), giving an increasing rate of precipitation of 1.82 mm/10a for the entire area. Regionally, precipitation showed the fastest rate of increase (2.14 mm/10a) in northern Xinjiang, while southern Xinjiang and the Hexi Corridor were essentially the same (1.61 mm/10a).

Winter saw rising precipitation trends at 68 (91.89 %) of the 74 (Fig. 2.31), with precipitation values increasing at 22 sites (32.35 %), which was not at the level of statistical significance (although others did reach $P<0.05$, under the Mann–Kendall test). The mean precipitation increased 1.16 mm/10a, which was significantly lower than in spring, summer, and autumn. Regionally, northern Xinjiang showed the fastest increasing rate in precipitation (2.21 mm/10a), followed by southern Xinjiang (0.59 mm/10a) and the Hexi Corridor (0.38 mm/10a) (Table 2.9).

Correlations between elevation, longitude, latitude, AO, AAO, SO, NAO and seasonal precipitation trends were also investigated in our study (Table 2.10). We found that spring precipitation is significantly correlated with latitude ($P<0.01$), summer precipitation is strongly correlated with elevation and longitude ($P<0.05$), autumn precipitation is not significantly correlated with longitude, latitude or elevation, and winter precipitation is significantly correlated with longitude, latitude and elevation ($P<0.05$).

2.3.1.3 Decadal Variability in Precipitation

In the 1960s, the mean precipitation was 134.34 mm in China's arid northwest and showed an increasing trend through the 1970s at 60.81 % of the meteorological stations. The sites showing precipitation increases were located mainly in the north

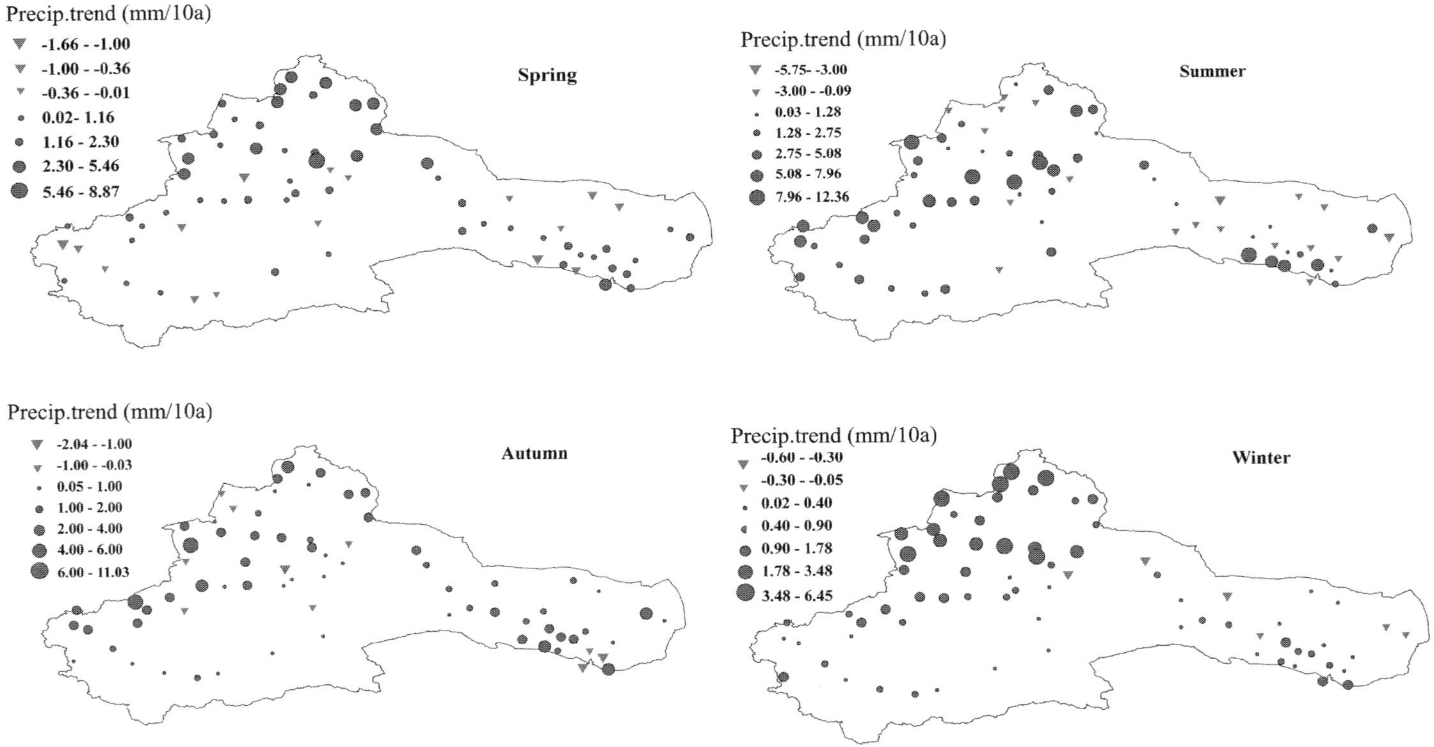

Fig. 2.31 Seasonal precipitation trends (1960–2010) in the arid region of northwest China

Table 2.9 The annual and seasonal trend in precipitation over different regions

Region	Spring (mm/10a)	Summer (mm/10a)	Autumn (mm/10a)	Winter (mm/10a)	Annual (mm/10a)
ARNC	1.29	2.51*	1.82*	1.16**	6.07**
Northern Xinjiang	2.41	2.67	2.14*	2.21*	9.20**
Southern Xinjiang	0.55	3.80*	1.64	0.59	5.35*
Hexi Corridor	0.86	0.99	1.64	0.38	3.96

$^*P<0.05$; $^{**}P<0.01$

Table 2.10 The relationship between seasonal precipitation trends and elevation, longitude, latitude and atmospheric circulation

Item	Spring	Summer	Autumn	Winter
Longitude	0.01	−0.34*	−0.13	−0.30*
Latitude	0.41**	−0.03	0.14	0.58**
Elevation	−0.18	0.35*	0.10	−0.33*
SOI	−0.18	−0.03	0.18	−0.20
AAOI	−0.02	0.04	−0.10	−0.04
NAOI	0.09	0.06	−0.17	0.14
AOI	−0.02	0.12	−0.05	0.09

$^*P<0.05$; $^{**}P<0.01$

and south slopes of the Tianshans and in the Hexi Corridor. Overall, precipitation increased the fastest in the Hexi Corridor at a rate of 8.23 %, while increasing only 1.12 % in southern Xinjiang and 0.36 % in northern Xinjiang (Fig. 2.32).

In the 1980s, the increasing rate in precipitation was 3.91 %, which is higher than the 1970s. Precipitation at 80.39 % of the weather stations showed notable increases, whereas 74 % of weather stations in the Hexi Corridor registered decreases. Northern Xinjiang exhibited the fastest precipitation rate increase (8.91 %), followed by southern Xinxiang (8.49 %). However, precipitation in the Hexi Corridor showed a decreasing trend at a rate of 3.65 % (Fig. 2.32).

In the 1990s, the increasing rate of precipitation was the fastest (6.99 %) compared with the other decades under study, and precipitation at 72.97 % of the meteorological sites increased. Southern Xinjiang showed the fastest increasing rate (15.46 %) followed by northern Xinjiang (7.63 %), while the Hexi corridor trailed far behind (1.61 %).

During the period 2000–2010, the mean precipitation charted a continuous increasing trend, but the increasing rate of 4.31 % was less than in the 1990s. Overall, precipitation at 60.81 % of the weather stations increased, while regionally, precipitation in northern Xinjiang showed the fastest increasing rate (5.08 %), followed by the Hexi Corridor (4.84 %) and southern Xinjiang (1.83 %) (Fig. 2.32).

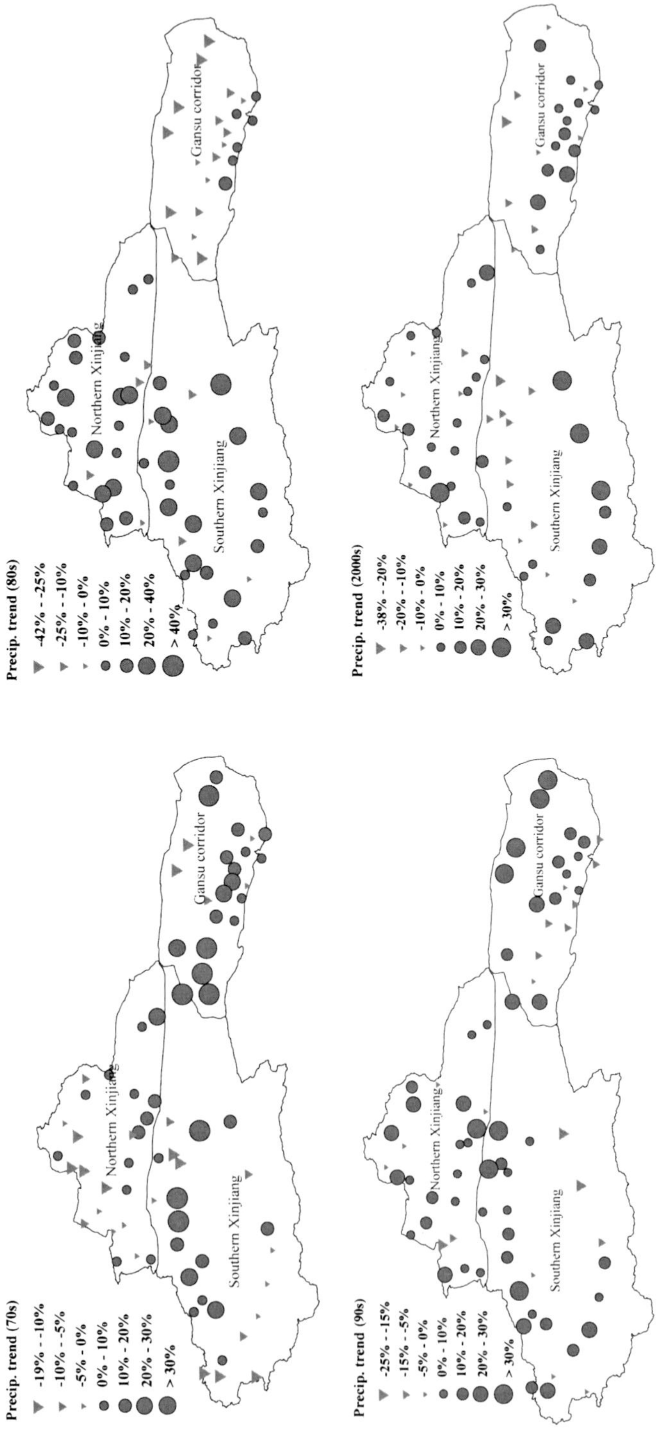

Fig. 2.32 Precipitation trends (1960–2010) in different decades in the arid region of northwest China

Table 2.11 Results of Mann–Kendall test for annual and seasonal precipitation and temperature

Parameter	Annual	Spring	Summer	Autumn	Winter
Precipitation (mm/decade)	8.01[b] (172.38)	1.78 (43.93)	2.55[a] (78.48)	1.72[a] (35.50)	1.73[b] (13.80)
Temperature (°C/decade)	0.35[b] (7.20)	0.24[a] (9.18)	0.26[b] (17.75)	0.41[b] (7.40)	0.39[b] (–8.83)

The data in the parenthesis are the mean of the study area

[a] delineate the significance at 5 % significance level

[b] delineate the significance at 1 % significance level

2.3.1.4 Spatio-Temporal Distribution of Monotonic Trends

The statistical characteristics and trends of precipitation and air temperature in the study area were examined (Table 2.11). Spatial variations are observed for precipitation trends, with few stations showing significant changes at the 0.05 significance level, although annual precipitation showed a significant positive trend (8.01 mm/decade). On a seasonal basis, statistically significant positive trends ($P<0.05$) can be detected for all seasons except for spring, with winter experiencing the most significant precipitation increases. Annually, 56 % of the stations exhibited significant increasing precipitation trends at the 0.05 significance level.

Spatial trends for annual and seasonal precipitation at individual stations and on a basin scale are shown in Figs. 2.33 and 2.34. For spring, summer, autumn and winter, the percentages for all stations exhibited notable increasing trends of 6, 18, 19, and 36 %, respectively (Fig. 2.33). For annual precipitation, stations in North Xinjiang and in the middle and west of the south slope of the Tianhshan Mountains showed the most significant increasing trends (Fig. 2.33a), while the basins of the Aibuhu Drainage, northern North Xinjiang (the Wulungu, Jimunai and Irtysh Rivers), and the south slope of the Tianshan Mountains (the Weigan, Aksu and Kaidu Rivers) showed strong increasing trends at a significance level of 0.01 (Fig. 2.33a).

Seasonally, the majority of the stations showed no significant trends in spring (Fig. 2.33b), with only the basins in the Chenerchen River, the Kumu Tage Desert, the Hami Basin and the Wulungu River showing statistically significant trends (at a significant level of 0.05). In summer, 18 % of the stations demonstrated notable increasing trends at a 0.05 significance level, especially stations around the Tianshan Mountains and in western South Xinjiang (Fig. 2.33c). Additionally, the Kaidu River exhibited a large increase trend at the 0.01 significance level (Fig. 2.34c). In autumn, precipitation showed an increasing trend ($P<0.01$) in the Aksu River and the Bayi Basin (Fig. 2.34d), while in winter, precipitation changes were most apparent in North Xinjiang (Fig. 2.33e). Specifically, basins in North Xinjiang (except for the Ili and Wulungu Rivers), the Kaidu River and the Kumu Tage Desert exhibited the greatest statistical variations (Fig. 2.34e).

Thus, overall, we can see that precipitation in the study area was dominated by generally increasing trends, especially in North Xinjiang and the south slope of the Tianhshan Mountains. The reasons for the increase in precipitation are related to larger-scale circulation currents. The results of previous studies showed that water

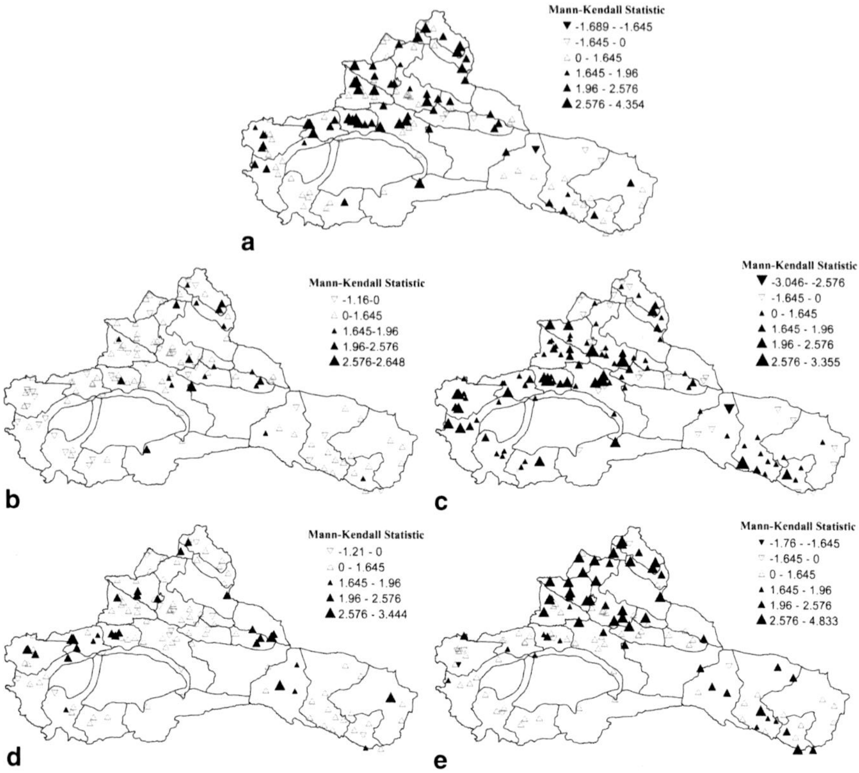

Fig. 2.33 Distribution of Mann–Kendall trends of annual and seasonal precipitation, **a** annual, **b** spring, **c** summer, **d** autumn, **e** winter

vapor from the Indian Ocean and the western Pacific has increased in recent years, bringing larger amounts of rainfall into China's arid region. However, the mixed pattern in the precipitation indicates that this change is a complex process affected both by regional atmospheric circulation and the local environment.

A detailed analysis of the relationships between trends in precipitation and elevation, and between precipitation and latitude/longitude show that precipitation is not significantly correlated with elevation (Fig. 2.35a) but is strongly correlated with latitude and longitude (Figs. 2.35b, c) ($P<0.05$).

2.3.2 Step Change Detection for Precipitation

Cumulative sum (CUSUM) charts of annual and seasonal precipitation for the study area (Fig. 2.36) showed that step changes for precipitation were identified, with the annual precipitation in 1991 at the 0.01 significance level. Annual precipitation prior to 1991 was 163 mm, whereas after 1991, it increased to 188 mm. Step changes

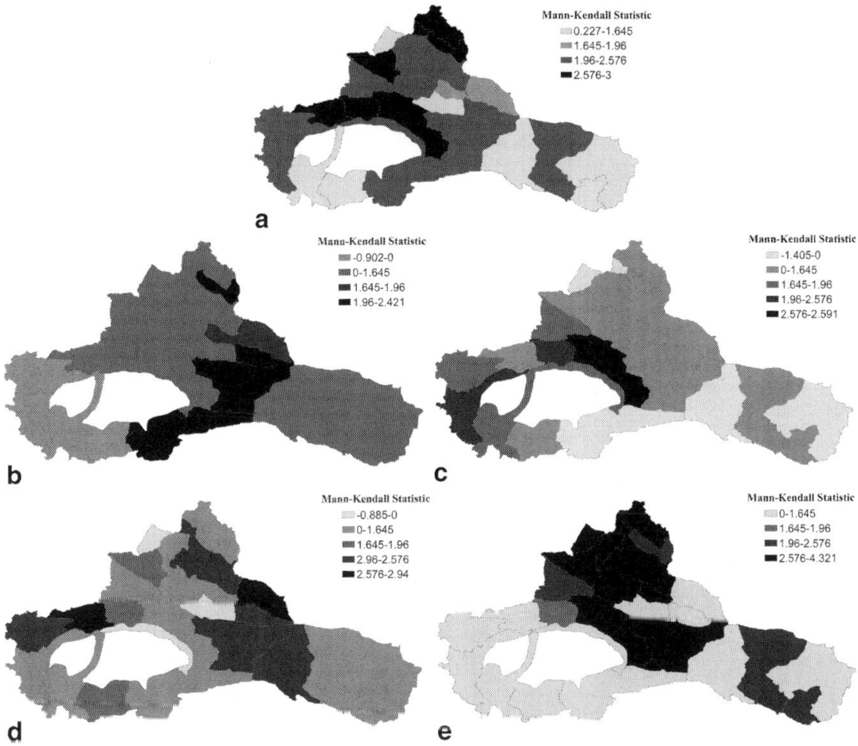

Fig. 2.34 Regional trends of precipitation at basin scale **a** annual, **b** spring, **c** summer, **d** autumn, **e** winter

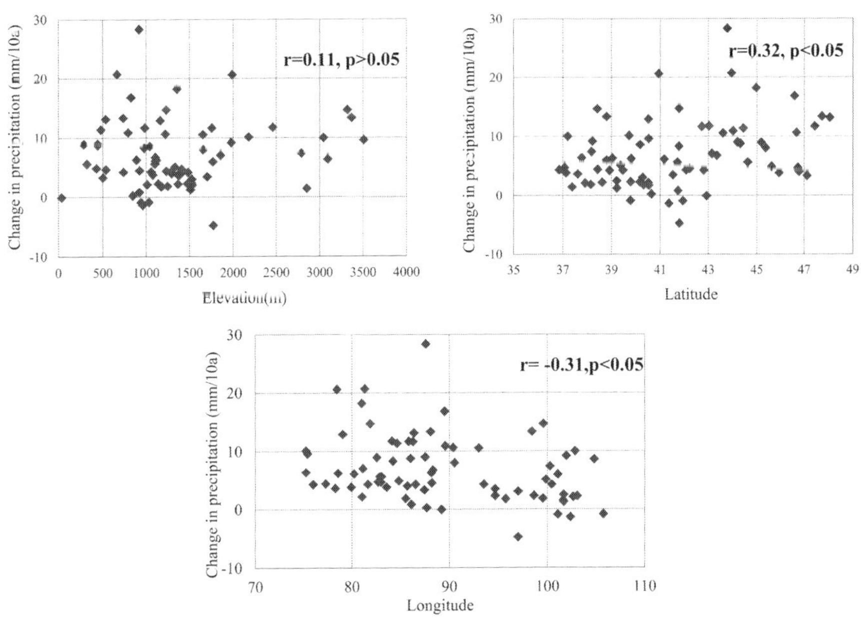

Fig. 2.35 **a** Relationships between the trends in precipitation and elevation, **b** latitude, **c** longitude

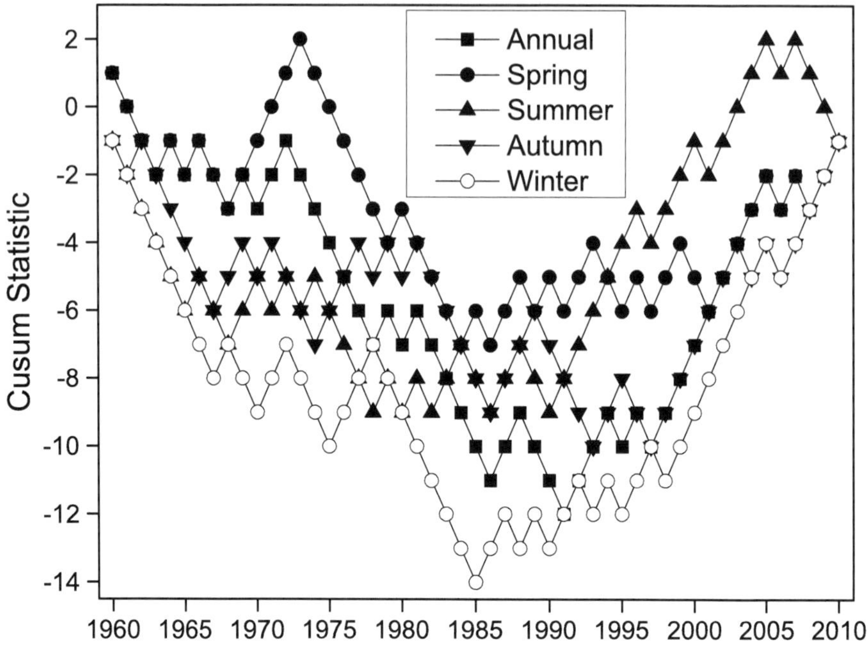

Fig. 2.36 Cumulative sum (CUSUM) charts for annual and seasonal precipitation

were also are observed in individual basins. About 12 basins showed step changes in 1986 at a significance level of 0.05, whereas only 1 basin showed changes in 1970, 1980 and 1991, respectively. On a seasonal basis, the change points can also be detected for all seasons. The change points are identified as 1986, 1990, 1997 and 1985 in spring, summer, autumn and winter, respectively; moreover, the precipitation in the later time series is higher than that in the earlier series. The step changes in individual basins on a seasonal scale show a mixed pattern, from which a uniform conclusion cannot reasonably be drawn.

The results of step change detection for precipitation (Fig. 2.37) showed that step changes were found in mountain and oasis areas in 1986 and 1987, respectively, but that no step change occurred in the desert area. For the five typical river areas, step changes in precipitation (Table 2.12) appeared in the southern slope of the Altai Mountains and the northern slope of the Tianshan Mountians in 1987 and 1992, respectively. However, no step change points were found in the northern slope of the Qilian and Kunlun Mountains or in the southern slope of the Tianshan Mountains.

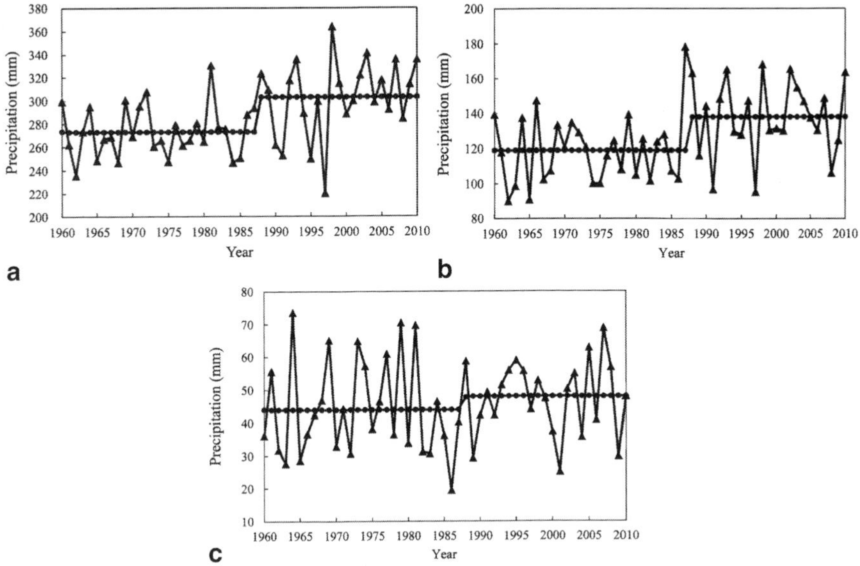

Fig. 2.37 Step change of precipitation in **a** mountains, **b** oasis, **c** desert areas

2.3.3 *Future Precipitation Trends*

2.3.3.1 Period Analysis

A variance analysis and an extrapolation method were employed to predict precipitation in the river areas. The results show a 22-year period as being the longest, which is predicted to occur in the northern slope of the Qilian Mountains. This will be followed by cycles of 20, 18 and 17 years in the northern slope of the Kunlun Mountains, the southern slope of the Altai Mountains and the northern slope of the Tianshan Mountains, respectively. The southern slope of the Tianshan Mountains, however, will experience a relatively short cycle of only 12 years. Overall, the precipitation periods for each river area are expected to show sizeable differences.

Further, the wavelet analysis method was used to analyze precipitation periods in the four headstreams of the Tarim River Basin (Fig. 2.38). Precipitation in the Aksu River basin had four weak periods of 4, 8, 18 and 21 years and one strong period of 11 years. Precipitation in 1962–1965, 1971–1974, 1981–1984, 1986–1991, 1997–2001 and 2001–2004 was in a positive phase, indicating that precipitation in these years was high. Precipitation in the Yarkand River basin had two weak periods of 4 and 14 years and one strong period of 9 years. Precipitation in 1962–1965, 1969–1972, 1975–1978, 1980–1983, 1984–1987, 1989–1992, 1994–1997 and 2000–2003 was in a positive phase, indicating that precipitation in these years was high. Precipitation in the Hotan River basin has four weak periods of 4, 5, 11 and 14 years and one strong period of 9 years. Precipitation in 1966–1968, 1971–1973,

Table 2.12 Precipitation trends and tests in typical river areas of the arid region of northwest China

Item	Average annual (mm)	Cv.	Trend	Zc	Step chang (year)	Primary period	Prediction model
NS of Qilian Mountians	365.69	0.120	↑	1.25	none	22**	$X(T)=392.68+1.024T+d_{t+T}$
SS of Altay Mountians	286.89	0.255	↑	1.96*	1987*	18**	$X(T)=329.456+1.03T+d_{t+T}$
NS of Tianshan Mountians	281.40	0.186	↑	2.64**	1992*	17*	$X(T)=244.449+0.822T+d_{t+T}$
SS of Tianshan Mountains	260.31	0.170	↑	0.75	none	12**	$X(T)=303.535+0.74T+d_{t+T}$
NS of Kunlun Mountains	57.03	0.340	↑	1.72	none	20**	$X(T)=62.556+0.249T+d_{t+T}$

X(T) is predicted value; T is period; d_{t+T} is the increment or decrement of period
SS southern slope, *NS* means northern slope
$^*P<0.05$;$^{**}P<0.01$

1979–1982, 1986–1988, 1989–1992, 1993–1995 and 2001–2004 was in a positive phase, indicating that precipitation in these years are high. Precipitation in the Kaidu River basin had four weak periods of 4, 9, 16 and 18 years and one strong period of 18 years. Precipitation in 1960–1970 and 1991–2002 was in a positive phase, indicating that precipitation in these years was high.

2.3.3.2 Future Precipitation Trends

The results of the linear regression coefficient test show that the period superimposed trend prediction model has a high simulation precision. Precipitation simulation reaches $P<0.05$ level of significance in the northern slope of the Tianshan Mountains, while that in the other river areas reach $P<0.01$, indicating that the model is suitable for precipitation forecast in the study area (Fig. 2.39).

In comparing average precipitation during the periods 1990–2010 and 1957–1989, we can see that precipitation increased the most in the southern slope of the Altai Mountains (Fig. 2.40) followed by the northern and southern slopes of the Tianshan Mountains, with precipitation increments of 55.41, 39.56 and 32.91 mm, respectively. Precipitation in the northern slopes of the Kunlun and Qilian Mountains increased least, with increments of only 11.33 and 10.77 mm, respectively. Overall, precipitation increased the most in northern Xinjiang, followed by southern Xinjiang and the Hexi Corridor.

In comparing average precipitation between 2011–2020 and 1990–2010, an increasing trend becomes apparent in the southern slope of the Tianshan Mountains and the northern slope of the Qilian Mountain, with respective increments of 29.29 and 18.54 mm. However, the northern slope of the Tianshan Mountains, the southern slope of the Altai Mountains and the northern slope of the Kunlun Mountains

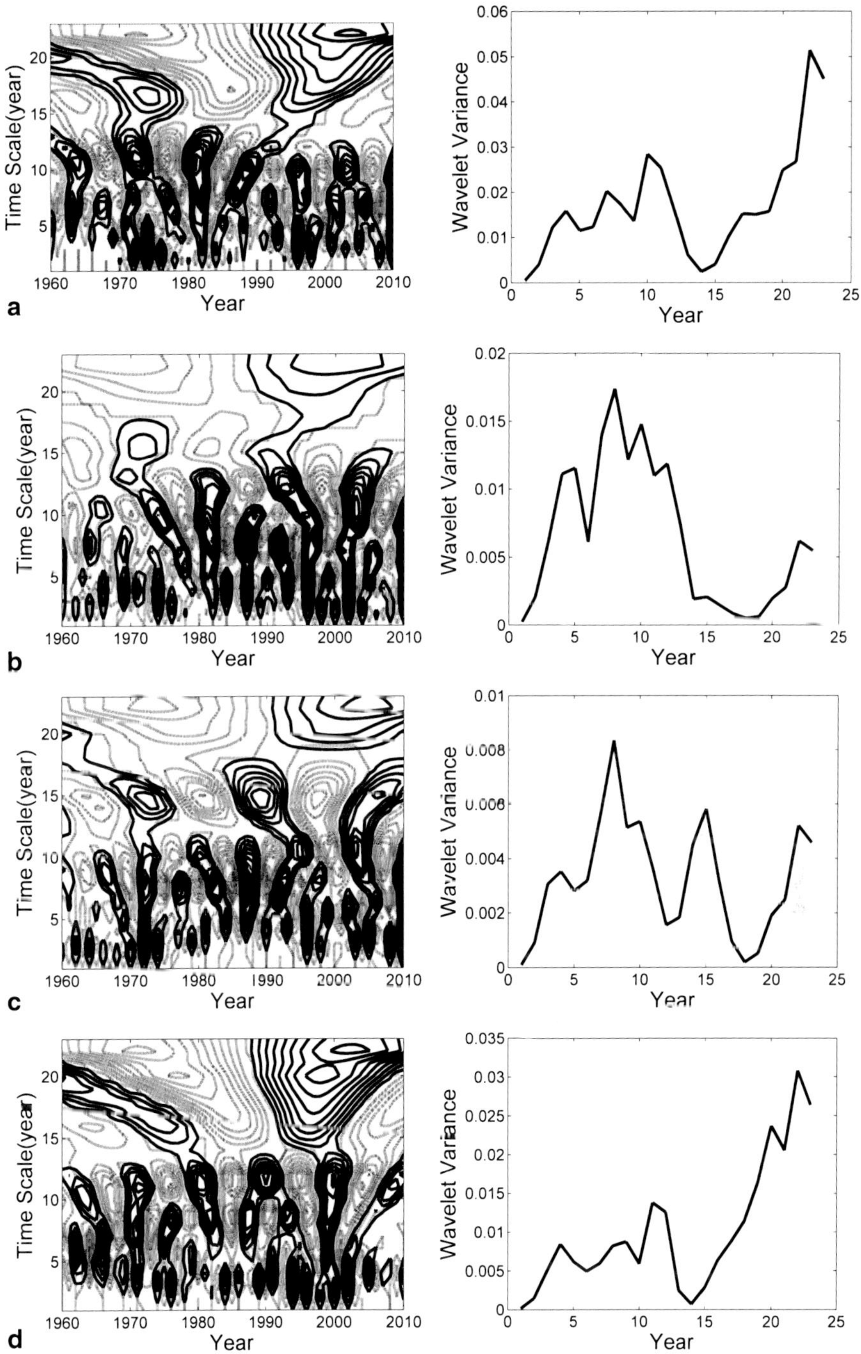

Fig. 2.38 Wavelet time-frequency distribution (*left*) and variance (*right*) of precipitation in the four headstreams of the Tarim River Basin **a** Aksu River; **b** Yarkant River; **c** Hotan River; **d** Kaidu River

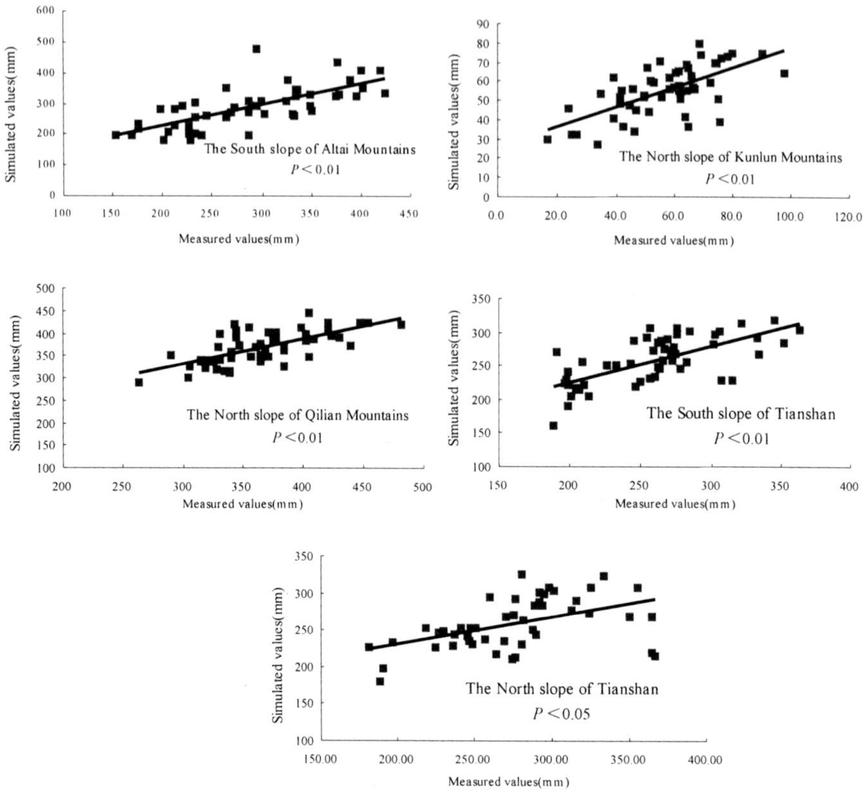

Fig. 2.39 Relations between simulated and measured values of precipitation

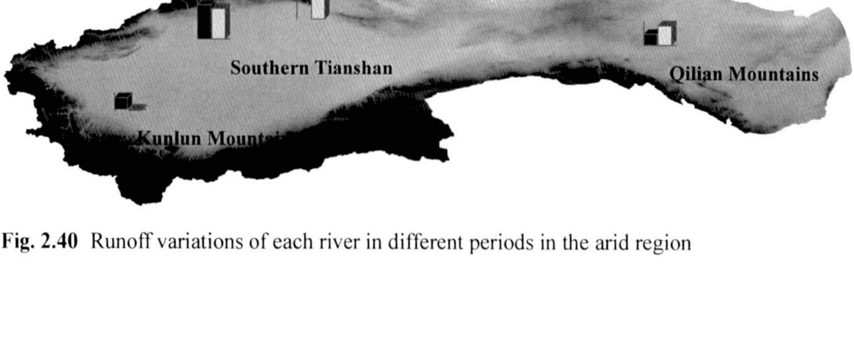

Fig. 2.40 Runoff variations of each river in different periods in the arid region

show a decreasing trend, with respective decrements of 76.32, 10.33 and 1.10 mm. The reason for decreasing precipitation may be similar to that for temperature.

2.4 Past Climate Change

Long-term climate variability is paramount for estimating the impact of climate on human activities and for predicting future climate (Rehman 2010). Because of their extensive spatial availability and high climatological sensitivity (Fritts 1976), tree rings are widely used to reconstruct past climate (Liu et al. 2006; Fang et al. 2009). Climate reconstruction using tree rings and instrumental records has revealed both fine- and coarse-scale climatic change (Moberg et al. 2005).

Previous studies in the area have mainly focused on temperature and precipitation changes and their impacts on water resources (Yuan et al. 2005; Jiang et al. 2009). However, little has yet been done in this research area with regards to the past 500 years of climate change using dendrochronology data. Detailed impact assessments require information about climatic variability at regional and national levels along with analyses not only of mean changes but also of trends (Hasanean and Basset 2006). In this research, we analyzed temporal climate variability in the study areas, placing short-term variability in a long-term context.

2.5 Temperature Variability

According to tree-ring reconstructions, the temperature in the Jing River (Fig 2.41, Yang et al. 2012) has undergone 7 warming and 7 cooling events over the past 500 years. Among the alternating warm and cold periods, three cold periods (1496–1544, 1623–1683, 1798–1850) were consistent with cold periods on the Tibetan plateau (Yang 2003) and matched glacier advancement in western China. However, only one of the cold periods (1469–1510), at Jinghe, was consistent with one of the three cold periods of the Little Ice Age in China, whereas three cold periods (1623–1683, 1798–1850, 1884–1922) were consistent with the surrounding areas in Xinjiang (Wang et al. 1998). From tree-ring reconstructions over the past five centuries, the temperature and climate change of Jinghe has not reflected the unconventional warming of recent decades. However, reconstruction of the temperatures of the past two centuries from the Qinling Mountain range TRW series showed a pronounced warming since the mid-twentieth century (Liu et al. 2009). Recent warming was shown to be unprecedented in the past 2,000 years from tree-ring chronologies in northwest Eurasia (Briffa et al. 2008) and western China (Holmes et al. 2009).

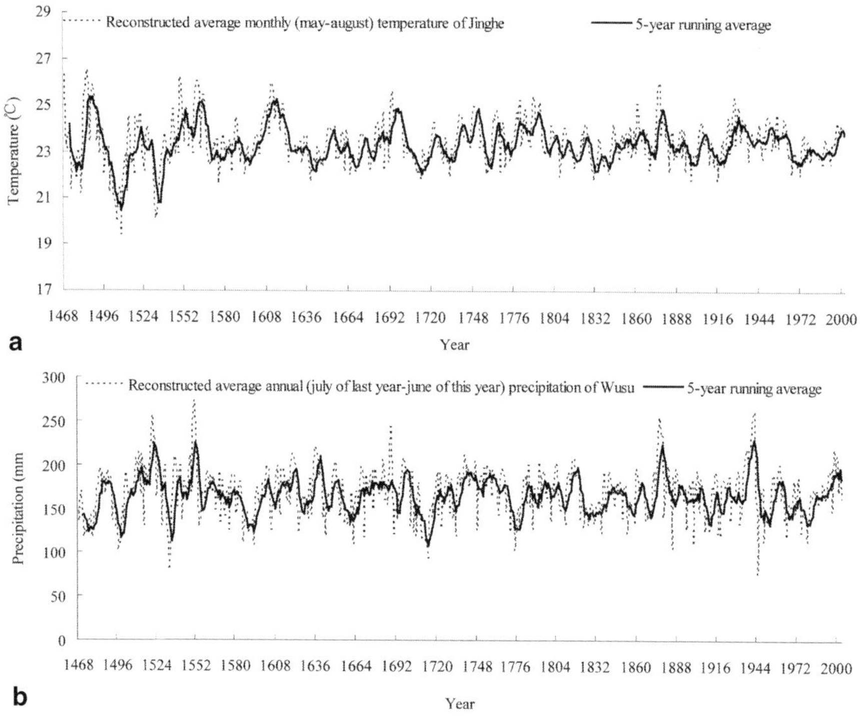

Fig. 2.41 Reconstructed **a** AMT of Jinghe, **b** ANP at Wusu in northern Xinjiang

2.6 Precipitation Variability

Tree-ring reconstructions of the precipitation at Wusu from the past 500 years (Fig. 2.41, Yang et al. 2012) have shown 9 wet periods and 9 drought events. Among the alternating wet and dry periods, wet periods were dominant prior to 1820, followed by mainly dry periods. In contrast, the occurrences of wet and dry periods in Shanxi (Li et al. 2006) were negatively correlated to those of Jinghe in northern Xinjiang.

2.7 Extreme Temperature and Precipitation

The maximum temperatures of recent years are lower than those of the past 500 years, while recent minimum temperatures are higher. Similarly, the maximum precipitation of recent years is lower than that of the past 500 years, while minimum precipitation is higher (Table 2.13, Yang et al. 2012).

Table 2.13 Extreme temperature (T) and precipitation (P)

Item	Max T (May–August) °C	Min T °C	Max P (July of previous year to June of current year) (mm)	Min P (July of previous year to June of current year) (mm)
Past 500–year history	26.6	19.4	273	219
Recent 40 years	24.5	22.0	76	113

These findings suggest that northern Xinjiang is located within a unique geographic unit that is collectively representative of China and global patterns, but conforms to patterns established in the surrounding areas of Xinjiang and the Tibetan plateau. Furthermore, data for Jinhe and Wusu do not show abnormal contemporary warming and wetting trends, and recent temperature and precipitation variability does not exceed variability ranges in the 500-year period under investigation.

Summary

1. Over the past 50 years, the temperature in the arid region of Northwest China has shown a significant increasing trend ($P<0.01$), at a rate of 0.343 °C/10a. Regionally, temperature increases reveal strong regional differences. The temperature in northern Xinjiang increased the fastest, at a rate of 0.386 °C/10a, followed by increased temperature rates in the Hexi Corridor (0.352 °C/10a) and in southern Xinjiang (0.283 °C/10a). Among the three landscape types (mountains-oasis-desert), the desert has the highest rate of increase (0.360 °C per decade), followed by the oasis region (0.339 °C per decade) and the mountain area (0.325 °C per decade). In addition, analyses of seasonal variability exhibited an increasing rate of mean air temperature of 0.486 °C/10a, which is higher than that in spring, summer, or autumn. The increasing rate of temperature in northern Xinjiang is the highest (0.587 °C/10a) in winter, followed by the rate in the Hexi Corridor (0.486 °C/10a), while Southern Xinjiang has the lowest rate (0.385 °C/10a). The winter temperature of the arid region of northwest China has a strong and significant correlation with the Siberian High Index and China's carbon dioxide emissions.
2. The Mann–Kendall method was used to detect step changes in temperature. Results show that a step change probably occurred in 1988 ($P<0.01$) for regional annual air temperature, while step changes in 1998, 1996, 1987 and 1986 ($P<0.05$) were detected in spring, summer, autumn and winter for the whole region, respectively. In selecting time-series analysis of variance extrapolation to analyze temperature change periods, we found that each river basin exhibited changes around every 17 to 20 years.
3. Mean precipitation has been calculated at about 153.37 mm for the past 50 years, with the increasing rate of mean annual precipitation measuring 6.07 mm/10a and a precipitation step change occurring in 1991. There are some regional differences in precipitation increases. For instance, the northern Xinjiang has the highest rate of increase (9.20 mm/10a), followed by southern Xinjiang (5.35 mm/10a) and the Hexi Corridor (3.96 mm/10a). In addition, from our

analysis of seasonal variability, we found that summer-time precipitation had the largest rate (2.51 mm/10a), followed by autumn and spring, and that winter had the smallest rate (1.16 mm/10a).

4. The variance analysis for precipitation indicated that the 22-period in the northern slope of the Qilian Mountains was the longest, followed by the northern slope of the Kunlun Mountains, the southern slope of the Altai Mountains and the northern slope of the Tianshan Mountains, with cycles of 20, 18 and 17 years, respectively.
5. Climate reconstruction using tree rings and instrumental records have revealed both fine- and coarse-scale climatic change. According to tree-ring reconstructions, the temperature in the Jing River has undergone 7 warming and 7 cooling events over the past 500 years, while tree-ring reconstructions of precipitation at Wusu have shown 9 wet periods and 9 drought events.

References

Abouabdillah A, Oueslati O, De Girolamo AM, Lo Porto A (2010) Modeling the impact of climate change in a mediterranean catchment (Merguellil, Tunisia). Fresenius Environ Bull 19:2334–2347

Ashrafi K, Shafiepour M, Ghasemi L et al (2012) prediction of climate change induced temperature rise in regional scale using neural network. Int J Environ Res 6(3):677–688

Bengtsson L (1997) The numerical simulation of climatic change. Ambio 26:58–65

Beniston M, Rebetez M (1996) Regional behavior of minimum temperature in Switzerland for the period 1979–1993. Theor Appl Climatol 53:231–243

Briffa KR, Melvin TM, Vaganov EA, Grudd H et al (2008) Trends in recent temperature and radial tree growth spanning 2,000 years across northwest Eurasia. Philos Trans R Soc B 363:2271–2284

Brohan P, Kennedy JJ, Harris I, Tett SFB, Jones PD (2006) Uncertainty estimates in regional and global observed temperature changes: a new data set from 1850. J. Geophys Res-Atmos 111:D12106. doi:10.1029/2005JD006548

Chen ZQ, He YS, Yan JP (1984) Arid climate and the regulation of the hydrothermal resources in the Northwest China. J Desert Res 14(4):108–112

Crowley TJ (2000) Causes of climate change over the past 1000 years. Science 289(5477):270–277

D'Arrigo R, Jacoby G, Wilson R, Panagiotopoulos F (2005) A reconstructed Siberian High index since A.D.1599 from Eurasian and North American tree rings. Geophys Res Lett 32:L05705. doi:10.1029/2004GL022271

Fang KY et al (2009) Drought variations in the eastern part of northwest China over the past two centuries: evidence from tree rings. Clim Res 38:129–135

Fritts HC (1976) Tree rings and climate. Academic Press, New York

Gong DY, Ho CH (2002) The Siberian High and climate change over middle to high latitude Asia. Theor Appl Climatol 72(1–2):1–9

Guo J (2010) Research on the effects of climate change on water cycle and water resources. Wuhan University, Ph. D. Thesis

Hasanean HM, Basset HA (2006) Variability of summer temperature over Egypt. Int J Climatol 26:1619–1634

He YQ, Lu AG, Zhang ZL, Pang HX, Gu J (2005) Seasonal variation of regional warming-up structure across China in the past half century. Clim Res 28:213–219

Holmes JA, Cook ER, Yang B (2009) Climate change over the past 2,000 years in western China. Quat Int 194:91–107

Intergovernmental Panel on Climate Change (IPCC) (2007) Climate change 2007: the physical science basis. In: Solomon S et al (ed) Contribution of working group I to the Fourth Assessment Report of the intergovernmental panel on climate change. Cambridge University Press, Cambridge, pp 996

Jiang FQ, Li XM, Wei BG, Hu RJ (2009) Observed trends of heating and cooling degree-days in Xinjiang Province, China. Theor Appl Climatol 97:349–360

Li Q, Liu Y, Cai QF, Sun JV et al (2006) Reconstruction of annual precipitation since 1686 A. D. from Ningwu region, ShanxiI province. Quat Sci 26(6):999–1006

Li QH, Chen YN, Shen YJ et al (2011) Spatial and temporal trends of climate change in Xinjiang, China[J]. J Geogr Sci 21(6):1007–1018

Li BF, Chen YN, Shi X, Chen ZS, Li WH (2012a) Temperature and precipitation changes in different environments in the arid region of northwest China. Theor Appl Climatol, doi:10.1007/s00704-012-0753-4

Li BF, Chen YN, Shi X (2012b) Why does the temperature rise faster in the arid region of northwest China? J Geophys Res 117:D16115. doi:10.1029/2012 JD017953

Li Dongliang, Peng Suqin, Yao Hui (1995) The climate characteristics of winter mean temperature in northwestern China. Atmospheric Sci 19(22):192–199

Li D, Wei L, Cai Y et al (2003) The present facts and the future tendency of the climate change in Northwest China. J Glaciol Geocryol 25(2):135–142

Liu X, Yin ZY, Shao X, Qin N (2006) Temporal trends and variability of daily maximum and minimum, extreme temperature events, and growing season length over the eastern and central Tibetan Plateau during 1961–2003. J Geophys Res 111:D19109

Liu Y, Linderholm HW, Song H, Cai Q et al (2009) Temperature variations recorded in Pinus tabulaeformis tree rings from the southern and northern slopes of the central Qinling Mountains, central China. Boreas 38:285–291

Lu AG (2009) Impacts of global warming change on patterns of temperature in China. J Mt Sci 6:405–410

Mahlstein I, Knutti R (2010) Regional climate change patterns identified by cluster analysis. Clim Dynam 35(4):587–600

Mariotti L, Coppola E, Sylla M B, et al.(2011) Regional climate model simulation of projected 21st century climate change over an all-Africa domain: Comparison analysis of nested and driving model results[J]. J Geophys Res Atmos 116 (D15, 16)

Moberg A, Sonechkin DM, Holmgren K, Datsenko NM et al (2005) Highly variable northern hemisphere temperatures reconstructed from low- and high-resolution proxy data. Nature 433:613–617

Moriondo M, Bindi M, Fagarazzi C et al (2011) Framework for high-resolution climate change impact assessment on grapevines at a regional scale. Reg Environ Chang 11(3):553–567

Panagiotopoulos F, Shahgedanova M, Hannachi A, Stephenson DB (2005) Observed trends and teleconnections of the Siberian high: a recently declining center of action. J Clim 18(9):1411–1422

Pierce DW, Barnett TP, Santer BD et al (2009) Selecting global climate models for regional climate change studies[J]. Proc Natl Acad Sci USA 106(21):8441–8446

Rehman S (2009) Study of Saudi Arabian climatic conditions using Hurst exponent and climatic predictability index. Chaos Soliton Fract 39(2):499–509

Rehman SFQ (2010) Temperature and rainfall variation over Dhahran, Saudi Arabia, (1970–2006). Int J Climatol 30:445–449

Rehman S, Siddiqi AH (2009) Wavelet based hurst exponent and fractal dimensional analysis of Saudi climatic dynamics. Chaos Soliton Fract 40(3):1081–1090

Ren GY, Xu MZ, Chu ZY, Guo J, Li QX, Liu XN, Wang Y (2005) Changes of surface air temperature in China during 1951–2004 (In Chinese with English abstract). Clim Envir Res 10(4):717–727

Ruelland D, Ardoinbardin S, Billen G et al (2008) Sensitivity of a lumped and semi-distributed hydrological model to several methods of rainfall interpolation on a large basin in West Africa. J Hydrol 361(1/2):96–117

Shi YF, Shen YP, Hu RJ (2002) Preliminary study on signal, impact and foreground of climatic shift from warm-dry to warm-humid in Northwest China. J Glaciol Geocryol 24(3):219–226

Shi YF, Shen YP, Li DL (2003) Discussion on the present climate change from warm-dry to warm-wet in northwest China. Quat Sci 23(2):152–164

Shi YF, Shen YP, Kang E, Li DL, Ding YJ, Zhang GW, Hu RJ (2007) Recent and future climate change in northwest china. Clim Chang 80(3–4):379–393

Shi YF, Zhang XS (1995) Impact of climate-change on surface-water resource and tendency, in the future in the arid zone of Northwestern China. Sci China 38(11):1395–1408

Segev U (2010) Regional patterns of ant-species richness in an arid region: the importance of climate and biogeography. J Arid Environ 74(6):646–652

Stahlschmidt ZR, DeNardo DF, Holland JN, Kotler BP, Kruse-peeples M (2011) Tolerance mechanisms in North American deserts: Biological and societal approaches to climate change. J Arid Environ 75(8):681–687

Sullivan C (2001) The potential for calculating a meaningful water poverty index. Water Int 26(4):471–480

Sun F, Roderick ML, Farquhar GD, Lim WH, Zhang Y, Bennett N, Roxburgh SH (2010) Partitioning the variance between space and time. Geophys Res Lett 37:L12704. doi:10.1029/2010GL043323

Tebaldi C, Smith RL, Nychka D et al (2005) Quantifying uncertainty in projections of regional climate change: A Bayesian approach to the analysis of multimodel ensembles[J]. J Clim 18(10):1524–1540

Vuille M, Bradley RS (2003) 20th century climate change in the tropical Andes: observations and model results. Clim Chang 59:75–99

Wang SU, Ye JL, Gong DY (1998) Climate in China during the Little Ice Age. Quat Sci 1:54–64

Wang JS, Fei XL, Wei F (2008) Further study of temperature change in northwest China in recent 50 years. J Desert Res 28(4):724–732

Xu Y, Ding YH, Zhao ZC (2003) Scenario of temperature and precipitation changes in Northwest China due to human activity in the 21~ (st) Century. J Glaciol Geocryol 23(5):327–330

Yang B (2003) Climate history of the Tibetan plateau during the last two millennia. Adv Earth Sci 18(2):285–291

Yang Y, Chen Y, Li W, Yu S, Wang M (2012) Climatic change of inland river basin in arid area: a case study in northern Xinjiang, China. Theor Appl Clim 107:143–154

Yuan YJ, Xie GH, Wei WS, Zhang JB (2005) Similarities and differences between summer temperature changes in Tianshan mountainous region and southern and northern Xinjiang. Meteorol Sci Technol 33(2):152–155

Zhang LY, Jiang ZL (1992) The genesis of the arid climate of northwestern China. Arid Land Geogr 15(2):1–12

Zhang XG, Li XQ (1982) Some characteristics of temperature variation in China in the present century. Acta Meteorol Sin 20(2):198–208

Zhang Q, Hu YQ, Cao XY et al (2000) On some problems of arid climate system of Northwest China. J Desert Res 20(4):357–362

Zhang Q, CJ Zhang CJ, Bai HX, Lin L, Sun LD, Liu DX, Wang JS, Zhao HY (2010a) New development of climate change in Northwest China and its impact on arid environment. J Arid Meteorol 28(1):1–7

Zhang XQ, Sun Y, Mao WF, Liu YY, Ren Y (2010b) Regional response of temperature change in the arid regions of China to global warming. Arid Zone Res 27(4):592–599

Chapter 3
Hydrologic System in Northwest China

Yaning Chen, Baofu Li, Zhongsheng Chen and Yuting Fan

Abstract Both the temperature and the precipitation in China's arid northwestern zone have increased, eliciting corresponding changes in hydrological processes in the region's inland basins. This chapter analyzes the characteristics of runoff and its components, the main findings are: (1) Runoff increased significantly at stations around the Tianshan Mountains. (2) The digital filtering method was used to separate baseflow from surface flow, after which the baseflow index (BFI) was calculated and analyzed. We find that baseflows of the four headstreams have increased considerably over the past 50 years. The baseflow and BFI showed obvious seasonal variations: The lowest baseflow and BFI typically occurred in December and January, and both increased gradually until reaching maximum values in August or July. And precipitation had a significant impact on runoff, whereas temperature strongly affected baseflow. In addition, in the Tizinafu River, the contribution of ice/snowmelt water varied from 25.96 to 68.87 % for spatial characteristics, and from 28.31 to 65.43 % for seasonal characteristics. The mean of the ice/snowmelt percentage is 43 %, which meant that ice/snowmelt water was the main supplying water source. (3) Using the data from 1960 to 2010, future runoff amounts were predicted. Some results can be concluded as follows: Runoff in the Aksu, Yarkand, and Hotan rivers will be low in 2010–2011 but will experience continued growth in 2017–2028.

Keywords Linear trends of runoff · Step change of runoff · Components of runoff · Isotope separation

---

Y. Chen (✉) · B. Li · Z. Chen · Y. Fan
State Key Laboratory of Desert and Oasis Ecology, Xinjiang Institute of Ecology and Geography, Chinese Academy of Sciences, No. 818 South Beijing Road, 830011 Urumqi, Xinjiang, China
e-mail: chenyn@ms.xjb.ac.cn

B. Li
e-mail: libf@ms.xjb.ac.cn

Z. Chen
e-mail: chenzhs@ms.xjb.ac.cn

Y. Fan
e-mail: fanyt@ms.xjb.ac.cn

Y. Chen (ed.), *Water Resources Research in Northwest China*,
DOI 10.1007/978-94-017-8017-9_3,

3.1 Introduction

Water is the foundation for the composition, development, and stability of oasis ecosystems in arid regions. It determines the evolution of the ecological environment, including the two contrary processes of oasis development and desertification, and is also the key ecological factor (Chen et al. 2007, 2009). Global warming and changes in precipitation patterns caused by CO_2 emissions and other human activities alter the availability of water resources and in turn affect natural ecosystems. Research results from the past 50 years have revealed that both the temperature and the precipitation in China's arid northwestern zone have increased (Shi et al. 2007), eliciting corresponding changes in hydrological processes in the region's inland basins. This relatively recent trend is characterized by a continual increase in temperature and precipitation, added river runoff volumes, increased lake water surface elevation and area, and elevated groundwater levels. These changes have led to increased water resources, providing immediate relief to the local water shortage.

Climate change has also caused the accelerated retreat of glaciers, which are important natural water reservoirs for the delta ecosystems in inland China. This phenomenon has raised widespread concerns worldwide and has become a hot topic in related academic fields. Although the future trends of changes in hydrological processes and water resources in China's arid region are not yet fully understood, these issues are very important for managing water resources in the river basin and deserve urgent attention.

Hypothesis testing to estimate long-term climate change trends is helpful to understand the inherent mechanism of hydrological process. In this section, two types of general trends are investigated: the monotonic trend and step change (Zhao et al. 2008). While it is difficult to gain a thorough understanding of the non-linear mechanism of any individual hydroclimatic process (Xu et al. 2009, 2011), such processes can be evaluated theoretically to determine if they comprise an ordered deterministic system, an unordered random system, or a chaotic dynamic system, and whether change patterns of periodicity or quasi-periodicity exist. Specific to the series of climate changes that have occurred in the arid/semi-arid regions of western China, such inquiries may be designed to determine if these changes represent a localized transition to a warm and wet climate in response to global warming, or merely reflect a centennial periodicity in hydrological dynamics.

To date, many researchers have explored the nonlinearity of runoff processes using various non-linear analytic methods and models, including wavelet (Labat et al. 2012; Melesse et al. 2010) and fractal theory (Xu et al. 2009). However, as we are still far from having an in-depth understanding of non-linear runoff characteristics in China's arid Northwest, this forms the focus of our present paper. In our study, we attempt to explain the chaotic dynamical characteristics of runoff by using correlation dimension analysis. We also investigate the long-term correlation characteristics of runoff by using the R/S analysis method and runoff duration by using wavelet analysis.

3.2 Characteristics of Runoff

3.2.1 Study Area, Data and Methods

3.2.1.1 Study Area

Arid regions account for approximately 30% of China's total land mass and are mainly situated in the sparsely populated Northwest. The largest arid zone (34°–50°N, 73°–108°E) includes the Uygur Autonomous Region of Xinjiang, Gansu Province, western Inner Mongolia, the northern Hui Autonomous Region of Ningxia and northern Qinghai Province (Fig. 3.1; Wang et al. 2013). In this vast region, almost all of the endorheic drainage basins are located in the western Helan Mountains in Ningxia and the western Wushao Mountains in Gansu Province, covering an area of 2.53×10^6 km^2. These basins are the Tarim Basin, the Qaidam Basin, the Badain Jaran Desert, and the Tengger Desert. Other endorheic drainage basins are also located in northern Xinjiang as well as in the desert region of Alxa Banner in western Inner Mongolia and in the Gobi desert of the Hexi Corridor in Gansu Province (Liu et al. 2010). Dominated by continental arid conditions and being only slightly impacted by East Asian monsoons, the region is controlled by a typical inner-continental land mass climate, which features wide-ranging temperatures, low precipitation, and low humidity.

3.2.1.2 Data

Monthly and yearly temperature and precipitation data covering the study area were provided by the National Climate Center (NCC) of the China Meteorological Administration (CMA). For this area, 84 stations passed the internal homogeneity check of the China National Meteorological Center (CNMC), including the moving *t* test (Peterson et al. 1998), the standard normal homogeneity test (Alexandersson 1986), and the departure accumulating method (Buishand 1982). Stations installed after 1960 or those with data gaps were excluded. As a result, 76 meteorological stations for 51 years (January 1960 to February 2011) were selected.

Another dataset that includes monthly and yearly air temperature, precipitation, and runoff was provided by the Hydrological Bureau of the Xinjiang region and Gansu Province. In this study, data from 122 hydrological stations and covering the period from 1965 to 2003 were selected. However, we only used 65 hydrological stations for precipitation analysis, 46 stations for air temperature analysis, and 52 stations for runoff analysis due to missing data in the other stations. Interpolations were applied, since most hydrological stations lacked data covering the period 2004–2010. A linear regression equation was adopted to analyze the relationship between the hydrological series and climate series from nearby meteorological stations. Couples with R-square above the level of 0.3843 (0.01 significant level) were deemed sufficient for this extension. The homogeneity test was also applied to these interpolated climate series, and all of the extended climate series passed this test.

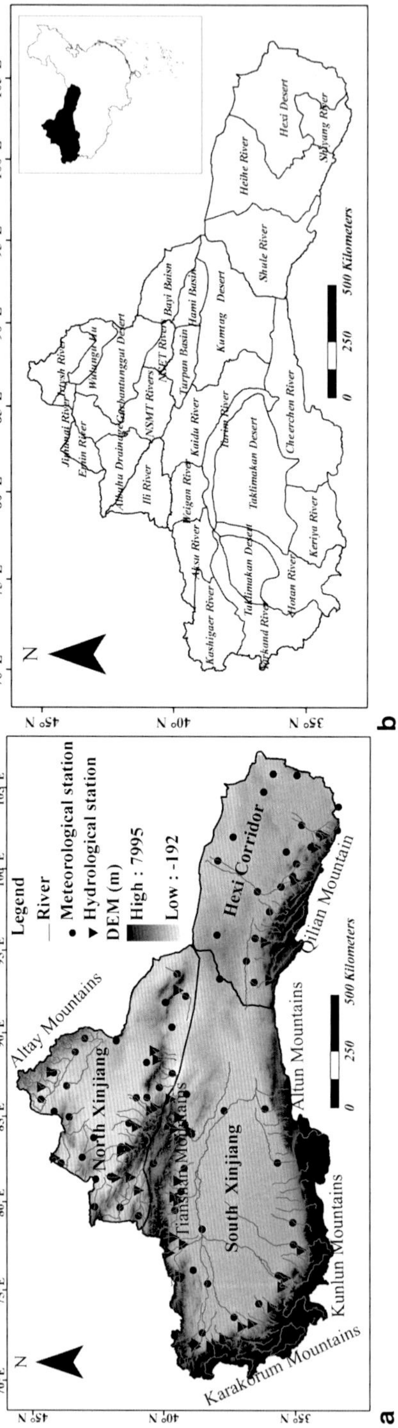

Fig. 3.1 **a** Distribution of meteorological station and hydrological station; **b** hydrological basin and drainage basin

3.2.1.3 Methods

In this study, the nonparametric Mann–Kendall (M–K) method (Mann 1945; Kendall 1975) is used to detect possible trends in hydrological extremes. The results of the M–K test are heavily affected by the serial correlation of the time serial correlation, so we adopt the Yue and Pilon method (Yue et al. 2002) to remove the serial correlation.

To achieve a better understanding of the non-linear characteristics of runoffs, a correlation dimension method was used to show their chaotic dynamical characteristics and wavelet analysis was used to obtain a specific period within a time series, after which an R/S analysis method was employed to study the long-term correlation characteristics. In this section, we introduce correlation dimension and R/S analysis.

(1) Correlation Dimension The correlation dimension method is usually applied to analyze a runoff series and check whether or not a river exhibits chaotic dynamic characteristics (Sivakumar 2007). The method's principle is as follows. Consider *X(t)*, the time series of annual runoff, and suppose it is generated by non-linear dynamic systems with m degrees of freedom. In order to restore the dynamic characteristics of the original system, it is necessary to construct an appropriate series of vectors $X^{(m)}(t)$ with delay coordinates in the m-dimensional phase space, according to the basic ideas initiated by (Grassberger and Procaccia 1983):

$$X^{(m)}(t) = \{X(t), X(t+\tau), \ldots, X(t+(m-1)\tau)\} \quad (3.1)$$

where m is called the embedding dimension and τ is an appropriate delay.

For a set $\{X_i / i = 1, 2, \ldots, N\}$ of points on the attractor, and using the G–P method, the correlation-integrals are defined in order to distinguish between stochastic and chaotic behaviors:

$$C(r) = \lim_{N\to\infty} \frac{1}{N^2} \left\{ \text{Number of pairs } (i,j), \text{whose distance}^{|X_i - X_j| < r} \right\}$$

where r is surveyor's rod for distance.

The correlation-integrals are computed as follows:

$$C(r) = \frac{1}{N_R^2} \sum_{j=1}^{N_R} \sum_{i=1}^{N_R} \Theta(r - |X_i - X_j| < r) \quad (3.2)$$

where N_R is the number of reference points taken from N, and N is the number of points $X^{(m)}(t)$. The relationship between N and N_R is $N_R = N - (m-1)\tau$. $\Theta(x)$ is Heaviside function defined as

$$\Theta(x) = \begin{cases} 0 & x \le 0 \\ 1 & x > 0 \end{cases} \quad (3.3)$$

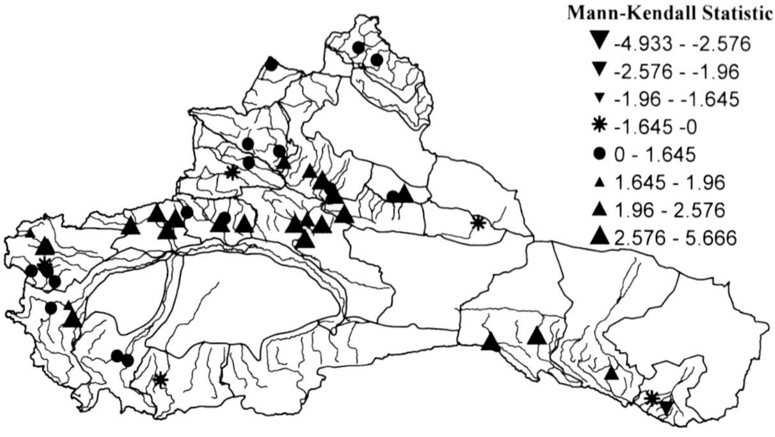

Fig. 3.2 Spatial distribution of annual trends for runoff

The expression counts the number of points out of the data set which are closer than radius r or within a hypersphere of radius r, and then divides it by the square of the point total (because of normalization). As $r \to 0$, the correlation exponent d is defined as:

$$C(r) \propto r^d \tag{3.4}$$

Apparently, the correlation exponent d is thus given by the slope coefficient of $\log(C(r))$ versus $\log r$. According to $(\log r, \log C(r))$, d can be obtained by the least squares method (LSM) in a log-log grid. To detect the chaotic behavior of the system, the correlation exponent has to be plotted as a function of the embedding dimension. If $X(^m)(t)$ is purely random (e.g., white noise), the correlation exponent increases with the embedding dimension without reaching the saturation value. If there are deterministic dynamics in the system, the correlation exponent reaches the saturation value, which means that it remains approximately constant with an increase of the embedding dimension.

3.2.2 Linear Trends in Runoff

(1) Overall Changes in Runoff Figure 3.2 (Wang et al. 2013) illustrates the overall spatial distribution of statistically significant trends for annual and seasonal runoff. As we can see, most hydrological stations show increasing trends. Annual runoff increases significantly at the stations around the south slope of the Tianshan Mountains, especially for the basins of the Aksu, Kaidu, and Shule Rivers. Some stations located in the Weigan, Kashigaer and Yarkand Rivers also display significant increasing trends ($P<0.05$), whereas strongly decreasing trends are observed in the Shiyang River.

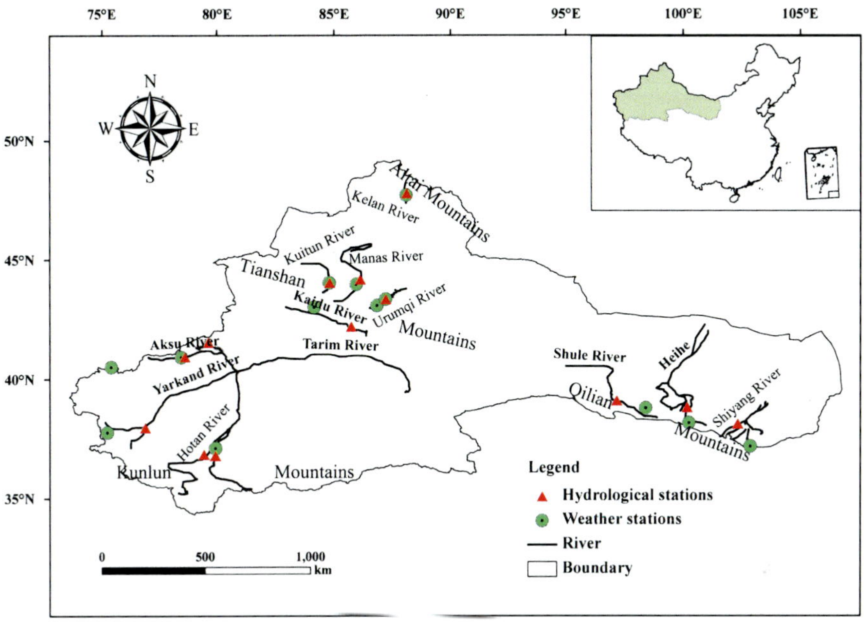

Fig. 3.3 The hydrological stations of rivers in the arid region of northwest China

(2) Runoff Trends in Typical Areas In this study, the research objects are eleven rivers flowing through five river areas of three zones in China's arid Northwest region, including northern Xinjiang (the south slope of the Altai Maintains and the north slope of the Tianshan Mountains), southern Xinjiang (the north slope of the Kunlun Mountains and the south slope of the Tianshan Mountains) and the Hexi Corridor (the north slope of the Qilian Mountains) (Fig. 3.3; Li et al. 2012). To minimize the impact of human activity and clearly delineate regional climate changes, data from the selected hydrological stations represents runoff from mountain passes in each river. The mean value of the hydrological stations' data represents runoff.

As the data clearly indicates, runoff in each river area shows an increasing trend (Fig. 3.4; Li et al. 2012). On the south slope of the Tianshan Mountains, the increasing trend is statistically significant at $P<0.01$, as are the trends related to the north slope of the Qilian Mountains and the north slope of the Tianshans ($P<0.05$). However, the increasing trends on the south slope of the Altai Mountains and the north slope of the Kunlun Mountains are insignificant. Over the past 50 years, the average increasing rate of runoff in each river area was 1.03×10^8 m^3 per decade. Specifically, the increasing rates of runoff on the north slope of the Kunlun Mountains and the south slope of the Tianshan Mountains are the fastest, at 4.25×10^8m^3/10a and 2.90×10^8m^3/10a, respectively, followed by those on the north slope of the Qilian Mountains and the north slope of the Tianshans, at 0.42×10^8m^3/10a, 0.26×10^8m^3/10a, respectively. Meanwhile, the increasing rate of runoff on the south slope of the Altai Mountains is the slowest, measuring only 0.15×10^8m^3/10a (Fig. 3.4).

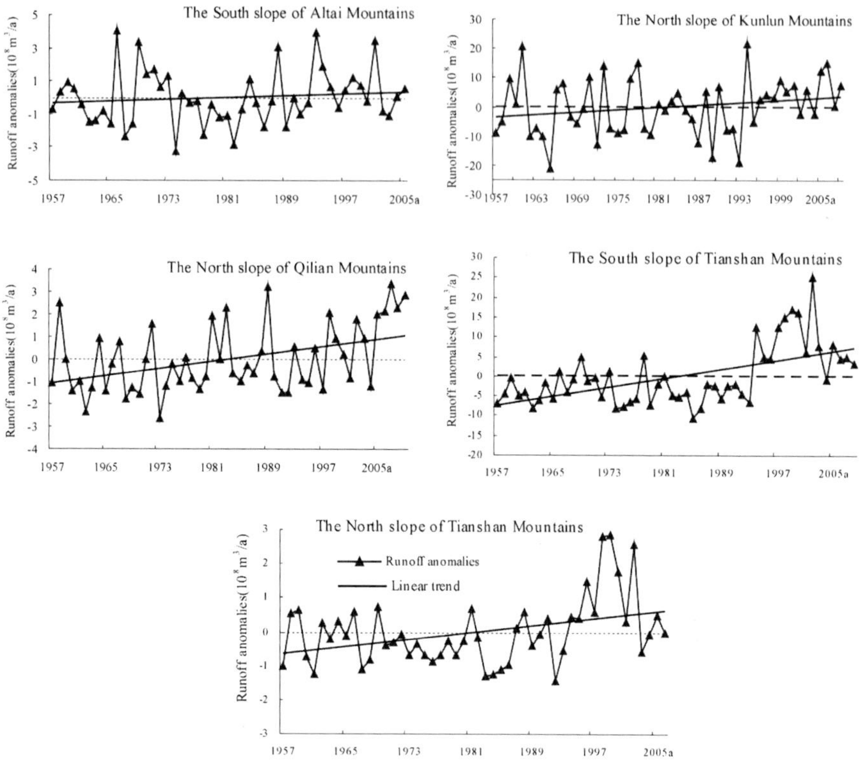

Fig. 3.4 Runoff anomalies trends in typical river areas of the arid region of northwest China (Li et al. 2012)

Since 1990, runoff from most of the mountains in the region has notably increased. Compared to the 1957–1989 timeframe, runoff from the south slope increased the fastest post-1990 (18.12%) (Fig. 3.5; Li et al. 2012), followed by runoff from the north slope of the Tianshans and the south slope of the Altais, which increased 12.56 and 10.26%, respectively. However, runoff on the north slope of the Qilian Mountains and the north slope of the Kunlun Mountains saw only slight increases of 9.43 and 8.12%, respectively. In general, then, runoff in southern Xinjiang increased the fastest, followed by runoff in northern Xinjiang, with runoff in the Hexi Corridor increasing at the slowest pace.

(3) Runoff Trends in Typical Rivers In Xinjiang, as can be seen from Fig. 3.6, runoff in the Hotan River exhibits a decreasing tendency, while the linear tendencies of the runoff in other three rivers (the Aksu, Kaidu, and Yarkand) show upward trends in the past half-century. Nevertheless, runoff change rates vary from region to region. Prior to the 1990s, runoff rate are lower than average, whereas after 1990, runoff rates are notably higher than average in the Aksu, Kaidu, and Yarkand Rivers.

The surface runoff mean velocities in the Aksu, Kaidu, Yarkand and Hotan Rivers were 76.8, 9.1, 65.9 and 44 m^3/s, respectively, while the increasing rates were

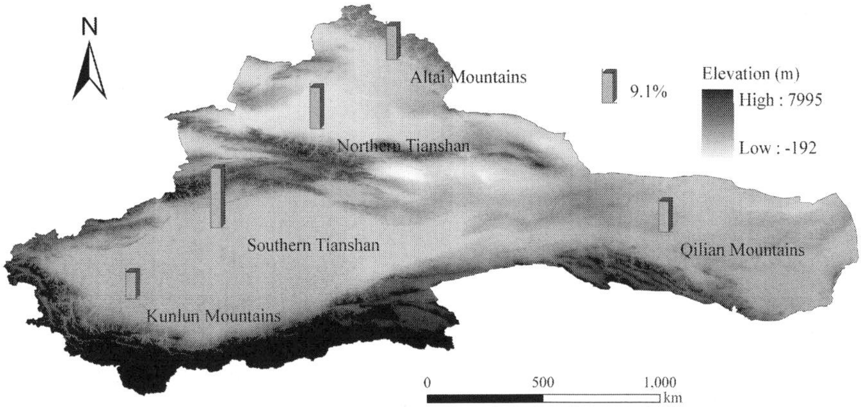

Fig. 3.5 Runoff variations of each river in different periods (before 1990 and after 1990) in the arid region of northwest China (Li et al. 2012)

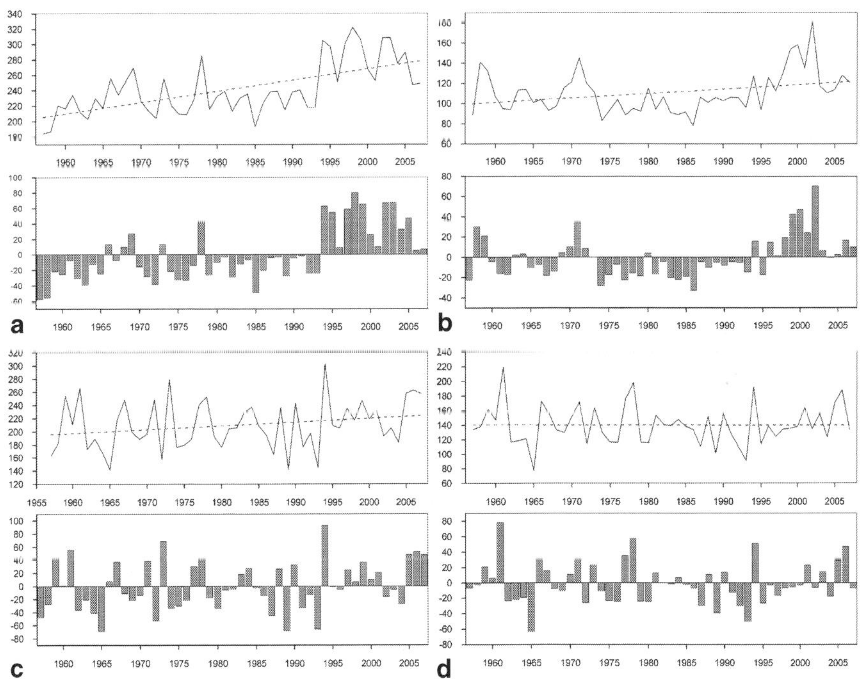

Fig. 3.6 The changes of runoff (m^3/s) in Aksu river (**a**), Kaidu river (**b**), Yarkand river (**c**), and Hotan river (**d**) in Xinjiang

0.48, 0.08, 0.19 and -0.03 m^3/a, respectively. The runoff trends in the Aksu and Kaidu Rivers were significant, at $P<0.01$, but the increasing trend of runoff in the Yarkand River was not significant, nor was the decreasing trend of runoff in the Hotan River.

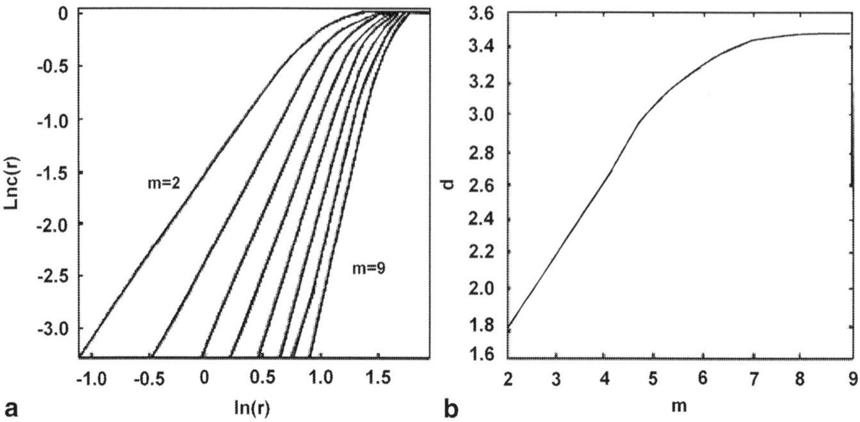

Fig. 3.7 The wavelet approximations for annual runoff in the Hotan River at different time scales. **a** The diagram of *lnC(r)* versus *ln(r)*. **b** The correlation exponent (d) versus embedding dimension (m)

In the Hexi Corridor, from 1960 to 2009, runoff in the Heihe and Shule Rivers showed upward trends, with mean increasing rates of $0.77 \times 10^8 m^3/10a$ and $0.93 \times 10^8 m^3/10a$, respectively. These trends were strong at the $P<0.01$ level of significance (not shown), while the runoff in the Shiyang River signaled a decreasing trend, with a mean increasing rate of $0.1 \times 10^8 m^3/10a$ indicating that its trend was significant at the $P<0.05$ level. Furthermore, Hurst index (H) analysis (H>0.5) indicated that runoff changes in the Heihe, Shule and Shiyang Rivers in the Hexi Corridor will continue to follow current trends.

3.2.3 The Non-linear Characteristics of Annual Runoff

To better understand the non-linear characteristics of annual runoff in the Aksu, Yarkand and Hotan Rivers, we employed a two-step comprehensive analysis method. In logical sequence, the correlation dimension method was used to show chaotic dynamical characteristics, after which the R/S analysis method was employed to study long-term correlation characteristics.

(1) Chaotic Dynamical Characteristics In Fig. 3.7, we used the Hotan River's time series of annual runoff to reconstruct the phase space and then calculated the correlation dimension of attractor. Figure 3.7a shows the relationship between $\ln C(t)$ and $\ln(r)$ for the annual runoff in the Hotan River, with a different embedding dimension m. Using the least square method (LSM), we calculated the slope coefficient of $\ln C(r)$ versus $\log r$, i.e., the correlation exponent for embedding dimension m = 1, 2, …. Figure 3.7b reveals the gradual saturation process of the correlation exponent. As we can see, the correlation exponent increases with embedding dimension m, and a saturated correlation exponent, i.e., the correlation dimension of attractor (D), was obtained when $m \geq 6$.

Table 3.1 The correlation dimensions for the annual runoff processes in the Hotan, Yarkand and Aksu rivers

Rivers	Hotan river	Yarkand river	Aksu river
Embedding dimension (m)	7	7	7
Attractor dimension (D)	3.22	3.21	3.21

Table 3.2 Hurst exponents for the annual runoff processes in the three headwaters of the Tarim River (Xu et al. 2009)

River name	1957–1972	1973–1988	1989–2002
Aksu river	0.76	0.60	0.91
Yarkand river	0.64	0.47	0.73
Hotan river	0.74	0.64	0.52

Through the above procedure, the correlation dimensions of attractor for the Hotan, Yarkand and Aksu Rivers were formulated at 3.22, 3.21 and 3.21, respectively (Table 3.1). The fact that none of these correlation dimensions is an integer indicates that the annual runoff processes in all three headwaters are chaotic dynamical systems that are highly sensitive to initial conditions. Since the value of the index is above 3 for the three rivers, at least four independent variables are needed to describe the dynamics of the annual runoff process in each river.

(2) Long-term Correlation Characteristics Our studies (Xu et al. 2009) revealed long-term temporal trends for annual runoff in the Tarim River at a time scale of 16 years. Over the past few decades, the annual runoff tended to increase in the Aksu and Yarkand Rivers but decrease in the Hotan River (Chen et al. 2009). Using the R/S analysis method, we calculated the Hurst exponents for annual runoff in each river during each period, as shown in Table 3.2. For the Aksu River, the Hurst exponents in the three periods 1957–1972, 1973–1988 and 1989–2002 are 0.76, 0.60 and 0.91, respectively. Since they are all greater than 0.50, this suggests that the annual runoff process in these rivers has a long-enduring characteristic in each period. In other words, the next period will have the same trend as the preceding one.

For the Yarkand River, the Hurst exponent for the period 1957–1972 equals 0.64, also indicating an enduring characteristic for the runoff. In the second period (1973–1988), however, the Hurst exponent (with a value of 0.47) is less than 0.50, suggesting an anti-persistence characteristic for the annual runoff process. In the last period (1989–2002), the Hurst exponent (with a value of 0.73) is greater than 0.50, again implying an enduring characteristic for the annual runoff process.

The Hurst exponents in the three periods for the Hotan River are 0.74, 0.64 and 0.52, respectively. Because all of these readings are greater than 0.50, a long-enduring characteristic is implied for the annual runoff process of each period. That the Hurst exponent for the last period is slightly greater than 0.5 indicates that the annual runoff of the Hotan River is likely to show a slight decrease over the next 14 years (after 2002).

3.2.4 Step Changes In Runoff

The Mann–Kendall–Sneyers test was applied to detect the step change point of the annual runoff series over the past 50 years. Figure 3.8 shows the computed

Fig. 3.8 Step change point of runoff in Aksu river (**a**), Kaidu river (**b**), Yarkand river (**c**), and Hotan river (**d**)

probability series of the step change point years. The results show that abrupt changes in runoff in the Aksu, Kaidu, Shule and Shiyang Rivers occurred in 1991, 1995, 1998 and 1995 (at the 0.05 significance level), respectively. The step change points of runoff in the Heihe River occurred in 1985, 1999 and 2004. No step change points of runoff were found for the Yarkand or Hotan Rivers.

3.2.5 Periodicity of Hydrological Variables

(1) Aksu River Wavelet variances (Fig. 3.9) indicate that the temperature in the Aksu River underwent weak periods of 3 and 11 years, and strong periods of 18 and 22 years. The 18-year period is considered the most reliable measurement due to the length of the data being only 50 years. According to the real part of the wavelet transformation coefficient at an 18-year phase (Fig. 3.9), the timeframes from 1964–1974, 1986–1997, and 2000–2002 were negative phases that exhibited lower temperatures. In contrast, other timeframes were positive phases that registered higher temperatures.

The same analysis of a precipitation time series of 4, 8, 18 and 21 years showed a weaker period, with the strongest period being 11 years. Seen from the perspective of

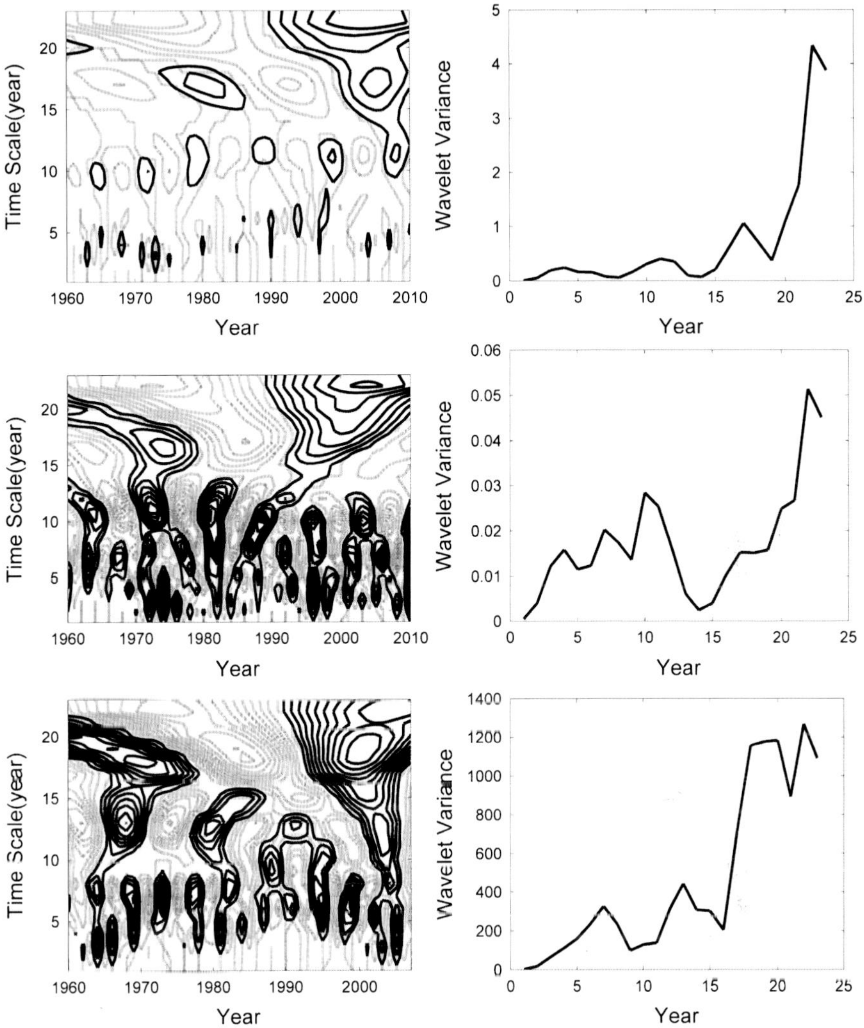

Fig. 3.9 The time-frequency distribution of wavelet coefficient and variance for temperature, precipitation and runoff in Aksu River, respectively (form *top* to *bottom*)

the 11-year time scale for Morlet wavelet transform coefficients, the relatively high periods of annual rainfall occurred in 1962–1965, 1971–1974, 1981–1984, 1986–1991, 1997–2001 and 2001–2004. A wavelet variance of runoff showed 6- and 14-year weak cycles, with the strongest period being an 18-year cycle (the same as for temperature). The wavelet coefficients of the real part of an 18-year cycle from 1970 to 1980 and from 2000 to 2007 show a positive phase, indicating high runoff in these periods. Thus, there are strong periods of 18 years for both temperature and runoff, while precipitation has a strong period of 11 years. This shows that runoff changes in the Aksu River have a strong correlation with temperature, which has a synchronized period.

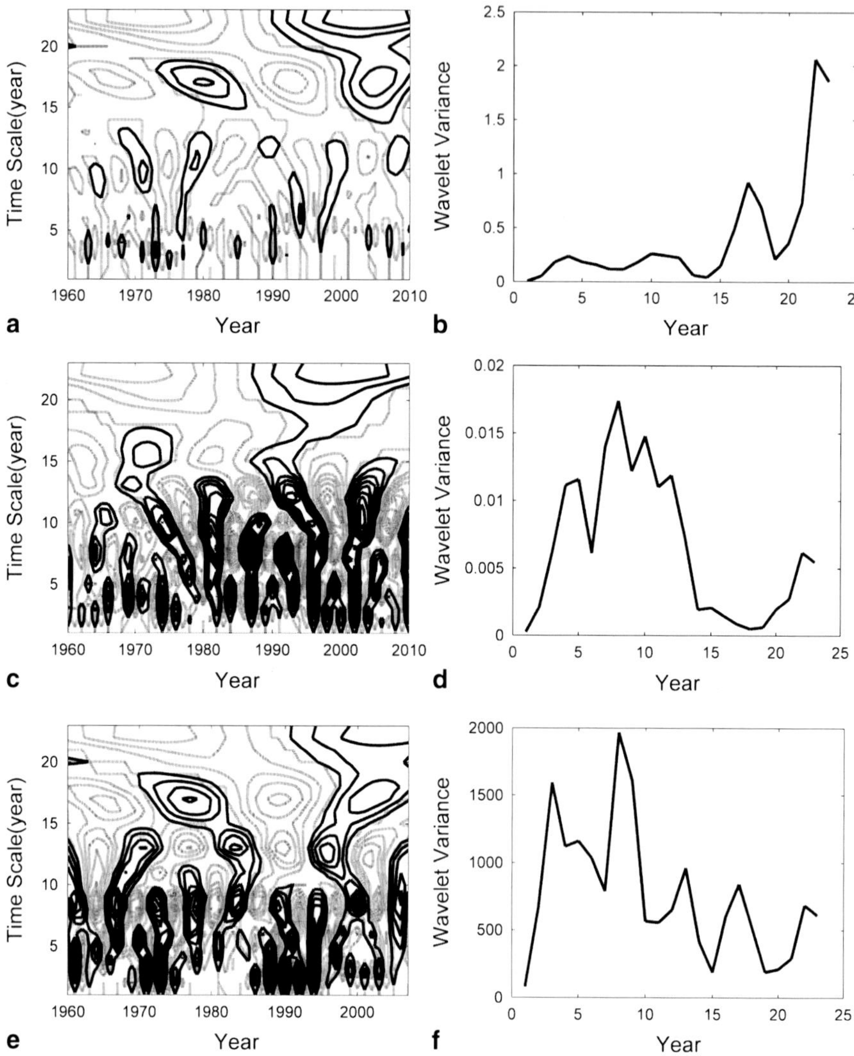

Fig. 3.10 The time-frequency distribution of wavelet coefficient and variance for temperature, precipitation and runoff in Yarkand River, respectively (form *top* to *bottom*)

(2) Yarkand River Wavelet variances (Fig. 3.10) indicated that weak 4- and 11-year temperature cycles were observed in the Yarkand River Basin. However, there were also strong periods of 17 and 22 years, with 19 years being the dominant period (again taking into consideration that the length of the recorded data is only 50 years). The wavelet coefficients at the real part of the 17-year cycle from 1973–1983 and 1997–2007 showed a positive phase, indicating higher temperatures in these intervals. For precipitation, weak cycles occurred in 4-, 5-, 11-, and 14-year timeframes, with the strongest cycle being 9 years. The timeframes from

1962–1965, 1969–1972, 1975–1978, 1980–1983, 1984–1987, 1989–1992, 1994–1997 and 2000–2003 belong to the wet period, with wavelet coefficients showing a positive phase. A weak cycle of runoff was observed at 4-, 14-, and 17-year phases, while the strong cycle was the same as for precipitation: 9 years. The coefficients at the real part of the 9-year cycle from 1962–1965, 1969–1972, 1975–1978, 1980–1983, 1984–1987, 1989–1992, 1994–1997 and 2001–2004 show a positive phase, indicating high runoff during these periods.

The wavelet analysis of the Yarkand River indicated that precipitation and runoff have strong periods of 9 years while temperature has a strong period of 17 years. This shows that runoff changes are synchronized with precipitation. The positive phase of the real part of the runoff is consistent with precipitation, which indicates that precipitation has a greater influence on runoff than temperature.

(3) Hotan River The main peaks of wavelet variances occurred at points on the scale equal to 4, 11, and 17, and 22 years. The first peak in wavelet variance, which corresponds to the scale a = 17, shows that the period at about 17 years oscillated the strongest and is the first main period. According to the real part change process of the wavelet transformation coefficient of different time scales in Fig. 3.11, we can conclude that 1970–1980 is a positive phase, as it registers a higher temperature. In contrast, weaker periods were found at 4 and 14 years for precipitation. The dominant period is 9 years, and the timeframes 1966–1968, 1971–1973, 1979–1982, 1986–1988, 1989–1992, 1993–1995 and 2001–2004 are positive phases that measured abnormally high amounts of rain. The energy change of the 9-year time-scale is the strongest, and the periods 1962–1968, 1969–1972, 1971–1974, 1979–1982, 1984–1987, 1989–1992, 1996–1999 and 2001–2004 are positive phases characterized by high runoff.

For time-frequency and variance distribution, precipitation and runoff have strong periods of 9 years while temperature has a strong period of 17 years. This shows that changes in runoff and precipitation are strongly synchronized. Moreover, the consistency of the positive phase also indicates that precipitation has a greater effect on runoff than temperature

(4) Kaidu River The variations indicated that temperature had weak periods of 7, 16 and 18 years. The 18-year period was the dominant one, with relatively strong periods of high annual temperature occurring in 1964–1974 and after 1994. Weak phases were observed for precipitation at 4-, 9-, and 16-year timeframes. The strongest cycle was the 18-year phase, with higher precipitation being observed from 1960 to 1970 and from 1991 to 2002. Wavelet variance of the runoff shows that within a dominant 18-year period, a relatively positive phase occurred in 1960–1971 and 1995–2001. In addition, the contour map is not completely closed, indicating that this runoff will continue to trend upward for some time.

Overall, temperature, precipitation, and runoff show a strong period of 18 years, indicating that runoff in the Kaidu River is affected both by temperature and precipitation, with the effects of precipitation being more obvious (Fig. 3.12).

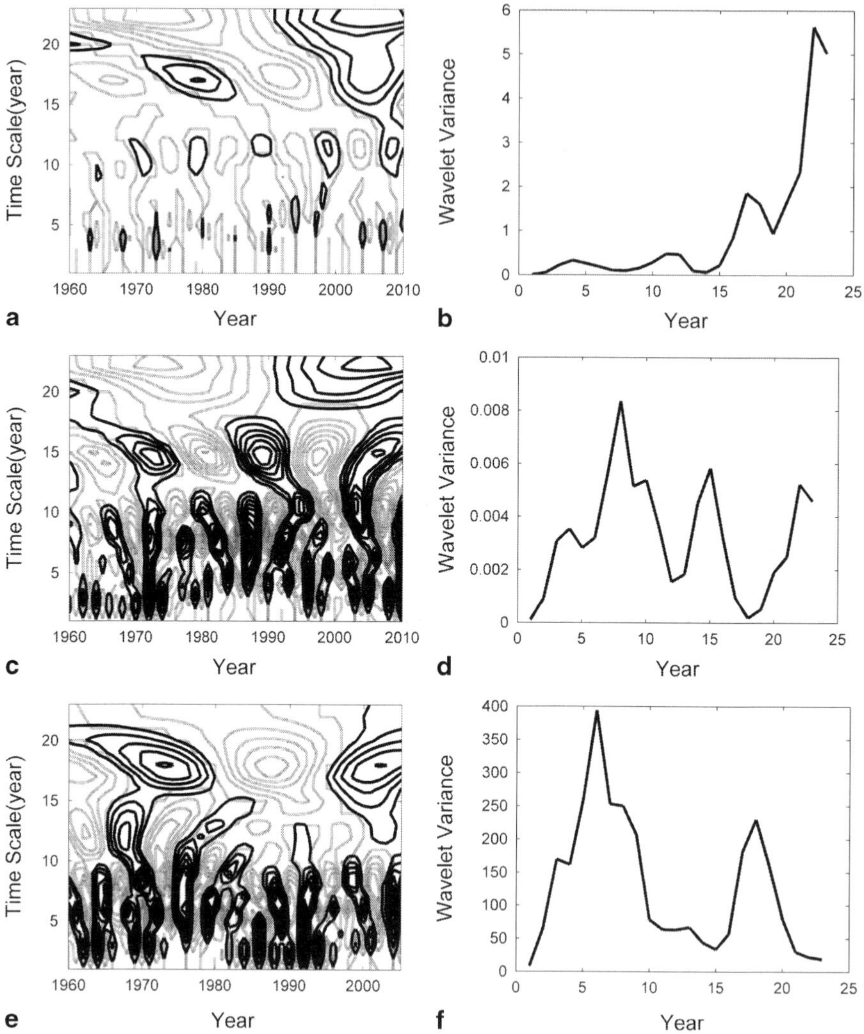

Fig. 3.11 The time-frequency distribution of wavelet coefficient and variance for temperature, precipitation and runoff in Hotan River, respectively (form *top* to *bottom*)

3.2.6 Conclusion

1. Affected by changes in precipitation and air temperature, runoff increases significantly at stations around the Tianshan Mountains, especially those near the south slope of the mountains and the Shule River. Few rivers show decreasing trends, and only the Shiyang River revealed significant negative trends.
2. Step changes are detected in a few stations for runoff series; the years with step changes also show non-uniform changes. Hence, the step changes of runoff

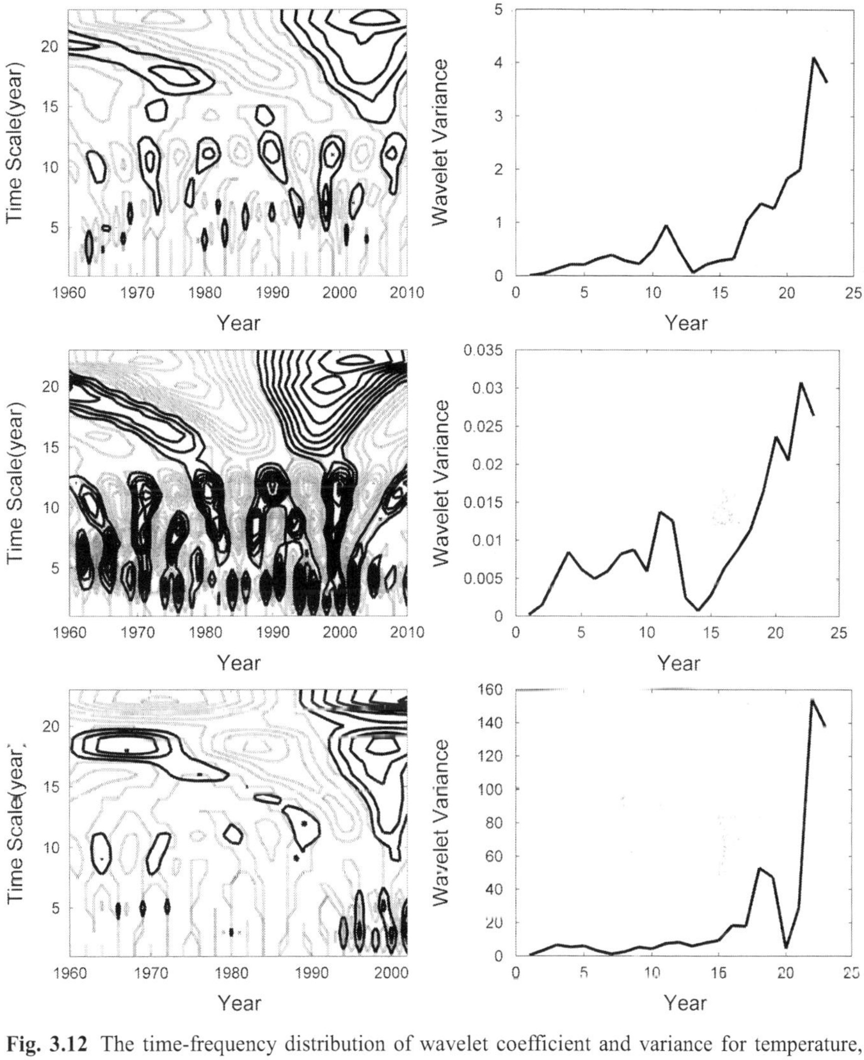

Fig. 3.12 The time-frequency distribution of wavelet coefficient and variance for temperature, precipitation and runoff in Kaidu River, respectively (form *top* to *bottom*)

exhibit a complicated pattern, with different hydrological stations displaying different step change points.

3. Changes in temperature and runoff in the Aksu River Basin reveal a strong consistency for a period of 18 years, while changes in runoff and precipitation in the Yarkand River and the Hotan River Basins show a similar 9-year phase, and temperature and precipitation in the Kaidu River have some synchronization with runoff and the same 18-year periodicity. Thus, we can initially speculate that the increase in runoff is mainly caused by snow and ice melting in the Aksu River, while the increase in accelerated snowmelt (caused by higher temperature

and precipitation levels) is responsible for increased runoff in the Kaidu River. For the rivers co-developed in the Kunlun Mountains (the Yarkand and Hotan Rivers), runoff is linked to precipitation.

4. The annual runoff processes of the three headwaters of the Tarim River are complex non-linear systems with fractal as well as chaotic dynamics. The correlation dimensions of attractor for the annual runoff process in the Hotan, Yarkand and Aksu Rivers are 3.2227, 3.2118, and 3.2092, respectively, none of which is integral. This means that the annual runoff processes in all three headwaters are chaotic dynamical systems that are sensitive to initial conditions. The time series of annual runoff in the Tarim River's three headwaters presents a long-term correlation characteristic. The Hurst exponents in the period from 1989 to 2002 indicate that the annual runoff in the Aksu and Yarkand Rivers will probably show an increasing trend, but that, in the Hotan River, the runoff is likely to show a slightly decreasing trend after 2002.

3.3 Characteristics of Runoff Components

Streamflow can be divided into quickflow and baseflow. Quickflow is normally from surface runoff, while baseflow is from shallow and deep groundwater. Typically, baseflow is not very sensitive to rainfall but is rather more associated with discharge from groundwater. The baseflow thus does not only supply the river's runoff during the rainless season, but also recharges groundwater during the wet season.

Because environmental isotopic tracers have the advantage of tracing, they are ideal for relating flow pathways and water residence time to river water than other common techniques used in classical hydrology. In alpine catchments, the change of meltwater in the total runoff is a direct response of water resources to climate change. Hence, calculating how much ice-melt water is in the total runoff can help evaluate the effects of climate change (Kong and Pang 2011). Moreover, as the stable isotope ratios of hydrogen and oxygen of water samples can provide essential information about water dynamics within a given watershed (Ruck and Mayer 2007), it is helpful to use tracers of stable isotope hydrogen or oxygen to identify water sources, since different water sources act within a river in different seasons. In hydrograph separation analysis, the two-component model was used to compute the contributions of the river water's various components.

3.3.1 Data And Separation Methods

(1) Baseflow Separation Typically, quickflow recession is fast and steep, whereas baseflow recession is slow and gentle in the flow process. Therefore, baseflow can be separated and estimated based on the measured flow hydrograph. Two types of techniques disaggregate daily streamflow into quickflow and baseflow. One type is graphical separation, such as the line segmentation method and the recession curve method the other is automatic methods that use computer programs to mimic manual separation methods such as the techniques of smoothed minima, HYSEP and recursive digital filtering (RDF). Although graphical separation methods are traditional and can be applied to different situations, they are subjective and not suitable for large data sizes. Automatic methods have been widely employed for streamflow separation with long time series. In this study, we selected the RDF technique due to its lower parameter requirements and relatively high accuracy (Hafzullah 2009). We also implemented the traditional recession curve method to provide a performance reference.

The RDF technique, which was originally used for signal analysis and processing, became popular in hydrology literature for baseflow separation (e.g., Eckhardt 2008). The RDF method was found to approximate well the baseflow sequence calculated by the manual separation technique using matching strips (Nathan and McMahon 1990). RDF smoothes sharp peaks in the fast component of the streamflow, so that the separated flow represents the delayed component of the streamflow, i.e., the baseflow. The RDF process can be represented as

$$Q_d(t) = \beta Q_d(t-1) + (1+\beta)/2\left[Q(t) - Q(t-1)\right] \tag{3.5}$$

where Qd (m^3/s) is the filtered quick streamflow, Q (m^3/s) is the total streamflow, t is the time step (day), and β is a filter parameter. The filtered baseflow Qb can then be obtained by:

$$Q_b(t) = Q(t) - Q_d(t) \tag{3.6}$$

The best results were obtained when β was in the range 0.90–0.95, with an optimal value of 0.925 (Nathan and McMahon 1990). The algorithm separates baseflow from total stream flow by passing the filter over the streamflow record three consecutive times (forwards, backwards, and forwards again). Each filter pass produces three time series estimates of baseflow, with the last being the projected baseflow time series. The justification for the use of this method rests on the fact that filtering out high-frequency signals is intuitively analogous to the separation of low frequency baseflow from the higher frequencies of quickflow (Nathan and McMahon 1990). The output of the filter is constrained, such that the separated flow cannot take negative values and is not greater than the total flow.

The recession curve method is the main traditional graphical separation method. It has been used to separate surface flow from total flow, and the derived groundwater flow is considerably higher during flood events than during dry seasons (Hafzullah et al. 2009). The recession curve method has the advantage of flexible adjustment according to basin characteristics, but it is tedious and time-consuming for baseflow separation with a long time series.

In this study, the recession curve method was implemented as a performance benchmark to the RDF. The method can be expressed as:

$$Q_t = Q_0 K^{-t} \tag{3.7}$$

where Q_t (m^3/s) is the discharge at any tie step t, Q_0 (m^3/s) is the discharge at the initial time step, and K is the recession constant.

The BFI is one of the most important low flow indexes. It is the long-term ratio of baseflow volume to total streamflow volume, and is affected by a number of climatic and topographic factors (Antonia and Paolo 2008). Therefore, it can be used to analyze the correlation between baseflow and climate factors.

BFI is calculated as a ratio between baseflow volume and total streamflow volume for a time period, and can be expressed as:

$$BFI = \frac{\int_{t_1}^{t_2} Q_b(t)dt}{\int_{t_1}^{t_2} Q(t)dt} \tag{3.8}$$

where t_1 and t_2 are starting and ending time points, respectively. In our study area, winter baseflow is less related to winter precipitation. Snowfall is accumulated from December to February and supplies baseflow during the spring snowmelt season. To study the variation of winter baseflow, stable BFI (SBFI) is defined as the ratio of winter baseflow volume to the total yearly streamflow volume, and is expressed as:

$$SBFI = \frac{\int_{t_D}^{t_F} Q_b(t)dt}{\int_{t_D}^{t_N} Q(t)dt} \tag{3.9}$$

where t_D, t_F and t_N are the months of December, February, and November, respectively. The recession curve method was applied to the Kaidu River for a comparison with the RDF method. The calculated BFI difference between the two methods varies with β, and reached its minimum when β is 0.925.

(2) Isotope Separation In the recharge areas of the Tizinafu River, ice/snowmelt plays a key role. It is important to determine the temporal and spatial distribution of ice/snowmelt percentage in runoff in order to utilize water resources rationally. In most previous research work with isotopic tracers, runoff is usually separated into different water types, such as two runoff components (event water and pre-event water sources) or three runoff components (event water, pre-event water, and the full mixture of precipitation from previous events). Both the two- and three-component method can be described as a uniform equation:

$$Q_t = \sum_{m=1}^{n} Q_m, Q_t C_t^j = \sum_{m=1}^{n} Q_m C_m^j, j = 1, \ldots, k \tag{3.10}$$

where Q_t is total runoff discharge, Qm is the discharge of component m, and C_m^j is the tracer j incorporated in the component m. In reference to isotope hydrograph separation, one of the tracers should be a kind of isotope.

Table 3.3 Catchment and baseflow characteristics of the four headwater streams

River	Drainage area	Average elevation	Average baseflow	Standard deviation	Median baseflow
	km^2	m	10^8 m^3	10^8 m^3	10^8 m^3
Aksu River	35,871	7,435	3.99	1.05	3.58
Hotan River	48,870	6,000	1.91	0.37	1.79
Yarkand River	48,100	5,395	12.31	1.25	12.45
Kaidu River	22,000	4,679	10.79	2.29	10.54

In the Tizinafu River mountain watersheds, river water is composed of ice/snowmelt, precipitation and groundwater. However, the groundwater comes only from ice/snowmelt and local precipitation without external groundwater flow into this area, so the runoff can be separated into the two components of ice/snowmelt and full precipitation mixture. Therefore, based on the mass balance of water and a tracer mass balance, we can use only oxygen isotopes to study the ice/snowmelt percent of the Tizinafu River streamflow.

The model requires several assumptions, as follows:

(1) C_m^j should be constant during the calculation period, such as a rainfall runoff process.
(2) C_m^j should be different between each component.
(3) A steady-state model adequately represents watershed conditions.

3.3.2 *Baseflow*

Table 3.3 presents the catchments and estimated baseflows of the four headstreams, as well as their characteristics. The annual average SBFIs of baseflow are 0.13 (Aksu River), 0.08 (Hotan River), and 0.23 (Yarkand River) to the upstream Tarim River, and 0.36 (Kaidu River) to the downstream Tarim River. The calculated baseflow increasing rates of the four rivers are 14.1, 6.9, 20.1 and 42.6 %, respectively (Fig. 3.13). The SBFIs of the Aksu and Hotan Rivers show continuously increasing trends, whereas the Yarkand River's SBFI fluctuates, and the Kaidu River denotes a decreasing trend (Fig. 3.13).

Table 3.4 presents the estimated baseflows and corresponding SBFIs in different decades. In terms of the relationship between baseflow amplitude and SBFI, each river has unique characteristics. From the 1980s to the 1990s, the rising amplitude of the Aksu River's baseflow is not proportional to the SBFI; from the 1970s to the 1980s, the declining amplitude of the Hotan River baseflow is not proportional to the SBFI; around 1970 and in 2000, the Yarkand River's baseflow increased but the SBFI decreased; the Kaidu River's baseflow first decreased, and then increased, while the SBFI has been steadily declining (Table 3.4). It is speculated that climatic factors have played a role in the relationship between baseflow amplitude and SBFI.

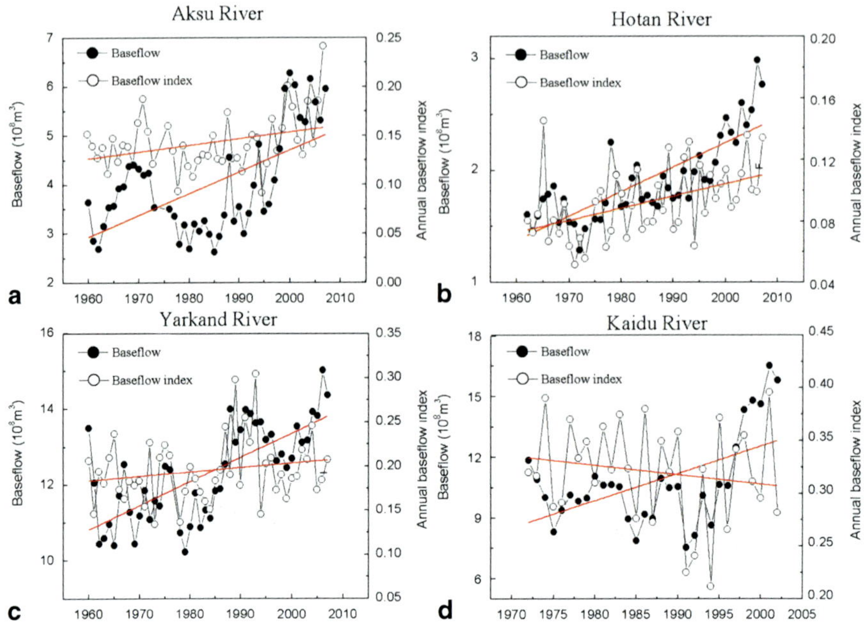

Fig. 3.13 Variation of annual baseflow and baseflow index for the four headstreams of the Tarim River. **a** Aksu River. **b** Hotan River. **c** Yakand River. **d** Kaidu River

Table 3.4 Annual variations of baseflow and BFI for the four headwater streams

River	Baseflow and BFI	1960–1969	1970–1979	1980–1989	1990–1999	2000–2008
Aksu	Qb (10^8 m^3)	3.61	3.63	3.19	4.06	5.75
	BFI (%)	13.37	14.42	13.12	13.49	17.62
Hotan	Qb (10^8 m^3)	1.66	1.65	1.80	1.96	2.54
	BFI (%)	8.65	7.52	9.21	9.80	10.95
Yarkand	Qb (10^8 m^3)	11.39	11.44	11.94	13.30	13.70
	BFI (%)	18.73	18.38	19.81	20.50	20.16
Kaidu	Qb (10^8 m^3)	/	10.06	9.93	10.79	/
	BFI (%)	/	33.23%	32.94%	30.05%	/

Qb is the baseflow, *BFI* is the baseflow index

Over the past 50 years, the SBFIs of the Hotan, Yarkand and Kaidu Rivers were 8.0–30.1 % of the streamflow in dry years (annual precipitation of less than 40 mm), and 10.8–37.0 % in wet years (annual precipitation greater than 100 mm) (Fig. 3.14). If the baseflow of a river is relatively constant, when the precipitation-supplied runoff increases, the proportion of baseflow in the entire stream flow should decrease. However, the SBFIs of the Hotan, Yarkand and Kaidu Rivers appeared generally larger in wet years than in dry years (Fig. 3.14). Tetzlaff and Soulsby (2008) concluded that BFI is larger in dry years than in wet years, based

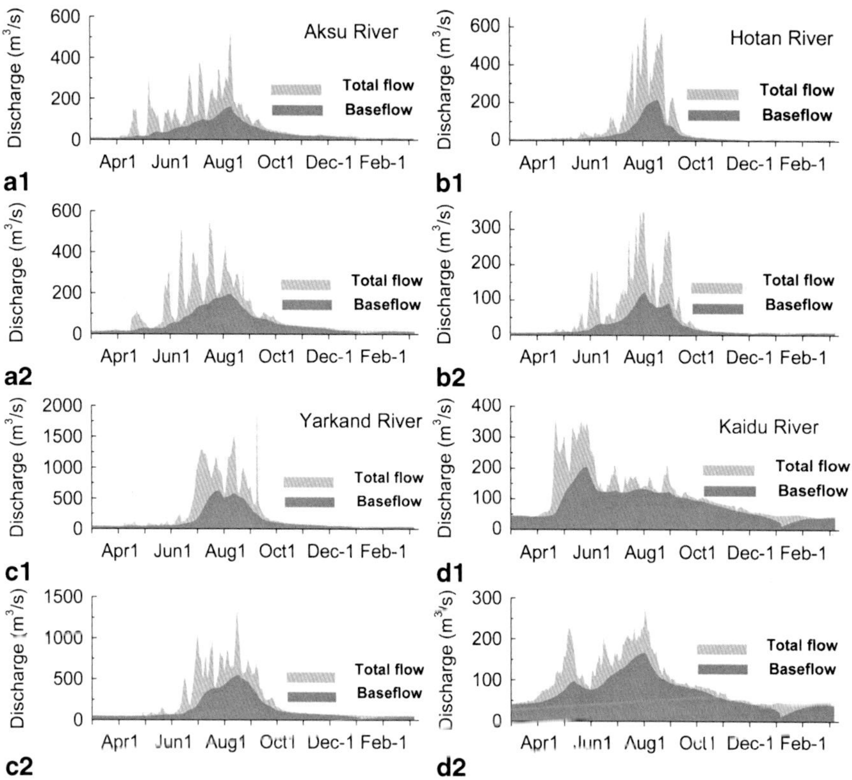

Fig. 3.14 Baseflow separation at the four headstreams of the Tarim River for typical dry and wet years. **a1** Aksu River in a dry year 1986 with SBFI = 12.7 %. **a2** Aksu River in a wet year 1996 with SBFI = 12.5 %. **b1** Hotan River in a dry year 1985 with annual SBFI = 8.0 %. **b2** Hotan River in a wet year 1987 with SBFI = 10.8 %. **c1** Yakand River in a dry year 1978 with SBFI = 13.0 %. **c2** Yakand River in a wet year 2007 with SBFI = 21.5 %. **d1** Kaidu River in a dry year 1980 with SBFI = 30.0 %. **d2** Kaidu River in a wet year 1981 with SBFI = 37.0 %

on a study in mountainous regions underlain by crystalline basement rocks and metamorphic formations with small aquifer potentials. In that kind of area, because of the limited storage in superficial deposits and thin soils, precipitation becomes surface runoff and enters streams directly instead of via deep percolation. In our study area, we see an opposite scenario, i.e., larger SBFI in wet years. An explanation for this uncommon phenomenon is that the increasing glacial meltwater had raised the level of baseflows in the rivers during those wet years, which is an outcome of warm-wet conditions. In the past 48 years, the warming-wetting trend in the Tarim River Basin was a regional response to global climate change, with the 1990s being the decade with the highest annual temperatures (Chen et al. 2009). Baseflows of the Aksu, Hotan and Kaidu Rivers began to increase in the 1990s, and baseflows of the Yarkand River began to increase in the late 1990s.

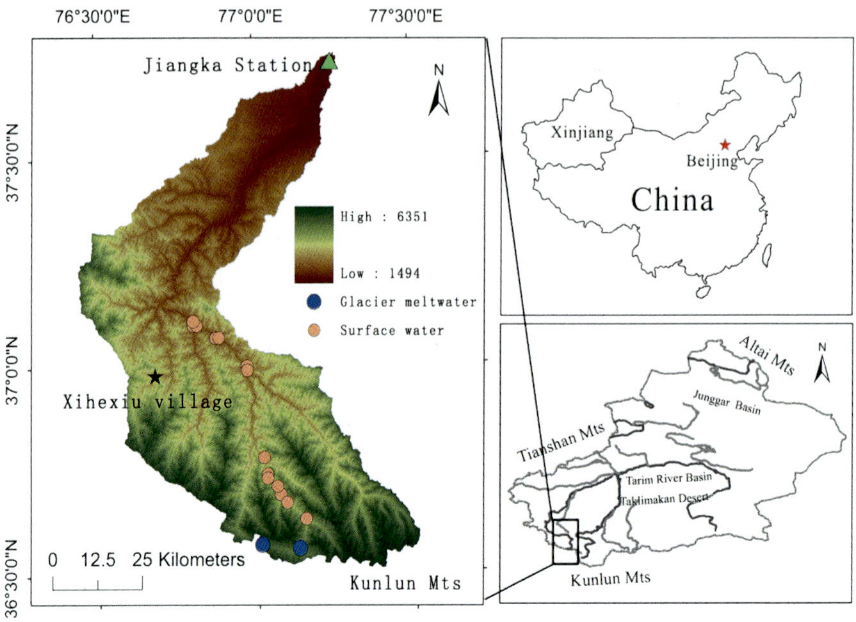

Fig. 3.15 Location map of the Tizinafu River and water sampling locations referred to in this paper

3.3.3 Ice-melt Water

The Tizinafu River, located in southeast Kashi and running a 335-kilometer course, originates in the northern slope of the Kunlun Mountains. From the southeast to the northeast, it flows through Yecheng, Zep, and the Yarkand agricultural irrigation area, where part of the river joins the Yarkand River. The streamflow of the Tizinafu River is 7.71×10^8 m^3/year and the average volume is 24.4 m^3/s, with a maximum flow of 519 m^3/s and a minimum flow of 0.42 m^3/s. Flood periods occur from May to September, dry periods from December to February. The Tizinafu River flows through the mountain valley; its river bed features rock, gravel and sand, and its downstream has an alluvial fan constitution. There is one hydrological station along the river and 25 small reservoirs built in the basin. Since the Tizinafu is supplied mainly by mountain snowmelt water, spring water and rain water, it has a higher water quality and contains little pollution. These features make it not only appropriate for human and animal consumption, but also for irrigation. Due to increasing temperatures, the melting of permafrost along the glacier front may also contribute a certain amount of water to the groundwater and increase baseflow in the rivers (Fig. 3.15).

As shown in Tables 3.5 and 3.6, the significant difference between the oxygen isotopes of different components is good enough to use the above tracer model. The $\delta^{18}O$ of ice/snowmelt Ci in the two tables refer to the mean value of the original 6

Table 3.5 Spatial characteristics of ice-snowmelt in the Tizinafu River streamflow

Stations	Cp (‰)	Cr (‰)	Ci (‰)	Estimated fractions (%)
Upstreams	−3.34	−10	−13.01	68.87
Kudi		−9.25		61.12
Mazhadaban		−8.19		50.16
Halastan		−7.22		40.12
Jiangka		−7.12		39.09
Downstreams		−7.02		38.06
Tributary		−6.87		36.50
Xihexiu		−6.81		35.88
Yuzimenleke		−5.85		25.96
Mean of the ice-snowmelt percentage (%)			43.97%	

Table 3.6 Temporal characteristics of ice-snowmelt in the Tizinafu River streamflow

Month	Cp (‰)	Cr (‰)	Ci (‰)	Estimated fractions (%)	Month	Cp (‰)	Cr (‰)	Ci (‰)	Estimated fractions (%)
January	−3.34	−7.18	−13.01	39.74	July	−3.34	−9.06	−13.01	59.18
February		−6.91		36.95	August		−6.98		37.63
March		−6.81		35.89	September		−7.21		40.00
April		−8.87		57.22	October		−6.97		37.51
May		−8.77		56.18	November		−6.08		28.31
June		−9.67		65.43	December		−6.46		32.24
Mean of the ice-snowmelt percentage (%)				43.86%					

sampling data. Because of the altitude's effect on isotope in this study area, all $\delta^{18}O$ values in precipitation weighted by rainfall were averaged, to get Cp, i.e., the $\delta^{18}O$ of the full mixture of precipitation from previous events. The $\delta^{18}O$ values in river water were averaged by station and by month, respectively, to get Cr in Tables 3.5 and 3.6.

The results of runoff separation according to different hydrological gauging locations are shown in Table 3.5. According to the table, the mean percentage of ice/snowmelt runoff is 43.97 %, which was evaluated by using the average of six ice/snowmelts $\delta^{18}O$. This is only 25.96 %, using the average from Yuzimenleke (at the lower section of the river) and enriched isotopic singular values. However, the mean percentage of ice/snowmelt runoff in the tributaries upstream is 68.87 %. This markedly significant variation of 42.91 % was caused by ice/snowmelt isotopic fractionation during melting, impeding an accurate separation of streamflow. From this, we can see that the error is large when ice/snowmelt contributes a dominant fraction of the streamflow. The ice/snowmelt percentages are less than the mean percentage at the stations with elevations lower than 2,500 m because in these lower areas, large ice/snowmelts infiltrate into the ground as lateral groundwater flow mostly discharges into rivers at mountain outlets. In light of the warming and drying climate in the region, more attention must be paid to the ice/snowmelt factor if the area's scarce water resources are to be rationally utilized.

As shown in Table 3.6, there are large diversities of ice/snowmelt percentages among the various months, mainly resulting from uneven distribution of rainfall. Ice/snowmelt percentages are lower than average from November to January and higher than average from May to July. In winter, rainfall amounts are minimal, whereas in the spring and summer, with high air temperatures, more ice/snowmelt is infiltrated into groundwater as lateral flow, delaying its discharge into the river. Because of these two factors (minimal rain in winter and groundwater infiltration in spring and summer), the percentage of ice/snowmelt is the largest in summer, with almost 60 % of the annual runoff in the warm months. Conversely, in winter, precipitation is abundant and ice/snowmelt is minimal, so the lowest percentage (28.31 %) of the annual runoff is recharged by ice/snowmelt in November.

3.3.4 Conclusion

1. Over the past 48 years, the increased rates of baseflow for the Aksu, Hotan, Yarkand, and Kaidu Rivers are 14.1, 6.9, 20.1, and 42.6 %, respectively. However, these baseflow increases do not coincide with changes in SBFI. The SBFI of the Hotan, Yarkand, and Kaidu rivers are 8.0–30.1 % of the streamflow in dry years, and 10.8–37.0 % in wet years. The increase of baseflow as a result of increasing glacial meltwater leads to the uncommon scenario of increasing precipitation and a larger BFI occurring in the same year.
2. Using hydrograph separation, we find that the Tizinafu River is a glacier-dependent river. From the glacier front to Jiangka station, ice/snowmelt water recharges the river, decreasing with elevation. Upstream, ice/snowmelt water accounts for 68.87 %, although it drops to less than 26 % downstream. For temporal characteristics of ice/snowmelt in the streamflow, we find that ice-melt water accounts for almost 60 % in late spring and summer, and less than 35 % in early winter. Therefore, the nature of Glacier Rivers determines the climate sensitivity of watersheds fed by them. Glacier Rivers like Tizinafu River are relevant to flood control and agriculture water division. In water resources management, both climate change and the replenishment of rivers should be emphasized.

3.4 Future Changes in Runoff

3.4.1 Study Area, Data And Methods

The study area in this section analyzes seven typical rivers: the headwaters of the Tarim River (the Aksu, Yarkand, Hotan and Kaidu Rivers), and the inland rivers in the Hexi Corridor (the Heihe, Shiyang, and Shule Rivers). For these rivers, the runoff time series from 1957 to 2009 were used. To predict streamflow using sensitivity analysis and the statistical downscaling model STAR (only for the Kaidu River), this study used the time series of average temperature and precipitation.

The relationships among air temperature, precipitation and runoff were used to simulate future change trends in runoff. As well, wavelet analysis was used to reveal the periodicity and non-linear trend of runoff and its related climatic factors. Based on the results of wavelet analysis, the quantitative relation between the runoff and climate factors (temperature and precipitation) was formulated with the regression analysis method.

On regional and global scales, the climate is generally warming, but changes in precipitation levels appear to be more complex and diverse. Therefore (see Sect. 4.4.2), this study analyzed typical river runoff variations under different scenarios. To do this, the statistical downscaling model STAR was also used in tangent with the above methods to predict meteorological factors such as precipitation, relative humidity and wind speed after setting the warming scenarios. According to the previously built multi-regression equations between runoff in Dashan Kou and Huangshuigou, the annual average temperature, and the annual precipitation regression model, combined with STAR downscaling model predictions under different scenarios in 2015, 2025 and 2035, runoffs were predicted under different scenarios.

3.4.2 *Predictions Using Continuous Wavelet Decomposition*

3.4.2.1 Aksu River

For the Aksu River, air temperature, precipitation and runoff periods are 18 a, 11 a, 18 a, respectively. Figure 3.16 shows the continuous wavelet decomposition curve of air temperature, precipitation and runoff wavelet coefficients in the Aksu River. Table 3.7 shows the appropriate equations for temperature, precipitation and runoff wavelet coefficients that have passed the significance test of 0.05. By applying these equations, temperature trends lower in the Aksu River in 2010–2015, after which the temperature slowly rises to a high temperature phase in 2018–2028. During 2010–2014 and 2026–2028, precipitation will be reduced but will increase again in 2018–2018. In 2010 to 2016 and 2017–2017, runoff will reduce.

3.4.2.2 Yarkand River

For the Yarkand River, air temperature, precipitation and runoff periods are 17 a, 9 a, and 9 a, respectively. Table 3.8 shows the appropriate equations for temperature, precipitation and runoff wavelet coefficients and Fig. 3.16b shows projections of the river's annual runoff and climate factors of wavelet coefficients. Temperatures in the periods 2010–2016 and 2026–2028 are expected to be relatively low, while those in 2016–2026 will be high. During the 2010–2025 timeframe, there will be only moderate levels of precipitation, which will give way to a rainy period from

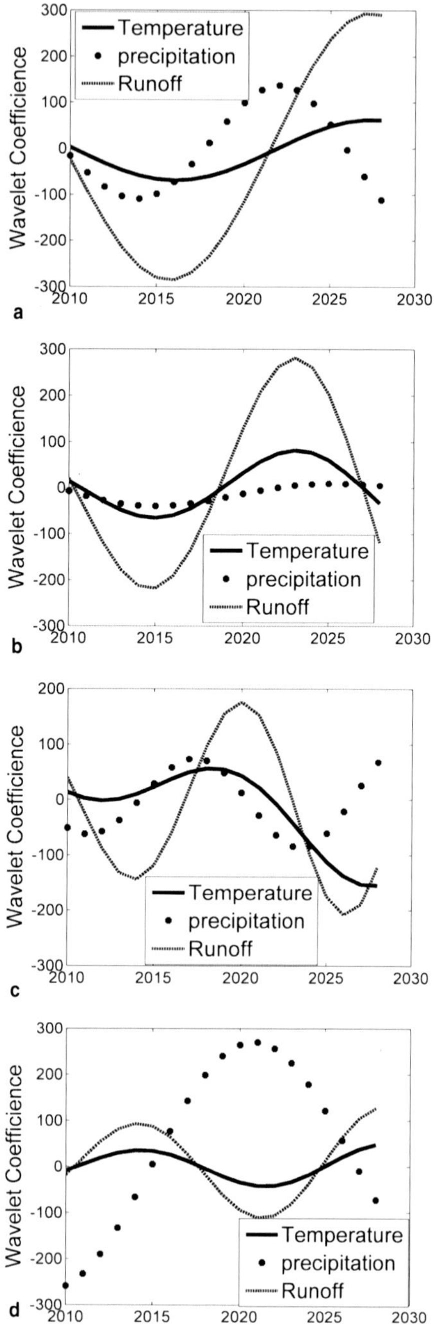

Fig. 3.16 Projections of Akesu river annual runoff and climate factor of wavelet coefficients in Aksu river (**a**), Yarkand River (**b**), Hotan River (**c**), and Kaidu river (**d**)

Table 3.7 Fitting equation of temperature, precipitation and runoffwavelet coefficients in Aksu river

Type	Periodic (a)	Fitting equation	R^2	P
Temperature	18	$Y=3.59\sin(0.21x-163)+6.29\sin(0.3x+166)$	0.97	<0.001
Precipitation	11	$Y=809\sin(0.37x-240)+822\sin(0.3x+257)$	0.92	<0.001
Runoff	18	$Y=26.1\sin(0.22x-192)+33.3\sin(0.2x+173)$	0.91	<0.001

Table 3.8 Fitting equation of temperature, precipitation and runoffwavelet coefficients in Yark-and river

Type	Periodic(a)	Fitting equation	R^2	P
Temperature	17	$Y=39.4\sin(0.37x-233)+41.3\sin(0.3x+6.36)$	0.96	<0.001
Precipitation	9	$Y=25.4\sin(0.18x-119)+20.5\sin(0.2x+91.7)$	0.97	<0.001
Runoff	9	$Y=226\sin(0.36x-221)+233\sin(0.37x+19.8)$	0.83	<0.001

Table 3.9 Fitting equation of temperature, precipitation and runoffwavelet coefficients in Hotan river

Type	Periodic(a)	Fitting equation	R^2	P
Temperature	17	$Y=137\sin(0.11x-22.4)+130\sin(0.11x-100)$	0.78	<0.001
Precipitation	9	$Y=368\sin(0.51x+250)+369\sin(0.5x-244)$	0.93	<0.001
Runoff	9	$Y=132\sin(0.49x-223)+140\sin(0.5x+15.6)$	0.87	<0.001

2025 to 2028. During the 2010–2018 and 2026–2028 timeframes, runoff will decrease, but from 2019–2026, it will increase.

3.4.2.3 Hotan River

For the Hotan River, air temperature, precipitation and runoff periods are 17 a, 9 a, and 9 a, respectively. Table 3.9 shows the appropriate equations for temperature, precipitation and runoff wavelet coefficients in the Hotan River. From 2010 to 2018, temperature and runoff will lower, but in 2018–2027, they will increase. Precipitation in 2010–2018 will decrease, and then it will start an increasing trend (Fig. 3.16c).

3.4.2.4 Kaidu River

For the Kaidu River, air temperature, precipitation and runoff periods are the same as 18a. Table 3.10 shows the appropriate equations for temperature, precipitation and runoff wavelet coefficients in the Kaidu area. During 2010–2018 and 2025–2028 (Fig. 3.16d), temperatures will be high, and from 2016 to 2026, precipitation is expected to increase. During the 2011–2018 and 2025–2028 timeframes, annual runoff will be higher than usual.

Table 3.10 Fitting equation of temperature, precipitation and runoffwavelet coefficients in Kaidu river

Type	Periodic(a)	Fitting equation	R^2	P
Temperature	18	$Y=32.6\sin(0.43x-270)+33.3\sin(0.43x+316)$	0.98	<0.001
Precipitation	18	$Y=189\sin(0.23x-180)+85.6\sin(0.31x-39)$	0.99	<0.001
Runoff	18	$Y=149\sin(0.43x-277)+149\sin(0.44x+309)$	0.97	<0.001

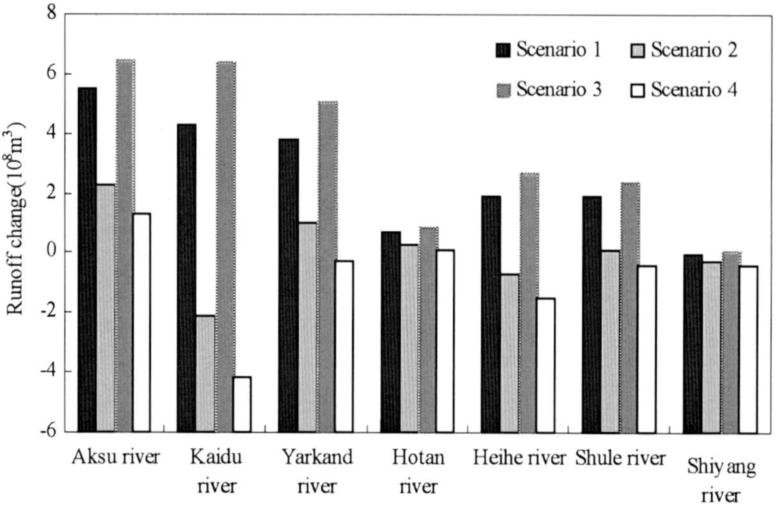

Fig. 3.17 The typical river runoff variation under different scenarios

3.4.3 Runoff Variations of Typical Rivers Using Sensitivity Analysis

The four scenarios are as follows: compared to the average temperature and precipitation for the past 50 years, the typical river area temperature increased by 0.5 °C and precipitation increased by 30 mm (Scenario 1); compared to the average temperature and precipitation for the past 50 years, the typical river area temperature increased by 0.5 °C and precipitation decreased by 30 mm (Scenario 2); compared to the average temperature and precipitation for the past 50 years, the typical river area temperature increased by 0.5 °C and precipitation increased by 50 mm (Scenario 3); and compared to the average temperature and precipitation for the past 50 years, the typical river area temperature increased by 0.5 °C and precipitation decreased by 50 mm (Scenario 4).

In Scenarios 1 and 3, runoff amounts in the Aksu, Kaidu and Yarkand Rivers increase substantially, in the range of $3.8\times10^8\sim6.5\times10^8$ m^3 (Fig. 3.17). Runoff amounts in the Heihe and Shule River also increase, but to the lesser degree of only $1.9\times10^8\sim2.7\times10^8$ m^3. Runoff is even less in the Hotan River (less than 1×10^8m^3), and the Shiyang river runoff under Scenario 1 will actually reduce by 0.02×10^8 m^3 and increase only slightly under Scenario 3 (0.07×10^8 m^3).

Table 3.11 STAR validation: observations and projections, both for the validation period from 1986 to 2010

	T(Bayinbuluke)		P(Bayinbuluke)		T(Baluntai)		P(Baluntai)	
	Obs	Sim	Obs	Sim	Obs	Simed	Obs	Sim
Spring	−1.4	−0.8	39.9	40.3	8.8	8.7	34.3	28.0
Summer	10.2	10.2	198.0	182.6	18.6	18.3	170.2	123.6
Autumn	−2.1	−1.4	39.2	45.0	7.4	6.6	27.2	31.7
Winter	−21.9	−21.6	10.9	8.2	−6.3	−7.0	1.8	1.1

In Scenario 2, runoff will be reduced in the Kaidu, Heihe and Shiyang Rivers, but the Akesu River will experience the highest increase in runoff (2.3×10^8 m^3) followed by the Yarkand River. In contrast, the Hotan River will experience the lowest increase in runoff at just 0.7×10^8 m^3. Under Scenario 4, only the Aksu and Hotan river runoff will increase slightly (1.3×10^8 and 0.1×10^8 m^3, respectively). The Kaidu runoff will be reduced (up to 4.2×10^8 m^3) as will the runoff of the Heihe and other rivers (the Yarkand, Shule and Shiyang) in the general range of $0.3 \times 10^8 \sim 0.4 \times 10^8$ m^3.

Only the sensitivity of the Shiyang River runoff to temperature is negative and the lowest (−0.06), followed by the Heihe River runoff, while the temperature sensitivity of the other rivers' runoff is high. For instance, the sensitivity of the Hotan River runoff to precipitation is the lowest (0.007), followed by the Yarkand and Aksu Rivers, whereas the sensitivity of the other rivers' runoff to precipitation is relatively high.

If the climate in China's arid Northwest continues its current trend of increasing warmth and humidity, runoff in the Aksu, Kaidu and Yarkand Rivers will greatly increase. However, if the climate instead turns warm and dry, then the runoff in the Kaidu and Heihe Rivers runoff will see larger reductions.

3.4.4 *Forecasts of Future Runoffs Combined with the STAR Model*

3.4.4.1 Temperature and Precipitation Forcasts Under Different Scenarios

The temperature will increase by 1.3–2.1 °C in 2020, 1.5–2.8 °C in 2030, and 2.3–3.3 °C in 2050, respectively, while the national average annual precipitation will increase 5–7 % in 2050. Based on the above-mentioned report regarding climate change in China, this section assumes that temperatures in the Kaidu-Kongque River will increase 0 °C (Scenario 2), 0.5 °C (Scenario 1), 1.0 °C (Scenario 3), 1.5 °C (Scenario 4), and 2.0 °C(Scenario 5). The statistical downscaling model STAR will used to predict meteorological factors such as precipitation, relative humidity and wind speed after setting the warming scenarios.

This section selects the seasonal temperature and precipitation in the Bayinbuluke and Baluntai weather stations to test the STAR model. The series in 1961–1985 is the training period, and the series in 1986–2010 is the model validation period. The results (Table 3.11) show that the measured value and the simulated value are consistent (except for summer precipitation in Baluntai), so the STAR model was used to predict the weather factors (temperature and precipitation).

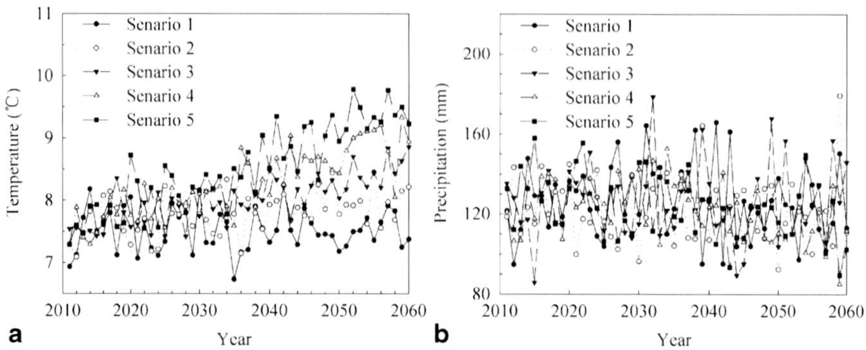

Fig. 3.18 The change trends of annual mean temperature and annual precipitation in the Kaidu-kongque River Basin under different scenarios

Table 3.12 Annual mean temperature and annual precipitation of Bayinbuluke and Baluntai in 2015, 2025 and 2035 under different scenarios

		Scenarior 1		Scenarior 2		Scenarior 3		Scenarior 4		Scenarior 5	
		T	P	T	P	T	P	T	P	T	P
Bayinbuluke	2015	−3.1	296.0	−4.7	259.0	−4.6	217.0	−4.3	250.0	−1.2	286.0
	2025	−4.5	244.0	−2.1	266.0	−5.0	339.0	−3.9	208.0	−3.6	234.0
	2035	−5.8	247.0	−4.6	211.0	−2.8	247.0	−5.0	312.0	−1.8	229.0
Baluntai	2015	6.8	227.0	9.0	134.0	7.9	150.0	8.1	135.0	7.9	301.0
	2025	7.4	139.0	8.8	163.0	8.4	217.0	8.7	193.0	8.6	196.0
	2035	6.9	202.0	8.8	213.0	8.2	185.0	8.3	152.0	8.2	183.0

Future precipitation levels show decreasing trends in Scenarios 4 and 5 (Fig. 3.18), while precipitation in the other scenarios is not significant. Moreover, the mean temperatures for months 2 and 10 in Scenario 2 are lower than in other scenarios, and the monthly precipitation does not vary significantly in the different scenarios for the Kaidu-Kongqu River. The annual average temperature in five scenarios is 7.5, 7.8, 8.0, 8.3 and 8.5 °C, and the annul average precipitation is 126.6, 123.2, 126.3, 122.0 and 123.6 mm, respectively. We selected the years 2015, 2025 and 2035 for planning years when forecasting water supply and demand. Table 3.12 shows the annual precipitation and temperature in 2015, 2015 and 2025 for the Bayinbuluke and Baluntai weather stations.

3.4.4.2 Forecasts of Future Runoffs Under Different Scenarios

According to the previously built multi-regression equations between runoff in Dashan Kou and Huangshuigou, the runoff was predicted under different scenarios. Here, the annual average temperature and the annual precipitation regression model are combined with STAR downscaling model predictions under different scenarios in 2015, 2025 and 2035. The results (Table 3.13) indicated that the runoff

Table 3.13 Prediction results of annual runoff in 2015, 2025 and 2035 under different scenarios

		Scenario 1	Scenario 2	Scenario 3	Scenario 4	Scenario 5
Dashankou	2015	38.45	33.04	29.75	32.93	40.41
	2025	32.19	37.47	38.90	30.13	32.67
	2035	30.45	29.35	34.91	36.84	34.96
Huangshuigou	2015	2.96	3.01	2.65	2.62	4.16
	2025	2.35	3.21	3.54	3.45	3.44
	2035	2.75	3.69	3.15	2.87	3.12
Total	2015	41.41	36.05	32.40	35.55	44.57
	2025	34.54	40.68	42.44	33.58	36.11
	2035	33.20	33.04	38.06	39.71	38.08

will increase to 44.57×10^8 m^3 (2015) under Scenario 5, with a warming context. Moreover, the predicted minimum runoff is 32.40×10^8 m^3, which may occur in 2015 under Scenario 3.

3.4.5 Conclusion

1. Trend simulation was conducted using the significant period of the wavelet transform coefficients for temperature, precipitation and runoff. Runoff in the Aksu, Yarkand and Hotan Rivers will be low in 2010–2016 but will see continued growth in 2017–2028. However, in the Kaidu River, the annual runoffs in 2011–2018 and 2025–2028 will be high, while runoff water in the 2018–2025 years will flatten.
2. Under Scenario 1, the climate will become warm and wet (temperatures will increase by 0.5 °C and precipitation will increase by 30 mm), as it will under Scenario 3 (temperatures will increase by 0.5 °C and precipitation by 50 mm). However, streamflow in the Kaidu, Heihe and Shiyang Rivers will reduce in the warm and dry climate of Scenario 2 (temperatures will increase 0.5 °C but precipitation will decrease 30 mm). Under Scenario 4 (temperature increase of 0.5 °C and precipitation decrease of 50 mm), the runoff in the Aksu and Hotan Rivers will increase slightly, but the Kaidu River will reduce its streamflow the most (up to 4.2×10^8 m^3), followed by the Heihe River. The runoff in the other rivers (the Yarkand, Shule and Shiyang) will be reduced in the range of $0.3–0.4 \times 10^8$ m^3.
3. Predicted climate changes for the next 50 years were calculated using the STAR model to forecast annual average temperature and precipitation under different scenarios. Runoff was also forecasted based on established multi-regressions between annual runoff, annual mean temperature, and annual mean precipitation. The model forecasted that annual runoff will increase up to 44.57×10^8 m^3 in 2015 under Scenario 5. However, under Scenario 3, the runoff in 2015 will be just 32.40×10^8 m^3, based on a climate warming context.

Summary

1. Based on runoff data from 1960 to 2010, the streamflow characteristics of China's arid Northwest region were investigated using the Mann–Kendall method, the wavelet transform method, correlation dimensions, and R/S analysis. Runoff was affected by changes in precipitation and air temperature and increased significantly at stations around the Tianshan Mountains, especially those on the south slope and along the Shule River. Step changes related to runoff were also detected in a few stations, but these variations were generally non-uniform. Temperature and runoff changes in the Aksu River Basin were strongly consistent for 18-year periods; changes in precipitation and runoff in the Yarkand River and Hotan River Basins shared the same 9-year period; and temperature and precipitation in the Kaidu River were somewhat synchronized with runoff variances over the same periodic 18-year timeframe. The correlation dimensions of attractor derived from the time series of the annual runoff for the Hotan, Yarkand and Aksu Rivers are all greater than 3.0 and non-integral, implying that all three rivers are chaotic dynamical systems. The time series of annual runoff in each river also presents long-term correlation characteristics.
2. The digital filtering method was used to separate baseflow from surface flow, after which the baseflow index (BFI) was calculated and analyzed. The major findings of this study include: (1) Baseflows of the four headstreams have increased considerably over the past 50 years. The largest baseflow index (BFI) usually occurred in the wet years, but the change rate was irregular due to increasing recharge from snow and glacial meltwater. (2) The baseflow and BFI showed obvious seasonal variations: The lowest baseflow and BFI typically occurred in December and January, and both increased gradually until reaching maximum values in August or July. (3) The responses of runoff and baseflow to climatic factors differed substantially. Precipitation had a significant impact on runoff, whereas temperature strongly affected baseflow. (4) In the Tizinafu River, the contribution of ice/snowmelt water varied from 25.96 to 68.87 % for spatial characteristics, and from 28.31 to 65.43 % for seasonal characteristics. The mean of the ice/snowmelt percentage is 43 %, which meant that ice/snowmelt water was the main supplying water source.
3. Using the data from 1960 to 2010, future runoff amounts were predicted using continuous wavelet decomposition, sensitivity analysis, and the STAR model. Some results can be concluded as follows: (1) Runoff in the Aksu, Yarkand, and Hotan Rivers will be low in 2010–2011 but will experience continued growth in 2017–2028. In the Kaidu River, annual runoff in 2011–2018 and 2025–2028 will be high, constituting a wet period, whereas runoff in 2018–2025 will be flat, prompting a dry period. These predications were formulated using the significant periods of the wavelet transform coefficients for temperature, precipitation and runoff. (2) Under Scenarios 1 and 3, the climate becomes warmer and wetter (temperature increases 0.5 °C in both scenarios, while precipitation increases 30 mm in Scenario 1 and 50 mm in Scenario (3). According to sensitivity analysis, the streamflows in the Kaidu, Heihe and Shiyang Rivers will reduce under

Scenario 2, which predicts a temperature increase of 0.5 °C and a precipitation decrease of 30 mm. (3) Climate change for the next 50 years was calculated using the STAR model's forecast of annual average temperature and precipitation under various scenarios. Runoff in the planning year was also forecasted based on established multi-regressions between annual runoff and annual mean temperature. The results indicate that annual runoff will change significantly under all scenarios.

References

Alexandersson H (1986) A homogeneity test applied to precipitation data. J Climatol 6:661–675

Antonia L, Paolo V (2008) Baseflow index regionalization analysis in a mediterranean area and data scarcity context: role of the catchment permeability index. J Hydrol 355:63–75

Buishand TA (1982) Some methods for testing the homogeneity of rainfall records. J Hydrol 58:11–27

Chen YN, Li WH, Xu CC, Hao XM (2007) Effects of climate change on water resources in Tarim River Basin, northwest China. J Environ Sci (China) 19:488–493

Chen YN, Xu CC, Hao XM, Li WH, Chen YP, Zhu CG, Ye ZX (2009) Fifty-year climate change and its effect on annual runoff in the Tarim River Basin, China. Quatern Int 208:53–61

Eckhardt K (2008) A comparison of baseflow indices, which were calculated with seven different baseflow separation methods. J Hydrol 352:168–173

Grassberger P, Procaccia I (1983) Characterization of strange attractors. Phys Rev Lett 50:346–349

Hafzullah A, Ilker K, Ebru E (2009) Filtered smoothed minima baseflow separation method. J Hydrol 372:94–101

Hurst HE, Black RP, Simaika YM (1965) Long-term storage: an experimental study. Constable, London, pp 1–155

Kendall MG (1975) Rank-correlation measures. Charles Griffin, London, p 202

Kong YL, Pang ZH (2011) Isotope hydrograph separation in alpine catchments: a review. Sci Cold Arid Reg 3:86–91

Labat D, Espinoza JC, Ronchail J, Cochonneau G, de Oliveira E, Doudou JC, Guyot JL (2012) Fluctuations in the monthly discharge of Guyana Shield rivers, related to Pacific and Atlantic climate variability (Variations des débits mensuels des fleuves drainant le bouclier guyanais en relation avec la variabilité climatique atlantique et pacifique). Hydrol Sci J (Journal Des Sciences Hydrologiques) 57:1081–1091

Li BF, Chen YN, Chen ZS, Li WH (2012) Trends in runoff versus climate change in typical rivers in the arid region of northwest China. Quatern Int 282:87–95

Liu C, Chen Y, Xu Z (2010) Eco-hydrology and sustainable development in the arid regions of China preface. Hydrol Process 24:127–128

Mann HB (1945) Non-parametric tests against trend. Econometrica 13:245–259

Melesse A, Abtew W, Dessalegne T, Wang X (2010) Low and high flow analyses and wavelet application for characterization of the Blue Nile River system. Hydrol Process 24:241–252

Nathan RJ, McMahon TA (1990) Evaluation of automated techniques for base flow and recession analyses. Water Resour Res 26 1465–1473.

Peterson TC, Easterling DR, Karl TR et al (1998) Homogeneity adjustments of in situ atmospheric climate data: a review. Int J Climatol 18:1493–1517

Ruck L, Mayer B (2007) Isotope hydrology of the Oldman River basin, southern Alberta, Canada. Hydrol Process 21:3301–3315

Shi YF, Shen YP, Kang E, Li DL, Ding YJ, Zhang GW, Hu RJ (2007) Recent and future climate change in northwest China. Clim Change 80:379–393

Sivakumar B (2007) Nonlinear determinism in river flow: prediction as a possible indicator. Earth Surf Process Landf 32:969–979

Tetzlaff D, Soulsby C(2008) Sources of baseflow in larger catchments—using tracers to develop a holistic understanding of runoff generation. J Hydrol 359:287–302

Wang HJ, Chen YN, Li WH, Deng HJ (2013) Runoff responses to climate change in arid region of northwestern China during 1960–2010. Chin Geogra Sci 23(3):1–15. doi:10.1007/s11769-012-0000-0

Xu JH, Chen YN, Li WH, Ji MH, Dong S (2009) The complex nonlinear systems with fractal as well as chaotic dynamics of annual runoff processes in the three headwaters of the Tarim River. J Geogr Sci 19:25–35

Xu JH, Chen YN, Lu F, Li WH, Zhang LJ, Hong YL (2011) The nonlinear trend of runoff and its response to climate change in the Aksu River, western China. Int J Climatol 31:687–695

Yue S, Pilon P, Phinney B, Cavadias G (2002) The influence of autocorrelation on the ability to detect trend in hydrological series. Hydrol Process 16:1807–1829

Zhao FF, Xu ZX, Huang JX, Li JY (2008) Monotonic trend and abrupt changes for major climate variables in the headwater catchment of the Yellow River basin. Hydrol Process 22:4587–4599

Chapter 4
Response of Runoff to Climate Change

Yaning Chen, Zhongsheng Chen, Baofu Li and Qihu Li

Abstract Rising global temperatures have accelerated the water cycle, causing a redistribution of water resources that ranges from minor to extreme, arid areas are especially affected by this process due to the fragile nature of their eco-hydrology, which is more sensitive to climatic changes. The main effects of climate change to runoff are as follows: (1) The runoff in North Xinjiang has a strong positive relationship with precipitation, while that in the south slope of the Tianshan Mountains, the middle section of the north slope of the Tianshan Mountains and the Shule River has a strong positive relationship with air temperature; and there is a positive significant relation between summer runoff and 0 °C level height (FLH). (2) Human activities are presently the main driving forces behind runoff changes. In the Aksu, Kaidu, Shule and Heihe Rivers, the influence percentage of human activity on runoff is 90.4, 55.7, 63.1 and 78.8 %, respectively. (3) It was discovered that runoff in the Aksu and Yarkand Rivers is increasing with rising precipitation and temperature levels, and that runoff in the Hotan River is likewise increasing, but the rate of increase there is minimal. As runoff's response to temperature is more sensitive to precipitation changes, runoff in the Aksu and Yarkand Rivers will increase 1.4–7.0 % with every 1 °C rise in temperature.

Keywords Attribution of runoff change · Correlation analysis · 0 °C level height · Quantification of climate and human factors · Sensitivity analysis

Y. Chen (✉) · Z. Chen · B. Li
State Key Laboratory of Desert and Oasis Ecology,
Xinjiang Institute of Ecology and Geography,
Chinese Academy of Sciences, No. 818 South Beijing Road, 830011 Urumqi, Xinjiang, China
e-mail: chenyn@ms.xjb.ac.cn

Z. Chen
e-mail: chenzhs@ms.xjb.ac.cn

B. Li
e-mail: libf@ms.xjb.ac.cn

Q. Li
College of Geomatics,
Xi'an University of Science and Technology, 710054 Xi'an, China
e-mail: liqh@ms.xjb.ac.cn

Y. Chen (ed.), *Water Resources Research in Northwest China*,
DOI 10.1007/978-94-017-8017-9_4, © Springer Science+Business Media Dordrecht 2014

4.1 Introduction

Climate change and its consequent impacts on hydrology and water resources are hot topics for scholars and researchers around the world. During the past 50 years, rising global temperatures have accelerated the water cycle, causing a redistribution of water resources that ranges from minor to extreme. Arid areas are especially affected by this process due to the fragile nature of their eco-hydrology, which is more sensitive to climatic changes.

Numerous studies have recently been carried out on the impact of climate change on glaciers, agriculture, and water resources. As the relationship among regional temperature (0 °C height), precipitation and runoff requires further investigation, this chapter will focus on the relationship between climate change and mountain runoff of seven typical watershed areas (the Aksu River, the Yarkand River, the Hotan River, the Kaidu River, the Shule River, the Heihe River, and the Shiyang River) in order to gauge the effects of regional climate change on the water cycle system process. Human activities are also an important factor impacting hydrological distributions, so this chapter will attempt to distinguish hydrological influences caused by climate change and those resulting from human activities.

The sensitivity of natural runoff to climate change refers to the magnifying effect of one or more climate variables (such as temperature, precipitation, etc.) causing changes in the natural runoff. If the magnifying effect of only a minor climate change is large, then we can conclude that the hydrology and water resources systems are highly sensitive to climate change. Sensitivity indices reflect the adaptability of hydrology and water resources systems to climate change (Qing et al. 2005).

4.2 Reasons Underlying Runoff Changes

China's arid Northwest area is an ecologically fragile zone that is extremely sensitive to climate change. Water resources mainly come from natural precipitation in mountains and mountain snowmelt water that is greatly affected by temperature. For instance, climate change impacts glacier movements and snow distribution, and can also lead to increases in heavy precipitation days and flood frequency. In high latitudes and snowmelt recharge basins, increases in temperature cause increases in the rain/snow ratio. This accelerates the rate of spring snowmelt and shortens the snowfall season, leading to a faster, earlier and larger spring runoff. Streamflows in China's arid region have changed significantly, but the reasons are very complex. In this section, we will investigate changes in surface runoff.

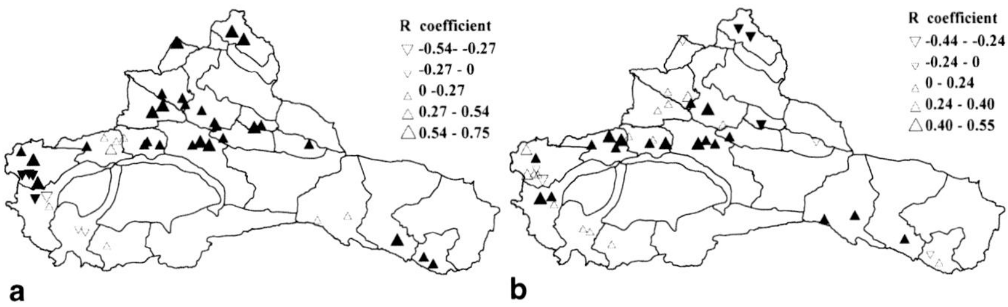

Fig. 4.1 Correlation between **a** annual runoff and precipitation, and **b** annual runoff and temperature at basin scale

4.2.1 Correlations Between Runoff and Climate Variables

4.2.1.1 Overall Relationships in the Northwest Arid Region

Figure 4.1a illustrates the strong correlation that exists between runoff and annual precipitation. As we can see from the figure, the correlation coefficients are extremely significant ($P<0.05$) for the Kaidu, Weigan, Heihe and Shiyang Rivers and for stations in North Xinjiang.

Correlations between runoff and air temperature are shown in Fig. 4.1b. The annual air temperature also has a significant positive influence on streamflow, especially in stations on the south slope of the Tianshan Mountains (the Kaidu, Weigan and Aksu Rivers) and the Yarkand, NSMT, and Shule Rivers. However, air temperature has a negative influence on the NSET and Irtysh Rivers.

Interestingly, annual runoff in the Aksu, Yarkand and Shule has significant positive correlation coefficients with annual temperature but no notable relationships with annual precipitation. This indicates that the runoff's main supply source is glacier meltwater. The Kaidu and Weigan Rivers are strongly affected by both annual temperature and annual precipitation, revealing that the runoff in these areas is supplied by both glacier and rainwater. Runoff changes in the Irtysh River are consistent with changes in annual precipitation but inconsistent with those of annual temperature, indicating that rainwater is the main supplier of this particular river.

4.2.1.2 Correlation Between Runoff and Climate Variables in Typical Rivers

(1) Trend Analysis of Climatic Factors and Runoff

① **Trend Analysis of Temperature** Figure 4.2 shows the temperatures of seven rivers' headstream areas, revealing varying degrees of increasing trends. The temperature of the Aksu River headstream area was relatively steady up until 1997, at which point it started to experience notable increases in temperature. Likewise,

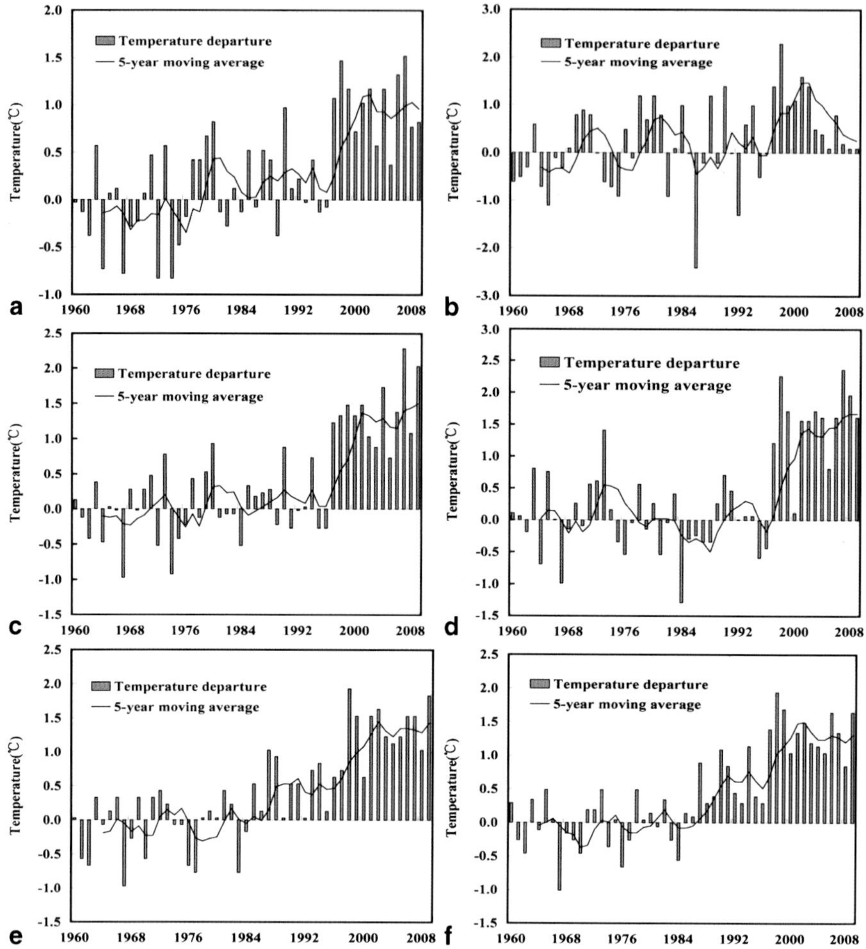

Fig. 4.2 The temperature departure of Aksu River (**a**), Yarkand River (**b**), Hotan River (**c**), Kaidu River (**d**), Shule River (**e**), and Shiyang River (**f**)

temperatures in the Yarkand headstreams area were in a steady state prior to 1997, and then increased. Despite a slight decline around 2002, temperatures there have remained in a positive anomaly state. Similar to the Yarkand, temperatures in the Hotan River headstream area were relatively stable before 1997, after which the temperatures increased substantially. Meanwhile, temperatures in the Kaidu River headstream area rose in 1973, fell for a while, and then rose again in 1996. Up to 1985, the Shule River area was relatively stable, after which it commenced a positive anomaly before entering a period of rapid warming in 1996. Temperatures in the Shiyang River experienced slight cooling in the 1960s, remained in a steady state during the 1970s and early 1980s, and then warmed up, showing significant down—steady—up trends. The temperature in the Heihe headstream area was stable

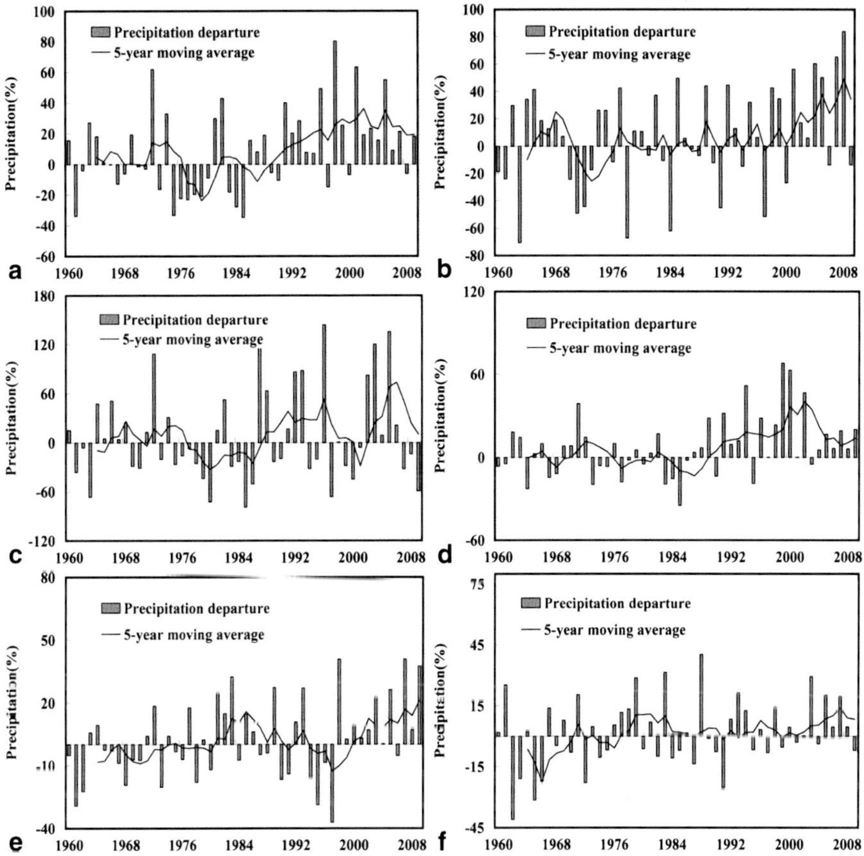

Fig. 4.3 The precipitation departure percentage of Aksu River (**a**), Yarkand River (**b**), Hotan River (**c**), Kaidu River (**d**), Shule River (**e**), and Shiyang River (**f**)

prior to 1987, with negative anomalies dominating. Temperatures then jumped after 1987, and since 1997 have maintained a positive anomaly state. Generally speaking, higher temperatures in headstream zones increase snowmelt runoff, which is bound to have some impact on hydrological conditions.

② **Trend Analysis of Precipitation Changes** Figure 4.3 shows precipitation anomaly percentage changes over the past 50 years for seven river headstreams. As we can see, the increasing trend for precipitation in the Hotan and Shiyang Rivers is not significant, but the trends for the other five rivers passed the significance test of 0.05. Precipitation in the Aksu River headstream showed a decreasing trend in 1985, followed by an increasing trend after 1985; since 1991, the precipitation for most years exceeds the 1960–1989 mean. Precipitation in the Yarkand headstream was in a steady state before 2000 and then increased significantly, showing an anomaly percentage of more than 80 % in 2008. The Hotan River headstream precipitation showed a decreasing trend until 1986, after which the trend disappears. In the Kaidu

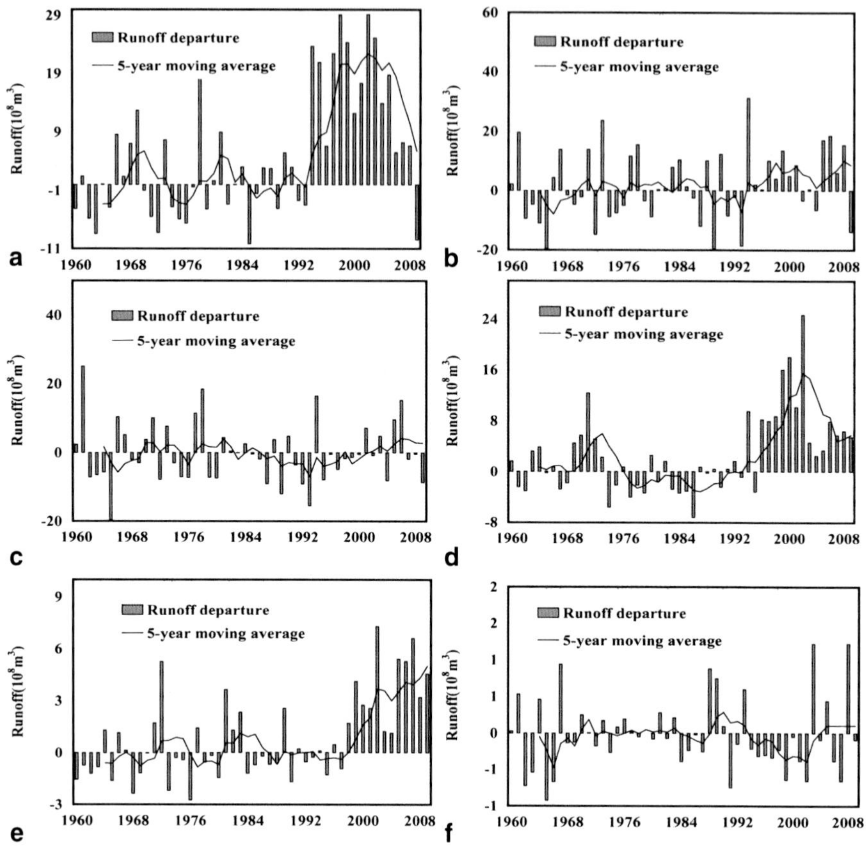

Fig. 4.4 The runoff departure of Aksu River (**a**), Yarkand River (**b**), Hotan River (**c**), Kaidu River (**d**), Shule River (**e**), and Shiyang River (**f**)

River headstream, precipitation changes were relatively stable before 1986 but increased significantly after 1986. The changes then experienced a decreasing trend from 2000 onwards, but are still higher than the 1960–1989 annual average. The Shule River headstream precipitation was relatively stable before 1998, after which a slight increase occurred. Precipitation in the Shiyang River headstream showed a decreasing trend in 1971, leading to a relatively stable period where there were more positive than negative anomaly years. The Heihe River headstream precipitation (not shown) underwent three stages of change: increases in 1960–1990, stability in 1990–2004, and significant increases in 2004–2009. Fluctuations in headstream precipitation will inevitably lead to precipitation runoff volatility, which is also bound to have some impact on hydrological conditions.

③ **Trend Analysis of Runoff Changes** Figure 4.4 shows runoff anomaly changes during the past 50 years. Runoff in the Aksu, Kaidu, Shule and Heihe Rivers increased significantly (a = 0.05), whereas runoff in the Yarkand and Shiyang Rivers showed a weak increasing trend and Hotan River runoff showed a slightly decreasing one. As the Aksu River runoff was greater than the standard period's mean after 1993, this proves that the Aksu River mountain outflow continued to rise during the 1993–2010 timeframe, which is the highest runoff period to date. During the dry season that stretched from 1960 to 1992, Yarkand River runoff showed only a weak rising trend. However, a closer look reveals that the runoff experienced unstable fluctuations in 1978–1986 and again in 1995–1991, and the rest of the period was also not stable. Hotan River runoff endured a dry season from 1963 to 1967, after which it was relatively stable from 1982 to 1988 and from 1998 to 2002. The runoff in the other years during the 50-year period is low. Meanwhile, runoff in the Kaidu River experienced a short wet season concentrated in 1993–2002, but its dry season was longer than that of the Aksu River, which lasted from 1960 to 1987. In addition, runoff changes for the Aksu were stable only in the short period of 1988–1991, but otherwise were close to average. Shule River runoff was in a steady state prior to 1997, with positive and negative anomalies alternately fluctuating, and then it entered into a wet season from 1998 to 2009, experiencing the highest runoff periods in the past 50 years. The Shiyang low runoff placed it in a strong dry season from 1960 to 1971, after which it alternated from wet to dry up to 1990 and then re-entered a dry season in 1990–2005. Heihe River runoff (not shown) was in a distinct dry season before 1974 (only 1964 and 1967 had a larger amount of water) and then began a significantly increasing trend and remained mainly wet throughout the 1980s and 1990s. Since 2005, it experienced 5 consecutive years of severely wet seasons.

(2) Cycle Analysis of Climatic Factors and Runoff

① **The Cycle Diagram Method** The cycle diagram method is suitable for recognition of short time series cycles. It assumes that the time series x_t was expanded to a Fourier series, which can then be represented as:

$$x_t = \sum_{i=1}^{k} (a_i \cos 2\pi f_i t + b_i \sin 2\pi f_i t) + \varepsilon_t \tag{4.1}$$

where f_i is frequency, t is time serial number, k is the number of periodic components (the main cycle and its harmonics number), and ε_t is the residual. When the frequency f_i is given, the above equation can be regarded as a multiple linear regression model. It can be proved with multiple regressions analysis theory that when $k \neq N/2$, the least squares estimation of undetermined coefficients b_i and a_i are:

$$\hat{a}_i = \frac{2}{N} \sum_{t=1}^{N} x_t \cos 2\pi f_i t \tag{4.2}$$

$$\hat{b}_i = \frac{2}{N} \sum_{t=1}^{N} x_t \sin 2\pi f_i t \tag{4.3}$$

where N is the number of observations. If N is even and when i = 0 or i = $N/2$, the estimates of the parameters a_i should be:

$$\hat{a}_i^* = \frac{1}{N}\sum_{t=1}^{N} x_t \cos 2\pi f_i t \tag{4.4}$$

The periodogram of the time series can be defined as:

$$I(f_i) = \frac{N}{2}(a_i^2 + b_i^2) \tag{4.5}$$

Where I(f_i) is the intensity of frequency f_i(i = 0, 1, 2, …, N − 1), f_i is set as the abscissa, and I(f_ii) is the ordinate. If the peak value could pass the Fisher test through a significant level of a, then the T = $1/f_i$ to which that peak value corresponds is the length of the period time series.

② Periodicity of Climatic Factors and Runoff

Aksu River Figure 4.5a illustrates the cycle of precipitation, temperature and runoff for the Aksu River headstream. At a significance level of a = 0.05, the precipitation cycle fluctuations are complex, showing cycles of 8, 6, 5 and 3 years. Additionally, temperature showed cycles of 25, 10, 7, 4 and 3 years, and runoff showed 4- and 25-year cycles. The precipitation, temperature and runoff of the Aksu River headstream experienced the same 25- and 4-year cycles, but precipitation and runoff are different. This indicates that runoff in the Aksu River headstream is mainly influenced by temperature, and that the effect of precipitation is weak.

Yarkand River Figure 4.5b shows the precipitation, temperature and runoff cycles of the Yarkand River headstream. Precipitation experienced cycles of 4 and 3 years, temperature changes emerged in multi-cycles of 25, 10, 5, 4 and 3 years, and runoff exhibited 6-, 3-, and 2-year cycles. From this, we can see that precipitation, temperature and runoff share a common cycle of 3 years, which suggests that runoff has a significant correlation with both precipitation and temperature. However, the periodogram analysis cannot discern the contribution of precipitation and temperature to runoff.

Hotan River Figure 4.5c showed the cycle of precipitation, temperature and runoff of Yarkand River headstream. Both precipitation and runoff showed complex cycle fluctuations with periods of 5, 3 and 2 years and 6, 4, 3, and 2 years, while temperatures showed cycles of 25, 10 and 2 years. The presence of a 3- and 2-year common cycle can be seen from runoff and precipitation, but there is only a 2-year common period for runoff and temperature. This indicates that although the Hotan River headstream runoff is driven by precipitation and temperature, precipitation plays the greater role.

Kaidu River Figure 4.5d shows the cycle of precipitation, temperature and runoff of the Kaidu River headstream. The fluctuations in precipitation and temperature are complex, charting 10-, 6-, and 3-year, and 25-, 8-, 5-, and 4-year cycles, respectively.

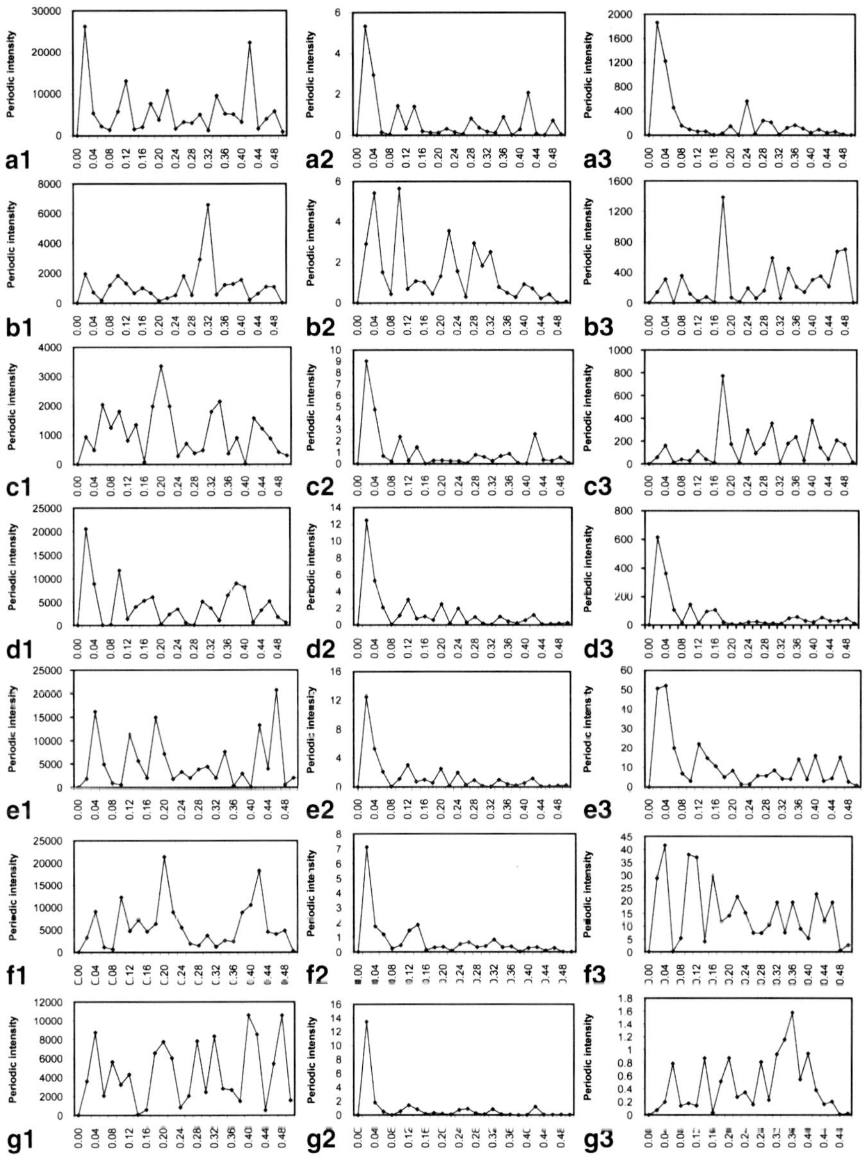

Fig. 4.5 The cycle of (**a1**, **b1**, **c1**, **d1**, **e1**, **f1**) precipitation, (**a2**, **b2**, **c2**, **d2**, **e2**, **f2**) temperature and (**a3**, **b3**, **c3**, **d3**, **e3**, **f3**) runoff. a Aksu River. b Yarkand River. c Hotan River. d Kaidu River. e Shule River. f Heihe River. g Shiyang River

Runoff reveals cycles of 25, 10, and 6 years. Thus, runoff and precipitation share 10- and 6-year cycles, while runoff and temperature share a common 25-year cycle. This shows that runoff in the Kaidu River headstream is influenced by the combined effects of precipitation and temperature, but that precipitation plays a greater role.

Shule River Figure 4.5e shows the cycle of precipitation, temperature and runoff of the Shule River headstream. Precipitation has complex cyclical fluctuations of 25, 8, 5, 3 and 2 years and temperature changes in periods of 7 and 3 years, while runoff keeps cycles of 25, 8 and 2 years. From this, we can see that runoff and precipitation have three common cycles of 25, 8 and 2 years, while runoff and temperature do not share a common cycle, indicating that the Shule River headstream runoff is mainly influenced by precipitation.

Heihe River Figure 4.5f illustrates the Heihe River headstream's precipitation, temperature and runoff cycles. Precipitation and runoff followed complex fluctuation periods of 5, 3 and 2 years and 10, 8, 6, 5, 3 and 2 years, while temperature showed cycles of 8 and 7 years. From this, we can see that runoff and precipitation share 5, 3, and 2-year common cycles, while runoff and temperature share common cycle of 8 years. This shows that while precipitation and temperature both have some impact on runoff, precipitation has the greater impact.

Shiyang River Figure 4.5g illustrates precipitation, temperature and runoff cycles for the Shiyang River headstream. Precipitation and runoff reveal slightly complex cycles of 5, 4 and 3 years and 7, 5, 4 and 3 years, respectively, whereas temperature chart 25- and 2-year cycles. Although runoff and precipitation both shared common 5, 4 and 3-year cycles, no common cycle was evident in temperature and runoff. This indicates that the Shiyang River headstream runoff changes are mainly due to the effects of fluctuations in precipitation, and that no obvious relationship exists with temperature.

(3) Correlation Analysis of Runoff, Temperature and Precipitation

Aksu River In order to quantitatively investigate the correlation of runoff, precipitation and temperature of the Aksu River, scatter plots were drawn between the annual runoff (Q) and annual precipitation (P), and between annual runoff (Q) and annual average temperature (T) (Fig. 4.6a). A statistical fitting method was carried out to establish a linear equation, and a statistical significance test was also performed. Linear fitting results show that there is a positive correlation between runoff and precipitation, and between runoff and temperature. The equations are as follows:

$$Q = 0.0537P + 65.323$$
$$Q = 8.7485T + 63.38$$

The multiple correlation coefficients (R2) of runoff and precipitation and runoff and temperature are 0.0765 and 0.2406, respectively. From this, we can see that the correlation of runoff and precipitation does not pass the significance test of 0.05 but that the correlation of runoff and temperature does pass the significance test of 0.01. The analysis results show that temperature change is the key factor inducing runoff changes.

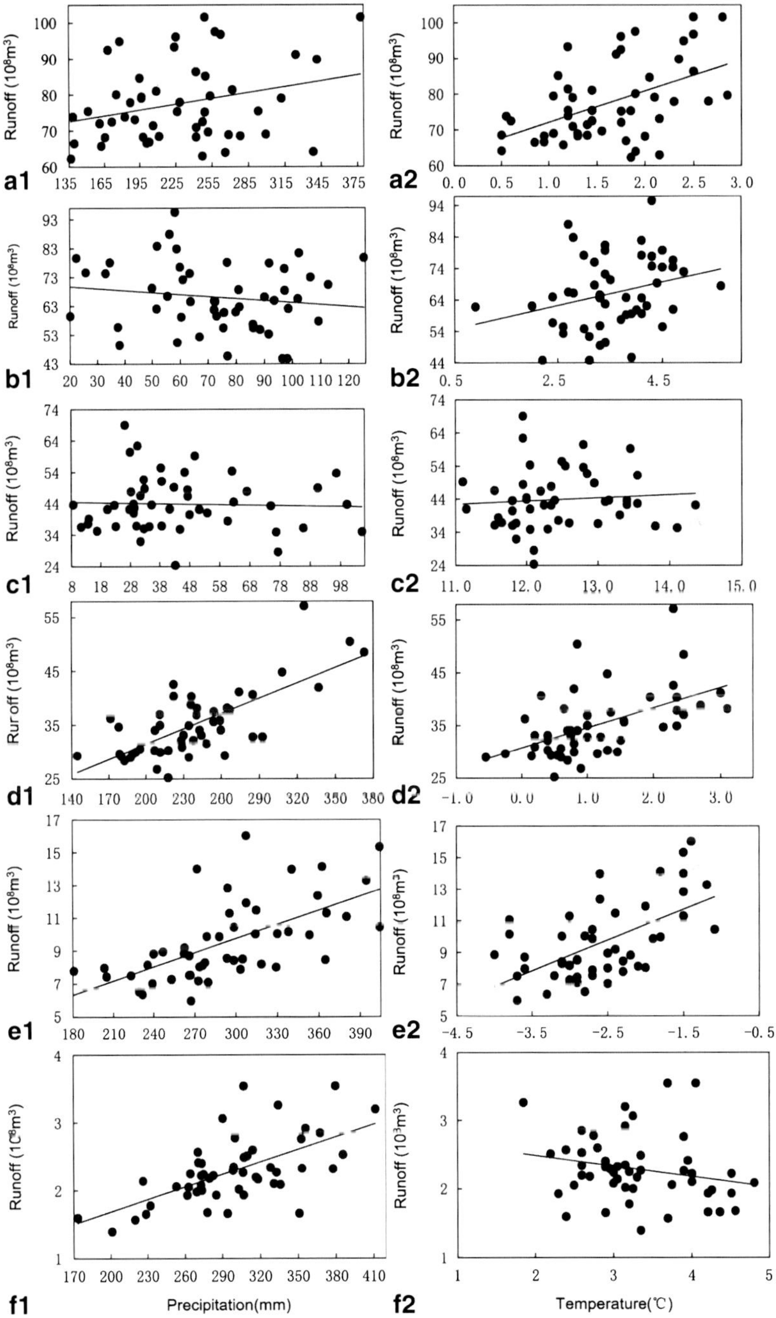

Fig. 4.6 Linear fitting of streamflow and precipitation (**a1**, **b1**, **c1**, **d1**, **e1**, **f1**), temperature (**a2**, **b2**, **c2**, **d2**, **e2**, **f2**) of the Aksu River Basin

In order to further reveal the relationship between runoff and climatic factors, we established a regression equation in which the annual precipitation and average temperature are the independent variables and the mountain outlet flow is the dependent variable:

$$Q = 0.0452P + 8.3613T + 53.7032$$

The multiple correlation coefficient (R2) of the regression equation is 0.2943. The F test shows that the F value of the regression equation is 9.802 and the p value is 0.0003 ($p<0.01$) under a confidence level of 99%. Hence, the regression equation has passed the 0.01 significance level of statistical tests, which indicates that the change of streamflow is the result of regional climate change.

Yarkand River In order to explore the quantitative correlation of runoff, precipitation and temperature in the Yarkand River, scatter plots were drawn between the annual runoff (Q) and annual precipitation (P), and between annual runoff (Q) and annual average temperature (T) (Fig. 4.6b). A statistical fitting method was carried out to establish a linear equation, and a statistical significance test was also performed. Linear fitting results show that there is a negative correlation between runoff and precipitation and a positive correlation between runoff and temperature. The equations are as follows:

$$Q = -0.069P + 71.169$$
$$Q = 3.722T + 52.941$$

The multiple correlation coefficients (R2) of runoff and precipitation and runoff and temperature are 0.0217 and 0.0782, respectively. Hence, the correlation of runoff and precipitation did not pass the significance test of 0.05, but runoff and temperature did. The results of the analysis show that temperature change is the key factor prompting runoff changes.

In order to further reveal the relationship between runoff and climatic factors, we established a regression equation in which the annual precipitation and average temperature are the independent variables and the outlet flow of the Yarkant River is the dependent variable:

$$Q = -0.0388P + 3.4475T + 56.7364$$

Here, the multiple correlation coefficient (R2) of the regression equation is 0.0846. The F test shows that the F value of the regression equation is 2.1719 and the p value is 0.1253 ($p>0.05$), which is below the confidence level of 99%. Therefore, the regression equation did not pass the significance test of 0.05, which indicates that changes in the streamflow cannot be explained by regional climate change. At the same time, we also noted that there is a weak negative correlation between annual runoff and annual precipitation and that this conclusion is obviously not true. As we all know, the original data contains a certain level of noise, which may affect the analysis results. Therefore, we performed wavelet decomposition of the original data to filter out the noise.

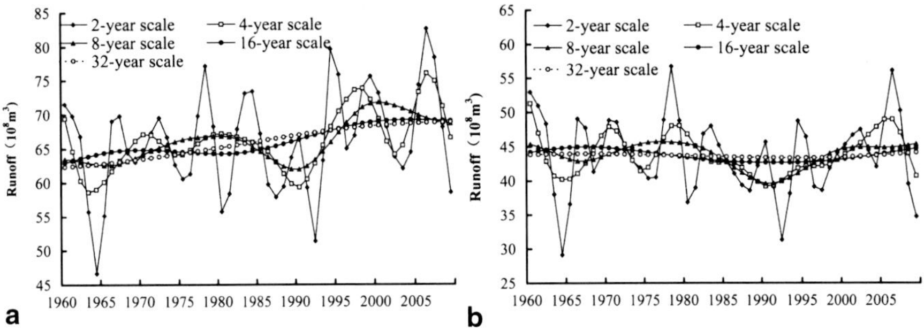

Fig. 4.7 The change trend of the streamflow of the Yarkand River and Hotan River under different time scales

Wavelet decomposition on different time scales was carried out for the runoff sequence of the Yarkand River using the Symmelt as basis wavelet and sym8 as the wavelet function. The results are shown in Fig. 4.7a.

As can be seen from the figure, the curve on the 2-year scale retains a greater amount of residual information compared with the original data. It was slightly flattened but fluctuated severely. An important message can be drawn from the curve: namely, that an increasing trend exists in the annual runoff of the Yarkand River in spite of the strong shock of ups and downs. The curve on the 4-year scale also retains substantial residual information and large fluctuations of four peaks and valleys. Further along, it is smoother compared to the 2-year scale curve, and the hidden increasing trend is more obvious. The curve on the 8-year scale also contains much less residual information and only one fluctuation compared with the 4-year scale curve. Overall, it is much smoother compared to the previous curve and the hidden increasing trend is decidedly more obvious. The curve on the 16-year scale contains only scant residual information and is smooth with a clear increasing trend. The curve on the 32-year scale basically contains no residual information and the increasing trend is obvious. The results of wavelet decomposition for the precipitation and temperature series are similar to those for runoff, but are not narrated here.

In order to reveal the relationship between runoff and climate factors from multi-time scales, further regression analyses based on wavelet denoising series should be performed. It can be seen that regression equations based on time scales can pass the 0.01 level significance tests, and that a 2 year scale denoising series can pass the 0.05 level significance test (Table 4.1). That runoff and precipitation still show a weak negative correlation in the 2-year scale denoising series is likely due to the 2-year scale wavelet decomposition having removed some noise but still retaining a substantial amount of residual information.

From the regression equations, we can see that annual runoff, precipitation and average annual temperature have positive correlations on the 4-, 8-, 16- and 32-year time scale wavelet compositions. With increasing time scales, the relationship becomes stronger. However, the relationship between runoff and temperature is stronger than the relationship between runoff and precipitation. A possible reason

Table 4.1 The regression equations of streamflow and precipitation, temperature under multi-time scales

Time scale	Regression equation	R2	F	*p*
2-year scale	Q=−0.05P+3.66T+56.81	0.1548	4.3038	0.0192
4-year scale	Q=0.08P+3.42T+48.25	0.2810	9.1833	0.0004
8-year scale	Q=0.11P+5.37T+39.41	0.8554	139.0479	0.0000
16-year scale	Q=0.12P+3.92T+43.21	0.9851	1,554.003	0.0000
32-year scale	Q=0.12P+4.63T+40.68	0.9992	29,132.820	0.0000

for this may be that snow (ice) runoff is the dominant water source in the Yarkand River. This further indicates that the changes in streamflow are the result of regional climate changes affecting Yarkand River runoff.

Hotan River In order to explore the quantitative correlation of runoff, precipitation and temperature in the Hotan River, scatter plots were drawn between annual runoff (Q) and annual precipitation (P), and between the annual runoff (Q) and annual average temperature (T). A statistical fitting method was carried out to establish a linear equation and a statistical significance test was also performed. Linear fitting results show that there is a negative correlation between runoff and precipitation and a positive relationship between runoff and temperature. The equations are as follows:

$$Q = -0.0161P + 44.556$$
$$Q = 0.9431T + 32.105$$

The multiple correlation coefficients (R2) of runoff and precipitation and runoff and temperature are 0.0019 and 0.0065, respectively. From these formulations, we can see that the correlations of runoff and precipitation and of runoff and temperature did not pass the significance test of 0.05.

In order to further investigate the relationship between runoff and climatic factors, we established a regression equation in which the annual precipitation and average temperature are the independent variables and the outlet flow of the Hotan River is the dependent variable:

$$Q = -0.0100P + 0.8724T + 33.4289$$

Here, the multiple correlation coefficient (R2) of the regression equation is 0.0072. The F test, under a confidence level of 99 %, shows that the F value of the regression equation is 0.1703 and the p value is 0.8439 ($p > 0.05$). Hence, the regression equation did not pass the significance test of 0.05. These unconvincing results are similar to those for the Yarkand River. Hence, we conducted wavelet composition of the original data to filter the noise. Figure 4.7b shows the wavelet composition results.

As we can see from the figure, the curve on the 2-year scale retains a greater amount of residual information compared with the original data. Although slightly smoothed, it nonetheless fluctuates severely. The curve on the 4-year scale also contains a large amount of residual information and sizeable fluctuations of three peaks and valleys. Further along, it smoothes out somewhat compared to the 2-year

Table 4.2 The regression equations of streamflow and precipitation, temperature under multi-time scales of the Hotan River

Time scale	Regression equation	R2	F	p
2-year scale	Q=−0.0379P+0.7683T+35.9875	0.0241	0.5806	0.5635
4-year scale	Q=−0.0235P+1.5077T+26.0975	0.1105	2.9204	0.0638
8-year scale	Q=−0.2071P+1.0705T+39.4678	0.6309	40.1626	0.0000
16-year scale	Q=0.3023P−2.8646T+65.8657	0.1316	3.5618	0.0363
32-year scale	Q=0.8362P−6.4011T+86.3073	0.9782	1054.61	0.0000

scale curves and the hidden increasing trend is also more obvious. From this, we can surmise that a decreasing trend in runoff is emerging but that fluctuations are still notable. The curve on the 8-year scale contains much less residual information and only one fluctuation compared with the 4-year scale curve. As well, it is much smoother compared to the previous curves and the hidden increasing trend is much more obvious. The curve on the 16-year scale contains only a small amount of residual information and is smooth with a clear increasing trend. The curve on the 32-year scale essentially contains no residual information and a weak decreasing trend can be seen. The results of wavelet decomposition of the precipitation and temperature series are similar to that of runoff and are not narrated here.

The regression equations for the 2-, 4-, and 16-year time scales do not pass the 0.05 level significance test; however, the equations for the 8- and 32-year time scales do pass the 0.01 level significance test (Table 4.2). Moreover, although runoff and precipitation still show a weak negative correlation in the 2-, 4-, and 8-year time scales, this is due to wavelet decomposition on these time scales removing some noise. Despite this, they still contain a large amount of residual information. Judging from the 16- and 32-year time scales, the annual runoff and precipitation was positively correlated, whereas runoff and temperature was negatively correlated. Hence, from every aspect, the relationship between runoff and temperature is stronger than that between runoff and precipitation. A possible reason for this may be that snow (ice) runoff contributes significantly to the Hotan River's streamflow.

Judging from the positive and negative correlations of runoff, precipitation and temperature, the causes underlying changes in the Hotan River runoff are highly complex. Nonetheless, from our findings, we can assume that these changes arise as a result of the regional impacts of climate change, which urgently needs to be studied on a larger scale.

Kaidu River In order to explore the quantitative correlation of runoff, precipitation, and temperature in the Kaidu River, scatter plots were drawn between the annual runoff (Q) and annual precipitation (P), and between annual runoff (Q) and annual average temperature (T) (Fig. 4.6d). A statistical fitting method was carried out to establish a linear equation, and a statistical significance test was also performed. Linear fitting results show a positive correlation between runoff and precipitation, and between runoff and temperature. The equations are as follows:

$$Q = 0.0946P + 12.531$$
$$Q = 3.8192T + 30.723$$

The multiple correlation coefficients (R2) of runoff and precipitation and runoff and temperature are 0.5179 and 0.2826, respectively. From these results, we can see that the correlations of runoff and precipitation and runoff and temperature have both passed the significance test of 0.01, and that the relationship between runoff and precipitation is higher than that between runoff and temperature. The analysis results show that temperature change is the key factor inducing changes in runoff.

In order to further investigate the relationship between runoff and climatic factors, we established a regression equation in which the annual precipitation and average temperature are the independent variables and the mountain outlet flow is the dependent variable:

$$Q = 0.0807P + 2.4485T + 13.0459$$

Here, the multiple correlation coefficient (R2) of the regression equation is 0.6229. The F test, under a confidence level of 99 %, shows that the F value of the regression equation is 38.815 and that the p value is 0.0000 ($p<0.01$). The regression equation has therefore passed the 0.01 significance level of statistical tests, which indicates that the changes in streamflow are the result of regional climate change.

Shule River In order to quantitatively investigate the correlation of runoff and precipitation and runoff and temperature of the Shule River, scatter plots were drawn between the annual runoff (Q) and annual precipitation (P), and between annual runoff (Q) and annual average temperature (T) (Fig. 4.6e). A statistical fitting method was carried out to establish a linear equation, and a statistical significance test was also performed. Linear fitting results show that there is a positive correlation between runoff and precipitation, and between runoff and temperature. The equations are as follows:

$$Q = 0.0286P + 1.1765$$
$$Q = 1.9465T + 14.648$$

The multiple correlation coefficients (R2) of runoff and precipitation and of runoff and temperature are 0.3866 and 0.3398, respectively. Hence, the correlation of runoff and precipitation and of runoff and temperature both passed the significance test of 0.01. The analysis results show that precipitation and temperature change are the key factors prompting runoff changes.

In order to dig further into the relationship between runoff and climatic factors, we established a regression equation in which the annual precipitation and average temperature are the independent variables and the mountain outlet flow is the dependent variable:

$$Q = 0.0217P + 1.3737T + 6.7796$$

The multiple correlation coefficient (R2) of the regression equation is 0.5335. The F test, under a confidence level of 99 %, shows that the F value of the regression equation is 26.879 and that the p value is 0.0000 ($p<0.01$). Hence, the regression equation has passed the 0.01 significance level of statistical tests, which indicates that changes in the Shule River's streamflow are caused by regional climate change.

Heihe River In order to quantitatively investigate the correlation of runoff and precipitation and runoff and temperature in the Heihe River, scatter plots were drawn between annual runoff (Q) and annual precipitation (P), and between annual runoff (Q) and annual average temperature (T) (Fig. 4.6f). A statistical fitting method was carried out to establish a linear equation, and a statistical significance test was also performed. Linear fitting results show that there are positive relationships between runoff and precipitation, and between runoff and temperature. The equations are as follows:

$$Q = 0.043P - 1.796$$
$$Q = 1.3237T + 17.071$$

The multiple correlation coefficients (R2) of runoff and precipitation and of runoff and temperature are 0.7142 and 0.0865, respectively. From this, we can see that the correlations of runoff and precipitation and of runoff and temperature passed the significance test of 0.01. Our analysis results reveal that temperature change is the key factor leading to changes in Heihe River runoff.

We further established a regression equation in which the annual precipitation and average temperature are the independent variables and mountain outlet flow is the dependent variable. Our intent is to reveal the relationship between runoff and climatic factors, as follows:

$$Q = 0.0422P + 0.2558T - 1.2170$$

Here, the multiple correlation coefficient (R2) of the regression equation is 0.7172. The F test, under a confidence level of 99 %, shows that the F value of the regression equation is 59.599 and that the p value is 0.0000 ($p < 0.01$). Therefore, the regression equation has passed the 0.01 significance level required of statistical tests, which indicates that changes in the Heihe River's streamflow can be explained by regional climate change.

Shiyang River In order to quantitatively study the correlation of runoff and precipitation and of runoff and temperature in the Shiyang River, scatter plots were drawn between annual runoff (Q) and annual precipitation (P), and between annual runoff (Q) and annual average temperature (T) (not shown). A statistical fitting method was carried out to establish a linear equation, and a statistical significance test was also performed. Linear fitting results show that there is a positive correlation between runoff and precipitation, and a negative correlation between runoff and temperature. The equations are as follows:

$$Q = 0.0062P + 0.454$$
$$Q = -0.1549T + 2.7997$$

The multiple correlation coefficients (R2) of runoff and precipitation and of runoff and temperature are 0.3778 and 0.0496, respectively. Furthermore, although the correlation of runoff and precipitation passed the significance test of 0.01, that of runoff and temperature did not pass the significance test of 0.05. The analysis results show that precipitation change is thus the key factor impacting runoff changes in the Shiyang River.

In order to further investigate the relationship between runoff and climatic factors, we established a regression equation in which the annual precipitation and average temperature are the independent variables and the mountain outlet flow is the dependent variable:

$$Q = 0.0063P - 0.1713T + 0.9921$$

The multiple correlation coefficient (R2) of the regression equation is 0.4384. The F test shows that the F value of the regression equation is 18.344 and that the p value is 0.0000 ($p<0.01$), which is below the confidence level of 99 %. Nonetheless, the regression equation has passed the 0.01 significance level demanded of statistical tests, which indicates that changes in the Shiyang's streamflow result from regional climate change.

As presented in Table 4.3, precipitation has a strong positive impact on the runoff of the Aksu and Kaidu Rivers but a considerably negative impact on the runoff of the Hotan River. In Yarkand River runoff, temperature plays a major role. When the impact of temperature is controlled, precipitation still exerts a strong positive impact on the runoff of the Aksu and Kaidu Rivers, and a considerable negative impact on Hotan River runoff, but its impact on the Yarkand runoff is only slight. When the impact of precipitation is controlled, the partial correlation between temperature and runoff in the Yarkand is weakened, and there is no apparent impact of temperature on the runoff of the other three rivers.

In terms of baseflow, temperature has a strongly positive impact on all four rivers. However, when the impact of temperature is controlled, precipitation has only a minor positive impact on the baseflows of the four rivers. When the impact of precipitation is controlled, the partial correlation between the temperature and baseflows of the Aksu River is weakened, but the partial correlations between temperature and baseflows of the Hotan, Yarkand and Kaidu Rivers are strengthened. Thus, while temperature exerts a strong positive impact on the baseflows of the Hotan, Yarkand and Kaidu Rivers, it has only a minor positive impact on the Aksu River's baseflow.

4.2.1.3 Conclusion

1. Affected by changes in precipitation and air temperature, runoff at stations around the Tianshan Mountains has significantly increased, especially in stations along the south slope of the Tianshan Mountains and in the NSET and Shule Rivers. Few rivers exhibit decreasing trends and only the Shiyang River showed significant negative trends. Step changes were detected in some stations with regards to runoff, but these changes were generally non-uniform in the arid sub-regions.

Table 4.3 Partial correlation analysis between flow variables (runoff, baseflow, and BFI) and climate factors (temperature and precipitation) for the four rivers

Item	Correlation and partial correlation	Aksu River	Hotan River	Yarkand River	Kaidu River
Runoff	Correlation P	0.410**	−0.393**	−0.147	0.515***
	Correlation T	0.23	−0.079	0.276*	0.255
	Partial correlation P	0.369**	−0.398**	None	0.468***
	Partial correlation T	None	None	0.266	None
Baseflow	Correlation P	0.252	0.144	0.221	0.099
	Correlation T	0.650**	0.680**	0.145*	0.572**
	Partial correlation P	0.275	0.259	0.269	0.045
	Partial correlation T	0.656	0.698***	0.212*	0.567***

*, ** and *** indicate the significance at the levels of $P<0.05$, $P<0.01$, $P<0.001$ (2-tailed), respectively

2. The rivers can be divided into three categories according to climate change: namely, the precipitation-temperature deterministic type (Yarkand River, Hotan River, Kaidu River, Heihe River), the precipitation deterministic type (Shule River, Shiyang River) and the temperature deterministic type (Aksu River). The data indicates that runoff in the Aksu River is mainly due to snow and ice melting; mountain runoff in the Yarkand, Hotan, Kaidu and Heihe Rivers is caused by precipitation and snowmelt due to temperature fluctuations; and the glacier on the Shule and Shiyang Rivers' upper reaches is limited but more sensitive to precipitation.
3. Huge regional differences exist in the correlation of runoff and precipitation and runoff and temperature in the above seven basins. The differences may be related to the recharge types of the streamflow. The correlation of runoff and precipitation is higher in the rivers where the proportion of glacier-melt runoff is low. However, in glacial-melt dominant rivers, the correlation of runoff and temperature is higher. The correlation of runoff and precipitation and of runoff and temperature is high in the rivers which are fed equally by glacier-melt and precipitation. This deep-seated mechanism needs further study.

4.2.2 Relationship Between Runoff and Snow Distribution

4.2.2.1 Data Source and Process

Six scenes were required to cover the Tarim River Basin. These are: h23v04, h23v05, h24v04, h24v05, h25v04 and h25v05. We selected 18,972 MYD10A1 images from July 4, 2002, to March 17, 2011, but discarded 106 as invalid data due to loss of sensor signals. We also selected 23,802 MOD10A1 data from February 24, 2000, to March 17, 2011, ignoring 438 unavailable data. The formats of both data are EOS-HDF (3 levels). The meanings of the classification codes in the MODIS product are shown in Table 4.4:

Table 4.4 Terrain classification coding in MODIS

Code	Represented features	Code	Represented features
0	No data	50	Cloud
1	Unclassified	100	Glacial lake
11	Night, polar data	200	Snow
25	No clouds	254	Invalid data
37	Lake	255	No data
39	Sea		

Table 4.5 Terrain re-classification coding in MODIS

Classification code	Re-classification code
100, 200	200 (Snow)
25, 37, 39	101 (land)
50	50 (cloud)
0, 1, 3, 4, 11, 254, 255	19 (invalid data)

As with other sensors, snow in MODIS is easily confused with clouds. This article uses MOD10A1 and MYD10A1 as the basic data source. Firstly, by using time lags between Aqua and Terra and by combining MOD10A1 and MYD10A1, we obtain a MOYD product that reduces the influence of the clouds. Then, by using daily data generated by a synthetic MOYD (N) product, we can effectively remove cloud affects and retention time resolution to identify the most appropriate time resolution for each basin.

In this article, the MOYD (N) based on the synthesis of MOD10A1 and MYD10A1 data have re-classified the classification criteria as follows (Table 4.5):

This section calculated the raster coded by 200 synthetic-based MOYD (N) products. The quality of the MOYDN snow products is an improvement over the Aqua (MYD10A1) and Terra (MOD10A1) products provided by NASA. The quality of data from Terra is higher than for Aqua with regards to snow identification, and the average recognition efficiency is about 0.5 % (and may even reach 1.5 % in April and October). The recognition efficiency of synthesized MOYD data is higher than Aqua (MYD10A1) and Terra (MOD10A1), with the lowest efficiency in August (0.4 and 0.9 %, respectively) and the highest in January (4.5 and 4 %, respectively). This indicates that the MOYD product can effectively improve recognition efficiency.

The climate factors related to snow are different in the southern slope of the Tianshans than in the northern slope of the Kunluns. In comparing the above three products, the results showed that, in the southern slopes of the Tianshan Mountains, the recognition efficiency for MOYD-products improved 2 and 5 % over Terra and Aqua products, respectively. Additionally, the Terra product is about 3 % higher than the Aqua product. In the northern slopes of the Kunlun Mountains, the Terra product has higher recognition efficiency than the Aqua one. For example, in April, Terra's percentage of difference in snow is 17 %, while for Aqua the percentage difference is 11 %, although both of them are less than the MOYD, which is 21 %. These differences indicate that MOYD is more valuable to use for discriminating snow. Nevertheless, even though the MOYD data improved the recognition rate of snow, the trend of the overall curve has not changed. Hence, we selected MOD10A1 for the analysis below (Fig. 4.8).

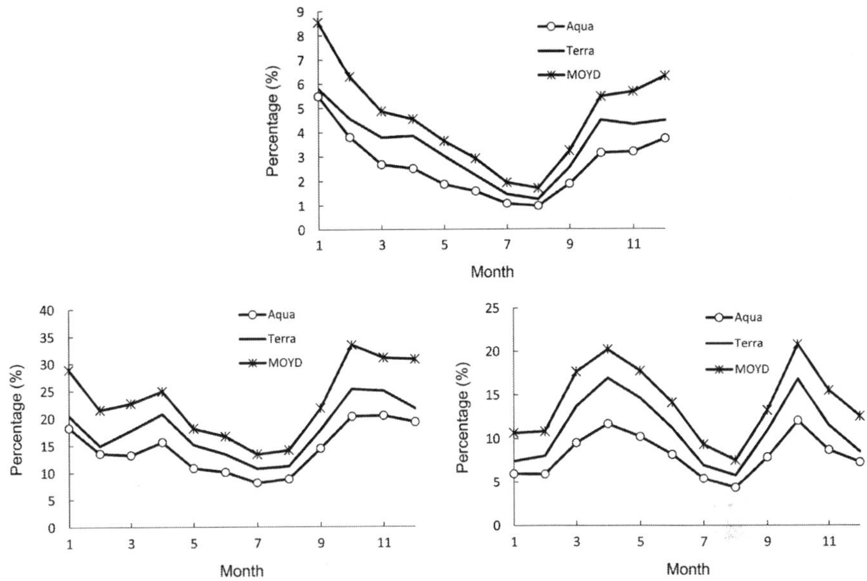

Fig. 4.8 Snow percentage in Tarim River Baisn, south slope of Tianshan and north slope of Kunlun Mountains

4.2.2.2 The Relationship Between Snow Cover, Runoff, and Precipitation

(1) The Relationship Between Snow and Runoff This section analyzes the trend of snow cover using MOD10A1 data, watershed runoff, and precipitation data. Taking into account the interference of clouds on the MOD10A1 data, the original data is synthesized as 10 days, with a 10-day synthesis maximum value to reflect the trends.

In Fig. 4.9, we can see that precipitation slightly precedes increased runoff during the flood season. This lag indicates either that runoff formed from precipitation requires some buffer time, or that precipitation reaches a certain level before it can produce significant effects on runoff. There is a strong correlation between reduction of snow cover and runoff in the flood season, which indicates that high temperatures lead to decreases in snow areas, which ultimately increases the runoff.

(2) The Relationship Between Runoff and Snow Areas Above the Snow Line Figure 4.10 illustrates that snow areas above the snow line in the summer flood season did not significantly decrease, but that from October to the following January, the snow area was significantly reduced. This indicates that the summer snow line was defined before it was reasonable do so, as snow cover showed no significant reduction in summer. However, because the snow area was decreased by October, we presume that snow contributed to runoff during the dry season.

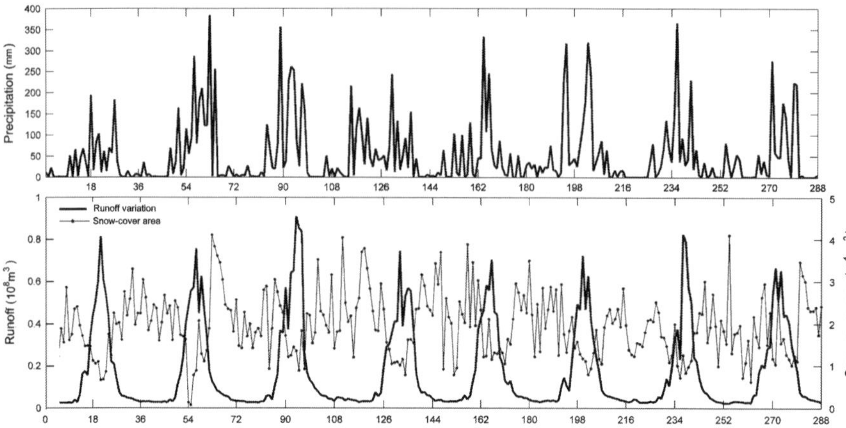

Fig. 4.9 Relationship between precipitation, snow cover and runoff in Aksu River

Even though mountain snow cover and precipitation time series from 2000 to 2007 is highly cyclical, a steady trend is not obvious. At the same time, the similarity between mountain snow cover and surface precipitation data is evident.

4.2.3 Response of Runoff to Atmospheric 0 °C Level Height in Summer

4.2.3.1 Study Area, Data, and Methods

(1) Study Area The arid region of Northwest China is sensitive to climate change, as the ecosystem is extremely weak and water resources originate mainly from mountain glacier-snow meltwater and precipitation. In addition, corresponding fluctuations in glacial movement and snow distribution are promoted by climate change, which has a greater influence on hydrological processes in this region (Shi et al. 2007; Wang et al. 2010). The 0 °C level (FL, also known as “freezing layer”) is an important characteristic in the lower layer of the high atmosphere that can reflect the temperature conditions of the mid-lower troposphere. In the troposphere, the temperature declines with rising height. When the surface temperature rises to above 0 °C, the FL will appear at a high atmosphere, of which, in meteorology, the corresponding altitude is the 0 °C level height (FLH). The rise or decline of the FLH will influence the fluctuation of mountainous surface 0 °C line to a certain extent, thereby influencing the glacier-snow meltwater processes on high mountains (Zhang et al. 2010). The glacial ablation areas in the Northwest are generally located at higher altitudes, usually in the mid-lower atmosphere, where melting mainly occurs in summer and most river runoff is directly related to glacial meltwater (Yang and Zeng 2001). Therefore, the rise and decline of summertime FLH will inevitably lead to changes in mountain runoff.

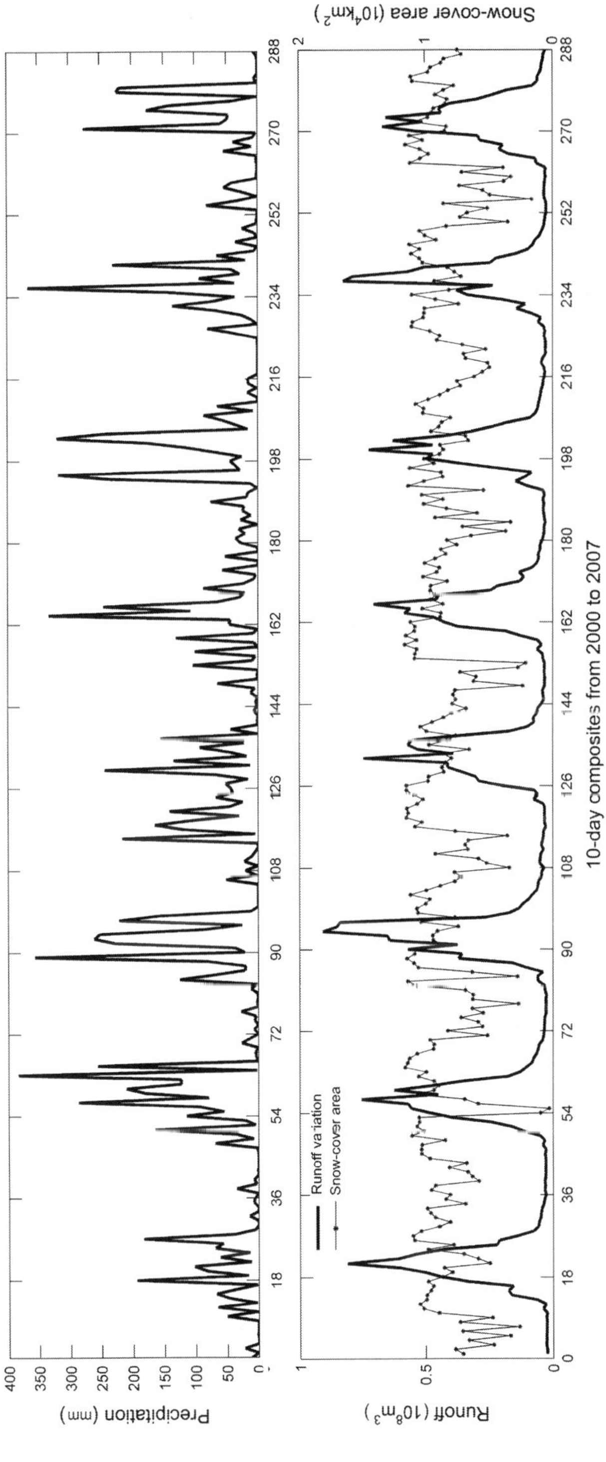

Fig. 4.10 Time series of precipitation, snow cover (above 4,200 m) and runoff in Aksu River

(2) Data Based on the topography and spatial distribution of existing data in China's arid Northwest, four typical regions were selected for the research, i.e., the northern slope of the Kunlun Mountains (NSKM), the southern slope of the Tianshan Mountains (SSTM), the northern slopes of the Tianshan Mountains (NSTM), and the northern slopes of the Qilian Mountains (NSQM) (Fig. 4.1). Summer FLH can be calculated based on daily sounding observations of eight sounding stations provided by the China Meteorological Administration from June to August within the 1960–2009 timeframe. The data shows that the summer FL is basically between two standard barospheres of 500 and 850 hPa. Based on these daily heights, temperatures in the three standard barospheres of 850, 700, and 500 hPa at 00:00 and 12:00 were selected as basic data. To ensure data quality, before calculating the FLH, the extreme decision method was used to amend any gross errors caused by operational errors in sounding observations. Additionally, a two-phase regression method was applied to homogenize the sounding heights and temperature sequences of the eight sounding stations, using NCEP reanalysis data as reference sequences. The intention here was to eliminate or reduce errors in the observing system.

The FLH was calculated by the following steps. First, the upper and lower standard barospheres were determined based on the locations of daily FL at 00:00 and 12:00, respectively. Then, a linear interpolation method (Bradley et al. 2009; Zhang and Guo 2011) was used to calculate the FLH each time, assuming that the temperatures between the two standard barospheres of 500 and 850 hPa have uniform changes in the vertical direction). Subsequently, the mean of two times was calculated to obtain a daily FLH. Finally, daily values from June to August were averaged to obtain the summer FLH. The linear interpolation method was used to calculate the FLH with the following equation:

$$H_i = \frac{H_j - H_k}{T_j - T_k}(T_i - T_k) + H_k \tag{4.6}$$

where variable H represents the height (m), T represents the temperature (°C), subscript i is the identifier of FL, and j and k are the identifiers of upper and lower standard barospheres of FL, respectively.

Summer runoff was sourced from the daily measured runoff data at the mountain hydrological stations of nine representative rivers from June to August during the period 1960–2009, which was less affected by human activities and thus able to represent the actual river runoff. The distribution of sounding stations, rivers, and hydrological stations is shown in Fig. 4.11.

(3) Methods Abrupt Change Point Analysis The MK non-parametric test is widely applied for determining the occurrence of abrupt change points of meteorological and hydrologic series. The method's advantages include the simplicity of its calculation, the confirmation of the start-time of the changes, and the identification of the area where the changes occur (Wei 2007). Let x1, …, xn be the data points. For each element xi, the numbers ri of elements xj precede it (j<i), such that xj<xi

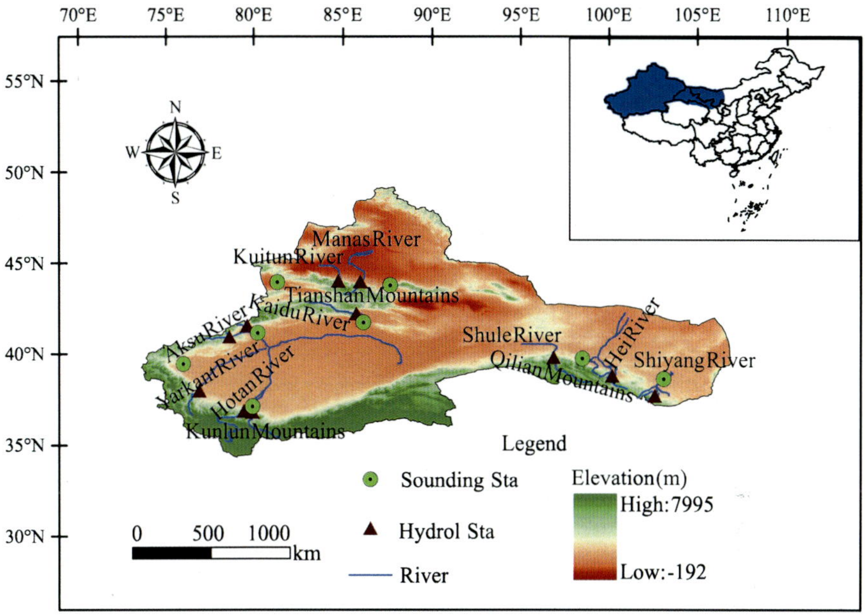

Fig. 4.11 The distribution of meteorological and hydrological stations in the ARNC

are computed. Under the null hypothesis (no abrupt change point), the normally distributed statistic S_k can be calculated via the following formula:

$$S_k = \sum_{i=1}^{k} r_i \quad (2 \le k \le n) \tag{4.7}$$

The mean and variance of the normally distributed statistic S_k can be given by the following formulas:

$$\bar{S}_k = E(S_k) = k(k-1)/4 \tag{4.8}$$

$$\mathrm{var}(S_k) = k(k-1)(2k+5)/72 \tag{4.9}$$

The normalized variable statistic UF_k is estimated as follows:

$$UF_k = \left(S_k - \bar{S}_k\right) \Big/ \sqrt{\mathrm{var}\left(S_k\right)} \tag{4.10}$$

The normalized variable statistic UF_k is the forward sequence, and the backward sequence UBk is calculated using the same equation but with a reversed series of data. When the null hypothesis is rejected (i.e., if any of the points in the forward sequence are outside the confidence interval), the detection of an increasing ($UF_k > 0$) or a decreasing ($UF_k < 0$) trend is indicated. The sequential version of the test used here enables the detection of the approximate time of occurrence of the trend by locating the intersection of the forward and backward curves of the test statistic. If the intersection occurs within the confidence interval, it indicates an abrupt change point.

Non-parametric Approach to Estimate the Climate Elasticity In relevant research on climate change's impacts on hydrological system to, elastic analysis of hydrological elements is often used to explore the sensitivity of the elements (Zheng et al. 2009; Ma et al. 2010). A vast number of climate elastic analysis methods exist, but the one based on a non-parametric method is best suited to our intent (Sankarasubramanian et al. 2001) to quickly and easily assess the influences of climate change on hydrological elements. In this study, the non-parametric approach to estimate the climate elasticity proposed by Zheng et al. (2009) was used for quantitative analysis on the sensitivity of asetival runoff to changes in summer FLH. Based on this method, the elasticity of runoff to changes in meteorological elements is defined as follows:

$$\varepsilon = \frac{\overline{X}}{\overline{Q}} \frac{\sum (X_i - \overline{X})(Q_i - \overline{Q})}{\sum (X_i - \overline{X})^2} \tag{4.11}$$

Where ε is the elastic coefficient, Q_i is the runoff, X_i is the value of the meteorological element, and $\overline{Q}$ and $\overline{X}$ are mean values of runoff and meteorological element series, respectively. The physical significance that relies on the meteorological element changed by 1% can induce the runoff to change by ε%. The greater the elastic coefficient, the more sensitive the runoff will be to changes in meteorological elements.

4.2.3.2 Changes in Summer FLH

In four typical regions of China's arid Northwest region, an upward trend of the summer FLH was shown on the southern slope of the Tianshan Mountains (SSTM), the northern slopes of the Tianshan Mountains (NSTM) and the northern slopes of the Qilian Mountains (NSQM) over the past 50 years, whereas a downward trend was shown on the northern slope of the Kunlun Mountains (NSKM) (Fig. 4.12). In these four typical regions, MK non-parametric test results (Table 4.6) revealed that the summer FLH on the NSTM and NSQM tended to increase significantly, with a rising rate of 2.05 and 2.28 m every year, respectively. Although an upward trend was shown on the SSTM, the 5% significance level was not passed. A significant decline was shown on the NSKM with a rate of 2.33 m every year. The significance of the changing trends of summer FLH was successively weakened on the NSKM ($|Z|=2.49$), the Tianshan Mountains ($|Z|=2.43$), and the NSQM ($|Z|=2.01$). In four typical regions, abrupt change points were related to changes

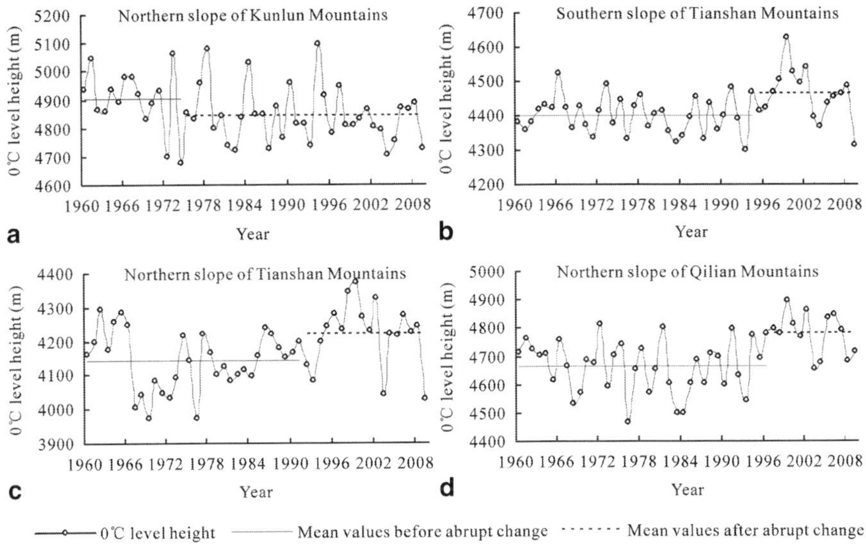

Fig. 4.12 Changes of the summer FLH for many years in four typical regions

Table 4.6 The test results of change trends and abrupt changes of the summer FLH in four typical regions

Typical regions	Trend test			Abrupt change points
	Trend rate (m/a)	*Z* values	Significance level	
NSKM	−2.33	−2.49	0.05	1974*
SSTM	1.21	1.91	–	1994*
NSTM	2.05	2.39	0.05	1991*
NSQM	2.28	2.01	0.05	1996*

* represents an abrupt change occurring at the 5% significance level

in summer FLH, but at different times (Table 4.6), i.e., in 1974 for the NSKM, which was significantly earlier than the SSTM (1994), NSTM (1991) and NSQM (1996). By comparing the time-interval prior and posterior to abrupt change points, the differences in mean values of the FLH are −77, 64, 78 and 243 m on the NSKM, SSTM, NSTM and NSQM, respectively, with change magnitudes increasing successively (Fig. 4.12).

Extremely complex geographical factors influence the above regional differences regarding changes in summer FLH. Firstly, since the changes can reflect circulation variations of cold and warm air masses high in the atmosphere, circulation characteristics unique to various regions are bound to greatly influence the FLH. Secondly, elevation is also an important factor, and different elevation locations correspond to different change degrees in FLH. Additionally, the underlying surface also plays a role in FLH changes.

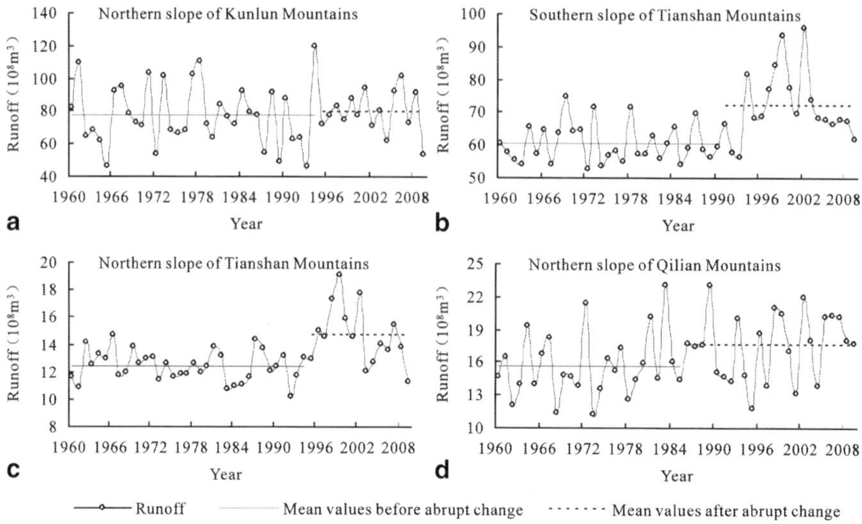

Fig. 4.13 Summer runoff changes for many years in four typical regions

Table 4.7 The test results of change trends and abrupt changes of the summer runoff in four typical regions

Typical regions	Trend test			Abrupt change points
	Trend rate (10^8 m^3/10a)	*Z* values	Significance level	
NSKM	0.12	0.28	–	1994
SSTM	3.33	3.36	0.01	1990*
NSTM	0.47	2.26	0.05	1994*
NSQM	0.76	2.29	0.05	1995*

* represents an abrupt change occurring at the 5 % significance level

4.2.3.3 Changes in Summer Runoff

Figure 4.13 shows the change sequence of summer runoff in four typical regions of the arid northwestern zones. Upward trends occur in summer runoff on the NSKM, the SSTM, NSTM, and NSQM over the past 50 years, of which the change trends of summer runoff are largely consistent on the SSTM and NSTM, with a high coincidence degree of curve change trends. According to the MK non-parametric test results (Table 4.7), all values of Z are greater than zero, indicating that the summer runoffs have generally increased. In this regard, noteworthy upward trends exist in the summer runoff on the SSTM, NSTM, and NSQK, with rising rates of 3.33×10^8, 0.47×10^8, and 0.76×10^8 m^3 every 10 years, respectively. However, summer runoff on the NSKM shows an insignificant upward trend, with a slight rise of 0.12×10^8 m^3 every 10 years.

Meanwhile, Z values can also reflect regional differences in summer runoff changes. The change trend of summer runoff in the Tianshan Mountains is the most significant, followed by NSQM and NSKM. According to MK non-parametric test results on abrupt changes (Table 4.7), abrupt changes in summer runoff occurred in 1990 (SSTM), 1994 (NSTM) and 1985 (NSQM). In comparing the time-intervals both before and after the abrupt change points, we find that the mean value of summer runoff increased by 11.76×10^8, 2.30×10^8 and 1.90×10^8 m^3, respectively (Fig. 4.13b–d). Although no abrupt change points occurred on the NSKM, a slight increase trend exists for summer runoff since 1994 (Fig. 4.13). Thus, 1994 can be considered the turning point for changes in summer runoff in that area (Chen et al. 2013).

Except for NSKM, the summer FLH in the other three typical areas share similar change trends to mountain runoff, indicating that a certain correlation exists between summer FLH and mountain runoff. The reduction in summer FLH on the NSKM is conducive to the accumulation of glacial material, resulting in a reduction in glacial meltwater runoff. As a result, in the warm and humid Northwest climate, insignificant changes occur in the summer mountain runoff on the NSKM, where glacial meltwater runoff accounts for 56.26 % (weighted by the proportions of glacial meltwater runoff of runoff components of the Hotan River and the Yarkand River). In contrast, with gradual increases in summer FLH in the Tianshan Mountains and NSQM, the melting of glaciers is accelerated, glacial meltwater runoff is increased, and summer mountain runoff shows a strong upward trend. It is noteworthy that the summer FLH and mountain runoff in these four typical sub-regions do not have exactly the same abrupt change points, which may be related to local natural conditions (such as topography and atmospheric circulation) and various social activities (such as the mode and intensity of human activities). However, abrupt change points for the summer FLH and mountain runoff are not the same in one typical region, indicating that changes in summer mountain runoff are not affected by a single factor, such as FLH, but result from the combined effects of multiple factors.

4.2.3.4 Changes in Summer Runoff in Response to Changes in Summer FLH

① Linear Fitting Between Summer Runoff and FLH (Chen et al. 2013) Through a comparative analysis, a better correlation can be found between summer runoff and changes in FLH in our study area. For a quantitative study on correlation, a scatter diagram was drawn on the summer runoff (Q) and FLH (H) in the four typical regions delineated above (Fig. 4.14). By using a statistical fitting method, linear equations were established with a statistical significance test on correlation.

The linear fitting between the summer runoff and FLH on the NSKM (Fig. 4.14) shows a significant positive correlation between them. The linear fitting equation is shown as follows:

$$Q_1 = 0.116H_1 - 483.459$$

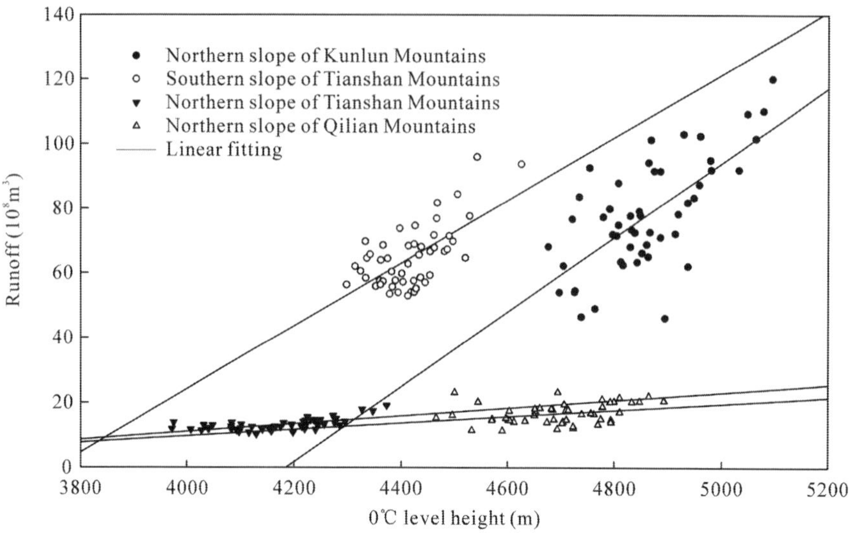

Fig. 4.14 The linear regression between the summer runoff and FLH in four typical regions

The linear fitting correlation coefficient R is 0.6862. Using the F-test, a correlation significance test was carried out on 50 samples of the summer runoff and the FLH on the NSKM. Confidence was gauged at 99 %, and $P = 0.0001 < 0.01$. The 1 % significance level of statistical test was passed, indicating that a very significant linear correlation exists between summer runoff and FLH.

As viewed from the linear fitting between the summer runoff and FLH on the SSTM and NSTM (Fig. 4.14), a positive correlation was also proven to exist. The linear fitting equations of the SSTM and NSTM are shown as follows:

$$\text{SSTM}: \quad Q_2 = 0.097H_2 - 365.502$$
$$\text{NSTM}: \quad Q_3 = 0.012H_3 - 36.754$$

The linear fitting correlation coefficients R are 0.6510 and 0.6354 on the SSTM and NSTM, respectively. Similarly, the F-test was used for a correlation significance test on 50 samples of the summer runoff and FLH. Confidence was measured at 99 %, and $P = 0.0001 < 0.01$ for the SSTM and NSTM. Both passed the 1 % significance level, indicating a very significant linear correlation between summer runoff and FLH in the Tianshan Mountains.

According to the linear fitting between the summer runoff and FLH on the NSQM (Fig. 4.14), a significant positive correlation was also found between them. The linear fitting equation is shown as follows:

$$Q_4 = 0.010H_4 - 29.656$$

Table 4.8 The statistics of rivers runoffcomponentsin four typical regions

Typical regions	Representative rivers	Runoff components[a] (%)		
		Glacial meltwater	Precipitation	Groundwater
NSKM	Yarkand River and Hotan River	56.26	21.08	22.66
SSTM	Aksu River and Kaidu River	33.71	38.37	27.92
NSTM	Manas River and Kuitun River	31.80	36.06	32.14
NSQM	Heihe River, Shule River and Shiyang River	15.13	35.70	49.17

[a] Runoff components for each of the four typical regions were calculated using the runoff components of the representative rives

The linear fitting correlation coefficient R is 0.3174. The F-test was used as a correlation significance test on 50 samples of summer runoff and FLH on the NSQM. The confidence percentage was 99 %, and $P = 0.0247 < 0.05$. The 5 % significance level was passed, indicating a significant linear correlation between summer runoff and FLH.

As we can see from the above, the correlation coefficients between summer runoff and FLH are reduced successively in the four typical sub-regions, which may be related to the type of runoff recharge. Moreover, it can also be seen that the recharge proportion from the glacial meltwater runoff of river runoff is reduced successively (Table 4.8), consistent with a weakening trend of the correlation between summer runoff and FLH. However, the recharge proportion from groundwater is increased successively, contrary to a weakening trend of the correlation between summer runoff and FLH. In other words, the greater the recharge proportion from glacial meltwater runoff and the smaller the recharge proportion from groundwater of river runoff, the higher the correlation between the summer runoff and FLH will be. In terms of significance, the recharge proportion from glacial meltwater runoff is only 15.13 % on the NSQM, which is less than that on the NSKM (56.26 %), SSTM (33.71 %) and NSTM (31.80 %). However, the recharge proportions from precipitation and groundwater are 35.70 and 49.17 %, respectively. The correlation between the summer runoff and FLH is less significant on the NSKM, SSTM, and NSTM. In summary, it can be concluded that regional differences in summer runoff in response to changes in summer FLH may, to a certain extent, be caused by river runoff components.

② **Sensitivity of Summer Runoff to Changes in FLH** The elastic coefficient can be used to characterize the sensitivity of river runoff to climate change. According to the equation, the elastic coefficients of the summer runoff to changes in summer FLH were calculated in four typical regions of China's arid Northwest. The purpose was to facilitate the sensitivity analysis between summer runoff and FLH. It can be seen that the elastic coefficients of the summer runoff to changes in summer FLH on the NSKM, SSTM, NSTM, and NSQM were 7.19, 6.63, 3.80, and 2.79, respectively (Fig. 4.15; Chen et al. 2013). In other words, a 1 % change in summer

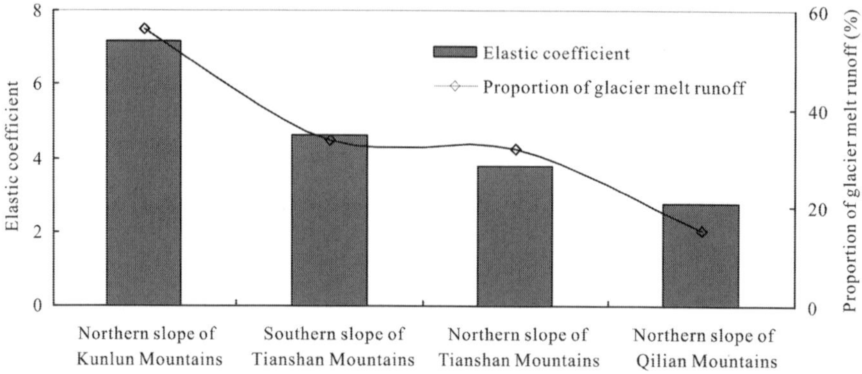

Fig. 4.15 The elastic coefficient and proportion of glacial meltwater runoff in four typical regions

FLH in these areas can induce their summer runoff to change by 7.19, 6.63, 3.80, and 2.79 %, respectively. It is noteworthy that although the elastic coefficient of the summer runoff to changes in summer FLH on the NSKM was 7.19 (ranking it first among the four areas), the summer FLH there showed a significant declining trend in 1960–2009, with an average height of 4,861 m. This was slightly higher than the minimum snow line altitude of 4,790 m in this zone. Accordingly, the temperature around the glaciers declined, the glacier melt slowed down, material accumulation increased, and glacial meltwater runoff was reduced, resulting in a slight increasing trend in summer runoff due to the effects of humidity.

By comparing these four zones, we found that the sensitivity of summer runoff to changes in summer FLH shows a successive weakening trend from south to north and from west to east. The differences in sensitivity may be related to the recharge proportions from glacial meltwater runoff. To verify this argument, correlation analysis is required of the elastic coefficients and recharge proportions from glacial meltwater runoff. Figure 4.15 shows the relationship among these elements. As can be seen, a high consistency exists, and the linear fitting equation is: $\varepsilon = 0.109G + 0.863$, where ε is the elastic coefficient, and G is the recharge proportion from glacial meltwater runoff. Hence, R = 0.9808. In this test, the 1 % significance level was passed, indicating a significant positive correlation between the sensitivity of summer runoff to changes in summer FLH and the recharge proportion from glacial meltwater runoff in the area under study, all of which confirms the validity of the foregoing argument.

In China's Northwest, the recharge proportion from glacial meltwater runoff in inland river basins is closely related to the distribution quantity and coverage of mountain glaciers in the upper reaches. The greater the quantity and coverage of the glaciers, the larger the recharge proportion from glacial meltwater runoff will be (Kang et al. 2002). In representative river basins on the NSKM, SSTM, NSTM, and NSQM, the area, reserves, and coverage of glaciers are successively reduced (Table 4.9); notable differences also exist in the contribution rates of glacial meltwater

Table 4.9 The distribution of glaciers covering four typical regions in ARNC

Typical regions	Area of river basins (km²)	Glaciers	Glacial area (km²)	Glacial reserves (km³)	Glacier coverage (%)
NSKM	129,100	6,472	10,652.29	1,190.81	8.25
SSTM	84,300	1,837	2,886.54	460.24	3.42
NSTM	51,000	1,105	808.65	44.47	1.59
NSQM	210,000	1,208	784.27	38.78	0.37

The data of glaciers, glacial area, and glacial volume were from the reference (Shi et al. 2005), whereas the area of river basins was from the reference (Chen 2010); the computing method of glacier coverage could be seen in detail from the reference (Shi et al. 2005)

runoff to river runoff. Combining the foregoing arguments regarding significant positive correlations between the sensitivity of summer runoff to changes in summer FLH and the recharge proportion from glacial meltwater runoff, we can see that different scales of glaciers are the key reason for regional differences in the sensitivity of summer runoff to changes in summer FLH (Chen et al. 2013).

③ **Response Mechanism of Summer Runoff to Changes in Summer FLH** The changes of mountain runoff in the area under investigation result from combined effects of multiple factors, but different factors have different contribution degrees to runoff changes. To explore this issue, under the premise of considering only the related indicators of meteorological elements, and after standardized processing of the summer FLH (H), surface temperature (T), precipitation (P), evaporation (E) and runoff (Q) data, multiple regression was carried out on H, T, P, E and Q. The intent was to obtain the regression equations on the NSKM, SSTM, NSTM and NSQM respectively, as follows:

$$Q_1 = 0.531H_1 + 0.400T_1 + 0.109P_1 + 0.014E_1 - 1.507e-06 \quad (\mathrm{F} = 18.562,\ \mathrm{N} = 50)$$

$$Q_2 = 0.511H_2 + 0.242T_2 + 0.416P_2 + 0.053E_2 - 7.415e-07 \quad (\mathrm{F} = 16.864,\ \mathrm{N} = 50)$$

$$Q_3 = 0.487H_3 + 0.247T_3 + 0.482P_3 + 0.043E_3 - 1.088e-06 \quad (\mathrm{F} = 16.195,\ \mathrm{N} = 50)$$

$$Q_4 = 0.820H_4 - 0.396T_4 + 0.580P_4 - 0.107E_4 - 9.438e-06 \quad (\mathrm{F} = 11.324,\ \mathrm{N} = 50)$$

It can be seen from the F value that the above four regression equations all passed the 1 % significance level of statistical testing. Based on these equations, it is clear that the FLH characterizing temperature change conditions in the mid-lower troposphere make the greatest contribution to changes in runoff. This indicates that the rise and decline of summer FLH has become a critical factor influencing changes in river runoff.

At the top of the Kunlun, Tianshan and Qilian Mountains, rain and snow are abundant, and a natural "solid reservoir" is formed by glaciers and permanent snow. This reservoir plays an important role in water resource composition and river runoff regulation. Within a global context, the mountain glaciers are exceedingly sensitive to climate change (Kang et al. 2002). With changes in near-surface temperature, mean summer FLHs on the SSTM, NSTM and NSQM over the past 50 years increased, respectively, from 4,462, 4,219, and 4,777 m before the abrupt change point to 4,398, 4,141, and 4,534 m after the abrupt change point. However, snow line heights were in the 3,850–4,390, 3,650–3,850, and 4,400–4,800 m range, respectively (Shi et al. 2005). As a consequence, the temperature around the glaciers rose and the glacial area under FL increased, resulting in accelerated glacial melt and retreat. Additionally, material accumulation was reduced and glacial meltwater runoff generally increased, as did river runoff. However, mean summer FLH on the NSKM was reduced from 4,916 m before the abrupt change point to 4,839 m after the abrupt change point, and the snow line height was around 4,790–6,260 m (Shi et al. 2005). In this area, temperatures around the glaciers actually declined, the glacial zone under the FL decreased, glacial melt slowed down, liquid precipitation was frozen, and water resources were reserved as snow, glaciers, and so on. Moreover, material accumulation increased, and the recharge from glacial meltwater runoff to river runoff was reduced. As a result, no significant changes occurred to summer mountain runoff in the two rivers on the NSKM, with the recharge proportion from glacial meltwater runoff being more than 50 %. Wang et al. (2008) used the functional model on glacial system change to analyze the sensitivity of China's glacial system to climate change and considered that the glaciers in the Kunlun Mountains belonged to a stable type, whereas those of the Tianshan and Qilian Mountains belonged to a sensitive type. These findings confirm the assumed response mechanism of summer runoff to changes in summer FLH to be reasonable and credible from a lateral perspective.

4.2.3.5 Conclusions and Discussion

Through analysis of the time series change trends of summer mountain runoff and atmospheric 0 °C level height and their correlations in four typical areas located in China's arid Northwest region during the 1960–2009 period, the main conclusions of this section are as follows:

1. In the past 50 years, a significant upward trend in summer FLH occurred on the NSTM and NSQM. Although an upward trend also occurred on the AATM, the 5 % significance level was not passed there, and a significant decline was shown on the NSKM at a rate of 2.33 m every year. In the four areas under study, summer runoff showed increasing trends. Specifically, the SSTM, NSTM and NSQM revealed strongly increasing trends, while the NSKM exhibited a slight increasing trend, at a rate of 0.12×10^8 m^3 every 10 years.
2. A strong positive correlation existed between summer runoff and FLH on the NSKM, SSTM, NSTM, and NSQM, with the elastic coefficients of summer runoff to changes in FLH pegged at 7.19, 6.63, 3.80, and 2.79, respectively. The correlation coefficient and elastic coefficient were successively reduced. Different recharge proportions from glacier meltwater runoff to rivers are a direct cause of

regional differences in correlation coefficient and sensitivity, whereas different glacial scales between regions are a root cause of such regional differences.

3. The rise and decline of summertime FLH has become a critical factor influencing changes in river runoff. Due to rises in summer FLH, temperatures around mountain glaciers will rise and glacial areas under FL will increase, resulting in accelerated glacial melt, an increase in material accumulation, and a decrease in glacial meltwater runoff. Conversely, declines in summer FLH will cause the temperature around glaciers in high mountains to decline and the glacial area under FL to reduce, resulting in mitigated glacial melt, increased material accumulation, and a decrease in glacial meltwater runoff, all of which influence changes in mountain runoff.

4.3 Quantification of Climate and Human Factors

4.3.1 Study Area, Data and Methods

The study area includes seven typical inland rivers (the Aksu, Yarkand, Hotan, Kaidu, Shiyang, Shule and Heihe Rivers), using annual runoff sequences from 1960 to 2009.

Time sequence mutation detection and identification Due to the impacts of human activities on the different strengths of the basin, the hydrological time series will show stage or trend changes. To eliminate impacts of climate change (such as precipitation and temperature) on runoff, the runoff series was divided into natural and human activity-impacted periods to quantitatively depict the stages of the hydrological changes caused by human activities and climate change.

Step jump is a statistical method to measure hydrological response to climate change and human activities. At present, the Pettitt method (1979) is widely used in the detection and identification of the step jump point of hydrological time series. It is grounded in non-parametric testing and has the advantage of simple calculation, clear change time, clear physical meaning, and the enhanced ability to identify the step change point. The Pettitt method is mainly based on the Mann–Whitney statistical functions $U_{T,N}$ and assumes that the two samples x_1, … x_t and x_{t+1}, … X_n are from the same sequence distribution. For continuous sequence, $U_{T,N}$ is calculated by the following equation:

$$U_{t,N} = U_{t-1,N} + \sum_{j=1}^{N} \mathrm{sgn}(x_t - x_j) \quad (t = 2, \ldots, N) \tag{4.12}$$

where:

$$\mathrm{sgn}(\theta) = \begin{cases} +1, & \theta > 0 \\ 0, & \theta = 0 \\ -1, & \theta < 0 \end{cases} \tag{4.13}$$

The test statistic calculated that the times of the first sample frequency exceeded those of the second sequence sample frequency. The null hypothesis of the Pettitt method has no step jump point. The statistic k(t) and probability (*P*) are calculated as follows:

$$k(t) = Max_{1 \le t \le N} \left| U_{t,N} \right| \tag{4.14}$$

$$P \cong 2\exp\left\{-6\left(K_N\right)^2 / \left(N^3 + N^2\right)\right\} \tag{4.15}$$

The step jump point can be used as the cut-off point to distinguish the natural period from the human activities period.

Method to Distinguish Climate Change and Human Activities If we consider the measured runoff in the natural period as a reference value, the measured runoff impact by human activities then consists of two parts: the human activities impact and the climate change impact. The effects of human activity and climate change on runoff are divided as follows:

$$\Delta Q_T = Q_{HR} - Q_B \tag{4.16}$$

$$\Delta Q_H = Q_{HR} - Q_{HN} \tag{4.17}$$

$$\Delta Q_C = Q_{HN} - Q_B \tag{4.18}$$

$$C_H = \Delta Q_H / \Delta Q_T \times 100\% \tag{4.19}$$

$$C_C = \Delta Q_C / \Delta Q_T \times 100\% \tag{4.20}$$

where ΔQ_T is the total change amount of runoff, ΔQ_H is the runoff impacted by human activities, ΔQ_C is the runoff impacted by climate change, Q_B is the runoff during the natural period, Q_{HR} is the measured runoff during the human activities period, Q_{HN} is the natural runoff during the human activities period, and C_H, C_C is the percentage of runoff impacted by human activities and climate.

4.3.2 The Judgment of the Natural and Human Activities Periods

The Pettitt method was used to test step changes for mountain outlet flow in seven rivers, with the results shown in Table 4.10. As we can see, runoff in the Aksu, Kaidu, Shule and Heihe Rivers experienced significant step changes ($P<0.05$). Specifically, a step change point was observed in 1993 in the Aksu, Yarkand and Kaidu, but

Table 4.10 Step changes of runoff time series

Rivers	Statistic *k(t)*	Step jump *t*	Probability*P*	Significance level *a*
Aksu River	422	1993	<0.0001	0.01
Yarkand River	198	1993	0.235	>0.05
Hotan River	91	1985	0.958	>0.05
Kaidu River	436	1993	<0.0001	0.01
Shule River	396	1997	<0.0001	0.01
Heihe River	312	1979	0.011	0.05
Shiyang River	167	1990	0.410	>0.05

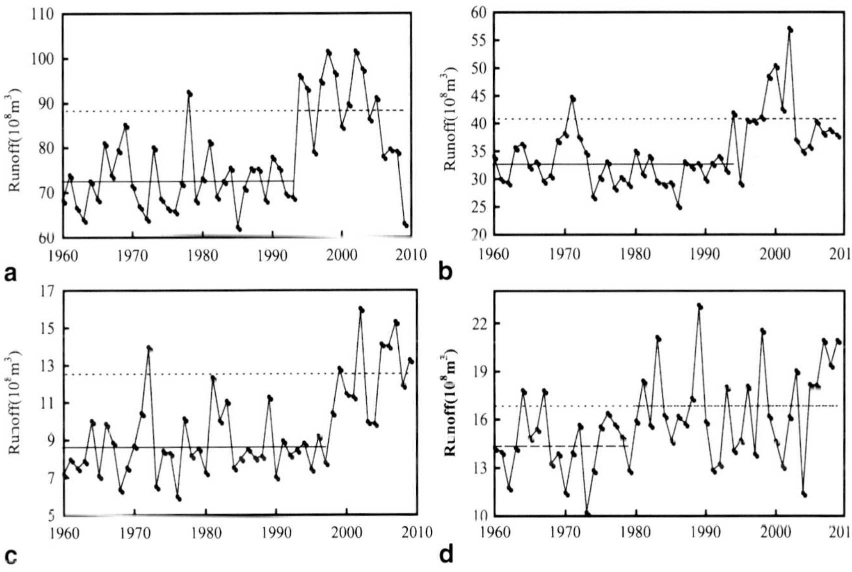

Fig. 4.16 Runoff before and after step changes in some typical rivers. **a** Aksu River. **b** Kaidu River. **c** Shule River. **d** Heihe River

the change in the Yarkand River was not significant. In the Shiyang River, runoff decreased in 1990 and then significant increased after 1997. Changes in the Heihe occurred the earliest (1979), followed by the Hotan (1985), although changes in runoff before and after the step change point were not significant for either river. Conversely, the runoff in the Aksu, Kaidu, Shule and Heihe underwent sizeable changes before and after the step change, increasing by 21.78, 26.48, 27.11 and 17.4 %, respectively (Fig. 4.16). From these data, we can see that step change years show major differences and are not consistent with the warm and wet phenomenon from west to east in the study area.

Due to the observed significant step jump changes in runoff in the Aksu, Kaidu, Shule and Heihe Rivers, we can see that these areas were strongly impacted by climate change and human activities. Based on the step change year, the natural change periods in the above rivers are, respectively, 1960–1993, 1960–1993, 1960–1997

Table 4.11 Runoff changes caused by climate change and human activities

River	Period	Obs (10^8 m^3)	Sim (10^8 m^3)	Variations (10^8 m^3)	Climate change		Human activities	
					Amount (10^8 m^3)	Percentage (%)	Amount (10^8 m^3)	Percentage (%)
Aksu	1960–1993	72.51	72.51					
River	1994–2009	88.3	74.02	15.79	1.51	9.6	14.28	90.4
Kaidu	1960–1993	32.36	32.36					
River	1994–2009	40.93	36.16	8.57	3.8	44.3	4.77	55.7
Shule	1960–1997	8.62	8.62					
River	1998–2009	12.55	10.07	3.93	1.45	36.9	2.48	63.1
Heihe	1960–1979	14.33	14.33					
River	1980–2009	16.83	14.86	2.5	0.53	21.2	1.97	78.8

and 1960–1979, while the human activity periods are 1994–2009, 1994–2009, 1998–2009 and 1980–200.

4.3.3 *The Effect of Climate Change and Human Activities on Runoff*

Multiple regression models were established in the Aksu, Kaidu, Shule and Heihe Rivers using hydrological and meteorological data before the step changes occurred. Models were used to predict the mean runoff in natural and change periods according to the longer time scale average state (Table 4.11). The results show that the measured runoff and simulated runoff are extremely consistent for the natural period, and so this regression model would be useful to calculate the impact of human activity on runoff. This section will consider the natural period as a reference to analyze the effects of human activities and climate change on runoff (Table 4.11).

Table 4.11 shows that: (1) measured runoff in the Aksu, Kaidu, Shule and Heihe Rivers have increased significantly; (2) the impacts of human activities and climate change vary in degrees from river to river due to differing scales of human activities and variations in differences in the magnitude of climate change; (3) the degree of human activities' influence on runoff is 90.4, 55.7, 63.1 and 78.8 %, respectively, in the Aksu, Kaidu, Shule and Heihe Rivers; and (4) human activities are the main cause of runoff changes in the above four rivers.

The term "human activities", as referred to in this section, is a relatively broad concept that includes activities that have a direct effect on water, such as the transformation of the basin's underlying surface, soil conservation and water conservancy. The term also encompasses activities that have an indirect effect on water, such as socio-economic structural adjustment.

This section considers general human activities and climate change as independent variables and quantitatively separates the influence of these. The results, however, are not accurate because the impacts caused by human activities and climate change are in reality not independent.

In the simulation of human activities and climate change impacts on runoff, numerous uncertainties exist. The first of these is the choice of a model, as researchers need to select a specific model according to their own data. The second uncertainty is data selection. As we know, meteorological data is only point data, and so the number of selected stations can cause different results. A third uncertainty is that basic geographical characteristics of runoff-generated effects are difficult to quantify. And finally, human activities and climate change are not independent variables, as was touched on above. The challenge, then, is constructing a reasonable model to accurately separate the ever-changing impacts and roles of human activities and climate change on runoff.

4.3.4 *Conclusions*

1. The Pettitt method was used to test step changes for mountain outlet flow in seven rivers, and step change points were observed in 1993 in the Aksu, Yarkand and Kaidu Rivers. The runoff in the Aksu, Kaidu, Shule and Heihe Rivers changed significantly before and after the step change, registering increases of 21.78, 26.48, 27.11 and 17.4 %, respectively. The step change year exhibits noteworthy differences, but these are not consistent with the warm and wet phenomenon from west to east in the study area.
2. The impacts of human activities and climate change occur to differing degrees in each of the seven rivers due to the different scale of human activities and variations in differences of magnitude of climate change. Human activities are the main cause of runoff changes in the Aksu, Kaidu, Shule and Heihe Rivers, with the degree of human activity influence on runoff measuring 90.4, 55.7, 63.1 and 78.8 %, respectively.

4.4 Sensitivity Analysis of Runoff to Climate Change

4.4.1 *Study Area, Data and Methods*

The sensitivity of natural runoff to climate change was calculated in the Aksu, Yarkand, Hotan, Kaidu, Shiyang, Heihe and Shule Rivers, which are considered typical inland rivers.

The runoff time series from 1957 to 2009 for these seven rivers were used. The time series of average temperature and precipitation for the rivers were also used to investigate the sensitivity of streamflow to climate change.

To assess the impact of climate change on runoff, we used the double mass curve method, the sensitive coefficient method and the elasticity theory formula to calculate the impact of precipitation changes on the Kaidu River's runoff.

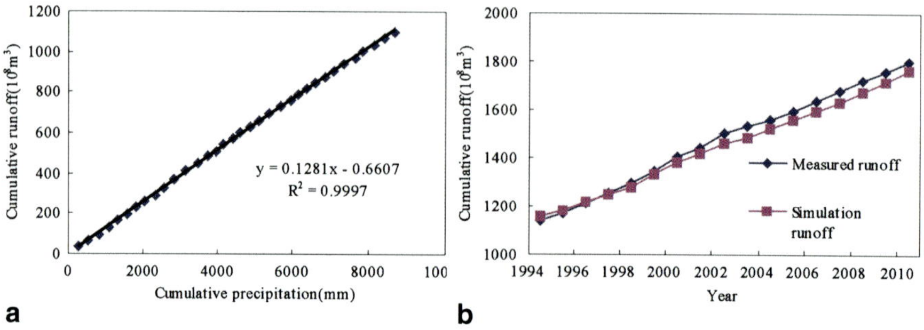

Fig. 4.17 **a** The relationship between cumulative precipitation and cumulative runoff, **b** the measured runoff and simulation runoff change during 1994–2010

According to previous studies (Chen et al. 2013), step change points of runoff occurred in 1993 and 1995, while step change point of precipitation occurred in 1993. We could thus divide the study timeframe into two periods—1960–1993 and 1994–2010—to analyze the impact of climate change on the Kaidu's runoff.

Figure 4.17a shows the relationship between cumulative precipitation and cumulative runoff for the first reference period (1960–1993). From this, we define the relationship as: $y=0.1281x-0.6607$ (y being cumulative runoff and x being the cumulative precipitation) with the coefficient of determination ($R^2=0.9997$), which indicates that the formula can be used to restore runoff changes in the second reference period (1994–2010). Figure 4.17b shows the measured runoff and simulation runoff change from 1994 to 2010. As we can see, the runoff simulation value is less than the measured values, indicating that other factors (i.e., temperature, human activities, etc.) are causing increases in surface runoff. In the 1994–2010 period, the average simulated runoff is 39×10^8 m^3, the measured average runoff is 41×10^8 m^3, and the simulation error is 4.9%. Meanwhile, according to the double cumulative curve formula, the contribution of precipitation changes to runoff increase is 76%.

We adopt the sensitivity analysis method proposed by Zheng et al. (2009) to explore the sensitivity of annual runoff to precipitation. This formula is as follows:

$$\Psi = \frac{\bar{X}}{\bar{Q}} \frac{\sum (X_i - \bar{X})(Q_i - \bar{Q})}{\sum (X_i - \bar{X})^2} \tag{4.21}$$

where X_i denotes the meteorological element, Q_i denotes the annual runoff, Ψ is the sensitivity coefficient, and $\bar{X}$ and $\bar{Q}$ are mean values of runoff and meteorological element over years, respectively. The results show that the sensitivity coefficient of annual runoff to precipitation is 0.84 and the contribution of precipitation changes to runoff increase is 55%.

Following the elasticity method, the formula is as follows:

$$\left(R_i - R_{1960}\right)/R_{1960} = \varepsilon\cdot\left(P_i - P_{1960}\right)/P_{1960} \tag{4.22}$$

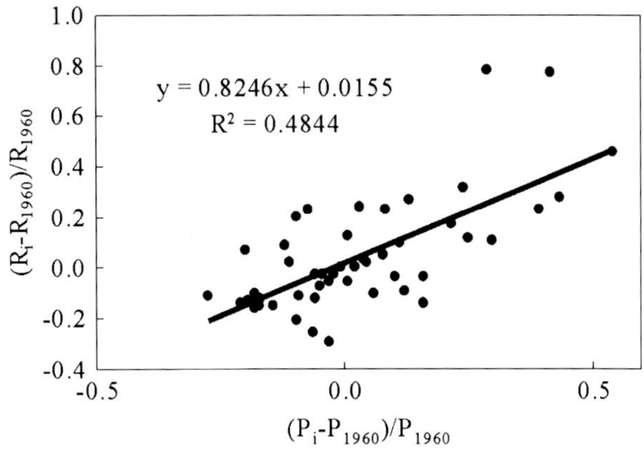

Fig. 4.18 Relationships between proportional changes of precipitation and runoff in Kaidu River

where ε is elasticity of runoff with respect to precipitation; R_i and P_i are yearly runoff and precipitation values, with $i = 1961, 1962, \ldots 2010$; and R_{1960} and P_{1960} are the runoff and precipitation in 1960, respectively. Therefore, ε can be regarded as the linear regression coefficient between $(T_i - T_{1960})/T_{1960}$ and $(C_i - C_{1960})/C_{1960}$.

To quantitatively identify runoff changes caused by precipitation changes in the past 50 years, the following formula is used to calculate the change rate of runoff:

$$\Phi = 100\% \times \left(\varepsilon \cdot \frac{(P_{1994-2010} - P_{1960-1993})}{P_{1960-1993}} \cdot R_{1960-1993} \right) \Big/ (R_{1994-2010} - R_{1960-1993}) \tag{4.23}$$

where $P_{1960-1993}$, $P_{1994-2010}$, $R_{1960-1993}$, and $R_{1994-2010}$ are the mean precipitation and run-off during the periods 1960–1993 and 1994–2010. Thus, Φ is the importance of the precipitation variation in run-off change. Using the above formula, we calculate 54 % as the importance of the precipitation variation in run-off change (Fig. 4.18).

The above analysis shows that the results of the elasticity methods and the sensitivity formula calculation are basically same. Nonetheless, Zheng et al. (2009) proposed the sensitivity calculation method and the theory of elasticity as being more suitable for the assessment of the impact of climate on runoff. Therefore, for the sake of simplicity, we use the sensitivity calculation method proposed by Zheng et al. (2009) to assess the effects of climate on runoff in the arid region of Northwest China.

The effects of climate change on hydrology and water resources systems are complex, but establishing a model to simulate the approximate prototype is still a reasonable and feasible way to investigate those impacts. Multivariate regressions among runoff depth (R), precipitation (P) and average temperature (T) were calculated (Fu and Liu 1991) and the control equation emerged as follows:

$$\log R = m \log P + n \log T + c_0$$

where P, T, R_i is average annual precipitation, average annual temperature, average annual runoff, respectively, and m, n, c is the regression coefficient.

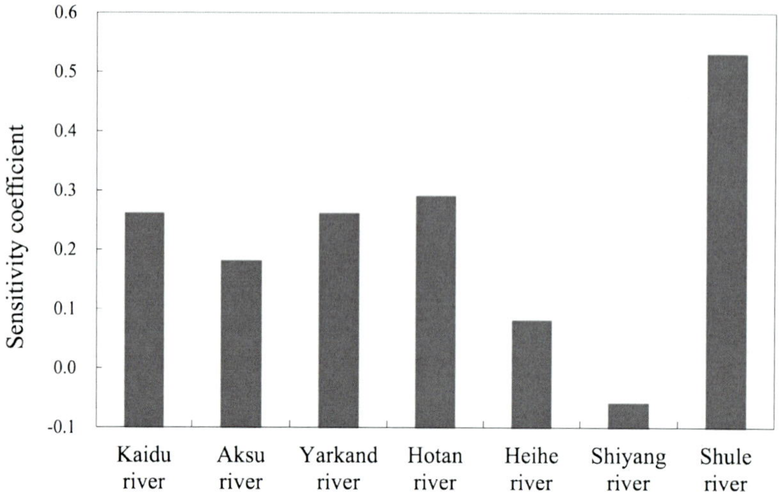

Fig. 4.19 Sensitivity coefficients of annual runoff to the temperature

We thus established the equations as follows for the headwater of the Tarim River:

Aksu River: $\text{LogR} = 0.29\log P + 0.214\log T + 2.254$
Yarkand River: $\text{LogR} = -0.0019\log P + 4.32\log T - 2.57$
Hotan River: $\text{LogR} = -0.00140.29\log P \pm 0.323\log T + 2.07$

The simulated flow fit well, with the measured flow based on the above equations. Moreover, the main peak and valley were in overall agreement with each other, and the correlation coefficients were 0.9, 0.86 and 0.75, respectively. We can see from the results of these three equations that runoff sensitivity to climate change has some credibility.

4.4.2 *Impact of Climate Change on Runoff of a Typical River*

Some results indicate that a step change occurred in the early 1990s for climate and runoff (Chen et al. 2013; Li et al. 2012, 2013). In this section, we divide the era under study into two periods (1960–1993 and 1994–2010) to assess the impact of climate change on the streamflow of different rivers.

4.4.2.1 The Sensitivity of Runoff to Temperature

Some studies suggest that a basic change occurred in the Northwest arid region's climate and runoff in the early 1990s (Chen et al. 2013). Thus, in order to facilitate comparison, the unified study period is divided into two stages of 1960–1993 and 1994–2010 to study the impact of climate change on changes in river runoff.

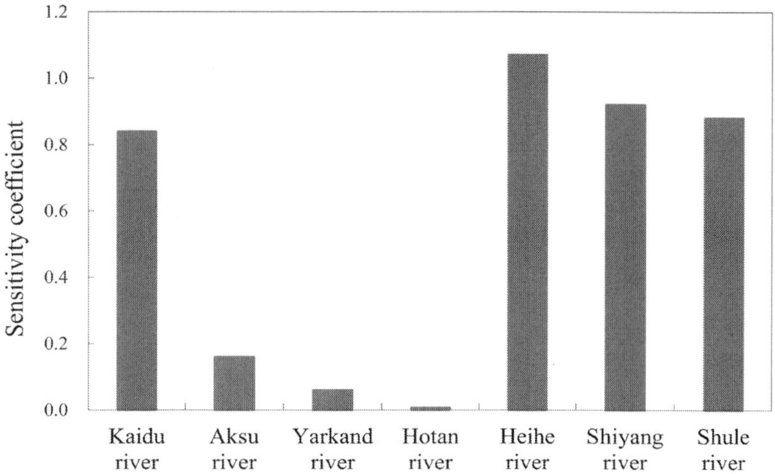

Fig. 4.20 Sensitivity coefficients of annual runoff to the precipitation and the change rates of annual runoff caused by the precipitation

The sensitivity of runoff to annual temperature in the Shule River and Tarim Basin is high (Fig. 4.19), with sensitive coefficients in the range of 0.18–0.53. However, sensitivity in the Heihe and Shiyang Rivers is very low, ranging between 0.08 and −0.06, respectively. From this, it can be seen that the sensitivity of the runoff of different rivers is related to annual air temperature. In addition, it is worth mentioning that the sensitivity of Shiyang River runoff to annual temperature is negative (−0.06), or, in other words, that a temperature rise of 1 % will reduce run-off by 0.06 %. This indicates that a temperature increase has a negative effect on run-off.

4.4.2.2 The Sensitivity of Runoff to Precipitation

The sensitivity coefficients of runoff to precipitation in the Heihe, Shiyang, Shule and Kaidu Rivers are high, ranging from 0.84 to 1.07. However, in the Hotan, Yarkand and Aksu Rivers, the coefficients are low, especially in the Hotan River, where runoff is 0.007 (Fig. 4.20). This indicates that if the sensitivity of river runoff to temperature is lower, the sensitivity of river runoff to precipitation will be higher, as in the Shiyang runoff.

4.4.2.3 Sensitivity Analysis of Runoff to Climate Change Based on Runoff Model

It is difficult to assess the effects of climate change on water resources due to the profound uncertainty of future climate change. An effective method to counter this uncertainty is to develop several possible climate change programs: assuming that future regional precipitation conditions have m possible change: P1, P2, … Pm, temperature status may change by: T1, T2, … Tn. Each possible position of precipitation and air temperature constitutes future regional climate scenarios to analyze differences in runoff characteristics under various programs (Fu and Liu 1991). This section assumes five temperature change statuses (+0, +0.5, +1, +2, and +3 °C) and five precipitation change statuses (−20 %, −10 %, +0 %, +10 %, +20 %), which can then be grouped into 25 climate scenarios. The sensitivity analysis results of the annual runoff are shown in Table 4.12.

The results showed that: (1) Runoff in the Aksu and Yarkand Rivers will increase with increased precipitation and temperature, and that runoff in the Hotan River will also increase with increased precipitation, but the rate of increase will be minimal. (2) Runoff response to temperature is more sensitive to precipitation changes, such that runoff in the Aksu and Yarkand Rivers will increase 1.4–7.0 %, respectively, with temperature increases of 1 °C; however, Hotan River runoff will be reduced by 2.9 % with the same temperature increase. (3) When there is a decrease in precipitation, the temperature compensation effect on runoff is more significant. (4) The strongest runoff reduction will occur when precipitation is decreased by 20 %.

4.4.3 Conclusion

1. Runoff is more sensitive to temperature in the Shule and Tarim Rivers, where sensitivity ranged between 0.18–0.53, while runoff in the Heihe and Shiyang Rivers exhibits a lower average annual temperature sensitivity of 0.08 and −0.06, respectively. Notable sensitivity coefficients were found in the Heihe, Shiyang, Shule and Kaidu Rivers, in values ranging from 0.84 to 1.07. However, sensitivity coefficients of runoff to precipitation for the Hotan, Yarkand and Aksu Rivers were very low, especially for the Hotan River.
2. Runoff in the Aksu and Yarkand Rivers will increase as precipitation and temperature increase, and runoff in the Hotan River will also increase with increased precipitation, but the rate of increase will be minimal. With changing temperatures, runoff is more sensitive to precipitation changes. So, for instance, runoff in the Aksu and Yarkand Rivers will increase 1.4–7.0 % with only minor temperature increases of 1 °C.

Summary

1. Based on temperature, precipitation and streamflow series, the causes underlying hydrological changes were discussed in the previous chapter. The results showed that: (1) the runoff in North Xinjiang (i.e., the Weigan, Heihe, and Shiyang Rivers) has a strong positive relationship with rainfall, while that in the south slope of the Tianshan Mountains, the middle section of the north slope of the Tianshan Mountains and the Shule River has a strong positive relationship with air temperature; and (2) there is a positive significant relation between summer runoff and 0 °C level height (FLH). Therefore, the rise and fall of the summer FLH is

Table 4.12 Sensitivity analysis of runoff to different climate senarioals in the headwater of Tarim River

Rain variation (%)	Temperature variation (℃)														
	Aksu River					Yarkand River					Hotan River				
	+0	+0.5	+1	+2	+3	+0	+0.5	+1	+2	+3	+0	+0.5	+1	+2	+3
−20	6.7	7.4	8.1	9.5	10.8	11.9	12.9	13.9	15.9	17.7	−5.4	−4.2	−3.0	−0.65	1.55
−10	3.1	3.8	4.5	5.9	7.1	8.2	9.3	10.2	12.1	13.8	−5.37	−0.413	−2.93	−0.63	1.6
0	0	0.7	1.4	2.8	3.9	0	6.4	7.0	8.8	10.5	0	−4.1	−2.9	−0.6	1.65
+10	−2.7	−2.1	−1.4	−0.1	1.1	−0.2	3.2	4.1	5.9	7.5	−5.34	−4.1	−2.9	−0.6	1.68
+20	−5.2	−4.5	−3.9	−2.6	−1.5	−2.3	0.7	1.6	3.3	4.9	−5.33	−4.09	−2.89	−0.59	1.69

a key factor inducing changes in summer runoff in arid regions of Northwest China.

2. The impacts of human activities and climate change on rivers vary according to the scale of the activities and variations in the magnitude of climate change. Human activities are presently the main driving forces behind runoff changes. In the Aksu, Kaidu, Shule and Heihe Rivers, the influence percentage of human activity on runoff is 90.4, 55.7, 63.1 and 78.8 %, respectively.
3. The sensitivity of natural runoff to climate change was investigated in several inland rivers. It was found that the sensitivity of river runoff to annual air temperature varies significantly. For instance, runoff in the Shule River and Tarim Basin showed extremely high sensitivity to temperature, while runoff sensitivity in the Heihe and Shiyang Rivers was very low with regards to temperature. For precipitation, the sensitivity coefficients of runoff to precipitation in the Heihe, Shiyang, Shule, and Kaidu Rivers are high, while in the Hotan, Yarkand and Aksu Rivers the same coefficients are low. We can assume from these findings that if a river's runoff has a low sensitivity to temperature, its sensitivity to precipitation will be higher. Runoff's sensitivity to climate change was also investigated. It was discovered that runoff in the Aksu and Yarkand Rivers is increasing with rising precipitation and temperature levels, and that runoff in the Hotan River is likewise increasing, but the rate of increase there is minimal. As runoff's response to temperature is more sensitive to precipitation changes, runoff in the Aksu and Yarkand Rivers will increase 1.4–7.0 % with every 1 °C rise in temperature.

References

Bradley RS, Keiming FT, Diaz HF (2009) Recent changes in freezing level heights in the Tropics with implications for the deglacierization of high mountain regions. Geophys Res Lett 36:L17701. doi:10.1029/2009GL037712

Chen X (2010) Physical geography of arid land in China. Science Press, Beijing, pp 226–289

Chen ZS, Chen YN, Li BF (2013) Quantifying the effects of climate variability and human activities on runoff for Kaidu River Basin in arid region of northwest China. Theor Appl Climatol 111:537–545

Fu G, Liu CM (1991) Estimation of regional water resource response to global warming—a case study of Wanquan Basin. Acta Geographica Sinica 58(3):277–288

Kang ES, Cheng GD, Dong ZC (2002) Glacier-snow water resources and mountain runoff in the arid area of northwest China. Science Press, Beijing, pp 97–124

Li BF, Chen YN, Chen ZS, Li WH (2012) Trends in runoff versus climate change in typical rivers in the arid region of northwest China. Quatern Int 282:87–95

Li BF, Chen YN, Shi X (2013) Temperature and precipitation changes in different environments in the arid region of northwest China. Theor Appl Climatol 112:589–596

Ma H, Yang D W, Gao B (2010) Impact of climate variability and human activity on streamflow decrease in the Miyun Reservoir catchment. J Hydrol 389:317–324

Qing DH, Ding YH, Su JL, Wang SM (2005) China climate and environmental change—the evolution and forecast of climate and environment. Science Press, Beijing, p 24, 121

Sankarasubramanian A, Vogel RM, Limburner JF (2001) Climate elasticity of streamflow in the United States. Water Resour Res 37:1771–1781

Shi YF, Liu CH, Wang ZT et al (2005) Concise glacier inventory of China. Shanghai Popular Science Press, Shanghai, pp 159–170

Shi Y F, Shen Y P, Kang E, Li D L, Ding Y J, Zhang G W, Hu R J (2007) Recent and future climate change in northwest china. Climatic Change 80:379–393

Wang SF, Kang SZ, Zhang L (2008) Modeling hydrological response to different land-use and climate change scenarios in the Zamu River basin of northwest China. Hydrol Process 22:2502–2510

Wang J, Li HY, Hao XH (2010) Responses of snowmelt runoff to climatic change in an inland river basin, northwestern China over the past 50 years. Hydrolog Earth Syst Sci 14:1979–1987

Wei FY (2007) Diagnose and forecast technology of modern climate statistics. China Meteorological Press, Beijing

Yang ZN, Zeng QZ (2001) Glacier hydrology. Chongqing Press, Chongqing

Zhang YS, Guo YJ (2011) Variability of atmospheric freezing-level height and its impact on the cryosphere in China. Ann Glaciol 52:81–88

Zheng HX, Zhang L, Zhu RR (2009) Responses of streamflow to climate and land surface change in the headwaters of the Yellow River Basin. Water Resour Res 45:W00A19

Zhang Q, Xu CY, Tao H (2010) Climate changes and their impacts on water resources in the arid regions: a case study of the Tarim River Basin, China. Stoch Env Res Risk Assess 24:349–358

Chapter 5
Glacier Change and Its Impact on Water Resources

Zhongqin Li, Meiping Sun and Puyu Wang

Abstract Glaciers play an extremely important role both on water resource and stabilization of river runoff in this northwest arid region of China. Urumqi Glacier No.1, located at the head water of Urumqi River is the monitored glacier with the longest continuous history of any glacier in China. The glacier has experienced remarkable shrinkage from 1962 to 2009. Its length has shortened by about 215.2 m (9.7 %) and its area has diminished by 0.304 km^2 (15.6 %). The cumulative mass balance of the glacier was − 13,693 mm, equivalent to 15.2 m of glacier ice, showing particular sensitivity to temperature change. Annual glacier melt runoff and discharge at gauging site have also exhibited increasing trends. The volume of glaciers in Xinjiang is about 23,611.2 × 10^8 m^3 water equivalent and rank first in the country. It is extremely sensitive of glaciers in Xinjiang to climate change. As temperatures rise, the melting of glaciers is accelerated and the meltwater runoff has increased to over 200 × 10^8 m^3, accounting for more than 25 % of the quantity of surface water 789 × 10^8 m^3. Due to serious melting, glacier area reduced by 11.7 % in the recent 30 years and the volume loss is more rapid. Changes of the spatial and temporal distribution of water resources and process of hydrologic cycle initiated by glacier changes will exercises a great influence on construction layout of Xinjiang and development patterns, which has attracted the common concern from the governments from central to local and people. It has been one of the important problems that needed to be solved

Z. Li (✉) · P. Wang
State Key Laboratory of Cryospheric Sciences/Tianshan Glaciological Station, Cold and Arid Regions Environmental and Engineering Research Institute, Chinese Academy of Sciences, 730000 Lanzhou, China

Cold and Arid Regions Environmental and Engineering Research Institute, Chinese Academy of Sciences, No. 320 Donggang Road, Lanzhou, China
e-mail: lizq@lzb.ac.cn

M. Sun
College of Geography and Environment Sciences, Northwest Normal University, 730070 Lanzhou, China

Northwest Normal University, No. 967 Anning Road, Lanzhou, China
e-mail: sunmeiping1985@163.com

P. Wang
e-mail: wangpuyu@lzb.ac.cn

Y. Chen (ed.), *Water Resources Research in Northwest China*,
DOI 10.1007/978-94-017-8017-9_5,

in the pace of development and construction of Xinjiang nowadays. For the Qilian Mountains, water stress and ecological problems are bottlenecks of social economic development. All of glaciers, seasonal snow and permafrost water resources account for the vast majority in the components of water resources. Among them, glacier is the most active factor because of its high sensitivity, river runoff regulate effect and feedback mechanisms of mountain climate. One thousand and seventy-eight glaciers in the middle of Qilian Mountains, including Heihe River Basin and Beidahe River Basin with the area of 420.55 km^2 are studied. Areas of these glaciers have reduced by 21.7 % in the past 50 years. It is estimated that glacier meltwater runoff account for over 10 % of river runoff and play an important role in the stabilization of river runoff, regulation of annual change and allocation.

Keywords Glacier change · Seasonal snow · Water resource · Hydrological response · Northwestern arid regions of China

Glaciers are important water resources sustaining social economic development and environment. The volume of glaciers in Xinjiang is about 23,611.2 × 10^8 m^3 water equivalent and rank first in the country, playing a vital role in the composition of water resources and regulation of river runoff. It is extremely sensitive of glaciers in Xinjiang to climate change. As temperatures rise, the melting of glaciers is accelerated and the meltwater runoff has increased to over 200 × 10^8 m^3, accounting for more than 25 % of the quantity of surface water 789 × 10^8 m^3. Due to serious melting, glacier area reduced by 11.7 % in the recent 30 years and the volume loss is more rapid (Li et al. 2010a). Changes of the spatial and temporal distribution of water resources and process of hydrologic cycle initiated by glacier changes will exercises a great influence on construction layout of Xinjiang and development patterns, which has attracted the common concern from the governments from central to local and people. It has been one of the important problems that needed to be solved in the pace of development and construction of Xinjiang nowadays. Similarly, in the Qilian Mountains, water stress and ecological problems are bottlenecks of social economic development. All of glaciers, seasonal snow and permafrost water resources account for the vast majority in the components of water resources. Among them, glacier is the most active factor because of its high sensitivity, river runoff regulate effect and feedback mechanisms of mountain climate. One thousand and seventy eight glaciers in the middle of Qilian Mountains, including Heihe River Basin and Beidahe River Basin with the area of 420.55 km^2 are studied in this chapter. Areas of these glaciers have reduced by 21.7 % in the past 50 years (Wang et al. 2011). It is estimated that glacier meltwater runoff account for over 10 % of river runoff and play an important role in the stabilization of river runoff, regulation of annual change and allocation.

This chapter concentrates on glacier changes in typical regions of Xinjiang and Middle Qilian Mountains and its impact on hydrology and water resources based remote sensing, geographical information systems and field survey.

5.1 Rapid Shrinkage and Hydrological Responses of Urumqi Glacier No.1 to Climate Change

Global warming has caused shrinkage of most glaciers and ice caps in the world over the last century, especially in recent decades (Dyurgerov and Meier 2000). Small glaciers are highly sensitive to changes in temperature and precipitation making them important indicators of climate change (Meier 1984; Oerlemans and Fortuin 1992; Dyurgerov 2002; IPCC 2007; Immerzeel et al. 2010). The presence of even a small amount of glacier cover in a catchment can influence streamflow variability on a range of time scales (He et al. 2010). Relationships between climate-induced glacier change and runoff yield have been the concern of several studies. e.g. Moore and Demuth (2001) related annual discharge from Place glacier in Canada to climatic variables. They found that the coefficients in a multiple regression equation varied through time, likely as a result of glacier changes resulting from climatic variability through the period of record. Braun and Escher-Vetter (1996) examined streamflow variability below a glacier in relation to change in mass balance. In that study, they discovered an upward trend in annual discharge from Vernagtferner, Bavaria, associated with a period of sustained negative net balance, during which the total glacier area decreased from 9.55 (1979) to 9.09 km^2 (1990). Aizen et al. (2007a) estimated the current glacier state and forecasted the potential impact of global and regional climate change on the glaciers and glacier runoff in the Tianshan. They attributed the decrease in glacier number, covered area, volume and glacier runoff to the drastically increase in temperature.

Since the beginning of the twentieth century, mountain glaciers in western China have experienced an overall retreat and thinning due to climate warming, and the corresponding hydrological regime has also exhibited remarkable changes especially over the past few decades (Yao et al. 2004; Liu et al. 2006). Glacier meltwater is an important water resource in the arid northwest China (Yang 1991; Stahl et al. 2008). Climate-induced water resource shortages has been posing an urgent problem in this arid and glacier meltwater supplied region (Shi and Zhang 1995; Shi and Liu 2000; Yao et al. 2004; Chen et al. 2006). Although a great deal of research has been focused on the hydrological effects of climate change, relatively little research has examined the effects on streamflow of interactions between climate variability and resulting glacier response. On the Urumqi Glacier No.1 catchment scale, Li et al. (2003b, 2010b) examined streamflow variability and found an increasing trend in annual discharge. Ye et al. (2005) studied the relationship among summer temperature, precipitation and glacier mass balance and runoff, they revealed that both summer precipitation and temperate were negatively correlated with mass balance and positively associated with runoff. Jing et al. (2006) investigated the glacier mass balance and recession, they concluded that Urumqi Glacier No.1 underwent accelerated recession and the speed of glacier flow has gradually declined since the 1980s.

Urumqi Glacier No.1, at the headwater of Urumqi River, eastern Tianshan, has the longest monitoring record of any glacier in China, covering the period

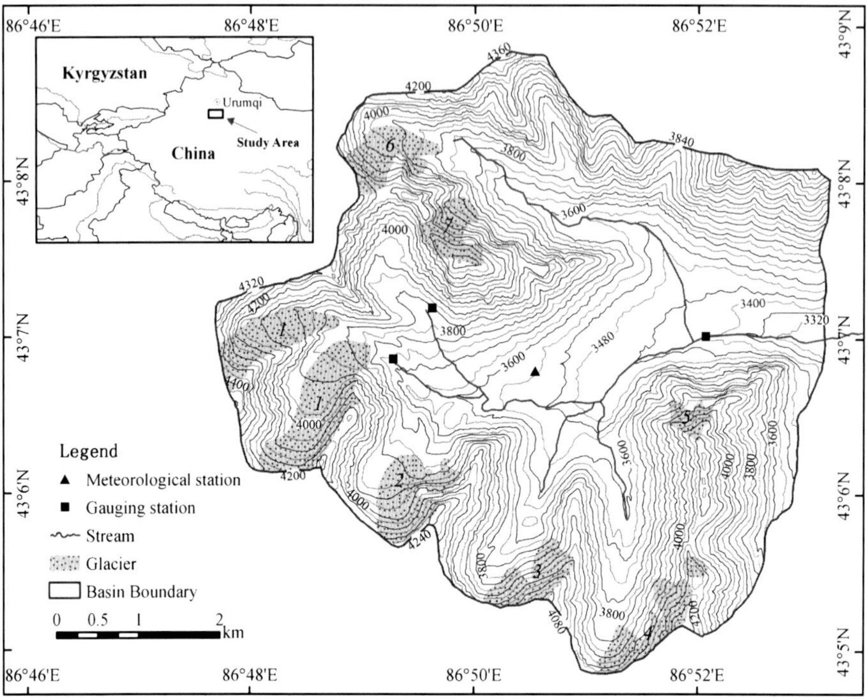

Fig. 5.1 Map of Urumqi River source region, showing locations of Urumqi Glacier No.1, hydrological and meteorological stations

1959–2008. This study presented analysis of the shrinkage history and runoff variations of the glacier in relation to recorded climatic variability over the past 50 years. We calculated the glacier hydrological balance, relating it to recent change in glacier mass loss. Additionally, the governing factor that affecting the change in discharge at glacierized catchment were explored, by comparing with an adjacent non-glacierized catchment. Finally we evaluated the possible runoff change at Glacier No.1 and Zongkong catchment in Urumqi River source region by the middle of the century (2041–2060) for different stages of glacier coverage, based on HBV hydrological model coupled with regionally downscaled climate change scenarios (RegCM3 regional model, IPCC SRES A1B scenario). This result would be useful for the utilization of water resources and development of good watershed management strategies in the arid region of northwest China.

5.1.1 Study Area

Urumqi Glacier No.1 is located at the headwater of Urumqi River, eastern Tianshan, China (43° 06′ N, 86° 49′ E) (Fig. 5.1). It is a northeast-facing valley glacier composed of east and west branches, which became separated into two small independent

Table 5.1 Main geographical and hydrometeorological features of the two catchments

	Glacier No.1	Empty cirque	Zongkong
Total area (km^2)	3.34	1.68	28.9
Glacierized area (km^2)	1.646	0	18.5
Elevation range (m a.s.l.)	3,740–4,486	3,805–4,301	3,405–4,486
Elevation of gauging station (m a.s.l.)	3,693	3,805	3,405
Mean annual discharge (10^4 m^3)	241.81	70.69	1,357.11
Mean annual temperature (°C)	−5.9	−6.3	−4.9
Mean annual precipitation (mm)	504	528	463

glaciers in 1993 due to continued glacier shrinkage (Li et al. 2003b; Ye et al. 2005). The glacier, an area of 1.646 km^2 and 2.23 km long in 2009, is the largest glacier of 7 glaciers distributed in the Urumqi river source region. Its altitude ranges from 3,756 to 4,486 m above sea level (a.s.l.) with an average elevation of 4,130 m a.s.l.

There are three gauging stations in the study area. Glacier No.1 gauging station was operated just below the glacier terminus since 1959. It covers a drainage area of 3.34 km^2, of which almost 50% was glacierized. In addition to this highly glacierized catchment, the Empty Cirque, a small non-glacierized but barren rocked catchment, was also hydro-analyzed in this study. Empty Cirque gauging station commencing in 1982 situated at the outlet of the basin. Zongkong gauging station, commencing in 1985, is situated at the outlet of the basin. Its drainage area is 28.9 km^2 of which 18.5% is glacierized. Its altitude ranges from 3,405 to 4,486 m a.s.l. This catchment represents a moderate glacieriezed area. In addition, Daxigou meteorological station is located at 3,539 m a.s.l., about 3 km downstream of Glacier No.1. It also has a long-term and excellent observed record. The main geographical and hydrometeorological characteristics of these two catchments are listed in Table 5.1.

5.1.2 Data Sets

The meteorological data during 1959–2008 have been collected at Daxigou meteorological station located at 3,539 m a.s.l., about 3 km downstream of the glacier (Fig. 5.1). Annual mean temperature and total precipitation data were used in this study to provide indices of regional climatic variability An advantage of using Daxigou data is that it has a long-term observed record; in addition, it can provide high-quality data for this research in view of it being national meteorological station.

The hydrological data were measured at Glacier No.1 and Empty Cirque gauging stations, which were established at 3,693 and 3,805 m a.s.l., respectively (Table 5.1). Unfortunately, the streamflow measurements at Glacier No.1 station were interrupted during 1967–1979, so the reconstructed data for this period using the meteorological data were employed in this study (Yang 1991). It should be noted that the observations at these two stations were carried out from May to September each year, over 95 % of annual runoff at the station occurs during the observation period, whereas for the rest of the year, the streams are mostly frozen.

Similarly, glacier mass balance data were available from 1959 to 2008, except for 1967–1979. The missing data during the period have been reconstructed using the meteorological data (Zhang 1981). Glacier length and area have been determined using glacier maps made in 1962, 1964, 1986, 1992, 1994, 2000, 2001, 2004, 2006 and 2009. These data mentioned above have been published in annual reports of the Tianshan Glacier Station since 1980.

5.1.3 Methods

(1) Statistical analysis Trend test and abrupt change analysis

The long-term trends of the meteorological and hydrological time series were investigated by linear regression, and standard t-test was used to determine the statistical significance of the trends. All the trend results in this paper were evaluated at the 99 %, 95 % confidence levels. Additionally, Mann-Kendall test was employed in the study in order to search for abrupt change in trends with time, the methodology was described in greater detail in many researches (Burn and Hag Elnur 2002; Lu et al. 2004; Chen et al. 2006; Liu et al. 2007; Zhao et al. 2008).

Regression analysis The relationship between hydrological and meteorological variables was explored by regression analysis. The analyses were conducted for mean monthly discharge at Glacier No.1 and Empty Cirque gauging sites. The calculated values of R^2 were tested for statistical validity at the 95 % confidence level.

(2) Water balance model The water balance of Urumqi Glacier No.1 was calculated to show how glacier mass loss contributes to runoff. The elements of water balance for a watershed depend upon watershed characteristics and processes. For a glacierized catchment, the formula can be presented as follows:

$$P - E - R = \Delta S \tag{5.1}$$

Where P is precipitation input from snow and rain; E is evaporation/sublimation output; R is runoff; ΔS is the change in the water storage of glacier and groundwater in the catchment (Woo et al. 1994). According to the experimental results of Kang et al. (1997), mean annual evaporation at Glacier No.1 was about 18 % of total annual precipitation. Altitudinal gradient of precipitation was 19 mm/100 m, and the maximum height in precipitation was about 4,000 m in the east branch and 4,200 m in the west branch. Generally, groundwater discharge within a glacier is negligible, and this error is small compared to the observed precipitation and glacier melt runoff (Mark and Seltzer 2003). So in the water balance calculation, the change in groundwater storage was not taken into account.

(3) RegCM3 regional climate model The regional climate model used in this study is ICTP RegCM3, which is based on the model proposed by Giorgi et al.

(1993a, b) and was upgraded by Pal et al. (2007). The RegCM3 uses the IPCC SRES A1B greenhouse gas emission scenario, which is a middle end scenario with CO_2 concentration of about 700 ppm by 2100, to simulate the future climate. The climate scenario outputs were taken from the study of National Climate Center, China Meteorological Administration (Gao et al. 2011), where climate modelling was carried out for the whole China and surrounding areas with a horizontal resolution of 25 km and 288 × 138 grid points. The regional model RegCM3 is extensively described by Pal et al. (2007) and detailed features of these simulations are described by Wu et al. (2011). The projected temperature and precipitation agree well with the observed ones, which can reproduce more detailed spatial distributions and amount of them in annual, winter and summer. Nevertheless, meteorological outputs from the climate model are not directly applied for our hydrological applications, because it can not represent local subgrid-scale features and dynamics, although the spatial scale of the RegCM3 is high (Giorgi and Francisco 2000; Chen et al. 2011). So we apply a simple statistical downscaling method to build a relationship between the observed data at Daxigou meteorological station and projected climate variables at four neighboring grid points. The future daily temperature and precipitation time series in A1B emission scenario are constructed by the following equation:

$$y = a_1x_1 + a_2x_2 + a_3x_3 + a_4x_4 + b \tag{5.2}$$

where y is a future time series of temperature or precipitation; x is a projected temperature or precipitation time series (using RegCM3); and a_1, a_2, a_3, a_4 and b are parameters.

(4) HBV Hydrological model For river discharge simulation, the HBV hydrological model of the Swedish Meteorological and Hydrological Institute (SMHI) is used. The original version of the HBV model was developed for runoff simulation and hydrological forecasts on Swedish lowlands (Bergström 1976) and later extended for the application in glaciated alpine catchments (Braun and Renner 1992). A wide variety of model applications covering the catchments of ranging from the Alps (Braun et al. 2000; Konz and Seibert 2010), Tian Shan (Hagg et al. 2006), Himalayas (Akhtar et al. 2009) and other regions (Wang et al. 2006) have demonstrated its robust performances and capabilities for solving different hydrological problems, such as runoff modeling regarding climate change impacts on the water availability of alpine river basins (Hagg et al. 2007), the forecast of extreme runoff events (Akhtar et al. 2008) and the development of good strategies for sustainable water management.

The HBV model is a semi-distributed, conceptual precipitation-runoff model using subbasins as the primary hydrological unit. It takes into account area-elevation distribution and basic land use categories (glaciers, forest, fields and lakes). Subbasins are considered in geographically or climatologically heterogeneous basins (SMHI 2006). It describes hydrological processes in a simplified way while its parameters have some physical meanings. The model consists of subroutines for snow accumulation and melt, soil moisture accounting procedure, routines for runoff generation and a simple routing procedure. It works on a daily time-step and has a very low data demand: daily mean air temperature and precipitation, long-term mean monthly potential evaporation and daily discharge for calibration. Parameters of the model in this study are calibrated using a manual calibration procedure. In order to assess the agreement between modelled and measured discharge, the Nash-Sutcliffe efficiency R^2 (Nash and Sutcliffe 1970) and the relative error RE are calculated:

$$R^2 = 1 - \frac{\sum_{i=1}^{i=N} \left[Q_s(i) - Q_o(i) \right]^2}{\sum_{i=1}^{i=N} \left[Q_o(i) - \bar{Q}_o \right]^2} \tag{5.3}$$

$$RE = 100 \frac{\sum_{i=1}^{i=N} \left[Q_s(i) - Q_o(i) \right]}{\sum_{i=1}^{i=N} Q_o(i)} \tag{5.4}$$

Where i is the time step, N is the total number of time steps, Q_s represents simulated discharge, Q_o is observed discharge and $\overline{Q_o}$ is the mean of Q_o over the calibration/validation period. A perfect model would result in an R^2 equal to 1 and the RE value should be close to zero.

5.1.4 Climate Change at the Headwater of Urumqi River

Annual mean air temperature and precipitation from Daxigou meteorological station during the study period 1959–2008 was shown in Fig. 5.2. Similarity in annual fluctuation trends of these two variables has been observed at the headwater of Urumqi river, whose characteristics were shown as follows.

(1) Temperature change Annual average temperature in the region during 1959–2008 showed a general warming trend (Fig. 5.2). From 1970s to the early 1980s, temperature was in a relatively low state. However, since the middle 1980s, it has been increasing continually and maintaining to be in high state especially after 1996. Annual average temperature reached to the maximum in 2007 (−3.8 °C) while the minimum occurred in 1984 (−6.7 °C). The change rate of annual temperature was 0.22 °C/10a at 0.01 significance level (Table 5.2), indicating that the temperature increased remarkably in the study area. This was mainly attributed to the strong rising of temperatures in four months, increasing at rates of 0.45 °C/10a in November, 0.30 °C/10a in September, 0.27 °C/10a in June and 0.25 °C/10a in July, respectively. For the temperature in the remaining months, a weak rising trend was present in January, while no trends were found for another months. In addition, the abrupt change tests revealed that a sharp warming trend for monthly and annual air temperatures mostly occurred in 1996. Comparison of the temperature during 1996–2008 (−4.4 °C) and 1959–1995 (−5.3 °C) indicated an increase of almost 1 °C, confirming the fact of climate warming in the region.

(2) Precipitation change Annual precipitation also showed an overall increasing trend in the period of 1959–2008 (Fig. 5.2). In line with the variations in annual

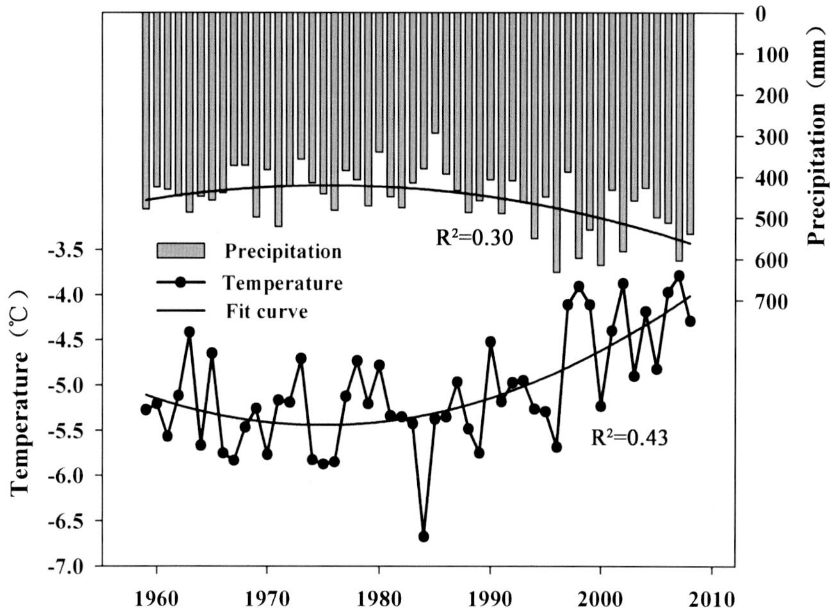

Fig. 5.2 Annual average air temperature and precipitation and their fitting curves for 1959–2008

Table 5.2 Statistically significant coefficients for monthly temperature, precipitation and discharge and year of abrupt change for 1959–2008

	Jan	Jun	Jul	Sep	Nov	Dec	Year
T	0.012*	0.027**	0.025**	0.03**	0.045**	No	0.022**
Y	1990	1997	1996	1989	1993	No	1996
P	No	No	0.937*	No	0.089*	0.055**	2.114**
Y	No	No	1991	No	1994	1987	1994
Q	–	0.005*	0.002**	No	–	–	0.003**
Y	–	1993	1994	No			1994

Months having statistically significant were only listed

T monthly mean air temperature (°C), *P* monthly precipitation (mm), *Q* monthly mean discharge (m^3 /s), *Y* year of abrupt change

** and *: statistical significance at 99 % and 95 % confidence levels, respectively

average temperature, annual precipitation has been in the similar changing trend. It has gone through two phases: a downward trend from 1960s to the late 1970s and an upward trend from the beginning of 1980s–2008. The 1970s was a period of relatively low precipitation, in which the driest years included 1973 (368.8 mm), 1970 (396.4 mm) and 1977 (399.2 mm), while the most humid year was 1996 (632.4 mm). The change rate of annual precipitation was 21.14 mm/10a at 0.01 significant level (Table 5.2), indicating that the increase of precipitation in the study area is evident. This was mainly due to the increase in precipitation for the months of July, November and December, whereas the other remaining months did not

Table 5.3 Glacier area change since 1962

Year	Area (East branch, West branch) (km^2)	Rate of area change (km^2/a)	Map scale and measurement method
1962	1.950		1:10,000, plane-table instrument survey (Wang et al. 1965)
1964	1.941	−0.005	1:10,000, aerial survey (Chen et al. 1996)
1986	1.840	−0.005	1: 5,000, photogrammetric survey (You 1988)
1992	1.833	−0.001	1:10,000, aerial survey (Chen et al. 1996)
1994	1.742 (1.115, 0.627)	−0.046	1:10,000, photogrammetric survey (Chen and Sun 1996)
2001	1.708 (1.097, 0.607)	−0.005	1:5,000, theodolite instrument survey (Jing 2004)
2006	1.677 (1.086, 0.591)	−0.006	1:5,000, electronic total station instrument survey (Zhou 2009)
2009	1.646 (1.069, 0.578)	−0.010	1:5,000, electronic total station instrument survey (Cao 2012)

exhibit clear trends. Abrupt change test showed that significant increasing trends in monthly and annual precipitation started in almost the same year, around 1994. From 1994 to 2008, annual average precipitation increased by 91 mm (17.5%), relative to the average value for 1959–1993, indicating that the annual precipitation increased significantly.

In summary, during 1959–2008, the climate regime shifted from cold-dry to warm-wet, which occurred around 1996. Comparing with the research in western Tianshan of China, our result was nearly 10 years later than that obtained by Shi et al. (2002), which may reflect the regional difference in temperature and precipitation.

5.1.5 Recent Shrinkage of Urumqi Glacier No.1

(1) Glacier area change The area of Urumqi Glacier No.1 has decreased by 0.304 km^2 (15.6%), from 1.950 km^2 in 1962 to 1.646 km^2 in 2009, which showed a remarkable shrinkage (Table 5.3). The change rate maximum of glacier area occurred in 1994 and arrived at −0.046 km^2/a. Comparatively speaking, the change rate after 1994 was weakly quicker than that before 1994. Except for climate warming reason, it may also be related to the increase in effective ablation area at the end of glacier after separating into east and west branches in 1993 (Li et al. 2003b; Wang et al. 2004).

(2) Glacier length change Urumqi Glacier No.1 has been retreating since there was observed record in 1962 (Fig. 5.3). The terminus retreat of east and west branches were respectively 195.0 m and 235.4 m, with a mean length of 215.2 m, during the period 1962–2009. Due to continued shrinkage, the glacier became completely separated into two small glaciers with a consequent sudden increase in retreat speed in 1993. The retreat speed of the west branch terminus apparently quickens more than that of the

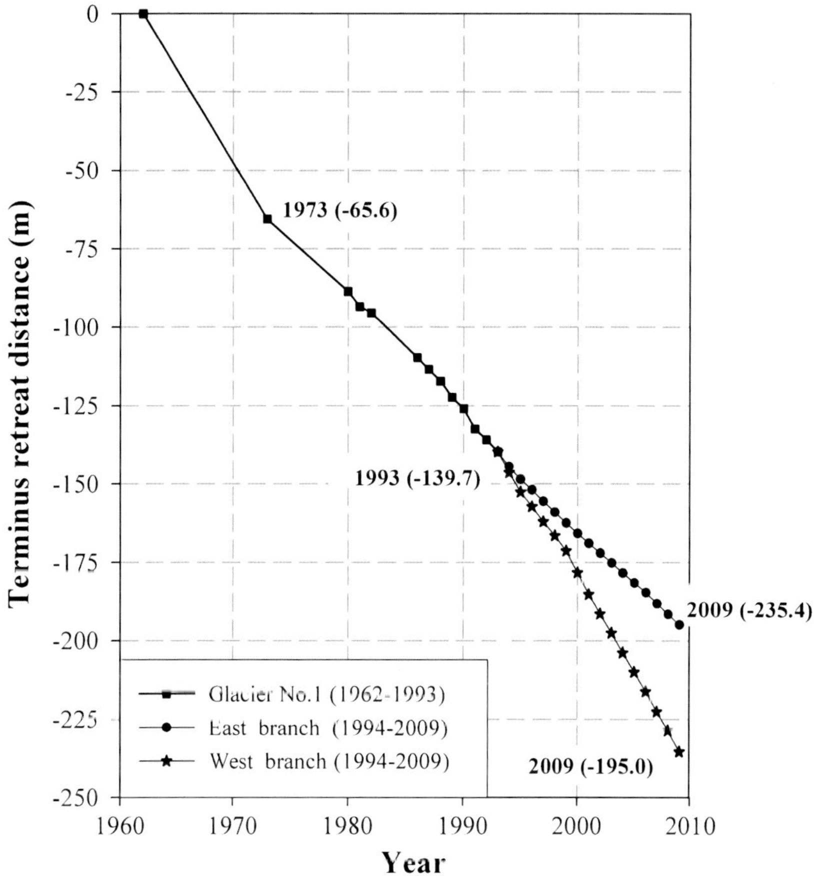

Fig. 5.3 Terminus retreat distance history since 1962 based on directly field measurements

east branch. The phenomenon may be related to that the west branch is of smaller dimensions than the east branch, and so is more sensitive to climate change. Additionally, the underlying topography of the west branch terminus is an abrupt ice ridge, which causes the ice not only to thin, but also to break up more easily (Jin et al. 2006).

(3) Glacier mass balance change Urumqi Glacier No.1 experienced a mostly negative mass balance through the measurement period, with a shift to more negative balance beginning in 1985, especially after 1996 (Fig. 5.4a). Consecutive negative mass balance has existed in the last decades, and the mass loss accounted for 61.6 % of the total over the past 50 years, of which the cumulative mass balance amounted to −13,693 mm, equivalent to 15.2 m of glacier ice, or $2{,}610 \times 10^4$ m^3 loss of the glacier volume. Accordingly, equilibrium line altitude of the glacier moved up 64 m for the period 1997–2008, compared to 4,042 m for 1959–1996 (Fig. 5.4b). It indicated that the large precipitation in the region was also hard to resist strong glacier

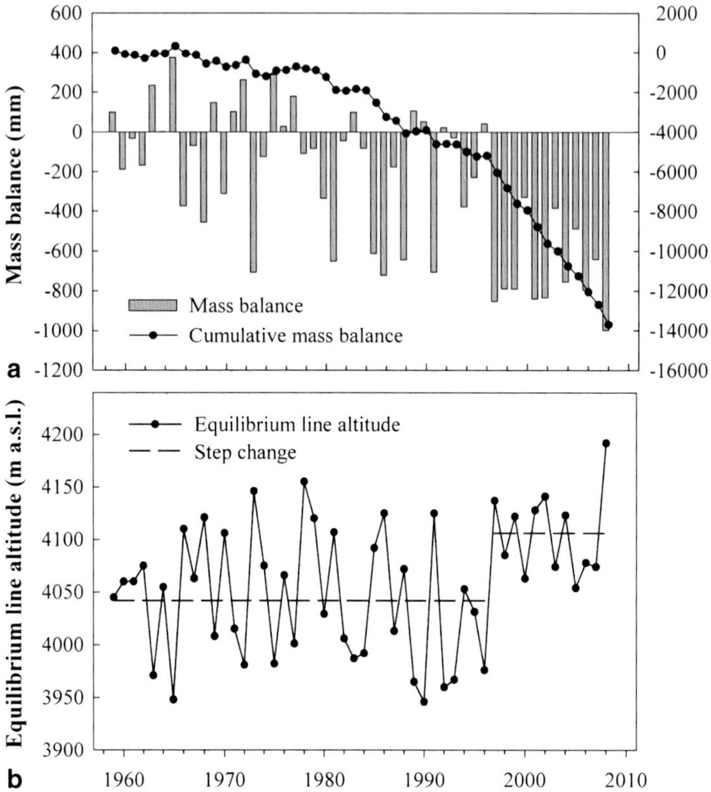

Fig. 5.4 **a** Annual cumulative mass balance and **b** annual equilibrium line altitude of Urumqi Glacier No.1 during 1959–2008

melting, although in glacier accumulation area, i.e. the mass balance of the glacier was particularly sensitive to temperature change.

5.1.6 Glacier Runoff Variation and Its Contribution to River Runoff

(1) Changes in water balance at Urumqi Glacier No.1 Based on water balance model, the annual glacier meltwater runoff of Urumqi Glacier No.1 during the period 1959–2008 was calculated (Fig. 5.5). As can be seen, glacier runoff varied with a significant upward while glacier storage showed an obviously decreasing trend. The opposite tendency indicated that the change in glacier storage led to continuously rising discharge amounts, such that the effect of declined glacier storage was a reduction in mass balance, resulting in increased meltwater production. In addition, the changes of precipitation and evaporation in the region were also evident, but their amplitudes of variation were far less than those of glacier runoff and storage.

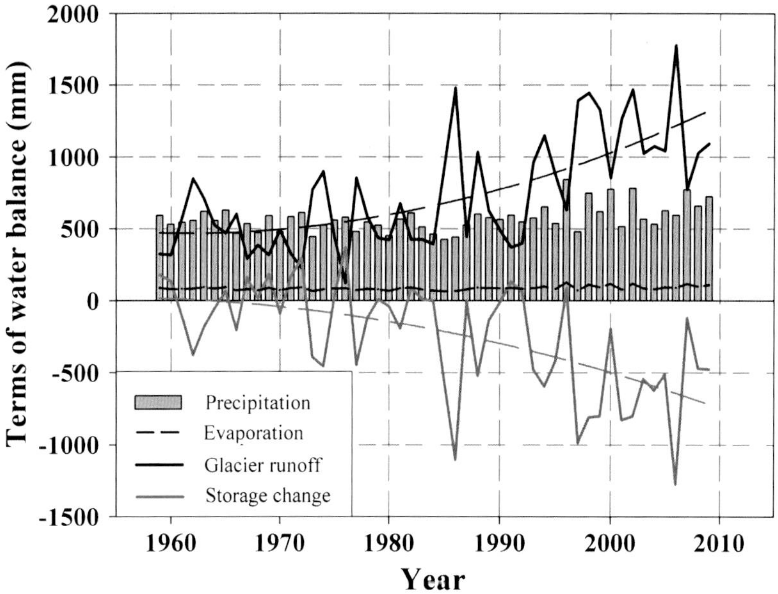

Fig. 5.5 Interannual changes of glacier runoff, glacier storage, precipitation and evaporation of Urumqi Glacier No.1 from1959 to 2008

So the catchment precipitation and evaporation have been relatively stable, although with increasing trends. The results revealed that glacier runoff variation was mainly controlled by the mass balance fluctuation. The change in glacier storage provides an estimate of the mean annual contribution of glacier melt to river runoff.

(2) Glacier runoff variation and its contribution to river runoff As shown in Fig. 5.6, it was obvious that glacier melt runoff and discharge at gauging site have been both in increasing trends, especially since the early 1990s. Abrupt change test also revealed that a steady rise in annual discharge took place in 1994. The effect of the changes can be demonstrated with step values of glacier runoff and its contribution to river discharge before and after 1994. The annual average glacier runoff during 1994–2008 was 197.8×10^4 m^3, increased by 88.0 %, compared with 105.2×10^4 m^3 for the period 1959–1993. Accordingly, the contribution rate to river flow went up sharply from 62.8 % to 72.1 %, implying that the variation amplitude of glacier runoff was greater than that of river discharge. Moreover, as indicated by correlation analysis, there was a strong relation ($R^2 = 0.975$) between the glacier runoff and river flow, indicating that glacier runoff has a substantially influence on the variability of river discharge in the region.

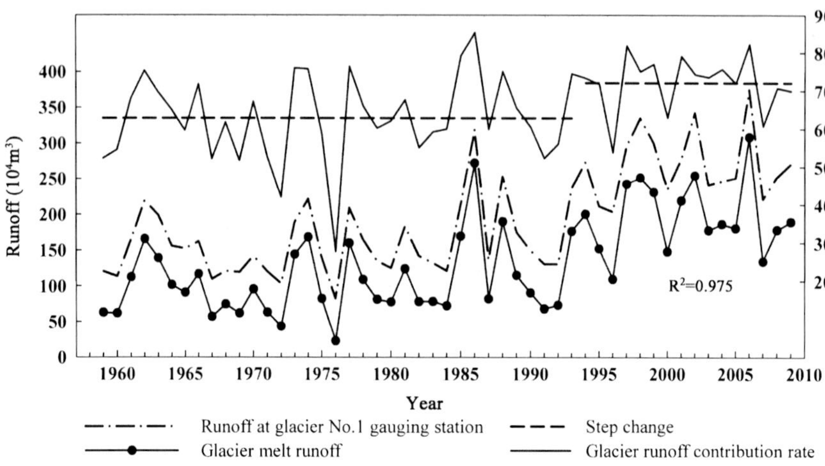

Fig. 5.6 Interannual change of glacier runoff and discharge at Glacier No.1 gauging station and the glacier runoff contribution to river discharge during 1959–2008

5.1.7 Factor Comparison of Affecting Runoff in Glacierized and Non-glaciered Catchments

The difference between the glacierized and non-glacierized catchments is that the runoff derived from the non-glacierized catchment is precipitation-dominated while the glacierized catchment is energy-dominated (Chen and Ohmura 1990; Liu et al. 2010). The main contributions to runoff in the study area are glacier or snowmelt and rainfall. Of a mean annual precipitation of 511 mm from 1959 to 2008, an average of 87.7 % was concentrated in the ablation season (May–September) and 67.2 % in the summer season (July–September). The monthly mean air temperature was usually higher than 0 °C, with average value of 2.6 °C in the ablation months (Fig. 5.7a). Due to the concurrence of ablation and precipitation in the region, the two tested sites (Urumqi Glacier No.1 and Empty Cirque catchments) have similar monthly discharge distribution patterns, except for a slight dissimilarity between June and July (Fig. 5.7b).

At the two catchments, runoff distributed unevenly during a year that occurred from May to September and reached their peaks in July. For Urumqi Glacier No.1 catchment, runoff increased sharply from June to July, at a greater magnitude of variation compared with that of Empty Cirque area, which was likely related to different supply of runoff. Discharge at Empty Cirque was mainly derived from rainfall after depleting snowmelt in late May or early June, whereas when there has substantially increased in the area of exposed ice at Glacier No.1 catchment, which would reduce the albedo and increase melt rate, and would also produce a large number of glacier meltwater.

Visually, comparing 5.7a with b, discharge at Glacier No.1 catchment was consistent with the major regime shifts in air temperature while discharge at Empty Cirque

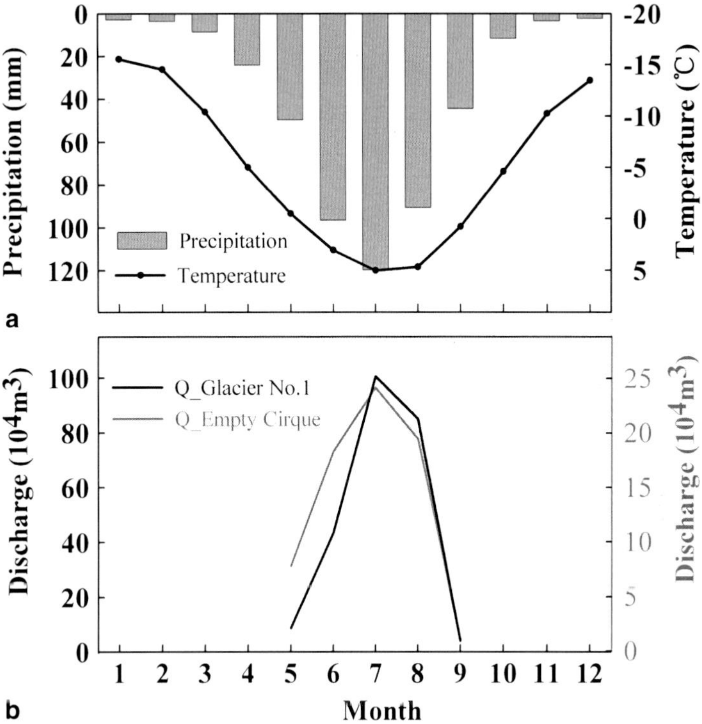

Fig. 5.7 Monthly distribution of **a** average air temperature, precipitation and **b** discharges of Glacier No.1 and Empty Cirque gauging stations over the period 1959–2008

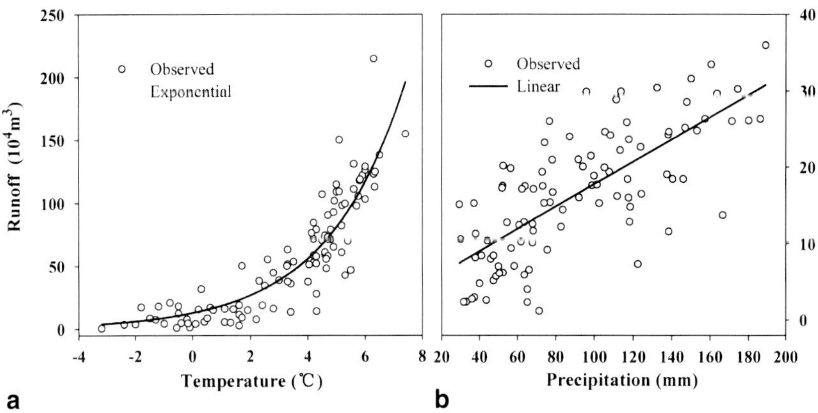

Fig. 5.8 Fitting charts between **a** discharge of Glacier No.1 gauging station and air temperature **b** discharge of Empty Cirque gauging station and precipitation

Table 5.4 Regression analysis of the dependence of discharge on air temperature and precipitation in the ablation period at the two tested sites

Catchment	Regression Equation	R^2	p	n
Glacier No.1	$Q=8.287e^{0.442T}$	0.762	<0.01	115[a]
Empty cirque	$Q=0.146P+3.196$	0.542	<0.01	114[a,b]

T, *P* denote monthly air temperature and precipitation from May to September, respectively
[a] Discharge data were incomplete for 1996
[b] 1990 was an outlier and was deleted from the analysis

Table 5.5 Changes in precipitation, temperature, glacier mass balance, glacier runoff and river discharge for the periods 1959–1993 and 1994–2008

Period	P (mm)	T (°C)	M (mm)	Q_{gd} (mm)	Q_g (10^4 m^3)	Q (10^4 m^3)
1959–1993	538	−5.3	−102	592	105.2	159.9
1994–2008	651	−4.5	−586	1,113	197.8	270.9
Change	113	0.8	−484	521	92.6	111
Change(%)	21.0			88.0	88.0	69.4

P, T, M, Q_{gd}, Q_g and Q are precipitation, temperature, glacier mass balance, glacier runoff depth, glacier runoff volume and river discharge at Glacier No.1 gauging station, respectively

catchment seemed to follow the variations of precipitation. Regression analysis and statistical tests proved that air temperature in ablation season was the governing factor affecting discharge at Glacier No.1 catchment (Fig. 5.8 and Table 5.4). As temperature increases, the resulting runoff exhibited a nonlinear increasing trend, showing that when air temperature was higher than 2 °C, the runoff would increase at an accelerated rate, indicating that runoff in the glacierized area was quite sensitive to temperature change. Nevertheless, for non-glacierized Empty Cirque catchment, precipitation was the important factor. 1 mm increase of precipitation during the melt season can lead to an increase of 0.146×10^4 m^3 runoff, the effect of precipitation on runoff should not be neglected. It is important to note that such an increase in discharge at Urumqi Glacier No.1 catchment can only be transient. As the glacier snout recedes, the decrease in glacier cover will eventually limit the total volume of meltwater generation, even if high temperatures are sustained.

5.1.8 Relationships Between Climate Change, Glacier Response and Runoff Yield

Compared with the results of abrupt change tests for air temperature, precipitation and runoff, it could be found that these change points mostly fell between 1991 and 1997.This strongly suggested an association between them. The relationships among climate change and resulting glacier response and runoff yield would be discussed below to provide a better explanation of their interactions.

Table 5.5 listed the average precipitation, air temperature, mass balance, glacier runoff and river discharge during the periods 1959–1993 and 1994–2008. Glacier

Table 5.6 Sensitive parameters and the optimal values of HBV for the two investigated areas

Parameter	Description	Glacier No.1	Zongkong	Unit
TT	Temperature limit for snow/rain	2.3	2	°C
DTTM	Value added to TT to give the threshold temperature for snowmelt	−3.3	−3.8	°C
CFMAX	Snow melt factor	2.7	3.4	$mm°C^{-1}d^{-1}$
GMELT	Glacier melt factor	4.5	5.1	$mm°C^{-1}d^{-1}$
FC	Maximum soil moisture storage	300	500	mm
PERC	Percolation from upper to lower response box	0.132	0.35	$mm\ d^{-1}$

runoff increased by approximately 521 mm (92.6×10^4 m^3), about 88.0%, while precipitation increased by 113 mm (21.0%), and there was a 0.8 °C increase in air temperature, and −484 mm decrease in glacier mass balance. This indicated that about 6.2% of glacier runoff came from increases in precipitation, and another 81.8% was contributed by the loss of glacier mass. River discharge increased by 111×10^4 m^3 (69.4%), of which approximately 21.0% was contributed by increase in precipitation, and the other 48.4% came from glacier runoff. It meant that the increased glacier runoff accounted for 69.7% of the increased river flow. The 0.8 °C increase of air temperature has caused a 484 mm annual glacier mass loss, which was compensated by a 113 mm increase in annual precipitation. This was equal to a change in mass balance of −746.3 mm/°C without increasing precipitation. By analyzing the river flow and mass balance, we found that a 100 mm change in mass balance could cause a fluctuation in river flow of 22.9×10^4 m^3. The total cumulative mass balance of the glacier over the past 50 years was −15.2 m, equal to a 3 486.0×10^4 m^3 contribution to river flow, which was about 14.4 times the annual discharge at Glacier No.1 gauging station. This strongly illustrated that the change of hydrological regime in the region was greatly linked to the change in glacier conditions and recent climate warming, which further emphasized the fact that it is of great importance of glacial meltwater to the local water resource.

5.1.9 *Modelling the Runoff Response in Glacier No.1 Catchment for Present-Day and Future Climate*

(1) Modelling with observed climate The model was calibrated using daily discharge observations of the period between 1997 and 2008 in the investigated area, validation runs were performed from 1986 to 1995. These two time slice not only include relatively dry and wet years, but also cover the sufficient observation data to calibrate and verify the model. For Glacier No.1 catchment, information about the glacier mass balance in addition to discharge were also considered in the calibration procedure to find the appropriate parameters of the snow and glacier routine. While for Zongkong catchment, measurement data of glacier mass balance are not available. In our study, prior knowledge to the studied area in combination with much experience gained in the parameter estimation in the previous HBV studies

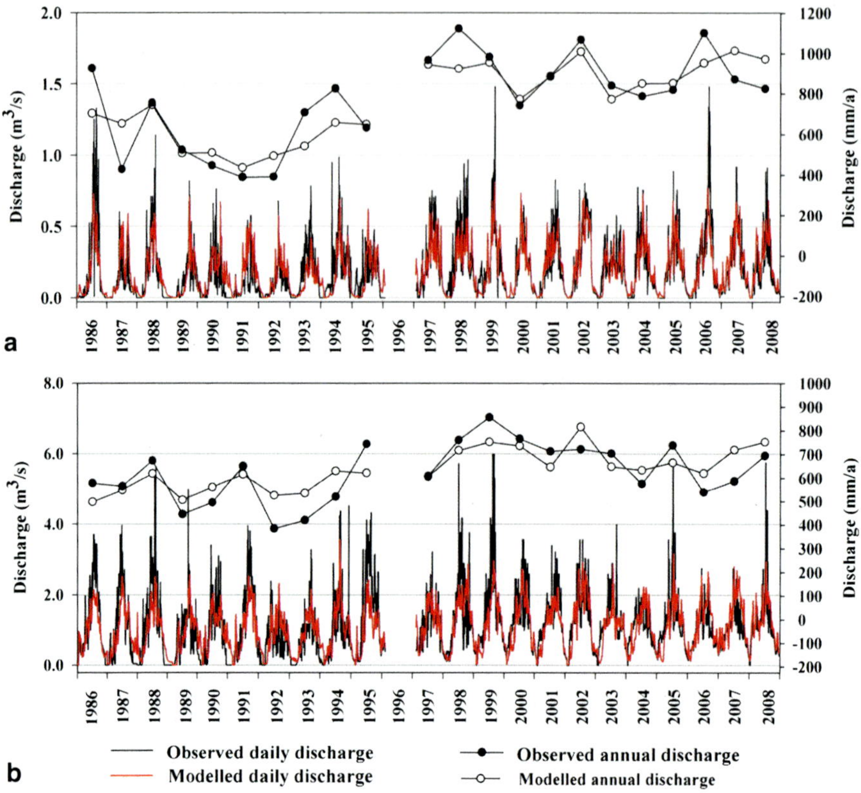

Fig. 5.9 Time series of modeled and observed discharge for **a** Glacier No.1 and **b** Zongkong and gauging stations, of which the daily discharge at the two stations are listed from May to September each year

Table 5.7 Efficiency criterions of modeling performance for the two investigated areas [R^2 is from Eq. (5.2), *RE* is from Eq. (5.3)

Catchment	Period	R^2_{mean}	R^2_{min}	R^2_{max}	*RE* (%)
Glacier No.1	Calibration (1997–2008)	0.81	0.65	0.91	−3.8
	Validation (1986–1995)	0.71	0.62	0.83	−5.4
Zongkong	Calibration (1997–2008)	0.77	0.74	0.87	−1.6
	Validation (1986–1995)	0.75	0.72	0.8	2.5

are together used to help us acquire the reasonable values of main parameters. During the calibration period, parameters TT, DTTM, CFMAX and GMELT are found to be most sensitive, followed by the parameters FC, PERC, and there is a strong interdependence among these parameters. The values of sensitive parameters after calibration are given in Table 5.6 for the two subbasins.

Figure 5.9 shows the time series of modeled and observed discharge during the calibration and validation periods for the two stations. The daily discharges are listed from

Table 5.8 Parameters of the statistical relationship between the observed and projected data

Variable	a_1	a_2	a_3	a_4	b	r	R^2
T	2.744	−2.923	0	0.897	−0.994	0.963	0.928
P	0	0.398	0	0	1.384	0.822	0.675

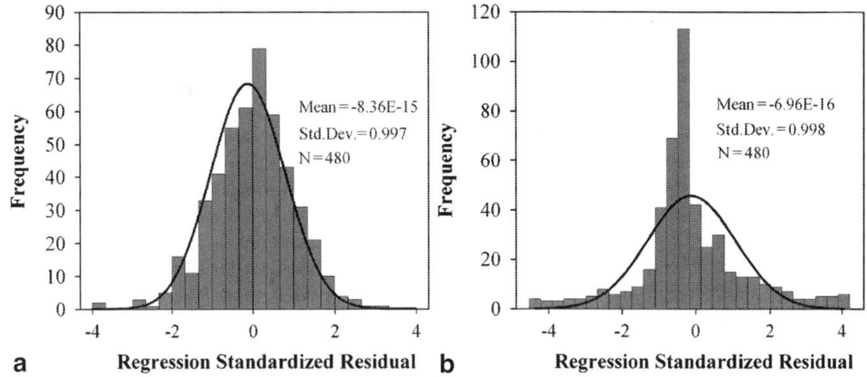

Fig. 5.10 Histogram of the residual of **a** temperature and **b** precipitation

May to September each year because over 95 % of the annual runoff at the stations occurs during the period, while for the rest of the year, the rivers are mostly frozen. As can be seen from the figure, the variation of the discharge during the calibration periods are in good agreement with the observed values but also during the validation periods the computed daily discharges, especially in Zongkong catchment, fit the observed ones. This is also confirmed by the higher R^2 and the smaller RE values, which are in the ranges of 0.71–0.81 and −3.8–5.4 %, respectively (Table 5.7). In addition, by comparing the annual sums of the simulated and observed discharges, the satisfactory model performance is also expressed by the correlation coefficient of 0.85 (Glacier No.1) and 0.77 (Zongkong). In the Glacier No.1 catchment there is a small underestimation of the discharge sum in 1986, 1993, 1994 and 1998 and an overestimation in 1987, 2007 and 2008. While in the Zongkong catchment, the simulated discharge is slightly larger than the observed total runoff in 1992, 2007 and 2008, and vice versa in 1986 and 1999. This was probably caused by the constant value in parameters for the whole simulation period. These, however, are within 6 % relative to the observed discharge, and are therefore acceptable. Furthermore, the annual change in the simulated mass storage of Glacier No.1 also coincides well with the measured values (see the next section, Fig. 5.3a). The correlation of simulations with observations gives a coefficient of 0.86, significant at the < 1 % level, which demonstrates that the HBV model is capable to simulate the discharge regime very well and is suitable in this study.

(2) Modeling results with future climate After the successful calibration and validation of the HBV model using the observed climate variables, the statistically

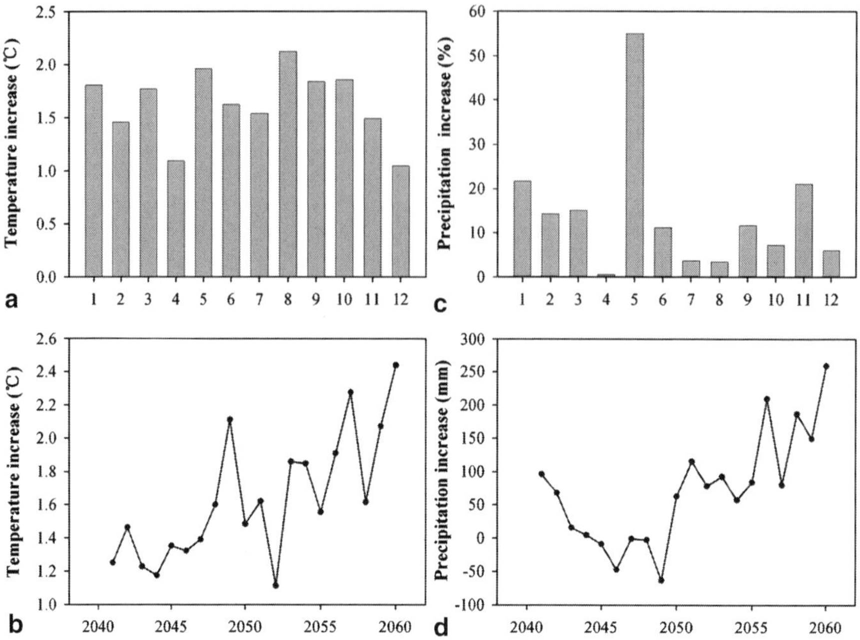

Fig. 5.11 Variations of **a**, **b** temperature and **c**, **d** precipitation derived from downscaling from RegCM3 under SRES A1B scenario for 2041–2060 relative to 2000–2008

downscaled temperature and precipitation of the RegCM3 are used to drive the model to estimate the variation of water resource in the future.

a. Changes of temperature and precipitation during 2041–2060 We carry out the linear regression for monthly meteorological data collected at Daxigou and the output from RegCM3 at four neighboring grid points for the period 1961–2000, the parameters of equation (5.1) are determined given in Table 5.8. The histogram of the residuals indicates that the data follows an approximately normal distribution (Fig. 5.10), so it can be seen that the downscaling method is valid. In addition, Correlation coefficients of temperature and precipitation reached 0.963 and 0.822, respectively, which implies that the time series of future climate are simulated in a closely agreement with the observed values and also demonstrates the statistical relationship has a good performance in this study.

Figure 5.11 shows the possible variation of temperature and precipitation projected by RegCM3 based on SRES A1B scenario for 2041–2060 relative to 2000–2008. These indicate an obviously increase in temperature and the increase is observed in all months. The temperature rise reaches the maximum of 2.0 °C in August and the lowest rise of 1.0 °C observed in December. The annual mean temperature rises up to 1.7 °C for the period 2041–2060. Similarly, precipitation is projected to increase although it gone through a downward trend in the first several years. From the monthly changes, the mean annual cycle of precipitation exhibits a stronger variability than the annual cycle of temperature. A more meaningful feature is that almost no change in

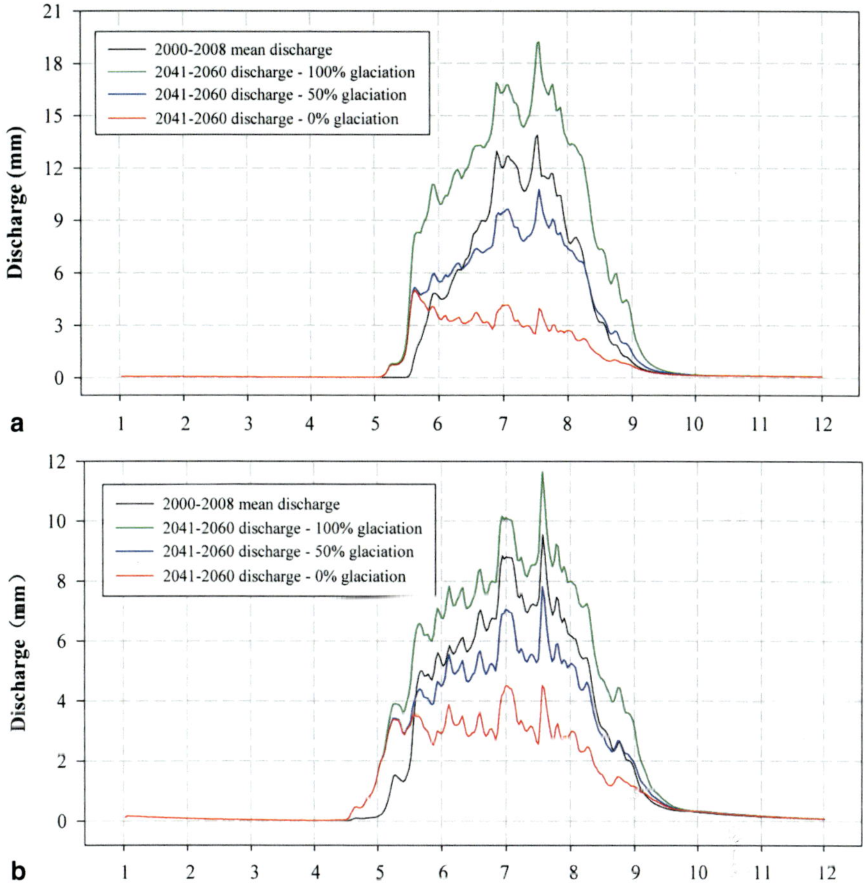

Fig. 5.12 Annual discharge cycle of **a** Glacier No.1 and **b** Zongkong catchments simulated by HBV for current climate (2000–2008) and future climate (2041–2060) for three stages of glaciations

precipitation is found in April but it reaches a maximum increase of 55 % in May. The annual precipitation will increase by 14.2 % by the middle of the century. The overall increase in temperature and precipitation seem to be consistent with those of Hagg et al. (2007) and Akhtar et al. (2008), who showed that the projected temperature and precipitation in Central Asian and Hindukush-Karakorum- Himalaya region would have an increasing trend.

b. Future annual discharge cycle Figure 5.12 shows the average monthly modelled results for the present (2000–2008) and future climate scenario (2041–2060, SRES A1B scenario) with three stages of glacier coverage: 100 % glacier, 50 % glacier and 0 % glacier. Table 5.9 presents the mean changes in future discharge for three glaciations stages relative to the present discharge in two catchments. The effects of above climate change on the river discharge are significant and the following can be observed.

Table 5.9 Mean change in future discharge (2041–2060) under A1B emission scenarios relative to the present discharge (2000–2008) for three glaciations stages and for two catchments

Catchments	Month	Present discharge	Change in future discharge (%)		
		(mm)	100 % glacier	50 % glacier	0 % glacier
Glacier No.1	5	11	106.1	64.8	61.4
	6	135	128.9	26.8	−16.5
	7	334	37.7	−44.8	−58.3
	8	301	53.0	−17.9	−72.3
	9	55	162.4	25.7	−51.1
	Annual	852	66.3	−8.8	57.5
Zongkong	5	28	183.2	140.2	135.6
	6	162	28.3	−14.0	−41.6
	7	230	17.0	−22.2	−53.6
	8	205	26.4	−19.1	−56.1
	9	65	64.3	1.5	−41.5
	Annual	719	33.3	−9.4	−40.4

100% glaciation There is an increase in river discharge throughout the melt season in the changed climate for the two basins. This mainly reflects that snowmelting starts half a month earlier and discharge rise towards its peak in summer months. The 100 % glacier extent predicts a significant discharge increase of 66.3 % in Glacier No.1 catchment and a relatively small increase of 33.3 % in Zongkong catchment. This indicates that a heavily glacierized catchment is likely to produce a greater discharge variation as compared to a less glacierized one. In addition, Compared to the present discharge, the rise magnitude of future discharge is found much higher in May, June and September. This is partly due to the fact that the two basins have a very small amount of discharge during the 3 months and a small absolute increase can lead to a larger perceptual discharge change relative to months of July and August when the highest peaks are observed. However, this case might not be occurred, it can thus be showed only a trend because future 100 % glacier extent is not realistic with climate change.

50% glaciation If the glacierized area is reduced by 50 %, snowmelt still begins half a month earlier and discharge is increased in May and September due to more frequent rainfall and prolonged time of snow and glacier melt caused by the regional climate warming. While in July and August the discharge decrease considerably, but the peak discharge for the two stations remains concentrated in the 2 months. From the year scale, this scenario predicts no obviously change with a slightly drop of 8.8 % in Glacier No.1 catchment and a small decrease of 9.4 % in Zongkong catchment, probably was the result of the glacier area decrease is nearly balanced out by the enhanced melt rates.

0% glaciation After complete disappearance of the glaciers, snowmelt still starts half a month earlier and there is still an increase in discharge in May. There is a marked decrease in discharge during June-September. In Glacier No.1 catchment, it predicts about a 57.5 % decrease in annual discharge, with the highest decline reaching 72.3 % in July. The most striking feature of these results is that there is a clear seasonal shift of peak discharge from mid-summer to late spring-early summer by the middle of the century. For the Zongkong catchment, there also predict a drastic decrease in the discharge (40.4 %). Similarly, the largest decline occurred in July and August when discharge will be reduced by 53.6 % and 56.1 %, respectively. This, on one hand, could indicate that glacial melt is major contribution to the total discharge for the two basins. There is neither forest nor any major lake present in the two subbasins and glaciers and fields (area without forest) are considered as the only two land use types in the hydrological model framework. Therefore, the effect of complete melting of glaciers on the hydrological cycle will depend on the degree of glaciations in the river basins and response of the river basins to climate change. On the other hand, the significant decrease in discharge may aggravate the crisis of water resources shortage in this region, especially in July and August when demand of water for agricultural production reaches its peak. The temperature increase will also lead to the augmentation in snowmelt runoff, which may arise more frequent and earlier spring runoff, and finally will provoke more changes in the runoff amount and timing and the allocation of the intraannual water resources in the catchment.

5.2 Glacier Changes and Their Impacts on Water Resources in Xinjiang

5.2.1 Background

(1) Glacier water resources in Xinjiang and its importance Glaciers are important water source for oases and for the sustainable development of the ecological environment, industry and agriculture in Xinjiang Uyghur Autonomous Region. There are 18,311 glaciers with a total area of 24,721.93 km^2 and approximately ice volume of 2,623.4711 km^3 (23,611.2 × 10^8 m^3 water equivalent) in Xinjiang, accounting for about 46.84 % of the total glacier volume in China being the first place (Lanzhou Institute of Glaciology and Geocryology 1986a, b, c, d; Shi 2005). Glaciers are distributed in Altai, Tianshan, Pamir, Karakorum and Kunlun Mountains, which are included in the Irtysh River, Junggar Interior River, Central Asia Interior River, Tarim Interior River, etc.

Glacier meltwater runoff contains the rainfall runoff produced by solid and liquid precipitation on the glacier surface and glacial runoff by the ablation of glacier ice. The runoff caused by seasonal snow in the non-glacier regions are called the snow runoff, not belonging to glacier meltwater runoff. Glacier meltwater runoff is the dynamic water resources involved in the river runoff cycle, representing the contribution of glaciers to the water resources. It can be calculated by the glacier mass balance method, glacial meltwater runoff modulus method, flow and temperature relations method, the comparative observation experiment and the glacier systems approach (Xie et al. 2006) and hydrological separation method. The glacial runoff was estimated to be about 563.3 × 10^8 m^3 (then revised to be 604.65 × 10^8 m^3) for the whole country and 188 × 10^8 m^3 for Xinjiang based on the glacier data before the mid-1980s (Yang 1991; Kang et al. 2000; Xie et al. 2006). In the recent 30 years, glacier meltwater increase remarkably due to the temperature rising. According to the glacier observation data in Xinjiang and runoff changes (Li et al. 2003b, 2009a), the increase of glacier meltwater is more than 10 % and thus the current glacier meltwater runoff is calculated to be over 200×10^8 m^3, accounting for more than 25 % of 789×10^8 m^3 the average quantity of surface water in Xinjiang for many years.

Acting as a huge alpine reservoir, glaciers can regulate river runoff. In the wet cold periods, glacier meltwater runoff reduces due to deficient calories, weak melting and increased accumulation of glaciers. In the arid dry periods, glacier meltwater runoff increases because of the increased fine weather and intensified glacier ablation. Moreover, glaciers can regulate the annual distribution of water resources. In the summer with little rainfall, rivers would not cutoff for the supply of glacier melting water. And for the storm, some of the rainfall would be conserved on the glacier as solid form to reduce the flood hazard. Most of river water resources in Xinjiang are stable and interannual variations of runoff is small. Generally the coefficient of variation (Cv value) is between 0.1 and 0.5, which decreases with the increase of glacier meltwater proportion, with the smallest value of 0.1–0.2 (Yang 1991).

The response of glaciers is very sensitive to climate change. Over the past hundreds years, the global average temperature increased 0.74 °C due to climate change (IPCC 2007). According to the observation data from meteorological stations, temperature in Xinjiang shows rising trend in the recent 50 years, especially for the last 10 years (1998–2007), which is the warmest period in Xinjiang. The average annual temperature increased 0.9 °C in Xinjiang, which is higher than the global average warming rate.

Affected by the impact of rising temperatures, global ice sheets and the number of most mountain glaciers has been in an obvious retreat state since the twentieth century. The Central Asian region is the most significant retreating of glaciers in the world. Most glaciers in Xinjiang are in the state of retreat since the 1950s, especially after the 1980s the retreat accelerated. The response to the climate warming is most sensitive for the glaciers with small size and low altitude, which is rapidly shrinking, even disappearing. Strong ablation of glaciers can produce large quantities of water in the short term, and is apt to produce unexpected hazards. However, ablation also leads to substantial drops in the solid water mass of the glacier and an eventual decrease in meltwater generation. Additionally, the regulation (balance between seasons) of river runoff can even be weakened. In the serious ablation period, higher amounts of glacier meltwater would lead to the outburst of glacial-dammed lake, causing sudden flood damage, which need special concerns and prevention.

According to IPCC (2007) and the prediction from Xinjiang Meteorological Bureau, the global temperatures will rise 1.2–3.8 °C in the next century. The annual average temperature in Xinjiang will increase in the range of 0.5–0.9 °C in the early twenty-first century (2001–2020) and 1.2 °C in the early twenty-first century period (2021–2030). The temperature will further increase in the mid and late twenty-first century. Such serious temperature rising will cause greater change of glacier snow water resources, and its impact on hydrology, water resources in Xinjiang concerns the national security, stability and socio-economic sustainable development, which is a scientific problem that needs urgent explaining.

Some previous studies have focused on the glacier changes in Xinjiang (Wu et al. 1983; Chen et al. 1996; Li et al. 2007b, c 2006; Shen et al. 2003; Liu et al. 2006; Xie et al. 2006). However, studies about the recent glacier change in Xinjiang are limited and cannot form overall cognition to the Xinjiang glacial fluctuation due to the lack of data. In addition, some studies are concentrated on the glacier changes in Central Asia (Aizen et al. 2007b; Bolch 2007). Nevertheless, although it focuses on the changes of water resources in the arid regions of Central Asia, the study on the glaciers is still very limited.

(2) Theory and method Length, area, ice thickness and volume are the key parameters for extracting glacier information, which can be determined by remote sensing or field survey approaches. However, it is more difficult to predict the future changes of glacier water resources in Xinjiang. The response mechanism of glaciers to the climate change is firstly made clear and the corresponding glaciology model is established to make simulation verification of the response process. Then the dynamic response processes to the future climate change and the state at different time scales are then simulated using the prediction scenarios of climate change. Glacier response to climate change included direct response and indirect response. The direct response is that the response of glacier surface and mass and energy balance to climate change, which can cause the immediate changes of glacier accumulation and ablation. It can be observed or calculated by mass balance model. The indirect response is the glacier dynamics response to climate change. Glacier is a huge ice body. The movement of glaciers changes the glacier morphology, causing the change of glacier water and thermal conditions. Therefore, glacier dynamics changes will indirectly affect the characteristics of glacial accumulation and ablation in various parts of glaciers. Glacier accumulation and ablation change caused by an indirect response is a slow process, leading to glacier change lag behind climate change. It is more difficult to simulate this process using glacier physical model, the parameter requirement of which are especially high.

Although there are several ways used to predict future glacier changes, for example, time extension method based on statistical model and the mass balance model, it cannot predict the dynamic process of glacier change and the error is large for the long time scales and non-steady-state glaciers (i.e. glaciers in the process of rapid melting). Internationally, glacier dynamics model is commonly used to make prediction of indirect response. Based on mass, energy and momentum conservation, glacier dynamics model can not only predict the geometry response process of the glaciers to the climate change in detail, but also the state of retreat under the climate change scenarios. It can realize the complete process from climate change to glacier mass balance change, dynamic response of glaciers, and finally glacier volume change that is glacier meltwater change (Oerlemans et al. 1998; Aðalgeirsdóttir et al. 2006; Li et al. 2007a).

Due to the high requirement of observed parameters for the glacier dynamics model (Li et al. 2007a), the study aimed at Urumqi Glacier No.1 and Urumqi River

Basin with complete information. In addition, a preliminary simulation analysis has been carried out for Qingbingtan Glacier No.72, located in the upstream of Aksu River in the Tarim Basin with long-term observation. Based on the simulation predicting results, sensitivity test is used to study the relation between changes of parameters and simulation results. These parameters include the glacial parameters, glacier's thermal mechanical parameters and water and heat conditions parameters, etc., containing as many relevant parameters of glaciers in Xinjiang as possible. Therefore, prediction can be used for more glaciers by parameter sensitivity test.

Based on the theory and method mentioned above, on the one hand this part aims to analyze the glacier changes in the last 30–50 years in Xinjiang, on the other hand to explain current glacier changes and their impacts on water resources in Xinjiang, taking the prediction simulation results and parameter sensitivity test of Urumqi Glacier No.1 and Qingbingtan Glacier No.72 as references.

It is important to note that the average area of glaciers in Xinjiang is 1.35 km^2. The amount of the large glaciers ($\geq$100 km^2) is more than the other regions around the country, but with a small proportion. The amount of small glaciers, with the area less than 1 km^2, accounting for more than 70 % of the total glaciers (Shi 2005). The study results of Urumqi Glacier No.1 with the area of 1.7 km^2 are of great reference value for studying small glaciers. The simulation work of Qingbingtan Glacier No.72 with the larger area of 5.62 km^2 is still in the exploratory stage. Due to lack of large-scale glaciers simulation, it is still difficult to make quantitative analysis on the future change of glaciers water resources in Xinjiang in value, process and time scales.

The distribution of glaciers and proportion of meltwater runoff are different in various water systems in Xinjiang. Therefore, the impact and form of future climate change on glacier changes and meltwater runoff are quite different. The main concern is about the future change of water resources caused by climate change, and their impact on all economic regions in Xinjiang. Taking this into account, five economic geography units are classified as follows, including Tarim River Basin, Rivers on the North Slope of Tianshan, Rivers in the East Xinjiang Basin, Ili River Basin and Irtysh River Basin.

5.2.2 Data

The used including observation data of four monitored glaciers, field survey data of eight glaciers, and remote sensing images and topographic maps used for the study of large-scale glacier changes. In addition, references such as "Glacier inventory III of China" (Lanzhou Institute of Glaciology and Geocryology 1986a, b, c, d), "Concise Chinese glacier inventory" (Shi 2005) and "Glacier Water Resources in China" (Yang 1991) are used to obtain the basic information of glaciers and dynamic water resources in Xinjiang.

(1) In-site glacier observation and field survey data The long-term observation and study of glaciers is very limited. Urumqi Glacier No.1 is the longest moni-

Table 5.10 The investigated glaciers

River basin	Four in-site observation glaciers and time series	Eight investigated glaciers and survey period
Rivers on the North Slope of Tianshan	Urumqi Glacier No.1, 1959–2009; Kuitun Haxilegen Glacier No.51, 1998–2009	Kuitun Haxilegen Glacier No.48, 2005; Fan-Shaped Diffluence Glacier, 2009; Glacier No.4 of Sigong River, 2009
Tarim River Basin	Qingbingtan Glacier No.72, 2007–2009	Keqikekuzibayi Glacier, 2007; Qingbingtan Glacier No.74 2008, 2009; Tomor glacier, 2009
Rivers in the East Xinjiang Basin	Miaoergou Ice Cap, 2004–2009	Heigou Glacier No.8, 2009
Irtysh River Basin		Kanas Glacier in the Friendship Peak, 2009
Ili River Basin		

tored glacier in China with the history more than 50 years. Tianshan Glaciological Station (TGS) was established 3 km northeast of the glacier since 1959 by Chinese Academy of Sciences. In 1998, in order to study glacier changes in different regions of Xinjiang, TGS established "Xinjiang glacier observation network". As of 2009, observation glaciers have increased to six, including Urumqi Glacier No.1, Kuitun Haxilegen Glacier No.51, Miaoergou Ice Cap, Qingbingtan Glacier No.72, Fan-Shaped Diffluence Glacier and Kanas Glacier, which has formed a relatively complete glacier observation network over Tianshan and Altai Mountains in China. The biggest advantage of glacier observation is the long-term and detailed data series. In addition to regular observation of glaciers themselves, other related items have been added, such as hydro-meteorological observation, glacier dynamics parameters observation, physical and chemical process of snow to ice transformation, aiming at establishing a quantitative relationship between glaciers and climatic conditions. In this study four glaciers with longer in-site observation data series are selected (Table 5.10), study and future change simulation of which constitute the basic knowledge of glacier response to climate change in Xinjiang.

Glacier investigation began since 1959 in Xinjiang and so far the total number of the investigated glaciers is more than 30 all over the mountain ranges in Xinjiang (Wang et al. 1983; Su 1998; Cao 1993; Mountaineering and Expedition Team of Chinese Academy of Sciences 1985). The field survey data of individual glaciers provide reliable basis for the study of glacier changes despite the short observation time. In this chapter, all of the current data of the eight glaciers are from the investigation implemented by the authors, including the location of the glacier boundaries, mass balance, ice thickness, surface velocity, temperature, hydro-meteorological, etc. The glacier change can be acquired by the comparison of current glacier data with the previous study.

(2) Remote sensing images and topographic maps Changes of glacier area, length and terminus position on the regional scale can be obtained by comparing recent remote sensing images with the earlier topographic maps, which is an important complement to the observation of single glacier and widely used internationally.

Table 5.11 Information of remote sensing images

River basin	Study area	Remote sensing image	Resolution (m)	Period
Rivers on the North Slope of Tianshan	Urumqi River Basin	SPOT5; Multispectral	5	2,005.10
	Toutun River Basin	SPOT5; Multispectral	5	2,005.9
	Kuitun Region	SPOT5; Multispectral	10	2,004.10
	Bogda Peak Region	ASTER	15	2,006.9
Tarim River Basin	Tomor Region	SPOT5; Panchromatic	5	2,003.9
	Yarkand River basin	ASTER	15	2,002.9
Rivers in the East Xinjiang Basin	Miaoergou Region	SPOT5; Panchromatic	5	2,005.8
Irtysh River Basin				
Ili River Basin	Kukesu River Basin	ASTER	15	2,004.9

High resolution remote sensing images of the observed regions and other important regions have been purchased in order to obtain the current glacier information by the combination of remote sensing images and field survey. The aerial photographs and 1:50,000 topographic maps in period 1962–1972 have also collected to obtain the early glacier information. The interpretation of images is strictly according to the approach supplied by World Glacier Monitoring Service (WGMS) (Rau et al. 2005).

The snow cover is a common serious problem for remote sensing images in Xinjiang. Most glacier terminus is debris covered, causing great difficulties to the determination of glacier boundary. SPOT5 high-resolution remote sensing images and a small amount of ASTER images were selected for the (nearly) cloud-free conditions and for the ablation period when the extent of snow cover was minimal to reduce potential uncertainty in glacier boundary delineation due to snow cover, which were the best choices since 2000. A digital elevation model (DEM) was used for the orthorectification of images and determination of glacier boundary. To verify the reliability of the results, purposeful field surveys are carried out every year. Some special observation research has been taken to study the impact of glacier debris.

Changes of 1,800 glaciers in the past 26–44 years have been studied, including 1,543 glaciers in the Tianshan Mountains with the area of 2,979.63 km^2, accounting for 17% of the total number of glaciers (9,035) and 32% of the glacier area (9,925 km^2) in the Chinese Tianshan Mountains (Table 5.11).

5.2.3 Glacier Changes in Different Regions and Their Impacts on Water Resources in Xinjiang

The total area of the selected 1,800 glaciers reduced by 11.7% in the past 26–44 years, with the area reduction of 0.243 km^2 for the individual glaciers and the terminus retreat rate of 5.8 m/a. The area reduction rate is between 8.8 and 34.2% with the average area reduction of 0.092–0.415 km^2 and terminus retreat of 3.5–10.5 m/a.

The area reduction rate (reflecting the relative change) and the average area change (reflecting the absolute change) are shown in Fig. 5.13. Seen from Fig. 5.13,

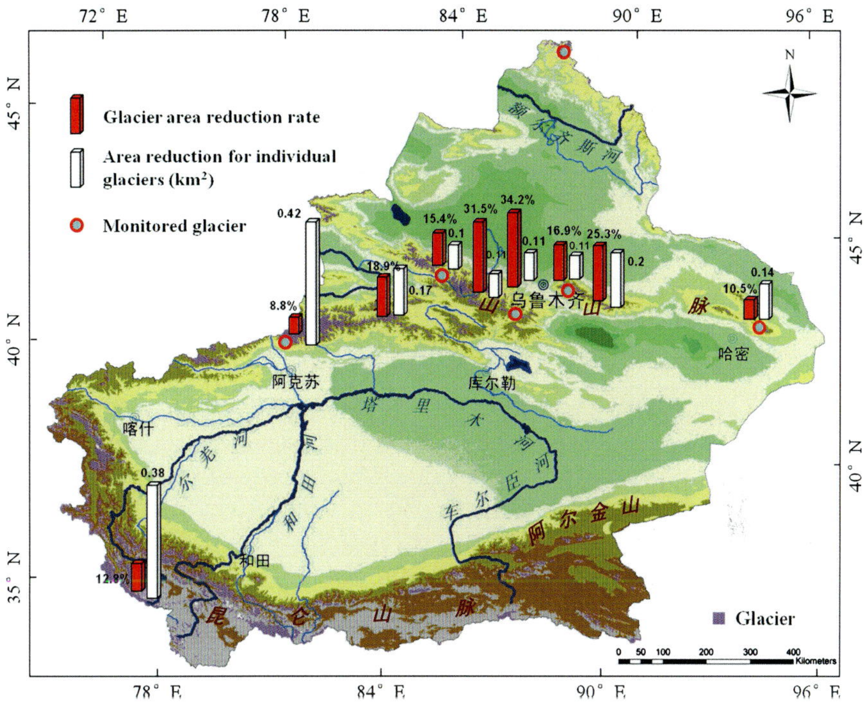

Fig. 5.13 Spatial characteristics of glacier variations in Xinjiang, North Western China

the spatial distribution of the two change indexes are different, which is closely related to the glacier size in the different regions. Because of the sensitive response of small glaciers to climate change, the relative area change is large for a small glacier dominated region but with small absolute value, such as Urumqi River Basin and Toutun River Basin. On the contrary, for the region with large glaciers, the relative glacier area change is small with large absolute value, such as Tomor Region and Yarkant River Basin. Glacier absolute changes also reflected the contribution to the river runoff. Actually glacier variation in different regions depends on not only the glaciers size, but also the water and thermal change (climate change), as well as glacier morphology, terrain conditions and orientation, etc. Even in the same region, similar hydrothermal conditions but different other conditions will still result in the differences of individual glacier changes. For the same reason, changes of the observed glaciers are quite different with the other individual glaciers and average value of glacier changes.

Terminus change is a valid parameter describing glacier change as it can be obtained easily. The terminus retreat rates are significant different in the different study areas within the range of 3.5–10.5 m/a, clearly related to glaciers area reduction rate and glacier length. For the glaciers with the same area reduction, the longer ones will retreat with large amount and vice versa. Moreover, although the glacier change will be influenced by the different time span, the glacier change is relatively slow before the 1980s. Therefore, the result will be in a good comparison covering the recent 30 years.

The glacier volume change can be estimated by the glacier area-volume empirical formula. However, this type of formulas will cause larger errors when calculating volumes of large-scale glaciers. Therefore, the glacier volume changes are not estimated in this chapter.

Shown as Fig. 5.13, the distribution of glacier changes and their impact on water resources in different regions will be discussed as follows.

(1) Rivers on the North Slope of Tianshan There are 3,399 glaciers in the Junggar Basin, the North Slope of Tianshan, with an area of 2,251 km^2 and ice volume of 137.35 km^3. The annual glacier meltwater runoff is calculated to be about 16.89×10^8 m^3, accounting for 13.5 % of river runoff (125×10^8 m^3). The river of this region can be divided into two categories by the amount of glacier meltwater. One category is primarily distributed small glaciers with the area less than 1 km^2, several within 2–5 km^2 and glacier meltwater runoff accounted for 6–20 %, including rivers in north slope of Bogda, Urumqi River, Toutun River, Santun River, Taxi River, Jinghe River, etc. The other category is Manas River, Horgos River, Anjihaihe, etc. The glacier area in this basin is more than 200 km^2 with the glacier meltwater accounting for 35–53 % of river runoff (Lanzhou Institute of Glaciology and Geocryology 1986a, b, c, d; Yang 1991; Shi 2005; Xie et al. 2006).

Urumqi Glacier No.1 and Kuitun Haxilegen Glacier No.51 are two monitored glaciers of this region. Urumqi Glacier No.1 had been retreating since 1959 and accelerated since the 1980s. Glacier area decreased by 14 % in 1962–2006, from 1.95 to 1.68 km^2, with an average reduction rate of 0.0061 km^2/a. Because of the glacier retreat, the two branches of the glacier have separated into two individual ones in 1993. The terminus retreat rate was 4.5 m/a in 1959–1993 and in 1994–2004 the East branch and West Branch retreated at the rate of 3.5 and 5.8 m/a, respectively. The glacier thickness of the mainstream line of Eastern Branch thinned by 0–30 m in 1981–2006 with the thickness reduction of the lower larger than the upper. Corresponding with the glacier retreat, the glacier meltwater runoff had substantial increased. The average glacier meltwater runoff of Urumqi Glacier No.1 was 508.4 mm/a, and reached to 936.6 mm/a in 1986–2001 with an increase of 84.2 % (Li et al. 2009b, 2003b, 2007). The observation data showed that the greatly increase of the glacier meltwater runoff is the result of the rapid climatic warming since the 1980s.

It is indicated that the accelerated ablation is very remarkable for Kuitun Haxilegen Glacier No.51 since the first observation in 1998, although the ablation extent was less than Urumqi Glacier No.1. Glacier area has been reduced from 1.558 km^2 (1964) to 1.356 km^2 (2004), with the shrinkage of 0.202 km^2 or 13 % and the terminus retreat rate of 3.9 m/a. The average annual terminus retreat rate was 2 m/a in 1964–2006, while in 1999–2006, the rate reached to be 5.1 m a^{-1}, representing an increase of 1.5-fold. The velocity of glacier surface showed a decreased tendency, indicating the glacier thinning. The terminus retreat rate of the neighboring Haxilegen

Glacier No.48 is no different than Haxilegen Glacier No.51, but the area change is relatively smaller for Haxilegen Glacier No.48, which is the result of its larger area.

Massive unfrozen glacier meltwater was still existed in the firn under 10 m glacier accumulation area in October, indicating that the glacier was in the state of rapid ablation because of its low cold reserves and weak resistance to climate warming. Seen from the data of Jiangjunmiao Hydrological Station downstream of the two glaciers, glacier meltwater runoff increased significantly since the 1990s.

In 1962–2006, the area of Fan-Shaped Diffluence Glacier and Glacier No.4 of Sigong River in the north slope of Bogda reduced by 7.1 and 10.6 %, with the average terminus retreat rate of 8.7 and 7.5 m/a, respectively. Compared to the investigation in 1981 (Wu et al. 1983), a large number of glacier lakes appeared in 2009, revealing the increase of glacier melting.

The combination of remote sensing images and field surveys are used to study 588 selected glaciers of this region, which are distributed in the Urumqi River basin, Toutun River Basin, Kuitun River Basin and Mt. Bogda Region. There are 150 glaciers in the Urumqi River Basin. The total area decreased by 13.8 % from 48.667 km^2 in 1964 to 41.965 km^2 in 1992 (Chen et al. 1996), and to 32.052 km^2 in 2005 (34.2 %), with the average glacier area reduction of 0.111 km^2 and the terminus retreat of 5.0 m/a. Eleven glaciers among these had completely disappeared due to serious ablation. Toutun River Basin is close to Urumqi River Basin, of which is small glaciers dominated. One hundred and seventy-two glaciers are existed in the basin in 1964 and five small glaciers had disappeared by 2005. The glacier area reduced by 31.5 % during the past 41 years with the average reduction of 0.092 km^2 and terminus retreat of 4.8 m a^{-1} for the individual one. There were 167 glaciers in the Kuitun River Basin in 1964. The area had decreased by 15.4 % with the average glacier shrinkage of 0.095 km^2 and terminus retreat rate of 3.5 m/a from 1964 to 2004. Eleven glaciers had disappeared. The glacier area of the 99 glaciers in the north slope of the Mt. Bogda Region had reduced by 16.9 %, with the average glacier shrinkage of 0.107 km^2 and terminus retreat of 3.6 m/a.

The simulation prediction results show that Urumqi Glacier No.1 will continue to retreat until 2,180 even if the current climate conditions remain unchanged, while the glacier meltwater runoff will steadily decrease in the next century and be half as much as present to 2,075. According to the climate change scenarios in the IPCC Report (IPCC 2007), Urumqi Glacier No.1 will disappear in the next 70–90 years, and probably in 50 years in the condition of extremely warming. Glacier runoff will remain relatively stable in the next 40 years, and then rapidly decline. This tendency will continue until the glaciers disappear. Whether the glacier melting water runoff will increase, it depends on temperature increasing rate. If adopting measured data from the nearby meteorological station to simulate, the glacier meltwater runoff will increase in the next 30 years by a rate up to about 21 % of the current level, and then falling sharply until the glaciers disappear. Taken the simulation results on the 150 glacier in the Urumqi River Basin, it is found that meltwater runoff has a slightly increasing trend in the next 20–30 years, then a sharp weakening, until all the glaciers disappear. Glacier meltwater runoff changes in the Urumqi River Basin is faster than Urumqi Glacier No.1 due to that Urumqi Glacier No.1 is one of the largest glaciers in this river basin.

By the sensitivity analysis, a total of 7,250 glaciers in the Tianshan Mountains are likely to change and disappear faster than Urumqi Glacier No.1. The number, area (2,133 km^2) and volume (60.27 km^3) of these glaciers account for 80 %, 23 % and 6 % of all glaciers in the Chinese Tianshan Mountains respectively.

The factors causing these glaciers melting stronger than Urumqi Glacier No.1 are such as instance glacier size smaller and altitude lower than Urumqi Glacier No.1, etc. If more strict parameters are used to make sensitivity analysis, the number of glaciers changing faster than Urumqi Glacier No.1 will be less, but the credibility will enhance.

Despite there are some difference in the glacier relative changes in the north slope of Tianshan, the average change for the individual ones are almost the same (about 0.1 km^2). According to the research results and sensitivity analysis, glaciers smaller than 1 km^2 in the Rivers on the North Slope of Tianshan tend to disappear within the next 30–40 years and glaciers larger than 5 km^2 will be still in the tendency of melting. Therefore, there are large differences in the impact of climate change on glacier water resources in the different basins of this region. For the river basins with smaller glacier coverage degree, such as Urumqi River Basin and Toutun River Basin, glacier meltwater runoff will become insignificant. However, for the river basin with larger glacier coverage degree, glacier meltwater run off will continue to maintain certain amount.

(2) Tarim River Basin Tarim River Basin is the largest interior river basin in China, with the runoff from the surrounded mountains into the Tarim River. There are 11,665 glaciers with the area of 19,878 km^2 and volume of about 2,313 km^3 in China and 613 glaciers in Kyrgyzstan with the area of 2,120 km^2 and volume of 260 km^3. The estimate total amount of glacier meltwater account for more than 40 % of the total runoff, of which the glacier water runoff in Yarkant River, Yulong Kashgar River and Kumalik River at the headwater of Tarim River was as much as 50–80 % (Lanzhou Institute of Glaciology and Geocryology 1986a, b, c, d; Yang 1991; Kang et al. 2000; Shi 2005; Xie et al. 2006).

Qingbingtan Glacier No.72 located in the upper of Aksu River of the Tomor Region is the monitored glacier of this region. Aksu River is the main tributary of the Tarim River, the water supply of which accounts for 70 % of the Tarim River runoff. The area of Qingbingtan Glacier No.72 decreased from 7.27 to 5.62 km^2 (22.7 %) in 1964–2008. The terminus retreat rate reached to 41.0 m/a. It is thin at the glacier terminus with the glacier temperatures close to 0 °C. The thickness of debris cover has a close relation with glacier ablation. The glacier surface velocities are up to 70 m/a and the dynamic effects cannot be ignored. Compared to Urumqi Glacier No.1, the ablation and movement supplies of Qingbingtan Glacier No.72 are much more intensive, which is characterized by the temperate-glacier and are very sensitive to climate change.

Three glacier been surveyed recently is Qingbingtan Glacier No.74 (9.55 km^2), Keqikuzibayi Glacier (42.83 km^2) and Tomor Glacier (310.14 km^2). The area of the three glaciers decreased by 14.7 %, 4.1 % and 0.3 % in the period 1964–2009 and the terminus retreat rate was 30.0, 22.9 and 3.0 m/a respectively. The glaciers are in the instable state due to intense ablation. The glacier has become very thin at the terminus and the area-volume empirical formula established in the river basins cannot be applied any more. Tomor Glacier is a huge glacier of this region. Investigation found that changes of the area and terminus are relatively small but the thinning is very remarkable, owing to the debris cover at the terminus.

Seven hundred and forty glaciers located in the Tomor Region at the head water of the Aksu River and upstream of Yarkant River have been studied by remote sensing and field survey. There are 483 glaciers in the Tomor Region, located in the northern and southern slope of Haerketawu Mountain, area of which differed greatly with the largest one more than 300 km^2. The total area of glaciers reduced by 8.8 % from 2,267.708 km^2 in 1964 to 2,067.412 km^2 in 2003 with the area reduction of 0.415 km^2 for individual glacier and terminus retreat of 6.2 m/a. The upstream of Yarkand River contains 257 glaciers, with a total area of 768.633 km^2 in 1976. It reduced to 670.328 km^2 in 2002, accounting for 12.8 % of the value in 1976 with an average shrinkage of 0.383 km^2 and terminus retreat of 10.5 m/a.

Observation found that there are four typical features of glaciers' response to the climate change in the Tomor region: (a) the melting of glaciers was extremely serious. Although the relative area reduction of glaciers was smaller, it exhibited larger absolute loss, (b) many large glaciers decreased in the state of thinning. Generally, in terms of glacier melting, changes of glacier are determined by area (or terminus) and thickness change. However, the change rate of glacier area (or terminus) largely depends on thickness of debris cover on the glacier tongue, which delayed the terminus retreat to a certain extent. It can be confirmed in the fact that the thickness of glaciers had become very thin, (c) the temperature increasing has an important influence on the compound valley glaciers with lower latitude. The compound valley glacier had occupied a very major part in this region, most part of which was existed in the bottom of the valley with a lower terminus latitude, indicating more sensitive to the climate change and the volume accounting for over 70 % of the whole glacier, which is the main part of the glacier meltwater runoff. According to the dynamics simulation of Qingbingtan Glacier No.72, the tongue of glaciers will disappeared in the next about 10 years, leading to a greatly reduced runoff, while the remaining small amount of ice are distributed at high altitudes, although it can exist for a long time, the glacier meltwater is limited, (d) the debris cover delay the terminus retreat but the effect of decreasing the glaciers ablation are limited. These glaciers were deemed as "tomor type" previously with debris cover widely distributed showing strong protection to glaciers. However, it indicated that debris cover merely account for 14.9 % of glacier area by remote sensing analysis, mainly distributed at the glacier terminus. Field observations found that only a small part of the debris thickness is more than 6–10 cm, inhibiting glacier ablation. Most of the debris thickness is very thin, enhancing the glacier ablation. The four characteristics mentioned above show that the glaciers in the Tomor Region are rapidly melting and the rate is much faster than expected.

Due to lack of field surveys, the whole characteristics of glacier ablation in the Yarkand River Basin cannot be realized accurately. Remote sensing analysis shows that the area reduction for the individual glacier was slightly larger than in the Tomor Region, indicating the glacier melting is still very intensive. Comparing with glaciers in the Tomor Region, the debris cover of these glaciers is smaller, which is probably a reason leading to the larger glacier area reduction.

According to the glacier distribution characteristics and glacier melting changes, combined with the simulation of typical glacier, it is speculated that, if continued warming in the next 30–50 years, the glaciers meltwater runoff will maintain a certain level. However, the sensitivity of meltwater runoff to climate change will increase and low water level intensifies in the periods with low temperature. Since then, most glacier solid resources exhausted, leading to a sharp drop of glacier runoff and eventually at a lower level. Further observations are urgently needed due to lack of systemic observation data, short time series and prediction uncertainty of time scales in glacier change process. Overall, the glacier water resources play a pivotal role in the Tarim Basin. Glaciers are in the state of intense ablation at present and once melting away, water resources will have disastrous effects to this region.

(3) Rivers in the East Xinjiang Basin Eastern Xinjiang Turpan-Hami Basin is seriously water deficient. Four hundred and forty-six glaciers with an area of 252.7 km^2 and ice volume of 11.4 km^3 are distributed in the Harkeli Mountain, Balikun Mountain, southern slope of Bogda Mountain and Tianger Mountain, which are the important water resources for the Urumqi and Turpan-Hami Basin. The area of glaciers in this region decreased 11 % over the past 40 years. It is estimated that annual glacier meltwater runoff at present is over 25 % of the river runoff mentioned above.

Glaciers located in the southern slope of Bogda belong to Turpan Basin river system. A field survey of Heigou Glacier No.8 in the southern slope of Bogda was carried out in August 2009. By comparison of topographic maps in 1962 and field survey data in 1981, it was found that the glacier area decreased by 1.3 % from 5.71 to 5.63 km^2 with the average terminus retreat rate of 11.0 m a^{-1}. Due to the narrow glacier tongue and low latitude of the terminus, the thinning changes of the glacier mainly occurred at the terminus. The average thinning of the glacier mainstream line was 10 m during 1981–2009 (Wang 1991).

One hundred and four glaciers in the southern slope of Bogda were studied by remote sensing and field survey. Results showed that the total area reduced by 25.3 % with an average shrinkage of 0.198 km^2 and the terminus retreat of 4.5 m/a in 1962–2006. Changes of glaciers in the north slope of Bogda Peak were smaller than the south slope during the same period. Twelve glaciers had disappeared in the south and north slopes of Mt. Bogda. A heavy reduction in the capacity of the local karez system can be obviously related to glacier retreat.

Miaoergou Ice Cap, located in the water system of Hami Basin, was observed since 2004 (Li et al. 2007c). Observation showed that the glaciers area reduced from 3.64 to 3.28 km^2, by 0.360 km^2 or 9.9 % in 1972–2005. The largest average retreat rate was 2.3 m/a for Miaoergou Ice Cap. The largest average retreat rate increased to 2.7 m a^{-1} since 2005. Mass balance observation showed that there was a slight melting at the top of the glacier. The ice core data indicated accelerated glacier melting in the recent 20–30 years. The ice thickness decreased by 0–20 m in 1981–2007, mainly occurred in the lower part of the ice cap and the thinning is not obvious at the top. The glacier temperature was relatively low with the value of −8 °C at the bottom of 60 m to the glacier surface.

By the remote sensing images and field survey, there are a total of 75 glaciers studied in the Miaoergou-Yiwu River Basin, belonged to the southern and northern slope of Harlik Mountain, of which 50 glaciers in the southern slope and 25 in the northern slope. The average area of the glaciers in the southern slope is slightly larger than the northern slope.

The area of 75 glaciers decreased from 98.252 to 87.964 km^2 in 1972–2005, with an area loss of 10.5 %. The average area reduction for individual glacier was 0.137 km^2 and the terminus retreated at the rate of 5.0 m/a. Four glaciers had already disappeared.

Changes of the glaciers in the Miaoergou Region are relatively smaller than the other regions of Xinjiang, because the altitude of the preserved glaciers is higher and the average size is larger. However, the water resources system is vulnerable, highly depending on the glacier meltwater. Rapid changes of glaciers in the recent 20 years indicated that glaciers were in the state of accelerated melting. Based on the observation data of Hami Hydrological Bureau, the runoff of rivers without glaciers meltwater supply, such as Toudaogou River, had decreased in the background of temperatures increasing and a slight increase of precipitation in the recent years, indicating that precipitation increment cannot compensate for the increased evaporation. For the rivers with less glacier meltwater supply, such as Guxiang River, their runoff kept increasing before 2,000 and after that there was a trend of decreasing and slower increasing. Moreover the range of change in river runoff was enlarged and the difference of amount of water in flood and dry season was large. The dry season was also extended. These are likely to be caused by the diminished regulation effect. For the rivers with larger glacier meltwater supply, such as Yushugou River, although the runoff continues to increase, the increase rate has started to decrease. The runoff process reflected the impact of glacier changes on hydrology and its different stages, which is mainly characterized by the decrease of water resources.

In short, glaciers in the Mt. Bogda Region of Tulufan Basin water system are in rapid retreating, regardless of the southern and the northern slope, exerting a significant impact on water resources to the downstream Urumqi and Tulufan Basin. There is a trend of increasingly strong ablation of glaciers in the Miaoergou Region of Hami Basin water system glaciers, resulting in a significant impact on the water resources and its annual distribution. Overall, glaciers in the East Xinjiang Basin are accelerated melting, leading to worsening conditions of water resources.

(4) Ili River and Irtysh River Basin Ili River is one of the largest rivers in Xinjiang. A total of 2,373 glaciers are located in this river basin with the area of 2,022.66 km^2 and volume of 142.18 km^3. The glacier meltwater runoff is about 193.0×10^8 m^3, accounting for 19.2 % of the total runoff. Glaciers in the Ili River Basin with the average area of 0.85 km^2 belonging to a medium size are comparatively sensitive to the response of climate change.

Two hundred and ninety-three glaciers in the upstream of Akesu River were studied using remote sensing. The results showed that the total glacier area of 265.812 km^2 reduced to 215.510 km^2 (area loss of 18.9 %) in 1963–2004, with the area reduction of 0.172 km^2 for the individual glacier and terminus retreat of 7.0 m/a. Eleven glaciers had already disappeared. Although these results have not been checked by the field survey data, it demonstrates that glacier change of this region belongs to the middle level of Xinjiang. The contribution and influence of glaciers to the river runoff cannot be ignored.

A total of 403 glaciers with the area of 289.29 km^2 are distributed in the Irtysh River Basin, which lies in the southern foot of Altai, Xinjiang. The glacier meltwater runoff of Irtysh River is about 7.73×10^8 m^3, accounting for 7.7% of the total runoff of about 100.0×10^8 m^3. The average area of glaciers in the Irtysh River Basin was 0.72 km^2 and the altitude of its snow line is the lowest in China.

Glaciers are mainly distributed around the Friendship Peak, the main peak of the Altai Mountains. An investigation of Kanas Glacier was carried out in 1980, which is the largest glacier near the Friendship Peak (Wang 1983). A second investigation was taken in August 2009. It was found that snow in the accumulation area of Kanas Glacier was significant thinning and the elevation of the snow line increased at least 30 m since 1980 due to the relative low altitude of glaciers and their sensitive response to climate warming. Some caves and a huge hole are formed at the lower part of the glacier with the melting water surging out. Comparing with the previous topographic maps, the glacier area reduced from 30.13 km^2 in 1959 to 28.74 km^2 in 2009 with a area loss of 4.6% and terminus retreat rate of 16.4 m/a.

Although a substantial proportion of glacier meltwater supplies in the Ili River and Irtysh River, the snow runoff plays the leading role. The impact of climate change has already behaved, such as Kelan River, of which meltwater runoff dominated in the Irtysh River Bain. The maximum runoff of the river has arrived a month earlier from June in the 1950s–1960s to May at present. The earlier melting of snow can improve the water supply in spring, especially the water for the irrigations in spring for agriculture and crops growth. However, the decrease of runoff in summer, especially from July to August, will have a great impact on downstream agricultural production, fisheries, etc.

Overall, the impact of future glacier changes on water resources is limited, but will greatly weaken the regulation effects of the glacier meltwater runoff. Special attention should be given to the impact of climate change to snow runoff and its possible consequences.

5.3 Rapid Shrinking of Glaciers and Its Impact on Water Resources in the Middle Qilian Mountain Region

Mountain Glaciers are sensitive climate indicators and thus subject to monitoring of environmental and climate changes (Oerlemans 2005, 1994). As a temperate country at low latitudes, China has the most abundant mountain glacier resources, with 46,298 glaciers having a total estimated volume of 5,590 km^3 (Shi et al. 2000). Glacier runoff is the major contributor to water resources that are used to support the sustainable development of the environment, industry and agriculture in arid and

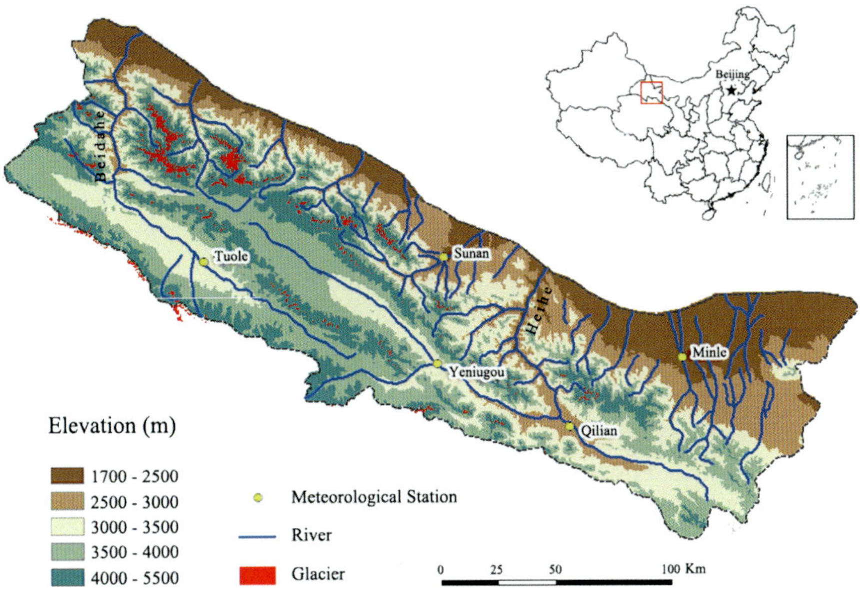

Fig. 5.14 Sketch map showing the study area and distribution of glaciers

semiarid regions, especially in northwest China (Yao et al. 2004). Recent evidence suggests an acceleration of glacier mass loss in several key regions in China with the climatic warming, such as Tibetan Plateau, Tianshan, Kunlun Shan (Li et al. 2010a, 2006; Shangguan et al. 2007; Ye et al. 2006; Jin et al. 2005; Yang et al. 2003; Fujita et al. 2000b), which has attracted wide attention.

Research on large-scale and long-time glacier fluctuations in the Qilian Mountain Region is limited, despite the fact that the glaciers in this region provide important water resources for local economic development. There has been some research work in the Qilian Mountains. From the late 1950s to the late 1980s, scientists from the Chinese Academy of Sciences (CAS) conducted some glacier expeditions in the region (Liu et al. 1992; Xie et al. 1985; Academia Sinica 1959). Liu et al. (2003) analyzed fluctuations of glaciers in the western Qilian Shan from the Little Ice Age to 1990. However, glaciers in the Qilian Mountain Region have experienced mass loss prominently owing to an increase in temperature of $0.5 \pm 0.2\,°C$ over the past 50 years in Northwest China (Li et al. 2003a). Therefore, it is necessary and urgent to do further research on recent glacier changes in the other parts of the Qilian Mountain Region.

Taking this into account, this part concentrates on the glacier changes in the middle Qilian Mountain Region and its impact to water resources. Glaciers, acting as a huge alpine reservoir that regulates annual runoffs, are a reliable water source for oases and for the sustainable development of the ecological environment, industry and agriculture in this region (Shen et al. 2001; Shi 2001; Yang 1991). According to the image interpretation method provided by World Glacier Monitoring Service

(WGMS), historical data (topographic maps, aerial photographs) are available for comparison with ASTER images in order to estimate changes of glaciers in the middle Qilian Mountain Region.

5.3.1 *Study Area*

The Qilian Mountains (36.5 °~39.5° N, 93.5 °~103 °E) are located on the north-eastern edge of the Tibetan Plateau. Their northern part is the Hexi Corridor of Gansu Province, and the southern part includes the Qaidamu basin and the Qinghai lake of Qinghai Province. The region of interest lies in the middle part of the Qilian Mountains, in the head water of rivers such as the Heihe and Beidahe (Fig. 5.14). The Heihe is the second largest inland river in the arid area of Northwest China, with the main stream extending 821 km in length. The drainage area, south of the Yingluo Gorge has catchments accounting for about 10,009 km^2 with an elevation ranging from 1,700 to 4,823 m a.s.l., characterized by high precipitation, low evaporation and low temperature (Chen et al. 2005; Lan et al. 2005; Gao and Li 1991).

According to the Glacier Inventory of China I (Qilian Mountains) (Wang et al. 1981), 1,078 glaciers were listed with a total estimated area and volume of 420.55 km^2 and 13.67 km^3, respectively, including 428 glaciers of the Heihe and 650 glaciers of the Beidahe. These data were compiled in 1979 on the basis of aerial photographs acquired largely in 1956, but partly in 1957 and 1966, and topographic maps based on the aerial photographs.

5.3.2 *Data and Methods*

(1) Data Processing and Uncertainty Assessment To compare recent changes in glacier area over the past ~ 50 years, 10 Advanced Spaceborne Thermal Emission and Reflection Radiometer (ASTER) images taken from June to September in 2003 and 2004 (9 in 2003; 1 in 2004) with the resolution of 15 m and 50 topographic maps (1:50,000) derived from aerial photographs acquired during 1956–1978 (40 in 1956; 3 in 1957; 5 in 1966; 2 in 1978) by the Chinese military geodetic service were analyzed. Therefore, the study period can be considered to be approximately from 1956 to 2003. All images were obtained for the (nearly) cloud-free conditions and for the ablation period when the extent of snow cover was minimal to reduce potential uncertainty in glacier boundary delineation due to snow cover, which were the best choices since 2000. A digital elevation model (DEM) with resolution of 30 m, provided by Environmental and Ecological Science Data Center for West China, National Natural Science Foundation of China, was used for the orthorectification of ASTER images using methodologies described by Toutin (2002) and Kääb (2004) and PCI Geomatica 9.1 Orthoengine software. Geocorrection and co-registration were established using ERDAS Imagine 9.0 software. The clearly distinguishable terrain features were selected from topographic maps that

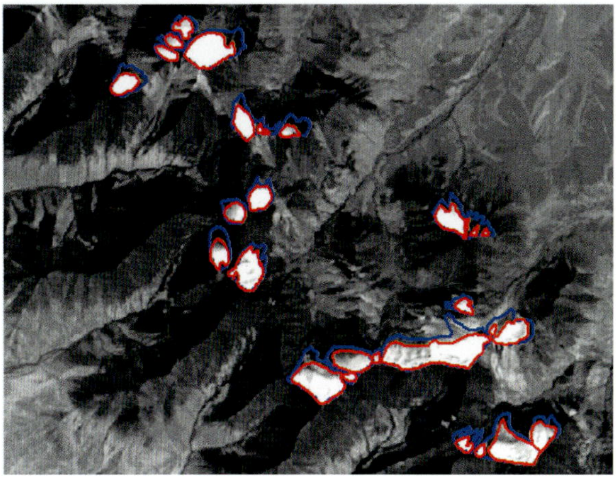

Fig. 5.15 Example of the glacier outlines extraction from ASTER images and topographic maps. *Red* glacier outlines are of 2003 and *blue* outlines of 1956

could be identified on each image. On average, 30–50 GCPs have been collected with the root mean square error (RMSE) value <1.0 pixel (15 m) in both of the x and y directions. All images and maps were presented in a Universal Transverse Mercator (UTM) coordinate system referenced to the World Geodetic System of 1984 (WGS84).

(2) Glacier Outline Extraction The glacier outlines are mapped manually with the DEM by commercial GIS software (ARCVIEW), using the topographic maps and ASTER images (Fig. 5.15), which is a useful tool for extracting detailed information from satellite imagery for glaciers (Raup et al.) particularly when mapping is conducted by the same person using a combination of different types of imagery (e.g. aerial photographs and ASTER) (Paul et al. 2002). Area and other parameters of the glaciers in the different periods can be computed from the extracted glacier polygons, resulting in a total sample of 910 glaciers. For some images and sections containing clouds or fresh snow cover, the outlines of some glaciers were not validated accurately, which were omitted. Supra-glacial debris cover is a factor reducing the accuracy of mapping as it contributes to the overall accuracy of the glacier outline. However, glaciers are debris-free in this particular case.

5.3.3 *Characteristics of Glacier Distribution*

The characteristics of glacier distribution in the study region are investigated by analyzing statistically the relations between topographic parameters (Fig. 5.16) and the glaciers using the extracted glacier polygon data in 2003. The relationship between area and number of glaciers and aspect shows clearly regional characteristics of glacier distribution. Seven hundred and twenty-one glaciers with areas of 245.79 km^2 which account for 79.0% of the total area locate in the three sectors northwest, north and northeast, similar to other regions of the Northern Hemisphere

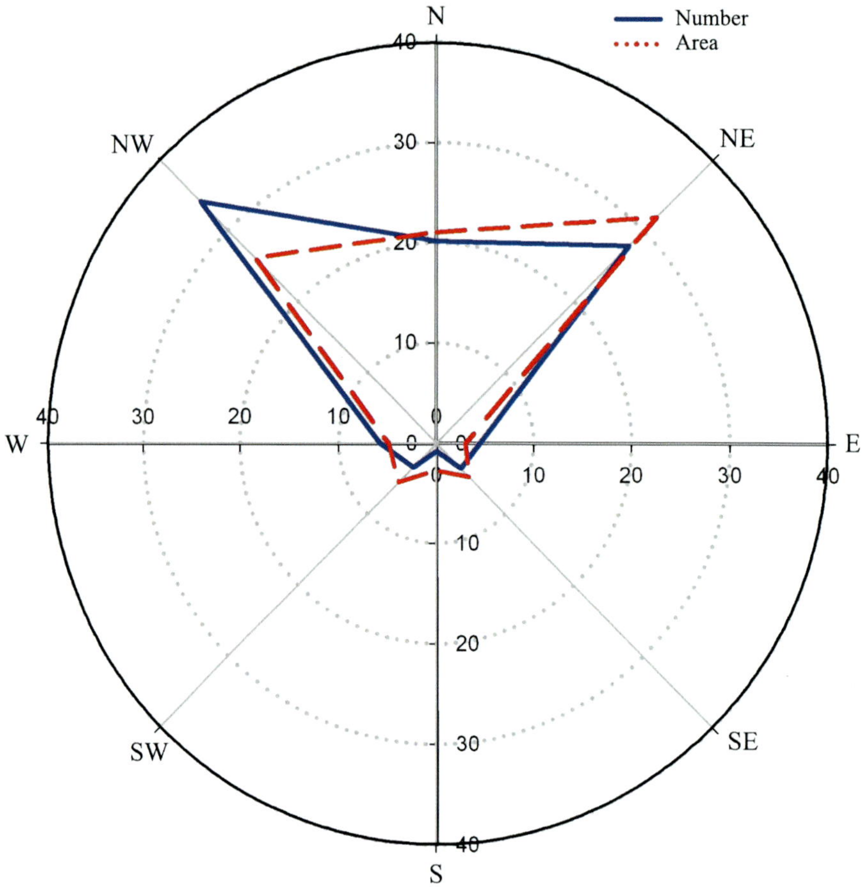

Fig. 5.16 Distribution of glacial number and area with different aspects in 2003/2004

(Evans 2006). The number of glaciers smaller than 1.0 km^2 in the three aspects occupies 93.1 % of the total value in 2003.

Figure 5.17 presents the distribution of glacier coverage in the middle Qilian Mountain Region, Heihe River Basin and Beidahe River Basin from approximately 1956–2003, according to the glacier size class (0.01–0.1 km^2, 0.1–0.5 km^2, 0.5–1.0 km^2, 1.0–5.0 km^2, > 5.0 km^2). In the middle Qilian Mountain Region, about 90.4 % of all glaciers in 2003 are smaller than 1.0 km^2 with a contribution of about 57.6 % to the total area, while only two glaciers in this sample are larger than 5.0 km^2 with a total area of 10.92 km^2. Moreover, about 9.3 % of the glaciers with the area of 1.0–5.0 km^2 account for 39.0 % of the total area. Glaciers with the area of 0.01–0.1 km^2, 0.1–0.5 km^2 and 0.5–1.0 km^2 account for 15.6, 56.9 and 18.0 % of the total numbers, respectively, and the summation of their areas contribute more than half of the total areas. Therefore, small glaciers (< 1.0 km^2) form a remarkable

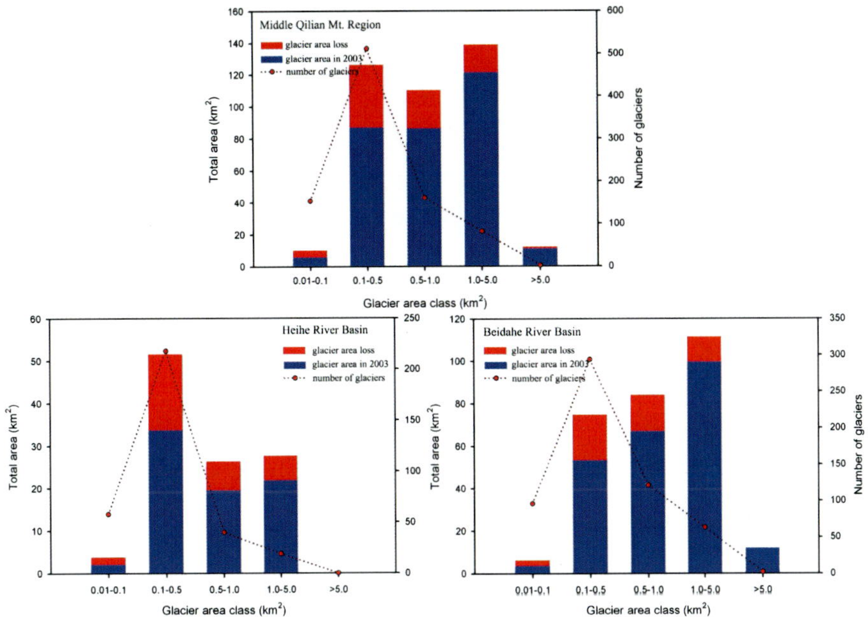

Fig. 5.17 Glacier area, number and area loss by glacier size class for 1956–2003 in the middle Qilian Mountain Region, including the Heihe River Basin and the Beidahe River Basin

Table 5.12 Glacier area changes between 1956 and 2003 in the middle Qilian Mountain Region

Basin	1956		2003		Glacier changes		
	Number	Area (km^2)	Number	Area (km^2)	Number	Area (km^2)	Percentage (%)
Heihe River Basin	335	109.35	322	76.94	−13	−32.41	−29.6
Beidahe River Basin	575	288.06	557	234.08	−18	−53.98	−18.7
Total	910	397.41	879	311.02	−31	−86.39	−21.7

part of the number and area cover in this region. In the Heihe River Basin, about 97.2 % of all glaciers are smaller than 1.0 km^2 with a contribution of about 82.5 % to the total area, while only 2.8 % in this sample are larger than 1.0 km^2 and contribute 17.5 % to the total area. No glaciers are larger than 5.0 km^2. In the Beidahe River Basin, small glaciers with area less than 1.0 km^2 occupy 57.2 % of the total area and glaciers in 1.0–5.0 km^2 size class has largest proportion of the total area (42.5 %).

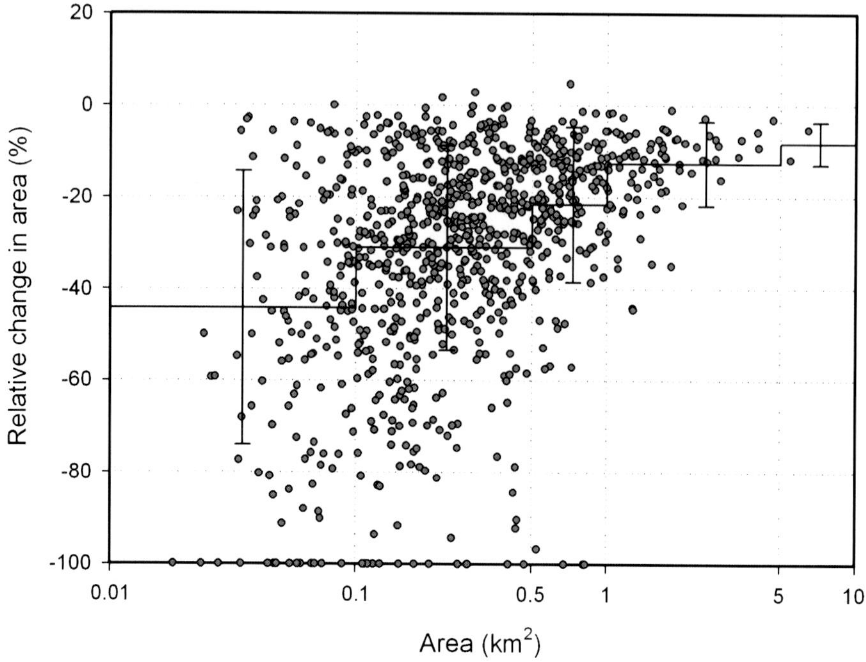

Fig. 5.18 Relative changes in glacier area of different sizes in the middle Qilian Mountain Region from 1956 to 2003. Mean values of glacier area change (*horizontal line*) together with standard deviation (vertical bars) are given for four area classes (in km^2: 0.01–0.1, 0.1–0.5, 0.5–1.0, 1.0–5.0, > 0.0

5.3.4 Changes in Glacier Area from 1956 to 2003

The glacier areas determined from the ASTER images allowed for analysis of the change in area between 1956 and 2003 (Table 5.12). The results show that the total area of the 910 studied glaciers had decreased by 21.7 % of the 1956 value from 397.41 to 311.02 km^2 with the area reduction of 0.10 km^2 for an individual glacier. Three glaciers, which enlarged their areas, with the largest increasing rate less than 5 %, were smaller than 1.0 km^2 and located on the lee side of the ridges. Accumulation on such glaciers depends primarily on snow drift and their behavior is determined by local conditions rather than climatic variations. All the glaciers in Colorado are of the same type (Raup et al. 2007b). The overriding importance of local conditions for the behavior of small glaciers has also been noted by Surazakov et al. (2007) for the Altai. The rate of glacier shrinkage was 29.6 and 18.7 % for the Heihe River Basin and the Beidahe River Basin, respectively. Thirty-one glaciers at lower elevations with area less than 1.0 km^2, which were present on the topographic maps, had disappeared completely by 2003, 16 of which were smaller than 0.1 km^2, 12 in 0.1–0.5 km^2 and the others in 0.5–1.0 km^2. Thirteen of the 31 disappeared glaciers were located in the Heihe River Basin and the others in the Beidahe River

Table 5.13 Comparison of glacier changes in the eastern, middle and western Qilian Mountain Region

Qilian Mountain Region	Study area	Period	Number of glaciers	Area change		Source
				km²	%	
Eastern	Lenglongling	1960s/1970s–2002	244	−29.85	−28.97	Zhao 2009
Middle	Middle Qilian Mountain Region	1956–2003	335	−86.39	−21.7	This study
Western	Western Qilian Mountain	1956–1990	1731	−151.9	−12	Liu et al. 2003

Basin. The average terminus retreat was 189 m with the reduction rate of 20.6 %. The glacier shrinkage shows obvious regional differences, which was strongly affected by the obvious differences in climatic conditions, due to the long distance from east to west of the middle Qilian Mountain Region. The annual temperature for the middle Qilian Mountain Region, including Beidahe River Basin and Heihe River Basin, showed increases with the range of 0.54~0.92 °C from 1950s to the present. The precipitation of the west was mainly affected by vapor transport of the Atlantic Ocean brought by the West Circulation, showing increased tendency, while the east was affected by the vapor transport of the Indian Ocean and Pacific Ocean brought by the Southeast and Southwest Monsoon, showing decreased tendency in the recent decades (Lan et al. 2004). The precipitation, determines glacier accumulation, for the Beidahe River Basin was larger than the Heihe River Basin, directly affecting the regional differences of the glacier shrinkage for the middle Qilian Mountain Region. Moreover, the proportion of glaciers smaller than 1.0 km^2 in the Heihe River Basin was higher (94.3 %) than the Beidahe River Basin (88.7 %).

The rate of shrinkage varied between glaciers of different size. The pattern of relative changes in glacier area between 1956 and 2003 is illustrated by Fig. 5.18. In 1956, 90.8 % of glaciers were smaller than 1.0 km^2 and occupied 62.0 % of the glacierized area. These glaciers accounted for 78.4 % of the total loss. This is a larger contribution than that observed in the Swiss Alps, where small glaciers accounted for 44 % of the total loss between 1973 and 1999 occupying 18 % of the glaciated area (Paul et al. 2004). Glaciers in the size class of 0.01–0.1 km^2 had lost 44.5 % of their total areas with the average area reduction of 0.03 km^2. Glaciers larger than 1.0 km^2 had only lost 12.3 % of their areas. However, the average area reduction reached to 0.22 km^2. In addition, the area reduction rate was 31.2 % (0.08 km^2) for 0.1–0.5 km^2, 21.7 % (0.15 km^2) for 0.5–1.0 km^2, 12.7 % (0.22 km^2) for 1.0–5.0 km^2 and 8.3 % (0.49 km^2) for >5.0 km^2, respectively. As for the area changes of glaciers in different size, the relative changes of small glaciers were usually higher than those of large ones, which exhibit larger absolute loss. This indicated that the small glaciers were more sensitive to climate change than large glaciers (Jóhannesson et al. 1989; Knight 1998; Nesje and Dahl 2000). The standard deviation of glacier area changes for glaciers in 0.01–0.1 km^2, 0.1–0.5 km^2, 0.5–1.0 km^2, 1.0–5.0 km^2 and >5.0 km^2 were 29.80, 22.35, 16.90, 9.22 and 4.66, respectively. The scatter of individual changes was strongly increasing towards smaller glaciers, which further demonstrated the particularly sensitive of small glaciers to climate change.

5.3.5 *Regional Differences in Glacier Area Changes in the Qilian Mountain Region*

While using remote-sensing data and geographical information systems are informative for understanding glacier changes over large regions, comparison was made with other glaciers in the Qilian Mountains of Northwest China. Two points emerged from the comparison studies: (1) glaciers in the Qilian Mountain Region are shrinking and (2) the rates of shrinkage vary in different regions as illustrated by Table 5.13. The largest shrinkage rate was in the middle Qilian Mountain Region, where glacierized area had declined by 21.7% during the past ~50 years, according to this study. The second was Lenglongling, eastern Qilian Mountain Region, with the shrinkage rate of 28.97% from 1960s/1970s to 2002 (Zhao 2009). The smallest decrease in glacier area was in the Western Qilian Mountain Region (12%) (Liu et al. 2003). Study of glacier changes in the western Qilian Mountain Region was limited from 1956 to 1990 and the reduction rate would probably change with the rapid temperature increasing, which can cause the comparison of results to be misleading; however, due to the lack of data, there may be no better way than to compare the area reduction rate for time periods that are as close as possible.

Furthermore, the shrinkage rate of glaciers in the Qilian Mountain Region appears to be faster than the other glaciers in the middle Chinese Tien Shan (13% in 1963–2000), west Kunlun Shan (0.4% in 1970–2001), Tarim Interior River basin (3.3% in 1960s/1970s–1999/2001) and Xinjiang Uyger Autonomous Region (11.7% in 1960s/1970s–2002/2006) in western China (Li et al. 2006, 2010; Shangguan et al. 2009, 2007). Mass balance observation of the Qiyi Glacier in the middle Qilian Mountains had further demonstrated this result. The Qiyi Glacier experienced positive mass balance of about 360 mm per year in the 1970s and 4 mm per year in the 1980s. However, the recent observation shows that mass balance of the glacier was −810 mm in 2001/2002 and −316 mm in 2002/2003, respectively, suggesting that the glacier was thinning dramatically (Pu et al. 2005). In the last 2–3 years the glacier has characterized by strong negative mass balance and its snow line has gone up steadily. Now the equilibrium line of the glacier has reached the highest altitude since the glacier has been observed. The intensive mass loss of the glacier in the recent years represents the sensitive response to the climatic warming. Against the background of global warming, it is expected that the glacier will experience negative mass balance, thinning, shrinking and rising snow line.

Glacier variation is driven by climate change, especially temperature and precipitation. It showed a significant temperature increase in the Qilian Mountain Region in the late 1980s, especially after 1990s, which basically agrees with the climate change in the northwest China, indicating obvious response to global warming (Jia et al. 2008). According to the study of Jia et al. (2008), the regional differences of climate changes in the Qilian Mountains were analyzed using the monthly temperature and precipitation data of the selected eight stations in the study area. The annual and summer temperature in the eastern, middle and western Qilian Mountain Region both showed a significant upward trend with a mean trend of 0.30 °C/10a, 0.33 °C/10a, 0.27 °C/10a for the annual temperature and 0.27 °C/10a, 0.27 °C/10a and 0.23 °C/10a for the summer temperature. The growth rate of the annual and summer temperature for the middle and eastern Qilian Mountain Region is relatively large, especially in the middle. The annual precipitation fluctuated with the mean rate of 10.0 mm/10a, 12.6 mm/10a and 12.2 mm/10a. In contrast, while the precipitation was almost fluctuating, the comparative rapid increase in temperature was an important factor resulted in the more intensive ablation of glaciers in the middle than the other areas of the Qilian Mountain Region.

Moreover, taking into account of the influence of the glacier size distribution, the average area of glaciers in the Lenglongling, middle Qilian Mountain Region and western Qilian Mountain Region was 0.41 km^2 (2002), 0.35 km^2 (2003) and 0.62 km^2 (1990), respectively. In contrast, glaciers in the middle Qilian Mountain Region were comparatively small and about 90.4 % of all glaciers are smaller than 1.0 km^2. Small glaciers in particular respond quickly and sensitively to the temperature increasing by retreating, which is another important reason for the strong glacier melt in the middle Qilian Mountain Region.

The strong ablation of the glaciers in the middle Qilian Mountain Region was also influenced by the location of the glaciers, elevation, length, orientation, type, configuration, velocity and other indexes (Evans and Cox 2010; Hodgkins et al. 2007). Debris cover (produced by englacial material brought to the ice surface or from falling rocks), a factor which can considerably reduce the ablation processes on a glacial surface, is nonexistent in the middle Qilian Mountain Region, and did not require consideration in this study. Chucca et al. (2005) pointed out that there seems to be a minimum response time (of very few years) between temperature and precipitation changes and the shrinkage of glaciers in the Maladeta massif (Central Pyrenees). As the lag time for a glacier to react to climatic fluctuations depends mainly on its size and its latitudinal location, climate changes, even if subdued, are felt faster and more precisely by smaller glaciers located in warmer settings than by larger glaciers in higher latitudes. As for the small elevation span for the glaciers in this study, the equilibrium line altitude will continue to rise with the climatic warming, and most of the glaciers will be in the ablation area. It is estimated that glaciers in the middle Qilian Mountain Region will experience more intensive shrinkage in the future.

5.3.6 Impact of Glacier Changes on Water Resources

Due to uneven area distribution of glaciers in different river basins, the contribution of glacier meltwater to river runoff is different. Glaciers in the north slope of Qilian Mountains are mainly distributed in the western river basin, meltwater supplies of which accounted for about 5 ~ 10 % (Yang 1991; He et al. 2008).

Climatic warming drives intensive glacier melting, glacier meltwater and its replenishment to the river runoff increasing. Glacier runoff will reduce along with the decrease of glacier area. Simulation results of Urumqi Glacier No.1 show that glacier runoff will increase with the temperature rising and decrease gradually when at its peak. Glacier size in the Heihe River Basin and Beidahe River Basin is small and mostly valley glaciers based, which is quite sensitive to climate change. The glacier area has reduced by 21.7% since 1950s. It can be speculated based on the simulation results of Urumqi Glacier No.1 that the peak period of runoff has already passed and the meltwater runoff are in the stage of decreasing rapidly.

However, glacier areas in the different river basins are different and supplies of meltwater to river runoff and annual allocation have large differences. In order to make quantitative analysis of this contribution, coupling research about glacier dynamic model and hydrological model at the basin scale are needed. To carry out this research, it is very important to obtain parameters about glacier morphology, mass balance, meteorology and hydrology of glaciated area. Shiyi Glacier in the Hulugou River Basin and Qiyi Glacier at the Beidahe River Basin are selected as reference glaciers to develop systematic observation research so as to study the impact of glacier changes to hydrology and water resources on river basin scale through the scale conversion between reference glacier and multiple glaciers in the river basin.

Summary (1) The Urumqi river source region has witnessed the fact of significant temperature upward of +0.22 °C/10a and an increasing trend in precipitation of +21.14 mm/10a. The climate regime shifted from cold-dry to warm-wet, occurring around 1996. Overall, there was a warm and wet tendency in the region during the past 50 years. Urumqi Glacier No.1 has experienced remarkably shrinkage from 1962 to 2009. Its length has shortened by about 215.2 m (9.7%) and its area has diminished by 0.304 km^2 (15.6%). The cumulative mass balance of the glacier was −13 693 mm, equivalent to 15.2 m of glacier ice, showing particular sensitivity to temperature change.

(2) Annual glacier melt runoff and discharge at gauging site have also exhibited increasing trends. A noticeable increase occurred in 1994. Before 1994, glacier runoff accounted for 62.8% of the total river flow, but after 1994, it rose to 72.1% probably due to the strong warming and increased precipitation since 1994.Temperature in the melt season was the governing factor determining the increase of runoff at the Urumqi Glacier No.1 catchment, while precipitation in the same period was the controlling factor in the non-glacierized Empty Cirque catchment. Remarkable mass loss of Urumqi Glacier No.1 led to continuously rising in glacier melt runoff and river flow. About 69.7% of the increased river flow was derived from the increased glacier runoff caused by loss of ice mass. Temperature rise and precipitation increase between 1959 and 2008 have had a combined effect impacting on glacier mass loss and runoff change.

(3) The climate input data of HBV model are taken from observed meteorological data and the downscaled data from output of the RegCM3. Using observed variables, the calibration and validation results show that the HBV model can reproduce the discharge reasonably well, except for the peak values, which are slightly underestimated at special years. Meanwhile, the variation of temperature and precipitation in the investigated area are also captured fairly well by the RegCM3, and there

is a projected general increase in temperature and precipitation by the middle of the century (2041–2060).

(4) In the changed climate and under three glacier coverage scenarios, snowmelt all begins half a month earlier and discharge all increase in May (up to 61.4–183.2 %) in the two catchments. This indicates that the temperature increase seems to cause more frequent rainfall and augmentation in snowmelt runoff, which may consequently increase the risk of flood disaster in spring season. Hundred percent glacier coverage scenario predicts a pronounced increase in river discharge in the two basins. The discharge will have a increase of 33.3 % in Zongkong catchment and it will rise up to 66.8 % in Glacier No.1 catchment by the middle of the century. While 50 % glacier coverage scenario predicts a slightly decrease in discharge. But 0 % glacier coverage scenario predicts a considerable decrease in discharge ranging from 40.4 to 57.5 % especially in July and August when demand of water for agricultural production reaches its peak. This, and thus, may aggravate the crisis of water resources shortage in this region. It should be noted that completely disappearance of the glacier will result in a clear seasonal shift of water resource from mid-summer to late spring-early summer by the middle of the century in Glacier No.1 catchment. It can be concluded that climate change and the induced glacier shrinkage will provoke changes in the runoff amount and timing and will finally result in the allocation of the intraannual water resources in the catchment.

(5) Xinjiang, Uygher Autonomous Region in north western China, possesses the biggest ice volume of the glaciers in China, which plays an extremely important role both on water resource and stabilization of river runoff in this vast arid region. During the past several decades, due to climate warming, the most glaciers in Xinjiang are in a state of rapid retreating. Therefore, the impact of the glacier recession on water resource has drawn a wide attention. Based on field observation and remote sensing technique, this study has revealed the variations of 1,800 glaciers during the past four decades and analyzed the potential influence of the glacier variations on the water resource in Xinjiang. As a result, the total area of the investigated glaciers has reduced 11.7 %. The average area of individual glaciers has reduced by 0.243 km^2, and the average retreat rate of that is 5.8 m/a. The glacier area reductions in different drainage areas range between 8.8 and 34.2 % for the total area, and 0.092–0.415 km^2 for the individual glacier. The potential impact of the glacier recession on water resource in future will be spatially different. For the Tarim River, the glacier runoff is estimated to maintain its current level or increase somewhat in next 30–50 years. Because the glacial runoff accounts for a large amount of the river's runoff, at the beginning, the increasing in glacier melt would enrich the river runoff. However, once the ice volume reduced to a certain value, a shortage of water resource in this region is inevitable. In the north slop of Tianshan, the glaciers with a size smaller than 1 km^2 are most likely to be melted away in next 20–40 years, and those larger than 5 km^2 are melting intensively. The impact of which on different river basin in this region would be different depending on the proportion of glacial runoff to the river runoff. In eastern Xinjiang, because

the number of the glaciers is small and also because the climate is extremely dry, the glacier retreating are causing the water shortage problem. For Ili River and Irtysh River, because they are dominant by snowmelt runoff, the impact of the glacier shrinkage and temperature rise would be limited on the quantity of the river runoff, but significant on the annual distribution of the river runoff. The snowmelt is important for these two river systems and needed to be particularly study.

(6) Glaciers in the middle Qilian Mountain Region decreased significantly in area from 1956 to 2003 and 21.7% of the area had been lost, including 29.6% in the Heihe River Basin and 18.7% in the Beidahe River Basin, primarily due to increasing summer temperatures. The shrinkage of the glaciers shows differences between glaciers of different size. The relative area reductions of small glaciers were usually higher than those of large ones, which exhibit larger absolute loss. This indicated that the small glaciers were more sensitive to climate change than large glaciers. Comparing with the other glaciers in the Qilian Mountain Region, the shrinkage of the glaciers in the middle Qilian Mountain Region was the largest. The regional differences of the glacier changes in both of the middle Qilian Mountain Region and the Qilian Mountain Region were greatly mainly influenced by the regional climate conditions, initial glacier area, elevation span, terminus elevation and other factors. In future, coupling research about glacier dynamic model and hydrological model at the basin scale are needed to make quantitative analysis of the supplies of meltwater to river runoff and annual allocation in the different river basins of Qilian Mountains.

References

Academia Sinica (1959) Report of the investigation of modern glaciers in the Qilian Mts., Investigation Team of Ice/Snow Utilization (Special Issue I). Science Press, Beijing. (In Chinese)

Ageta Y, Kadota T (1992) Predictions of changes of glacier mass balance in the Nepal Himalaya and Tibetan Plateau: a case study of air temperature increase for three glaciers. Ann Glaciol 16:89–94

Aizen VB, Aizen EM, Kuzmichonok VA (2007a) Glaciers and hydrological changes in the Tien Shan: Simulation and prediction. Environ Res Lett 2(4):045019

Aizen VB, Kuzmichenok VA, Surazakov AB et al (2007b) Glacier changes in the Tien Shan as determined from topographic and remotely sensed data. Global Planet Change 56:328–340

Akhtar M, Ahmad N, Booij MJ (2008) The impact of climate change on the water resources of Hindukush-Karakorum-Himalaya region under different glacier coverage scenarios. J Hydrol 355(1-4):148–163

Akhtar M, Ahmad N, Booij MJ (2009) Use of regional climate model simulations as input for hydrological models for the Hindukush- Karakorum- Himalaya region. Hydrol Earth Syst Sci 13:1075–1089

Aðalgeirsdóttir G, Jóhannesson T, Bjórnssonn H et al (2006) Response of Hofsjiikull and Southern Vatanj Okull, Iceland, to climate change. J Geophys Res 111(F03001). doi:10. 1029/2005. JF000388

Bergström S (1976) Development and Application of a Conceptual Runoff Model for Scandinavian Catchments, Department of Water Resources Engineering, Lund Institute of Technology, Bulletin Series A 52, Swedish Meteorological and Hydrological Institute, Norrköping, Sweden

Bolch T (2007) Climate change and glacier retreat in Northern Tien (Kazakhstan/ Kyrgyzstan) using remote sensing data. Global Planet Change 56:1–12

Braun LN, Escher-Vetter H (1996) Glacial discharge as affected by climate change. Interpraevent 1996-Garmisch Partenkirchen, 65–74

Braun LN, Renner C (1992) Applications of a conceptual runoff model in different physiographic regions of Switzerland. Hydrol Sci J 73(3):217–231

Braun LN, Weber M, Schulz M (2000) Consequences of climate change for runoff from Alpine regions. Ann Glaciol 31:19–25

Burn DH, Hag Elnur MA (2002) Detection of Hydrologic Trends and Variability. J Hydrol 255:107–122

Cao ZT (1993) Glacio-hydrological characteristics of Gozha glacier on south slope of the West Kunlun Mountians. J Glaciol Geocryol 15(4):582–589

Cao M (2012) Velocity and area of Urumqi Glacier No.1 in 2009. In: Tianshan Glaciological Station (eds) Annual Report of Tianshan Glaciological Station, vol 19. Cold and Arid Regions Environmental and Engineering Research Institute, CAS, Lanzhou. (In Chinese)

Chen JY, Ohmura A (1990) On the Influence of Alpine Glaciers on Runoff. IAHS Publ. 193 (Symposium at Lausanne 1990- Hydrology in Mountainous Regions. I-Hydrological Measurements; the Water Cycle), 117–125

Chen YW, Sun ZZ (1996) Changes in surface velocity and the tongue of Urumqi Glacier No.1. In: Tianshan Glaciological Station (eds) Annual Report of Tianshan Glaciological Station, vol 13. Cold and Arid Regions Environmental and Engineering Research Institute, CAS, Lanzhou. (In Chinese)

Chen JM, Liu CH, Jin MX (1996) Application of the repeated aerial photogrammetry to monitoring glacier variation in the drainage area of the Urumqi River. J Glaciol Geocryol 18(4):331–336. (In Chinese)

Chen Y, Zhang DQ, Sun YB et al (2005) Water demand management: A case study of the Heihe River Basin in China. Phys Chem Earth 30:408–419

Chen YN, Takeuchi K, Xu CC et al (2006) Regional climate change and Its effects on river runoff in the Tarim Basin, China. Hydrol Process 20(10):2207–2216

Chen J, Brissette FP, Leconte R (2011) Uncertainty of downscaling method in quantifying the impact of climate change on hydrology. J Hydrol 401:190–202

Chueca J, Julián A, Saz MA et al (2005) Responses to climatic changes since the Little Ice Age on Maladeta Glacier (Central Pyrenees). Geomorphology 68(3–4):167–182

Dikich AN, Hagg W (2004) Climate driven changes of glacier runoff in the Issyk-Kul basin, Kyrgystan. Zeitschrift für Gletscherkunde und Glazialgeologie 39:75–86

Dyurgerov MB (2002) Glacier Mass Balance and Regime: Data of Measurements and Analysis. Institute of Arctic and Alpine Research, University of Colorado, Boulder. Occasional paper, 55, 268–275

Dyurgerov MB, Meier MF (2000) Twentieth century climate change: evidence from small Glacier. Proceedings of the National Academy of the United States of America, 97, 1406–1411

Evans IS (2006) Local aspect asymmetry of mountain glaciation: a global survey of consistency of favoured directions for glacier numbers and altitudes. Geomorphology 73(1-2):166–184

Evans IS, Cox NJ (2010) Climatogenic north-south asymmetry of local glaciers in Spitsbergen and other parts of the Arctic. Ann Glaciol 51(55):16–22

Fujita K, Ageta Y (2000a) Effect of summer accumulation on glacier mass balance on the Tibetan Palteau revealed by mass-balance model. J Glaciol 46:244–252

Fujita K, Ageta Y, Pu JC et al (2000b) Mass balance of Xiao Dongkemadi glacier on the central Tibetan Plateau from 1989 to 1995. Ann Glaciol 31:159–163

Gao QZ, Li FX (1991) Case study of rational development and utilization of water resources in the Heihe River Basin. Gansu Science and Technology Press, Lanzhou, pp 1–228. (In Chinese)

Gao XJ, Shi Y, Giorgi F (2011) A high resolution simulation of climate change over China. Sci China Earth Sci 54(3):462–472. doi:10.1007/s11430-010-4035-7

Giorgi F, Francisco R (2000) Evaluating uncertainties in the prediction of regional climate change. Geophys Res Lett 27(9):1295–1298. doi:10.1029/1999GL011016

Giorgi F, Marinucci M, Bates G (1993a) Development of a second-generation regional climate model (RegCM2). Part I: boundary layer and radiative transfer processes. Mon Weather Rev 121:2794–2813

Giorgi F, Marinucci M, Bates G (1993b) Development of a second-generation regional climate model (RegCM2). Part II: convective processes and assimilation of lateral boundary conditions. Mon Weather Rev 121:2814–2832

Hagg W, Braun L (2005) The influence of glacier retreat on water yield from high mountain areas: comparison of Alps and Central Asian. In: de Jong C, Collins D, Ranzi R (eds) Climate and hydrology in mountain areas. John Wiley & Sons, Ltd, pp 263–275

Hagg W, Braun LN, Weber M et al (2006) Runoff modelling in glacierized Central Asian catchments for present-day and future climate. Hydrol Res 1:1–13

Hagg W, Braun LN, Kuhn M et al (2007) Modelling of hydrological response to climate change in glacierized Central Asian catchments. J Hydrol 332(1-2):40–53

He YQ, Pu T, Li ZX et al (2010) Climate change and Its effect on annual runoff in Lijiang Basin-Mt. Yulong Region, China. J Earth Sci 21(2):137–147

Hodgkins R, FOX A, Nuttall AM (2007) Geometry change between 1990 and 2003 at Finsterwalderbreen, a Svalbard surge-type glacier, from GPS profiling. Ann Glaciol 46:131–135

Immerzeel WW, van Beek LPH, Bierkens MFP (2010) Climate change will affect the Asian water towers. Science 328(5984):1382–1385

Intergovernmental Panel on Climate Change (IPCC) (2007) In: Solomon S, Qin D, Manning M, Chen Z, Marquis M, Averyt KB, Tignor M, Miller HL (eds), Climate Change 2007

Jia WX, He YQ, Li ZX et al (2008) The regional difference and catastrophe of climatic change in Qilian Mt. Region. Acta Geographica Sinica 63(3):257–269. (In Chinese)

Jin R, Li X, Che T et al (2005) Glacier area changes in the Pumqu river basin, Tibetan Plateau, between the 1970s and 2001. J Glaciol 51(175):607–610

Jing ZF (2004) Changes in surface velocity and the tongue of Urumqi Glacier No.1. In: Tianshan Glaciological Station (eds) Annual Report of Tianshan Glaciological Station, vol 16. Cold and Arid Regions Environmental and Engineering Research Institute, CAS, Lanzhou. (In Chinese)

Jing ZF, Jiao KQ, Yao TD et al (2006) Mass Balance and Recession of Urumqi Glacier No.1, Tien Shan, China, over the Last 45 Years. Ann Glaciol 43:214–217

Jóhannesson T, Raymond C, Waddington E (1989) Time-scale for adjustment of glaciers to changes in mass balance. J Glaciol 35:355–369

Kääb A (2004) Mountain glaciers and permafrost creep. In: Methodical research perspectives from Earth observation and geoinformatics technologies: habilitation Thesis, Department of Geography. University of Zurich, Switzerland, p 205

Kang ES, Shi YF, Yang DQ (1997) An Experimental Study on Runoff Formation in the Mountains Basin of the Urumqi River. Quat Sci 2:139–146

Kang ES, Yang ZN, Lai ZM et al (2000) Runoff of snow and ice meltwater and mountainous rivers. In: S Yafeng (ed) Glaciers and their environments in China—the present, past and future. Science Press, Beijing, pp 190–233

Knight PG (1998) Glaciers. Stanley Thornes Ltd, Cheletenham

Konz M, Seibert J (2010) On the value of glacier mass balances for hydrological model calibration. J Hydrol 385(1–4):238–246

Lan YC, Ding YJ, Kang ES (2004) Variations and trends of temperature and precipitation in the mountain drainage basin of the Heihe River in recent 50 years. Plateau Meteorol 23(5):723–727. (In Chinese)

Lan YC, Ding YJ, Liu JQ et al (2005) Change of water resources in mountainous area of Heihe River under global-warming scene. J Desert Res 25(6):863–868. (In Chinese)

Lanzhou Institute of Glaciology and Geocryology, Chinese Academy of Sciences (1986a) Glacier Inventory of China (III); Tianshan Mountains (Ili Biver Drainage Basin). Science Press, Beijing, pp 1-158. (In Chinese)

Lanzhou Institute of Glaciology and Geocryology, Chinese Academy of Sciences (1986b) Glacier Inventory of China (III); Tianshan Mountains (Interior Drainage Area of Scattered Flow in East). Science Press, Beijing, pp 11–83. (In Chinese)

Lanzhou Institute of Glaciology and Geocryology, Chinese Academy of Sciences (1986c) Glacier Inventory of China (III); Tianshan Mountains (Interior Drainage Area of Tarim Basin in Southwest. Science Press, Beijing, pp 1–187. (In Chinese)

Lanzhou Institute of Glaciology and Geocryology, Chinese Academy of Sciences (1986d) Glacier Inventory of China (III); Tianshan Mountains (Interior Drainage Area of Junggar Basin in Northwest). Science Press, Beijing, pp 1–206. (In Chinese)

Li DL, Wei L, Cai Y et al (2003a) The present facts and the future tendency of the climate change in Northwest China. J Glaciol Geocryol 25(2):135–142. (In Chinese)

Li ZQ, Han TD, Jing ZF et al (2003b) A summary of 40-year observed variation facts of climate and Glacier No.1 at headwater of Urumqi River, Tianshan, China. J Glaciol Geocryol 25(2):117–123. (In Chinese)

Li BL, Zhu AX, Zhang YC et al (2006) Glacier change over the past 4 decades in the middle Chinese Tien Shan. J Glaciol 52:425–432

Li HL, Li ZQ, Shen YP et al (2007a) Glacier dynamic models and their applicability for the glaciers in China. J Glaciol Geocryol 29(2):201–208

Li ZQ, Shen YP, Wang FT et al (2007b) Response of glacier melting to climate change–Take Urumqi Glacier No.1 as an example. J Glaciol Geocryol 29(3):333–342

Li ZQ, Wang FT, Zhu GC et al (2007c) Basic features of the Miaoergou flat-topped glacier in East Tianshan Mountains and its thickness change over the past 24 pears. J Glaciol Geocryol 29(1):61–65

Li HL, Li ZQ, Qin DH (eds) (2009a) Basic principles of glacial dynamic models and observation guide for corresponding parameters. China Meteorologlical Press, Beijing, pp 1–56

Li ZQ, Wang WB, Zhang MJ et al (2009b) Observed changes in streamflow at the headwaters of the Urumqi River, Eastern Tianshan, Central Asia. Hydrol Process doi: 10.1002/hyp.7431

Li ZQ, Li KM, Wang L (2010a) Study on recent glacier changes and their impact on water resources in Xinjiang, Northwestern China. Quat Sci 30(1):96–106. doi:10.3969/j.issn.1001-7410.2010.01.09. (In Chinese)

Li ZQ, Wang WB, Zhang MJ et al (2010b) Observed changes in streamflow at the headwater of the Urumqi River, eastern Tianshan, central Asia. Hydrol Process 24:217–224. doi:10.1002/hyp.7431

Li ZX, He YQ, Pu T et al (2010c) Changes of climate, glaciers and runoff in China's monsoonal temperate glacier region during the last several decades. Quat Int 218:13–28

Li ZQ, Li KM, Wang L (2010d) Study on recent glacier changes and their impact on water resources in Xinjiang, north western China. Quat Sci 30(1):96–106. (In Chinese)

Li ZQ, Li HL, Chen YN (2011) Mechanisms and simulation of accelerated shrinkage of continental glaciers: a case study of Urumqi glacier No.1 in eastern Tianshan, central Asia. J Earth Sci 22(4):423–430. doi:10.1007/S12583-011-0194-5

Liu CH, Song YP, Jin MX (1992) Recent change and trend prediction of glaciers in Qilian mountain. In: Lanzhou Institute of Glaciology and Geocryology Memoirs 7, Academia Sinica, pp 1–9. (In Chinese)

Liu SY, Sun WX, Shen YP et al (2003) Glacier changes since the Little Ice Age maximum in the western Qilian Shan, Northwest China, and consequences of glacier runoff for water supply. J Glaciol 49(164):117–124

Liu SY, Ding YJ, Shangguan DH et al (2006) Glacier retreat as a result of climate warming and increased precipitation in the Tarim river basin, northwest China. Ann Glaciol 43:91–96. doi:10.3189/172756406781812168

Liu JS, Wang SY, Huang YY (2007) Effect of climate change on runoff in a basin with Mountain Permafrost, Northwest China. Permafr Periglac Process 18:369–377

Liu Q, Liu SY, Zhang Y et al (2010) Recent shrinkage and hydrological response of Hailuogou glacier, a monsoon temperate glacier on the east slope of Mount Gongga, China. J Glaciol 56(196):215–224

Lu AG, He YQ, Zhang ZL et al (2004) Regional structure of global warming across China during the twentieth century. Climate Res 27:189–195

Mark BG, Seltzer GO (2003) Tropical glacier melterwater contribution to stream discharge a case study in the Cordillera Blanca, Peru. J Glaciol 49(165):271–281
Meier MF (1984) Contribution of small glaciers to global sea level. Science 51:49–62
Moore RD, Demuth MN (2001) Mass balance and streamflow variability at place glacier, Canada, in relation to recent climate fluctuations. Hydrol Process 15:3473–3486
Mountaineering and Expedition Term of Chinese Academy of Sciences (1985) Glacial and weather in Mt. Tuomuer District, Tianshan. Xinjiang People's Publishing House, Urumqi, pp 32–98
Nash J, Sutcliffe J (1970) River flow forecasting through conceptual models 1: a discussion of principles. J Hydrol 10:282–290
Nesje A, Dahl SO (2000) Glaciers and environmental change. Oxford University Press Inc., New York
Oerlemans J (1994) Quantifying global warming from the retreat of glaciers. Science 264:243–245
Oerlemans J (2005) Extracting a climate signal from 169 glacier records. Science 308:675–677
Oerlemans J, Fortuin JPF (1992) Sensitivity of glaciers and small ice caps to greenhouse warming. Science 258:115–117
Oerlemans J, Anderson B, Hubhard A et al (1998) Modeling the response of glaciers to climate warming. Climate Dyn 14(4):267–274
Pal JS, Giorgi F, Bi X et al (2007) Regional climate modeling for the developing world: the ICTP RegCM3 and RegCNET. Bull Am Meteorol Soc 88(9):1395–1409
Paul F, Kääb A, Maisch M et al (2002) The new remote sensing-derived Swiss glacier inventory: I. Methods. Ann Glaciol 34:355–361
Paul F, Kääb A, Maisch M et al (2004) Rapid disintegration of Alpine glaciers observed with satellite data. Geophys Res Lett 31:L21402. doi:10.1029/2004GL020816
Pu JC, Yao TD, Duan KQ et al (2005) Mass balance of the Qiyi Glacier in the Qilian Mountains: a new observation. J Glaciol Geocryol 27(2):199–204. (In Chinese)
Rau F, Mauz F, Vogt S et al (2005) Illustrated GLIMS glacier classification manual. http//nsidc.org/
Raup B, Racoviteanu A, Jodha S et al (2007a) The GLIMS geospatial glacier database: a new tool for studying glacier change. Global Planet Change 56(1–2):101–110
Raup B, Kääb A, Kargel JS et al (2007b) Remote sensing and GIS technology in the Global Land Ice Measurements from Space (GLIMS) Project. Computers Geosci 33(1):104–125
Sakai A, Fujita K, Duan KQ et al (2006) Five decades of shrinkage of the July 1st Glacier, Qilian Mountains, China. J Glaciol 52(176):11–16
Shangguan DH, Liu SY, Ding YJ et al (2007) Glacier changes in the west Kunlun Shan from 1970 to 2001 derived from Landsat TM/ETM+ and Chinese glacier inventory data. Ann Glaciol 46:204–208
Shangguan DH, Liu SY, Ding YJ et al (2009) Glacier changes during the last forty years in the Tarim Interior River basin, northwest China. Progress Nat Sci 19:727–732
Shen YP, Liu SY, Wang GX et al (2001) Fluctuations of glacier mass balance in watersheds of Qilian Mountains and their impact on water resources of Hexi region. J Glaciol Geocryol 23(3):244–250. (In Chinese)
Shen YP, Liu SP, Ding YJ et al (2003) Glacier mass balance change in Tailanhe River watersheds on the Tianshan Mountains and its impact on water resources. J Glaciol Geocryol 25(2):124–129
Shi YF (2001) Estimation of the water resources affected by climatic warming and glacier shrinkage before 2050 in West China. J Glaciol Geocryol 23(4):333–341. (In Chinese)
Shi YF (2005) Concise Chinese GLacier inventory. Shanghai Scientific Popularization Press, Shanghai, pp 17–188
Shi YF, Liu SY (2000) Estimation on the response of Glaciers in China to the global warming in the 21st century. Chin Sci Bull 45(7):668–672. (In Chinese)
Shi YF, Zhang XS (1995) Impact of climate change on surface water resource and tendency in the future in the Arid Zone of Northwestern China. Sci China Ser D 38(11):1395–1408
Shi YF, Huang MH, Yao TD et al (2000) Glaciers and their environments in China–the present, past and future. Science Press, Beijing. (In Chinese)

Shi YF, Shen YP, Hu RJ (2002) Preliminary study on signal, impact and foreground of climatic shift from warm-dry to warm-humid in Northwest China. J Glaciol Geocryol 24(3):219–226. (In Chinese)

SMHI (2006) Integrated hydrological modelling system (IHMS), Manual version 5.10, SMHI, Norrkoping

Stahl K, Moore RD, Shea JM et al (2008) Coupled modelling of glacier and streamflow response to future climate scenarios. Water Resour Res 44:W02422. doi:10.1029/2007WR005956

Su Z (1998) Glaciers and environment of the Karakorum-Kunlun Mountains. Science Press, Beijing, pp 10–123

Surazakov AB, Aizen VB, Aizen EM et al (2007) Glacier changes in the Siberian Altai Mountains, Ob river basin, (1952–2006) estimated with high resolution imagery. Environ Res Lett 2:045017. doi:10.1088/1748-9326/2/4/045017

Toutin T (2002) DEM from stereo Landsat 7 ETM+data over high relief areas. Int J Remote Sens 23:2133–2139

Wang ZT (1991) A discussion on the questions of development of Heigou Glacier No.8 at Bogda-peak region. J Glaciol Geocryol 13(2):141–158

Wang WY, Mi DS, Chen JM et al (1965) Glacial topographic map at the source of the Urumqi River, Tianshan (1:10000). In: An Studies on Glaciology and Hydrology on the Urumqi River, Tianshan, Laboratory of Glaciology and Geocryology, Chinese Academy of Sciences Institute of Geography (eds). Science Press, Beijing. (In Chinese)

Wang ZT, Liu CH, You GX et al (1981) Glacier Inventory of China I (Qilian Mountains). Lanzhou Institute of Glaciology and cryopedology, Academia Sinica, Gansu. (In Chinese)

Wang LL, Liu CH, Kang XC et al (1983) Fundamental features of modern glaciers in the Altay Shan of China. J Glaciol Geocryol 5(4):27–38

Wang NL, Jiao KQ, Li ZQ et al (2004) Recent accelerated shrinkage of Glacier No.1, Tien Shan, China. J Glaciol 50(170):464–466

Wang Puyu, Li Zhongqin, Gao Wenyu (2011) Rapid shrinking of glaciers in the middle Qilian Mt. Region of Northwest China, during the last ~50 years. J Earth Science, 22(4):539–548

Wang S, McGrath R, Semmler T et al (2006) The impact of the climate change on discharge of Suir River Catchment (Ireland) under different climate scenarios. Nat Hazard Earth Sys Sci 6:387–395

Woo MK, Yang ZN, Xia ZJ et al (1994) Streamflow processes in an Alpine permafrost catchment, Tianshan, China. Permafr Periglac Process 5:71–85

Wu GH, Zhang SP, Wang ZX (1983) Retreat and advance of modern glaciers in Bogda, Tianshan. J Glaciol Geocryol 5(3):143–152

Wu J, Gao XJ, Shi Y et al (2011) Climate change over Xinjiang Region in the 21st century simulated by a high resolution regional climate model. J Glaciol Geocryol 33(3):479–487. (In Chinese)

Xie ZC, Wu GH, Wang LL (1985) Recent fluctuations of glaciers in the Qilian Mountains. In: Lanzhou Institute of Glaciology and Geocryology Memoirs 5, Academia Sinica, pp 82–90. (In Chinese)

Xie CW, Ding YJ, Liu SY et al (2006) Variation of Keqikaer glacier terminus in Tomur peak during last 30 years. J Glaciol Geocryol 28(5):672–677

Xie ZC, Wang X, Kang ES et al (2006) Glacier runoff in China: an evaluation and prediction for the future 50 years. J Glaciol Geocryol 28(4):457–466

Yang ZN (1991) Glacier water resources in China. Gansu Science and Technology Press, Lanzhou, pp 1–158. (In Chinese)

Yang JP, Ding YJ, Chen RS et al (2003) Causes of glacier change in the source regions of the Yangtze and Yellow rivers on the Tibetan Plateau. J Glaciol 49(167):539–546

Yao TD, Wang YQ, Liu SY et al (2004) Recent glacial retreat in high Asia in China and its impact on water resource in Northwest China. Sci China Ser D 47(12):1065–1075

Ye BS, Yang DQ, Jiao KQ et al (2005) The Urumqi River source Glacier No.1, Tianshan, China: changes over the past 45 years. Geophys Res Lett 32:L21504. doi:10.1029/2005GL024178

Ye QH, Kang SC, Chen F et al (2006) Monitoring glacier variations on Geladandong Mountain, central Tibetan Plateau, from 1969 to 2002 using remote sensing and GIS technologies. J Glaciol 52(179):537–545

You GX (1988) Map of Glacier No.1 and Glacier No.2 at the source of the Urumqi River, Tianshan Mountains. Xi'an Cartographic Publishing House, Xi'an. (In Chinese)

Zhang JH (1981) Mass balance studies on Glacier No.1 in Urumqi River, Tianshan. J Glaciol Geocryol 3:32–40. (In Chinese)

Zhao ZM (2009) Monthly velocity of Glacier No.1 at the headwater of Urumqi River in 2006 and the changes in area. In: Annual Report of Tianshan Glaciological Station (vol 18), Tianshan Glaciological Station (eds). Cold and Arid Regions Environmental and Engineering Research Institute, CAS, Lanzhou. (In Chinese)

Zhao FF, Xu ZX, Huang JX et al (2008) Monotonic trend and abrupt changes for major climate variables in the headwater catchment of the Yellow River Basin. Hydrol Process 22:4587–4599

Zhou LQ (2009) A study of glacier changes and surface dust on lenglongling glaciers in the Qilian Mountains. Dissertation, Lanzhou University, Lanzhou. 1–57. (In Chinese)

Chapter 6
Spatiotemporal Variation of Snow Cover from Space in Northern Xinjiang

Xianwei Wang, Hongjie Xie and Tiangang Liang

Abstract Snow is an appreciable fraction of soil water recharge in the middle and high latitude areas, and snow cover and snow depth estimation is extremely important for regional climate change studies, agriculture and water source management. Traditional in situ measurments provide critical snow depth observations in limited areas and for ground control validation of remotely sensed estimations. Satellite-based snow measurements have revolutionized the monitoring of spatiotemporal variation of snow cover and snow depth in complex natural conditions at regional and global scales. This chaper introduces the algorithm of optical satellite snow cover detection and MODIS standard snow cover products, sumarizes recent studies on mitigating the cloud-blockage issues in MODIS snow cover products and their applications, and finally illustrates the spatiotemporal variatons of snow cover in the Northern Xinjiang, China by using the cloud-removed MODIS snow cover product. Together, snow cover days (SCD), snow cover index (SCI), snow cover onset date (SCOD) and snow cover end date (SCED) provide important information on the snow cover conditions and can be applied in any region of interests. These information would be critical for local government, such as land use planning, agriculture, live stock, and water resource management, for example, to mitigate snow-caused disasters and to plan for agriculture and industry water use. Long term availability of MODIS type of snow cover data for producing such datasets is key to study the connection between snow cover variation and climate change.

X. Wang (✉)
Center of Integrated Geographic Information Analysis,
School of Geography and Planning, and Guangdong Key Laboratory for Urbanization and Geo-simulation, Sun Yat-sen University,
No. 135 West Xingang Road, 510275 Guangzhou, China
e-mail: xianweiw@vip.qq.com

H. Xie
Laboratory for Remote Sensing and Geoinformatics,
Department of Geological Science, University of Texas at San Antonio,
San Antonio, TX 78249, USA
e-mail: Hongjie.Xie@utsa.edu

T. Liang
College of Pastoral Agriculture Science and Technology, Lanzhou University,
No. 222 South Tianshui Road, 730020 Lanzhou, China
e-mail: tgliang@lzu.edu.cn

Y. Chen (ed.), *Water Resources Research in Northwest China,*
DOI 10.1007/978-94-017-8017-9_6,

Keywords MODIS · Snow cover · Cloud removal · Northern Xinjiang

6.1 Introduction

Traditional in situ measurments at different climatic stations provide good snow depth observations in limited areas and are critical ground control for validating remotely sensed estimations of snow cover and snow depth. Satellite-based remote sensing snow measurements have revolutionized the monitoring of spatial and temporal distribution and variability of snow cover extent and snow depth in complex natural conditions at regional and global scales, such as those derived from the Interactive Multisensor Snow and Ice Mapping System (IMS), Scanning Multichannel Microwave Radiometer (SMMR), Special Sensor Microwave/Imager (SSM/I), Advanced Microwave Scanning Radiometer—EOS (AMSR-E), Geostationary Observational Environmental Satellite (GOES), Advanced Very High Resolution Radiometer (AVHRR), Moderate Resolution Imaging Spectraradiometer (MODIS), and Landsat (Hall et al. 2000).

Historical records of remotely sensed hemispherical snow cover maps date back to 1966, as being generated by the US National Ocean and Atmospheric Administration (NOAA) (Klein and Barnett 2003). Microwave remote sensing products, like SMMR, SSM/I, and AMSR-E, are used for global scale studies because of their high temporal resolution (daily) and with limited influence of cloud cover despite of low spatial resolution (25 km or coarser). Products derived from optical instruments using reflected solar radiation, such as AVHRR, MODIS, and Landsat, etc., have higher spatial resolution and are better for regional studies, but heavily depend on suitable weather conditions, especially clear sky (no clouds). The low temporal resolution (16 days) of Landsat data is an obstacle to its wide application in monitoring snow, even though it has much higher spatial resolution (30 m) than MODIS and AVHRR.

The Northern Xinjiang Uygur Autonomous Region is one of the three major snow distribution regions in China (Che and Li 2005; Huang and Cui 2006). Snow-related disasters frequently occur on a large scale in winter and spring, resulting in a large number of animal deaths and significant economic loss (Liu et al. 2003); meanwhile, snow is important water resources in this arid and semi-arid region. The limited ground observations revealed some temporal and minor spatial variations related to snowfall in this region over the last 50 years. However, all of the climatic stations are present in readily accessible and low elevation areas and may cause locational bias. The spatial variation of snow in mountainous and remote areas of this region may not be accurately represented in the current in situ data sets. Therefore, this chaper sumarizes the recent studies on the spatiotemporal variations of snow cover detected by the up-dated statellite technology and improved data processing methods.

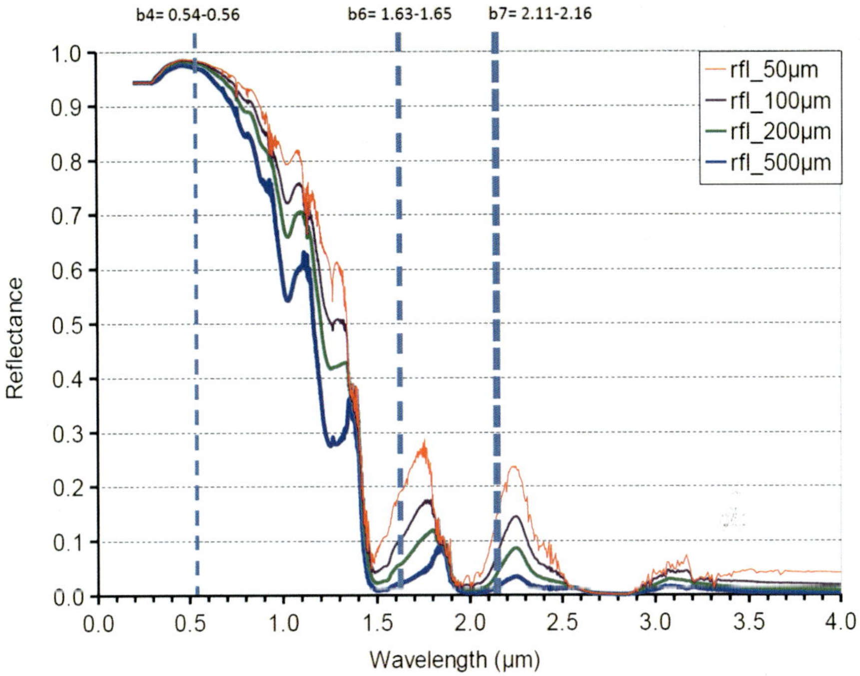

Fig. 6.1 Predicted spectral reflectance of dry snow surface under clear sky at the Summit station (72.5794 N, 38.5042 W) of Greenland. Snow density is 250.0 kg m^{-3}, and snow grain radiuses are 50, 100, 200 and 500 μm from top to bottom, respectively. Finer snow grain size has higher reflectance particularly in the near and mid infrared wavelength range. Detailed model and theory description are documented by Flanner and Zender (2006). The three *dashed lines* from *left* to *right* are the wavelength positions of MODIS band 4, band 6 and band 7, respectively, and the width of the line is proportional to the width of wavelength range

6.2 Principles of MODIS Snow Cover Products

Snow has high reflectance in the 0.3–0.7 μm (visible) wavelength range and low reflectance for wavelengths longer than 1.4 μm (near-infrared) (Fig. 6.1). Thus, snow cover is relatively straightforward to be detected using optical remote sensing. However, challenging does exist. First, reflectance in the visible wavelengths is limited by the surface illuminated from sunlight, and low illumination and darkness cause problems, resulting in a large part of high latitude areas in winter to be omitted.

Second, electromagnetic spectrum in the visible range can not pass through cloud, and thus can not reach the land surface under cloud; another problem due to cloud is that thin cloud has similar high relfectance with snow surface in the visible range, frequently causing confusion between thin cloud and patchy snow, although most cloud has higher reflectance in the near and mid-infrared wavelength range, which is used to distinguish snow from cloud (Frei et al. 2012). Third, vegetations, particularly the dense forest canopies, obstruct visible and near-infrared signal to reach the snow surface under canopies and reduce the snow surface albedo, making it difficult to accurately detect snow depth and extent (Klein et al. 1998; Nolin 2004). Last, surface heterogeneity, such as in mountainous areas and in the polar regions with wetlands and lakes, also makes it difficult to map snow using moderate resolution (500 m like MODIS or 1 km as AVHRR) visible and near-infrared imagery without high resolution (tens m) land surface data (Frei and Lee 2010).

6.2.1 MOD10A1/MYD10A1 Binary

The first MODIS standard snow cover products MOD10A1(Terra) and MYD10A1(Aqua) are a binary classification or hard classification, which can only tell whether it is covered by snow or not, but can not tell the snow depth and snow water equavilent (SWE). The MODIS snow-cover algorithm is based on the high reflectance of snow in the visible band (band 4=0.545–0.565 μm) and low reflectance in the near infrared bands (band 6=1.628–1.652 μm for Terra MODIS and band 7=2.105–2.155 μm for Aqua MODIS, due to the failure of band 6 sensor in Aqua MODIS) (Fig. 6.1). These two bands in the visible and near-infrared are used to calculate the normalized difference of snow index (NDSI) in Eq. (6.1a, b) (Hall et al. 1995).

MOD10A1:

$$NDSI = \frac{Band4 - band6}{band4 + band6} \tag{6.1a}$$

MYD10A1:

$$NDSI = \frac{Band4 - band7}{band4 + band7} \tag{6.1b}$$

Four masks are used in classifying a certain NDSI value as snow or not snow, including a dense forest stands mask, a thermal mask, a cloud mask, and an ocean and inland water mask. These masks have been incorporated in the snow cover algorithm to get the best estimation of snow. A pixel in a non-densely forested region is mapped as snow if its NDSI is ≥0.4 and reflectance in MODIS band 2 (0.841–0.876 μm) is >11 % and reflectance in MODIS band 4 is ≥10 %. The latter prevents pixels containing very dark targets such as black spruce forests, from being flagged as snow. The detailed algorithm and processing steps have been documented in several sources (Hall et al. 2002; Riggs et al. 2006).

6.2.2 MOD10A1/MYD10A1 Fractional

Accompanying with the binary snow cover products, MODIS also provides a fractional snow cover layer in the standard snow cover products (MOD10A1/MYD10A1), which are derived by using the empirical relationship between fractional snow cover and the NDSI in Eq. (6.2a, b) (Salomonson and Appel 2004, 2006; Rittger et al. 2013).

MOD10A1:

$$F_{SCA} = -0.01 + 1.45NDSI \tag{6.2a}$$

MYD10A1:

$$F_{SCA} = -0.64 + 1.91NDSI \tag{6.2b}$$

The coefficients were determined by comparing MODIS NDSI with fractional snow cover in a MODIS pixel derived from Landsat ETM+ images processed with the similar algorithm used for MOD10A1/MYD10A1 binary.

6.2.3 *MODSCAG*

Besides the MODIS standard snow cover products that are most popular in the snow community, MODSCAG (MODIS Snow Covered-Area and Grain size retrieval) algorithm has been developed to retrieve fractional snow cover as well as snow grain size based on spectral mixing analysis with MODIS spectral reflectance data (Nolin et al. 1993; Dozier and Painter 2004; Sirguey et al. 2009; Painter et al. 2009). Spectral mixture analysis is based on a set of simultaneous linear equations that are solved for the components of the pixel-averaged atmospherically corrected surface reflectance in Eq. (6.3), i.e., MODIS product MOD09GA/MYD09GA (Kotchenova and Vermote 2007).

$$R_{S,\lambda} = \sum_{K} F_k R_{\lambda,k} + \varepsilon_\lambda \tag{6.3}$$

Where $R_{s,\lambda}$ is the pixel-averaged MODIS surface reflectance in wavelength band λ; F_k is the fraction of endmember k (snow, vegetation, or soil); $R_{\lambda,k}$ is the reflectance of endmember k in wavelength band λ; and ε_λ is the residual error at λ for the fit of all endmembers. The least-squares fit to F_k can be solved by several standard methods. Analysis of residuals reveals spectral regions of poor modeling and can be used for separating near-degenerate spectra (Rittger et al. 2013).

In contrast to the algorithm of MODIS standard snow cover products that rely on absolute reflectance like in Eq. (1), MODSCAG uses the relative shape of the snow's spectrum, which is sensitive to the spectral reflectance of the snow fraction. Thus, MODSCAG allows the snow's spectral reflectance to vary pixel-by-pixel and can address the spatial heterogeneity that characterizes snow and its albedo in mountainous and patchy regions. Validation and comparison study shows that it is more accurate to characterize fractional snow cover than the MODIS standard fractional snow cover algorithm based on NDSI (Rittger et al. 2013).

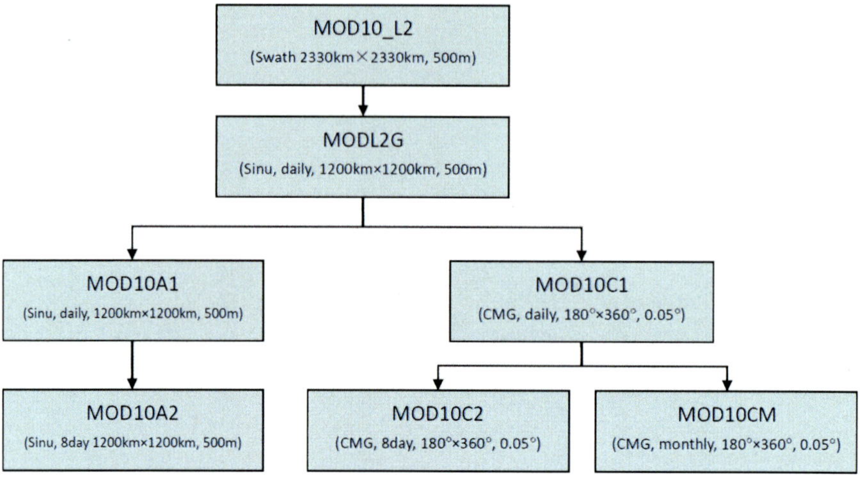

Fig. 6.2. Terra/Aqua MODIS standard snow cover products. Aqua MODIS snow cover product's names begin with MYD

6.3 MODIS Standard Snow Cover Products

Being launched on December 18, 1999, the NASA Earth Observing System (EOS) Terra satellite began collecting data on February 24, 2000 and passes the equator at about 10:30 a.m. The Aqua satellite, a counterpart of the Terra, was launched on May 4, 2002 and began collecting data on June 22, 2002. It passes the equator at about 1:30 p.m. Carried aboard the Terra and Aqua spacecrafts, MODIS standard snow cover products are produced as a series (Fig. 6.2), beginning with a swath (scene) product at a nominal pixel spatial resolution of 500 m with nominal swath coverage of 2,330 km by 2,030 km. The multiple swath observations at 500 m resolution of snow cover (MOD10_L2) are then projected onto a sinusoidal gridded tile (1,200 × 1,200 km) of MOD10L2G, which is further processed as a sinusoidal 500 m grid of daily (MOD10A1) and 8-day (MOD10A2) composite tile (1,200 × 1,200 km) products, a 0.05° global Climate Modeling Grid (CMG) daily product (MOD10C1), an 8-day product (MOD10C2), and a monthly product (MOD10CM) (Hall et al. 2002; Riggs et al. 2006; Wang et al. 2008).

Using a similar algorithm to Terra MODIS snow cover products, the Aqua MODIS also produces a parallel series of snow cover products that have a name beginning with MYD. In spite of the replacement of band 6 with band 7 in Aqua MODIS due to the failure of band 6 sensor in Aqua MODIS, both snow cover products (MOD10A1 and MYD10A1) have high agreement for cloud free snow cover classification (Wang et al. 2009).

MODIS standard snow cover products have high accuracy for clear-sky boservations and are widely applied in all kinds of studies around the world. In a validation of the MODIS swath snow product (MOD10_L2) and cloud mask in the Lower Great Lakes Regions, Ault et al. (2006) found that MOD10_L2 has very high accuracy under clear sky; highest errors occur for trace snow (as snow depth < 1 cm); and the accuracy increases as snow depth increases. In a study in the eastern part of Turkey, Tekeli et al. (2005) found that the worst match between ground observations and MODIS snow cover maps was only 21 % when the sky was cloud-covered on March 24, 2004. The high frequency of cloud cover seriously impairs the accuracy of MOD10A1. In a study at the Upper Rio Grande River Basin during the 2000–2001 snow year, Klein and Barnett (2003) identified that MOD10A1 has very high overall agreement (94 %) with in situ Snowpack Telemery (SNOTEL) observations under clear skies. Most of the disagreement between MOD10A1 and SNOTEL observations occurred at the beginning and end of the snow season when thin snowpack conditions were prevalent. The high agreement between MOD10A1 and SNOTEL observation was limited to clear skies. In addition, the MODIS cloud mask appears to frequently map edges of snow-covered areas as cloud, which is likely due to lower fraction of snow (patchy snow) or thin snow depth, and thus the cloud mask requires improvement in these transition zones (Klein and Barnett 2003; Tekeli et al. 2005; Riggs et al. 2006). The fractional snow cover contained in MOD10A1 and MOD10A2 (V005) is a better representation of these transition zones (Salomonson and Appel 2004). Compared with the snow product of National Weather Service National Operational Hydrologic Remote Sensing Center (NOHRSC), MODIS classified fewer pixels as cloud than NOHRSC and significantly greater amounts of snow in the presence of clouds for topographically complex, forested, and snow-dominated areas (Maurer et al. 2003; Klein and Barnett 2003). At the same basin studied by Klein and Barnett (2003), Zhou et al. (2005) also identified clouds as a major cause affecting the accuracy of MODIS snow classification.

6.4 Improvements on MODIS Standard Snow Cover Products

Annually near half of the MODIS daily snow cover images are blocked by cloud, which impedes wide applications of MODIS daily snow cover products. The MODIS 8-day product (MOD10A2) minimizes cloud cover and maximizes snow cover, thereby providing better input for snowmelt runoff models. It has higher classification accuracy for both snow and land than MOD10A1 through cloud suppression. In Northern Xinjiang, China, the MOD10A2 has high accuracies under clear sky condition when mapping snow (94 %) and land (99 %) at snow depth ≥4 cm, but a very low accuracy (< 39 %) for patchy snow or thin snow depth (< 4 cm) (Wang et al. 2008).

Although the 8-day MODIS snow cover composites (MOD10A2/MYD10A2) greatly reduce the cloud blockage compared to daily MODIS snow cover products (MOD10A1/MYD10A1), there are more than 10 % of the MOD/MYD10A2 images where cloud coverage is more than 20 % in a year, e.g., in Northern Xinjiang, China and on the Colorado Plateau, USA (Xie et al. 2009). Moreover, the reduction of cloud with MOD/MYD10A2 sacrified their temporal resolution from daily observation to 8-day composite. In addition, the 8-day MODIS snow cover products using fixed starting and ending dates each year. This loses flexibility in monitoring particular snowfall events especially when snow can easily melt away in a few days or even in a few hours within a day, making them difficult to timely monitor the snowfall and snowmelting, particularly in the snow accumulation and snowmelting periods. Thus, the cloud issues with MODIS snow cover products severely impede wide and proper applications of MODIS standard snow cover products.

Therefore, all kinds of approaches have been developed recently to mitigate negative impacts of cloud blockage with MODIS standard snow cover products. One solution is to separate the cloud-masked pixels into "snow" or "non-snow" using a ground truth-derived ratio approach proposed by Wang et al. (2008). But this algorithm can only correct the total snow cover area and does not allow the identification of individual cloud-masked pixels as either snow or non-snow. An alternative solution is to composite multiple daily snow cover images to produce a maximum snow cover image and thereby minimize short-term environmental contaminations, like cloud, aerosol, etc. (King et al. 1992; Liang et al. 2008a), with a reduced temporal resolution. For example, the MODIS 8-day composite product (MOD10A2) reduces the original temporal resolution of daily to 8 days. A third method is using a spatial filter to replace the value of a cloud-covered pixel with the majority of its neighboring pixels' value (Parajka and Blöschl 2008). In mountainous areas, a local time-related snowline is a practical approach for estimating the cloud-covered pixels as snow or snow free (Parajka et al. 2010). Comibing the MOD/MYD10A1 with microwave estimation like AMSR-E SWE is another suitable method to fill the cloud-covered pixels, although passive microwave observations have much coarser spatial resolution and are challenging to be applied in mountainous areas (Liang et al. 2008b; Gao et al. 2010b).

The two Terra and Aqua MODIS standard snow cover products are parallel using the almost identical instrument observations and algorihtms except Aqua MODIS uses the band 7 to replace the band 6 (see Eq. 1). Since clouds are always changing in both position and extent over a three-hour period (Hall and Riggs 2007), one place (or a group of pixels) which is covered by cloud in the morning (10:30 a.m., Terra) is likely to be clear in the afternoon (1:30 p.m., Aqua), and vice versa. So, the combination of Terra and Aqua MODIS snow cover products could, in most meteorological conditions, results in cloud-less daily snow cover images and more cloud-free multi-day snow cover composite images than any single type (Terra or Aqua) of multi-day image composites. Thus, it is desirable to generate a multi-day composite image based on both Terra and Aqua MODIS daily images with flexible starting and ending dates as long as cloud cover satisfying a user-defined criterion (Wang et al. 2009; Wang and Xie 2009; Xie et al. 2009).

Taking the northern Xinjiang, China and the Colorado Plateau, USA in the 2003–2004 hydrologic year as examples, Xie et al. (2009) developed an algorithm and automated scripts to produce multi-day Terra or Aqua, and Terra-Aqua (Terra-Aqua) snow cover image composites, with flexible starting and ending dates and a user-defined cloud cover threshold. The daily composite of Terra-Aqua MODIS snow cover products has mean cloud cover ~10–15% less than and mean snow cover

~1–4% larger than those of the daily Terra or Aqua product for the year examined. Using a cloud cover of 10% as a user-defined threshold for stopping the multiday combinations, the algorithm produced 152 and 162 multi-day Terra-Aqua composite images in the Northern Xingjiang and at the Colorado Plateau, an average of 2.4 and 2.2 days (composite period) per image, respectively. The multi-day Terra, Aqua, and Terra-Aqua composite products result in similar annual mean snow covers (~15% for the Colorado Plateau and ~30% for the northern Xinjiang) as those of the corresponding 8-day standard products, while they are ~3 times as those of the daily standard products. The multi-day Terra-Aqua composite product (average of 2–3 days composite) has much higher agreement with ground measurements than that of the standard daily products and has similar agreement as that of the standard 8-day products, but having many more images available as compared with the 46 images per year from the standard 8-day products. The new multi-day MODIS snow cover composite products (MODMYD10_MC) is a better representation of the snow cover than the standard 8-day MODIS Terra or Aqua composite alone (Wang et al. 2009). Therefore, MODMYD10_MC projects a significant contribution to the current MODIS snow cover product series and is further used to examine the spatiotemporal variation of snow cover in Northern Xinjiang, China.

Based on the daily combination of Terra and Aqua MODIS snow cover products, Hall et al. (2010) further generated a cloud-gap filled snow cover products that use previous or succeeding days' daily combination to replace the cloud-covered pixels on the current day combination to generate a near cloud-free daily snow cover image. Gao et al. (2010b) produced cloud free snow cover image and enchanced SWE product by using the daily combination of MOD/MYD10A1 and AMSR-E SWE products. We are improving our algorithm (Xie et al. 2009) to generate a daily cloud-free snow cover product, which first combines the MODIS daily Terra and Aqua snow cover products (MOD10A1 and MYD10A1) to form a daily cloudless snow cover product (MODMYD_DC), then replaces the cloud-covered pixels on the current-day MODMYD_DC with the previous-day MODMYD_DC till all cloud-covered pixels are cleared up, and finally generates a new daily cloud-free snow cover product (MODMYD_MC). The case study in Northeast China indicates that MODMYD_MC can be used to efficiently monitor the spatiotemporal variations of snow cover world around.

6.5 Spatiotemporal Variation of Snow Cover

Tianshan Mountains are the water tower and solid reservoir for the arid and semi-arid region in Mid Asia. The amount of rainfall and snowfall in the mountainous areas is several times higher than that in the surrounding plain, oasis and deserts. The rainfall and snow/glacier-melting runoff from the high mountains are critical recharge for the surrounding streams, lakes and groundwater. Thus, the spatiotemporal variations of snow cover are critical information for local water resource and agriculture management.

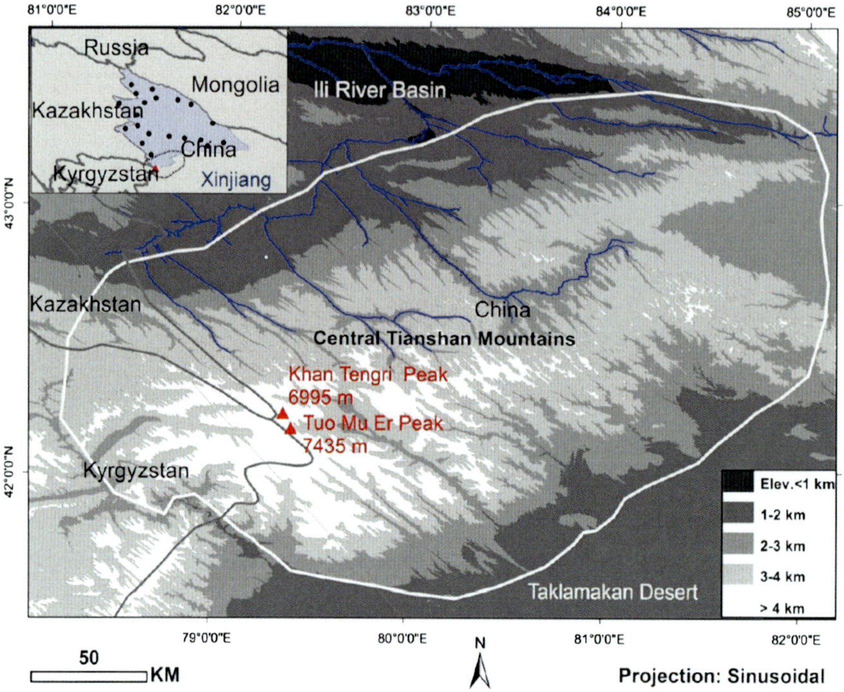

Fig. 6.3 Test areas showing elevation distribution on central Tianshan Mountains, indicating by the *white* polygon. The shaded area on the inset map is the northern Xinjiang, China, with *black* dots for climate stations of snow depth measurements. The Ili River basin is at the north foot of Central Tianshan Mountains. The Taklamakan Desert is in the south (Source: Wang and Xie 2009)

Using the multi-day MODIS snow cover composite product MODMYD10_MC developed in Xie et al. (2009), the purpose of this session is to examine the spatio-temporal variation of snow cover and its interaction with climage change in Northern Xinjiang, China and the adjacent countries under the umbralla of two MODIS tiles of h23v04 and h24v04 (Fig. 6.3). All new maps (SCD, SCOD and SCED) are produced for the entire region of the two tiles, but the Central Tianshan Mountains is illustrated for detailed analysis as an example. Background of the figure is the elevation map from NASA's SRTM (Shuttle Radar Topography Mission) with 90 m spatial resolution. The white polygon is used to constrain the geographic region in examining the overall spatiotemporal variations of snow cover conditions by the snow cover index (SCI). The majority of the materials are edited from Wang and Xie (2009).

6.5.1 Definition of SCD and SCI

Snow-covered days (SCD) are calculated as the total days with snow cover at a pixel within a hydrologic year starting from Spetember 1 through next August 31 using the multi-day MODIS snow cover composite product MODMYD_MC by Eq. (6.4).

$$SCD = \sum_{i=0}^{N} (D_{i2} - D_{i1}) \tag{6.4}$$

where N is the total number of available composite images within a hydrologic year; D_{i1} and D_{i2} are the beginning and ending dates of each image composite, respectively. For instance, the first image composite for the hydrological year of 2003–3004 is mod03244myd03245, which is a two-day composite from MODIS Terra and Aqua, $D_{i1}=244$, $D_{i2}=245$. If a pixel value on the image is equal to 200 (e.g. with snow), the SCD value at this pixel for this image is 2. Otherwise, if the pixel value on the image is not equal to 200, the SCD value at this pixel for this image is 0. Using all composite snow cover images within a hydrologic year, we can generate a SCD map for each year, a mean SCD map for all years from 2000 to 2006 and a SCD anomaly map for each year. Using a similar algorithm, a snow cover onset date (SCOD) in fall and a snow cover end date (SCED) map in spring in each year can be generated using the cloud-low MODMYD10_MC snow cover images (Wang and Xie 2009).

Using the SCD value derived in Eq. (6.4), a snow cover index (SCI, unit: km^2 day) in a hydrologic year is developed to represent the overall spatial and temporal signals of snow cover conditions in a certain region and defined as Eq. (6.5).

$$SCI = \sum_{i=0}^{N} A \times SCD_i \tag{6.5}$$

where A is the area of a pixel (in km^2; for MODIS, the actual spatial resolution is 463 m, so $A=0.214$ km^2); SCD(i) is the value (days) of SCD at pixel i within a hydrologic year; N is the total pixel number on an image.

6.5.2 Variations of Snow Cover

Compared to the SCD derived from in situ measurements at 20 climate stations in Northern Xinjiang (Figs. 6.3 and 6.4), MODIS-derived SCDs agree well (90 %) with in situ SCD and are relatively higher than in situ SCDs, except for the slightly lower values at three stations (ID 979, 971, and 965). The average MODIS SCD at the 20 stations is 9 days higher than the average in situ SCD, with exceptionally much higher values at the four stations (989, 984, 988 and 973) where the MODIS SCD is about 20 days more than in situ SCD. The overall higher MODIS SCD value

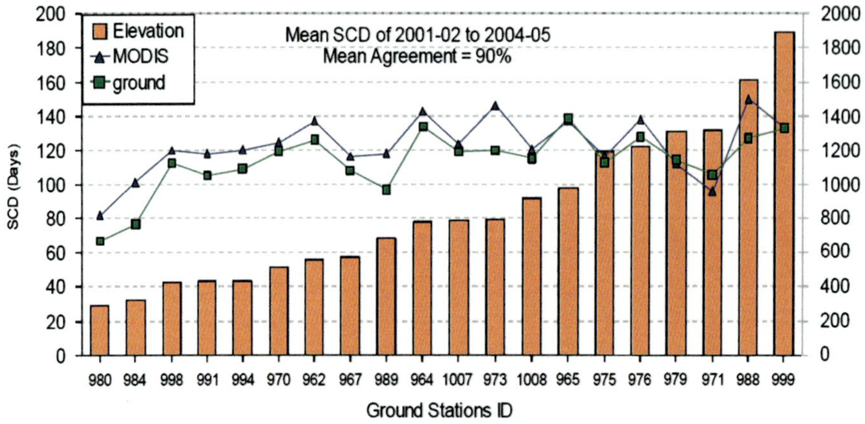

Fig. 6.4 Comparison of MODIS mean snow covered duration/days (SCD) and in situ observations of mean SCD at 20 climate stations with ascending order of elevation (*right* scale, m) in Northern Xinjiang, China, from 2001 to 2005. The mean agreement is the average Agreements defined in Eq. (6.3) for all 20 stations in the four hydrologic year from 2001–2002 to 2004–2005. (Source: Wang and Xie 2009)

is partly due to the multi-day composite algorithm that generates maximum days of snow cover during a composite period, resulting in more snow-covered days than the daily ground measurements within a year. Another contribution to this disagreement is the comparison between point ground measurements and areal observation (500 m × 500 m) of MODIS images. The third factor is the uncertainties caused by the cloud-blocked pixels, which are about 5 % of a multi-day composite image and are ignored in calculating the SCD. In addition, other factors related to the MODIS standard daily snow cover classification, such as patchy snow, haze, dry bright sand, transition zones between snow and land, and between cloud-covered and cloud-free area, trees and forest, shaded areas by cloud, mountains, trees and buildings, and so on, also contribute to the SCD overestimation. It is also found that the agreement does not have obvious association with elevation although the relative agreements at the two lowest elevation locations (ID: 980 and 984) are larger than other locations at the higher elevations.

Figure 6.5 demonstrates the mean SCD map within the six hydrologic years on the Central Tianshan Mountains. The SCD is grouped into seven classes on a 50-days interval basis (actual SCD stretches from a few days to 365 days at individual pixels). The SCD is larger than 50 days in most of the area. Only scattered pixels in the valley, forested areas and some desert areas have less than 50 days of snow cover duration. Compared to Fig. 6.3, the SCD map is closely associated with the elevation distribution. SCDs are generally less than 100 days where elevation is lower than 1 km at the bottom of Ili River Basin and in part of the desert with elevation of 1–2 km. Dominant areas, however, have SCD from 101 to 150 days, corresponding to the 1–2 km and part of 2–3 km elevation areas. Part of the 2–3 km elevation areas has SCD of 151–250 days. Where elevations are from 3 to 4 km, SCD spans

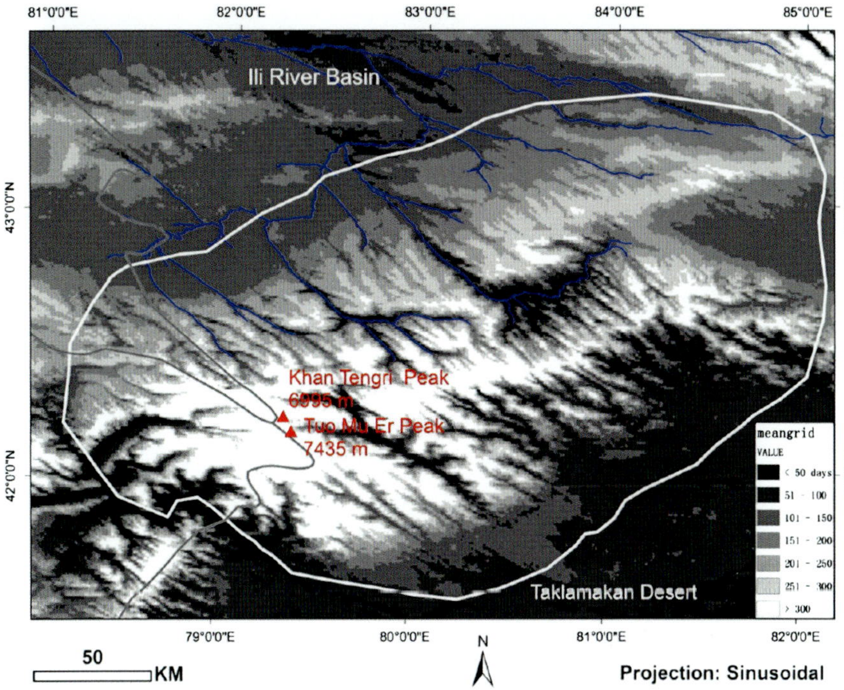

Fig. 6.5 Spatial distribution of mean snow covered duration/days (SCD) on the Central Tianshan Mountains within the six hydrologic years from 2000 to 2006. (Source: Wang and Xie 2009)

from 200 to 300 days. At the top of the mountains where elevations are higher than 4 km, SCD is generally larger than 300 days. As the challenge in demanding snow cover maps of high temporal and spatial resolution is concerned, MODIS provides unprecedented snow cover maps and now SCD maps for various applications.

Figure 6.6 displays the spatial distribution of annual SCD difference against the mean SCD of the six hydrologic years and the minimum snow cover (or perennial snow hereafter) in August. The differences of SCD in each year have also been grouped into seven categories, i.e., <−60 days, −60∼−31 days, −30∼−11 days, −10∼10 days (assumed as no change), 11–30 days, 31–60 days, and >60 days. One common feature of the six maps is that there is no obvious change on the top of the mountains surrounding the peak areas or the perennial snow cover. Compared to the mean SCD of the six hydrologic years, the SCD in the 2001–2002 and 2004–2005 hydrological years had overall higher values than the mean. In contrast, both 2003–2004 and 2005–2006 years had lower values than the mean in most of the areas, especially in 2005–2006 year.

Figure 6.7 shows the quantitative variation of overall snow cover conditions as indicated by SCI on the Central Tianshan Mountains. It has a similar pattern with Fig. 6.6 that 2001–2002 and 2004–2005 have higher SCD values than the mean SCD of the 6 years, resulting much higher SCI in the 2 years than other

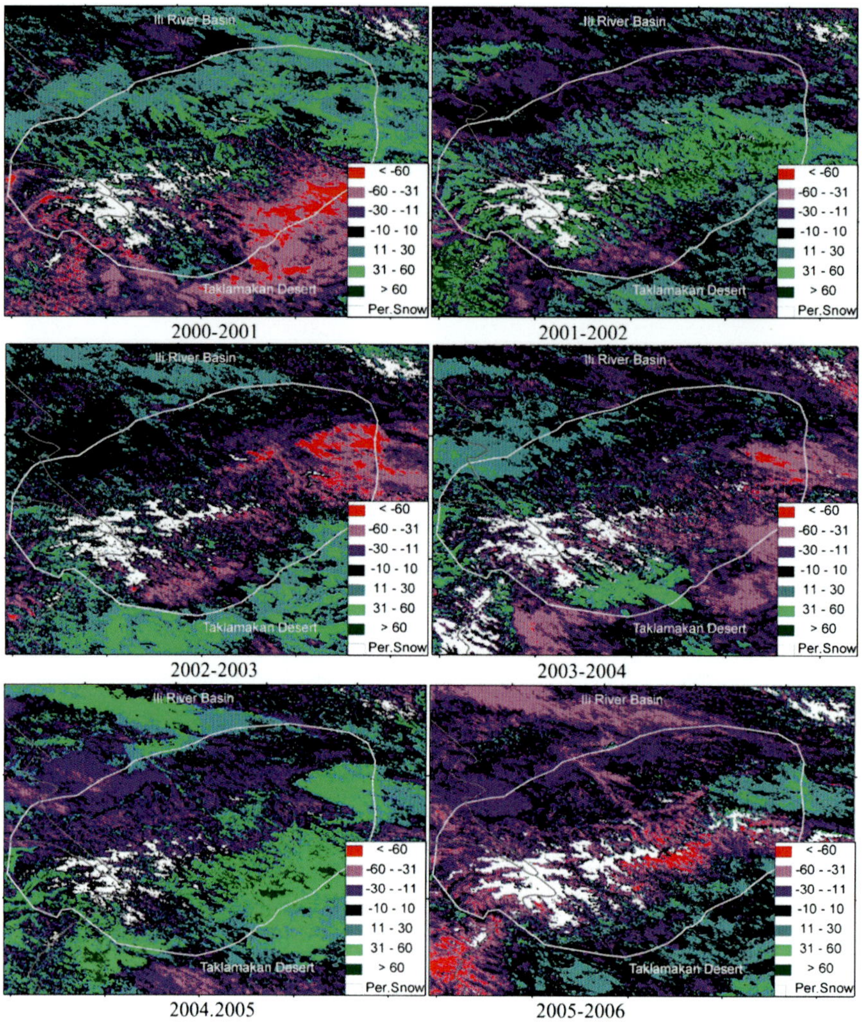

Fig. 6.6 Spatial distribution of SCD anomaly map in each hydrological year on the Central Tianshan Mountains. The mean SCD map of the 6 years was used as a basis. The *white* area represents the minimum snow cover or perennial snow in August of each year. (Source: Wang and Xie 2009)

years. The reverse patterns in Fig. 6.6 between 2000–2001 and 2004–2005, and between 2001–2002 and 2002–2003, are also quantified at a similar reverse pattern in Fig. 6.7. The 2 years in 2003–2004 and 2005–2006 have lower SCD distributions than mean SCD in Fig. 6.6, leading to similar lower SCIs than the other years. SCI in this short period (2001–2006) does not show any obvious trend. The longer records of snow cover products using NOAA's AVHRR data may provide a better change trend of snow cover in the recent 30 years. The continuing MODIS data is

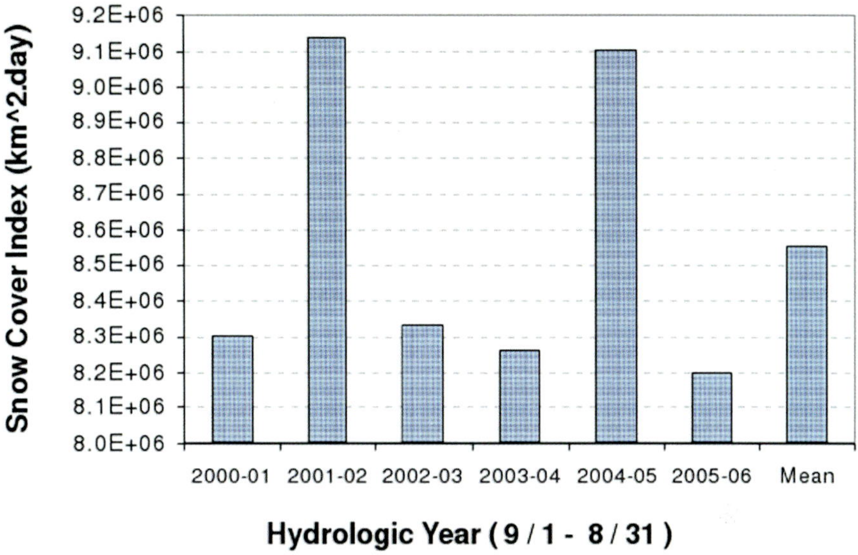

Fig. 6.7 Variation of snow cover index (SCI) on the Central Tianshan Mountains within the six hydrologic years from 2000 to 2006. (Source: Wang and Xie 2009)

also critical to monitor the change trend of the snow cover conditions at this region and other regions of the world.

Figure 6.8 illustrates the spatial distribution of snow cover onset dates (SCOD) and end dates (SCED) maps on the Central Tianshan Mountains in the fall and spring seasons of 2004 and 2005, respectively. The SCOD values are categorized into seven classes: before (<) September 15 (or O1), September 16–30 (O2), October 1–15 (O3), October 16–31 (O4), November 1–15 (O5), November 16–30 (O6), and after (> December 1 (O7). Similar to SCOD, the SCED values are also categorized into seven classes: after (>) May 1 (or E1), April 16–30 (E2), April 1–15 (E3), March 16–31 (E4), March 1–15 (E5), February 16–28 or 29 (E6), and before (<) February 15 (E7). While Figs. 6.5, 6.6 and 6.7 respectively give the general spatial distribution of the mean SCD in the 6 hydrologic years, the SCD anomalies against the mean SCD and the overall snow cover conditions (SCI) in each year, Fig. 6.8 furthern provides the spatial distribution of the specific dates when the snow cover begins and when the snow cover melts away at the scale of one MODIS pixel (500 m), respectively. Taking the 2004–2005 and 2005–2006 two hydolgic years as examples, Fig. 6.8 explains the reasons that formed the contrast of SCD variations in 2004–2005 and in 2005–2006 in the Ili River Basin, i.e., similar SCOD in 2004 Fall and in 2005 Fall, but SCED is in April 16–30 in 2005 Spring vs SCED mostly in March 1–15 in 2006 Spring, leading to 30–60 shorter SCD in 2005–2006 than that in 2004–2005. Similar patterns exist in the Southeast of Central Tianshan Mountains in these two hydrologic years (Wang and Xie 2009).

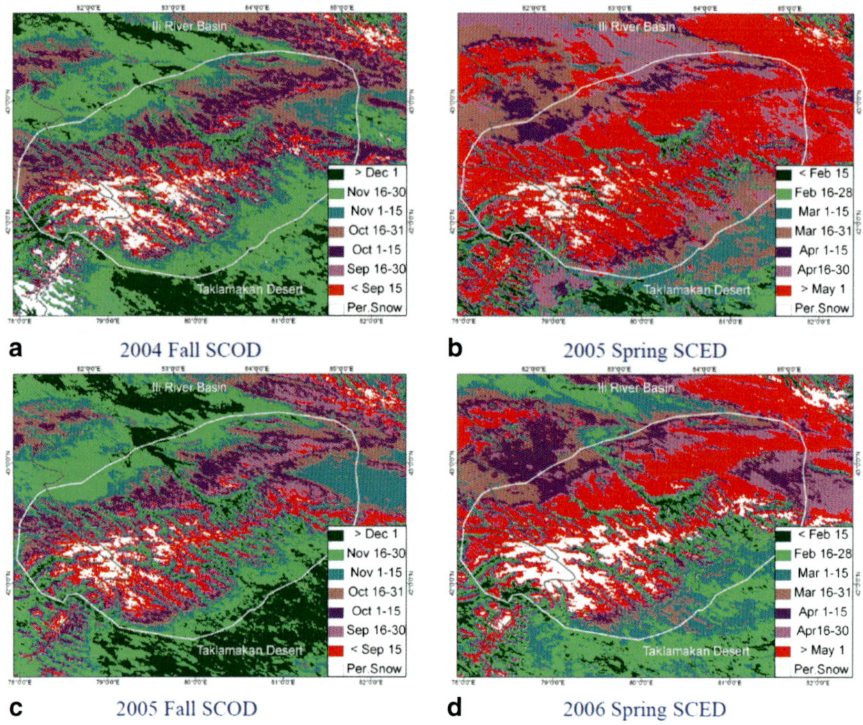

Fig. 6.8 Spatial distribution of snow cover onset dates (SCOD) and end dates (SCED) maps on the Central Tianshan Mountains in the fall and spring seasons of 2004 and 2005, respectively. The *white* area represents the minimum snow cover or perennial snow in August of each year. (Source: Wang and Xie 2009)

6.5.3 Variations of Perrenial Snow and Glacier

The spatial distribution of perennial snow is illustrated on Figs. 6.6 and 6.8. The perennial snow cover had similar spatial distribution pattern on the top of the high-elevation (>4 km) mountains in spite of yearly variation. The perennial snow cover in each year did not vary at the same trend with the overall snow cover duration. For example, on Fig. 6.6, although 2003–2004 and 2005–2006 has overall lower SCD, their perennial snow cover shown in August of 2004 and 2006 has larger areas than other years, especially in the August of 2004 when there only appeared snow cover at the southwest corner of the map in the 6 years. The 2004–2005 had overall higher SCD, but its perennial snow cover shown in the August of 2005 had the least snow cover extent in the six hydrologic years. This least snow cover extent is assumed to represent the maximum glacier distribution since it is difficult to distinguish perennial snow cover from glacier using the MODIS snow cover products. Perennial snow/glacier cover mainly distributes on the top of the mountains with elevation >4 km and has a spatial extent ~2,380 km^2 according to the least perennial snow cover map in the August of 2005. More accurate map of glacier distribution at this region can be obtained using Landsat or other higher resolution images (Li et al. 2004). Further study will examine the glacier coverage and its elevation (volume) change through integrating accurate laser elevation measurements (i.e. ICESat) (Wang and Xie 2009).

Summary and remarks Snow cover and snow depth estimation is extremely important for regional climate change studies, agriculture and water source management. Satellite-based snow measurements have revolutionized the monitoring of spatiotemporal variation of snow cover and snow depth in complex natural conditions at regional and global scales. This chaper introduces principles of optical satellite snow cover detection and the MODIS standard snow cover products, sumarizes recent efforts on mitigating the cloud-blockage issues in the optical snow cover detections and its applications, and finally illustrates the spatiotemporal variatons of snow cover in the Central Tianshan Mountains mostly in Northern Xinjiang, China by using the cloud-removed MODIS snow cover product.

Overall, MODIS standard snow cover products have high accuracy for clear-sky observations and are widely applied in all kinds of studies around the world. In contrast, MODSCAG uses the relative shape of the snow's spectrum and is more accurate to characterize fractional snow cover than the MODIS standard snow cover products (Rittger et al. 2013). However, the impact of this improvement for fractional snow cover on the total water resource budget is limited. Meanwhile, cloud blockage severely obstacles the wide application of MODIS daily snow cover products. All kinds of efforts have been done to mitigate the negetive effects of cloud cover.

Benefiting from daily twice observations of Terra and Aqua MODIS sensors and the parallel MODIS snow cover products in the morning and afternnoon, the flexible multi-day MODIS snow cover composite products MODMYD10_MC is a better representation of the snow cover than the standard 8-day MODIS Terra or Aqua composite alone, and is further used to examine the spatiotemporal variation of snow cover on the Central Tianshan Mountains using all kinds of snow mapping methods (Wang et al. 2009; Wang and Xie 2009).

While an SCD map gives the overall spatial distribution of snow cover duration, both SCOD and SCED maps provide the spatial distribution of the specific dates when the snow cover starts and when the snow cover melts away at the pixel scale of 500 m. Together, SCD, SCI, SCOD and SCED provide important information on the snow cover conditions and can be applied in any region of interests. This information could be critical for local government, such as land use planning, agriculture, live stock, and water resource management, for example, to mitigate snow-caused disasters and to plan for agriculture and industry water use. Long term availability of MODIS type of snow cover data for producing such datasets is key to study the connection between snow cover variations and climate change. For example, possible applications for this data are to study the relation between SCI and annual (hydrologic year) mean temperature change, to study the relation between SCI and El Nino-Southern Oscillation (ENSO), and to examine the time series change of annual mean SCD, SCOD and SCED for a region. Additional questions can also be addressed through the type of data presented. Does global warming reduce the mean snow cover days for a region? Does global warming have potential to forward shift or backward shift the mean snow cover onset date and/or mean snow cover melting date? How does the spring sand-storm or dust soil from the south desert impact the melting of snow cover?

References

Che T, Li X (2005) Spatial distribution and temporal variation of snow water resources in China during 1993–2002. J Glaciol Geocryol 27(1):64–67. (In Chinese)

Dozier J, Painter TH (2004) Multispectral and hyperspectral remote sensing of alpine snow properties. Ann Rev Earth Planet Sci 32:465–494

Flanner MG, Zender CS (2006) Linking snowpack microphysics and albedo evolution. J Geophys Res 111:D12208. doi:10.1029/2005JD006834

Flerchinger GN, Cooley KR, Ralston DR (1992) Groundwater response to snowmelt in a mountainous watershed. J Hydrol 133(3-4):293–300

Frei A, Lee S (2010) A comparison of optical-band based snow extent products during Spring over North America. Remote Sens Environ 114:1940–1948

Frei A, Tedesco M, Lee S, Foster JL, Hall DK, Kelly RE, Robinson DA (2012) A review of global satellite-derived snow products. Adv Space Res 50:1007–1029

Gao Y, Xie H, Yao T et al (2010a) Integrated assessment on multi-temporal and multi-sensor combinations for reducing cloud obscuration of MODIS snow cover products of the Pacific Northwest USA. Remote Sens Environ 114:1662–1675

Gao Y, Xie H, Lu N, Yao T, Liang T (2010b) Toward advanced daily cloud-free snow cover and snow water equivalent products from Terra-Aqua MODIS and Aqua AMSR-E measurements. J Hydrol 385:23–35

Hall DK, Riggs GA (2007) Accuracy assessment of the MODIS snow-cover products. Hydrol Process 21:1534–1547

Hall DK, Riggs GA, Salomonson VV (1995) Development of methods for mapping global snow cover using moderate resolution imaging spectroradiometer data. Remote Sens Environ 54:127–140

Hall DK, Tait AB, Foster JL, Chang ATC, Allen M (2000) Inter comparison of satellite-derived snow-cover maps. Ann Glaciol 31:369–376

Hall DK, Riggs GA, Salomonson VV et al (2002) MODIS snow-cover products. Remote Sens Environ 83:181–194

Hall DK, Riggs GA, Foster JL et al (2010) Development and evaluation of a cloud-gap-filled MODIS daily snow-cover product. Remote Sens Environ 114:496–503

Huang Z, Cui C (2006) Snow monitoring using EOS/MODIS data in Xinjiang, China. J Glaciol Geocryol 28(3):343–347. (In Chinese)

Kanneganti VR, Bland WL, Undersander DJ (1998) Modeling freezing injury in Alfalfa to calculate forage yield: II. Model validation and example simulations. Agron J 90:698–704

King MD, Kaufman YJ, Menzel WP, Tanre D (1992) Remote sensing of cloud, aerosol, and water vapor properties from the Moderate Resolution Imaging Spectrometer (MODIS). IEEE Trans Geosci Remote Sens 30(1):2–27

Klein AG, Barnett AC (2003) Validation of daily MODIS snow cover maps of the Upper Rio Grande River Basin for the 2000–2001 snow year. Remote Sens Environ 86:162–176

Klein AG, Hall DK, Riggs GA (1998) Improving snow cover mapping in forests through the use of a canopy reflectance model. Hydrol Process 12:1723–1744

Kongoli CE, Bland WL (2000) Long-term snow depth simulations using a modified atmosphere-land exchange model. Agric Forest Meteorol 104:273–287

Kotchenova SY, Vermote EF (2007) Validation of a vector version of the 6S radiative transfer code for atmospheric correction of satellite data. Part II.Homogeneous Lambertian and anisotropic surfaces. Appl Opt 46:4455–4464

Liang TG, Huang XD, Wu CX, Liu XY, Li WD, Guo ZG, Ren JZ (2008a) An application of MODIS data on snow cover monitoring in pastoral area: a case study in the Northern Xinjiang, China. Remote Sens Environ 112:1514–1526

Liang TG, Zhang XT, Xie HJ, Wu CX, Feng QS, Huang XD, Chen QG (2008b) Toward improved daily snow cover mapping with advanced combination of MODIS and AMSR-E measurements. Remote Sens Environ 112:3750–3761

Liu X, Liang T, Guo Z, Li L (2003) Assessment and monitoring of snow disaster effect on grassland livestock industry in the Aletai region using remote sensing technology. Acta Prataculturae Sinica 12(6):115–119. (In Chinese)

Maurer EP, Rhoads JD, Dubayah RO, Lettenmaier DP (2003) Evaluation of the snow covered area data product from MODIS. Hydrol Process 17:59–71

Nolin AW (2004) Towards retrieval of forest cover density over snow from the Multi-angle Imaging SpectroRadiometer (MISR). Hydrol Process 18:3623–3636

Nolin AW, Dozier J, Mertes LAK (1993) Mapping alpine snow using a spectral mixture modeling technique. Ann Glaciol 17:121–124

Painter TH, Rittger K, McKenzie C, Slaughter P, Davis RE, Dozier J (2009) Retrieval of subpixel snow covered area, grain size, and albedo from MODIS. Remote Sens Environ 113:868–879

Parajka J, Blöschl G (2008) Spatio-temporal combination of MODIS images-potential for snow cover mapping. Water Resour Res 44:W3406

Parajka J, Pepe M, Rampini A, et al (2010) A regional snow-line method for estimating snow cover from MODIS during cloud cover. J Hydrol 381:203–212

Riggs GA, Hall DK, Salomonson VV (2006) MODIS Snow Products User Guide to Collection 5, Online article. http://modis-snow-ice.gsfc.nasa.gov/userguides.html. Accessed 2 Jan 2007

Rittger K, Painter TH, Dozier J (2013) Assessment of methods for mapping snow cover from MODIS. Adv Water Resour 51:367–380

Rodell M, Houser PR (2004) Updating a Land Surface Model with MODIS-derived snow cover. J Hydrometeorol 5:1064–1075

Salomonson VV, Appel I (2004) Estimating fractional snow cover from MODIS using the normalized difference snow index. Remote Sens Environ 89:351–360

Salomonson VV, Appel I (2006) Development of the Aqua MODIS NDSI fractional snow cover algorithm and validation results. IEEE Trans Geosci Remote Sens 44:1747–1756

Sirguey P, Mathieu R, Arnaud Y (2009) Sub-pixel monitoring of the seasonal snow cover with MODIS at 250 m spatial resolution in the Southern Alps of New Zealand: methodology and accuracy assessment. Remote Sens Environ 113:160–181

Tekeli EA, Akyurek Z, Sorman AA, Sensoy A, Sorman AU (2005) Using MODIS snow cover maps in modeling snowmelt runoff process in the eastern part of Turkey. Remote Sens Environ 97:216–230

Wang X, Xie H (2009) New methods for studying the spatiotemporal variation of snow cover based on advanced combination products of MODIS Terra and Aqua. J Hydrol 371:192–200

Wang X, Xie H, Liang T (2008) Evaluation of MODIS Snow Cover and Cloud Mask and its Application in Northern Xinjiang, China. Remote Sens Environ 112:1497–1513

Wang X, Xie H, Liang T, Huang X (2009) Comparison and validation of MODIS standard and new combination of Terra and Aqua snow cover products in Northern Xinjiang, China. Hydrol Process 23:419–429. doi:10.1002/hyp.7151

Xie H, Wang X, Liang T (2009) Development and assessment of combined Terra and Aqua MODIS snow cover products in Colorado Plateau, USA and northern Xinjiang, China. J Appl Remote Sens 3(033559):1–14

Zhou L, Li H, Wang Q (2000) The basic characteristics of heavy snowstorm process and snow disaster distribution in eastern pastoral areas of Qinghai-Xizang Plateau. Plateau Meterol 19(4):450–458. (In Chinese)

Zhou L, Wang Q, Li H, Zhang H, Li J (2001) Study on real-time predictive assessment of snowstorm disaster in eastern pastoral area of Qinghai-Tibet Plateau. J Nat Disasters 10(2):58–65. (In Chinese)

Zhou X, Xie H, Hendrickxa MHJ (2005) Statistical evaluation of remotely sensed snow-cover products with constraints from streamflow and SNOTEL measurements. Remote Sens Environ 94:214–231

Chapter 7
Change of Potential Evapotranspiration and Its Implications to Water Cycle

Zhi Li, Yanjun Shen, Yaning Chen and Weihong Li

Abstract We analyzed the changes of potential evapotranspiration during the past 50 years, measured as the pan evaporation in the hyperarid region. We quantified the influence of water, radiation and aerodynamic attributions of the overall trend of potential evapotranspiration in this study region and analyzed the implications to hydroclimate and water balance in this region. This phenomenon implies that the region would face to more severe water scarcity in future. We also try to demonstrate the changes in water balance in this region will be facing an accelerated or enhanced cycle which implies that the increased river discharge and evapotranspiration will cause large challenge in water resources management. Both the ecosystems and social systems will be expected to facing with severe water problems under the climate warming scenarios.

Keywords Pan evaporation · Potential evapotranspiration · Water cycle · Water balance · Hyperarid region

Y. Chen (✉) · Z. Li · W. Li
State Key Laboratory of Desert and Oasis Ecology, Xinjiang Institute of Ecology and Geography, Chinese Academy of Sciences, No. 818 South Beijing Road, 830011 Urumqi, Xinjiang, China
e-mail: chenyn@ms.xjb.ac.cn

Y. Shen
Key Laboratory of Agricultural Water Resources, Center for Agricultural Resources Research, Chinese Academy of Sciences, 050021 Shijiazhuang, China

Center for Agricultural Resources Research, Chinese Academy of Sciences, No. 2 Huaizhong Road, Shijiazhuang, China
e-mail: yjshen@sjziam.ac.cn

Z. Li
e-mail: liz@ms.xjb.ac.cn

W. Li
e-mail: liwh@ms.xjb.ac.cn

Y. Chen (ed.), *Water Resources Research in Northwest China*,
DOI 10.1007/978-94-017-8017-9_7,

Evapotranspiration is a key process in the Earth's surface hydrological cycle and energy balance (Hupet and Vanclooster 2001), playing a key role in the global and regional hydrological processes under certain climate and landscape conditions. Since evapotranspiration is difficult to measure directly, traditionally measurements of pan and potential evaporation have been used to represent the evaporative demand of the atmosphere. The temporal variation of evaporation is an important indicator of climate change and the response of water resource and eco-environment to such change (Cannarozzo et al. 2006; Oguntunde et al. 2006; Liu et al. 2008). Water resources are limited in arid area, and drought is a serious threat to agricultural crops. Because of evapotranspiration is important in water and energy cycles, playing a key role in the global and regional hydrological processes under certain climate and landscape conditions, estimation of evapotranspiration is necessary for improving the planning, management, and efficient use of water resources (Gavilan and Castillo-Llanque 2009).

As the air temperature near Earth's surface was increasing at a rate of about 0.74 °C during the twentieth century (IPCC 2007), it is expected that evaporative potential of the atmosphere would increase as well. However, decreases in pan and potential evaporation have been widely reported over the past several decades (Peterson et al. 1995; Moonen et al. 2002; Roderick and Farquhar 2002; Roderick et al. 2009a, b; Gong et al. 2006; Burn and Hesch 2007; Zeng et al. 2007; Liu et al. 2009; Liu and Zhang 2011; Han et al. 2009, 2010; Shen et al. 2010). This has given rise to many studies on pan and potential evaporation trends and the attribution of their causes. Scholars have put forward various interpretations to explain this "paradox". Brutsaert and Parlange (1998) pointed out that the observed decline did not necessarily mean a declining actual evapotranspiration; in non-humid environments, decreasing pan evaporation may actually be a strong indication of increasing terrestrial evaporation. McVicar et al. (2012) suggested that the evaporative process is primarily driven by radiative and aerodynamic components, and thus while increase in air temperature in isolation may cause increases of pan evaporation, changes of other meteorological variables may have outweighed the effect of temperature.

Much effort, based on empirical meteorological data, has been made to investigate the dynamics of pan and potential evaporation in China (Liu et al. 2004; Zeng et al. 2007; Zhang et al. 2007; Shen et al. 2010). Liu et al. (2006) found close relationships between pan evaporation with diurnal temperature range and wind speed in North China; Liu et al. (2010b) found decreases in diurnal temperature range, sunshine duration and wind speed were the main attributing factors in the pan evaporation declines for China as a whole. However, most of the pan evaporation (E_{pan}) and potential evapotranspiration (ET_0) time series ended in 2001, and no detailed analysis has been reported on ET_0 trends and its spatial-temporal variability in the hyperarid regions of China (Shen et al. 2010). Because of pan evaporation track record data ended in 2001 in Northwest China, we use time series of meteorological data lasting 1958–2010 from 81 ground-based meteorological stations of the China Meteorological Administration to estimate the potential evapotranspiration, and the abrupt change in ET_0 in early 1990s was found. Then, we investigated the relationship with several climatic variables in the context of global climate change, and we

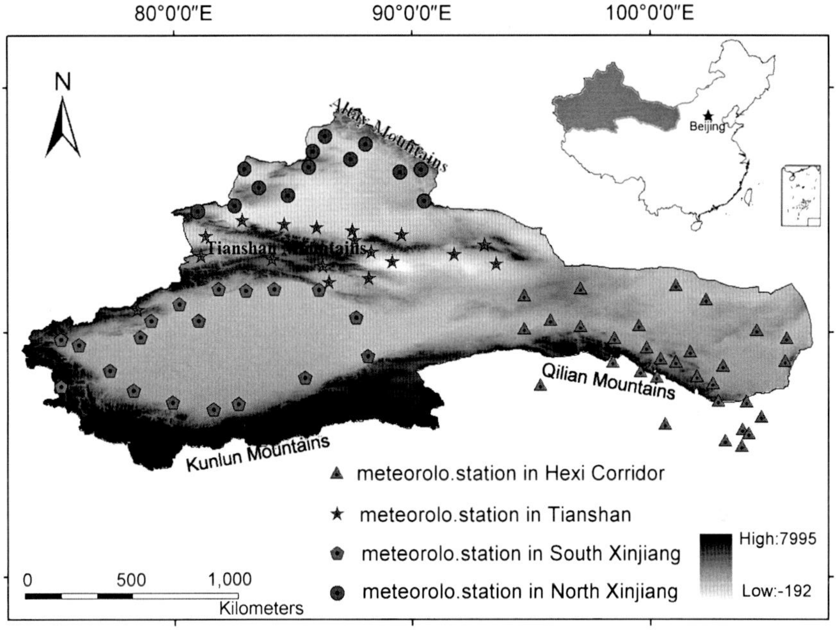

Fig. 7.1 Location of the arid region of northwest China

focused on the recent changes and attribution of potential evapotranspiration in arid region of Northwest China.

Study area and materials The arid region of Northwest China is located in the hinterland (35°–50°N, 73°–106°E) of the Eurasia continent, and far away from oceans, which causes the dry climate. It includes the entire Xinjiang Uyghur Autonomous, Hexi Corridor in Gansu, and west part of the Helan Mountains in Inner Mongolia in terms of admin. The region accounts for 24.5 % of the land area of China. The terrain of this region is featured by huge mountain ranges and vast basins or valleys. In the west part of the region, the terrain of Xinjiang is defined by three mountain ranges, including Altay Mountain, Tianshan Mountain and Kunlun Mountain (from north to south). Tianshan Mountain divides Xinjiang into two parts: Conventionally, the area south to the Tianshan Mountain is called South Xinjiang and the area north to this mountain is called North Xinjiang, the Tarim River, located in southern Xinjiang with a drainage area of 1.02×10^6 km^2, is the largest inland river in China. A narrow lowland, called Hexi Corridor lies in between Qilian Mountains in the Northwest of Gansu Province (Fig. 7.1).

Topographically, the altitude of the study area ranges from −192 to 7995 m above sea level. The lowest place in the study area is Aitin Lake in the Turpan depression and the highest place is in the Tibet-Qinghai plateau. The study area comprises desert depressions, mountains, high plateau and alluvial fans, where most oases and major human landscapes are located. The study area includes about 90 % of the deserts in China, including the Taklamakan desert, GurbanTongute desert, BadainJaran desert, Tengger desert, Qaidam desert, Kumtagh desert and the southwestern part of the Gobi desert.

This region has an extremely sparse population distribution, under-developing economy, and fragile ecological environment, mainly due to its arid climate (Ma et al. 2005; Chen et al. 2006). The climate is dominated by continental arid conditions because all the oceans are remote. Water is the key ecological factor in arid area, the oasis in Northwest China developed along the river. The climate of the arid region is typical of inner-continental land masses, with a wide temperature range, less precipitation and low humidity, annual precipitation in the study area ranges from less than 20 mm in the center of the Taklamakan desert to 500 mm in the Tianshan Mountains. The precipitation in most of the study area is below 200 mm. Exceptions are the Tianshan Mountains and the Qinghai-Tibet Plateau, where the annual precipitation exceeds 400 mm. Around one-third of the total area has an annual precipitation of less than 50 mm. Temperature also differs tremendously in the region between cold and hot. Xinjiang in the Northwest is affected by the Atlantic water vapor brought from Westerly Circulation, and does not belong to the monsoon region, the most northern edge of the monsoon can only reach the west of the Hexi Corridor.

The data used in this study were acquired from 81 ground-based meteorological stations of the China Meteorological Administration. These stations have complete records of almost all climatic factors from 1958–2010, daily observations on temperature (including minimum, maximum and average) at 2 m height, sunshine duration, wind speed measured at 10 m height, precipitation and relative humidity from 1958 to 2010 were used in analysis of potential evapotranspiration in this region. The pans are made of copper with a diameter of 0.2 m and depth of 0.1 m. They are positioned on a frame of 0.7 m above the ground and have a rim to keep birds away from drinking, monthly data on pan evaporation from 1955 to 2001 are also employed for the analysis.

7.1 Change of Pan Evaporation

7.1.1 Pan Evaporation Trends

Because of the observational data of pan evaporation afford by meteorological administration ended in 2001, firstly, we analyzed the trends in the pan evaporation relative to the average value for the period 1958–2001, in which the average annual pan evaporation was 2,213 mm in the study area. Figure 7.2 clearly showed that the pan evaporation in the study area has an obvious decreasing trend from 1960 to 1990, with sharply decreasing trend during 1960s–1970s, and an increasing trend since the 1990s.

The trends of pan evaporation observed at each station were examined in order to understand the spatial distribution of pan evaporation trends. Most of the

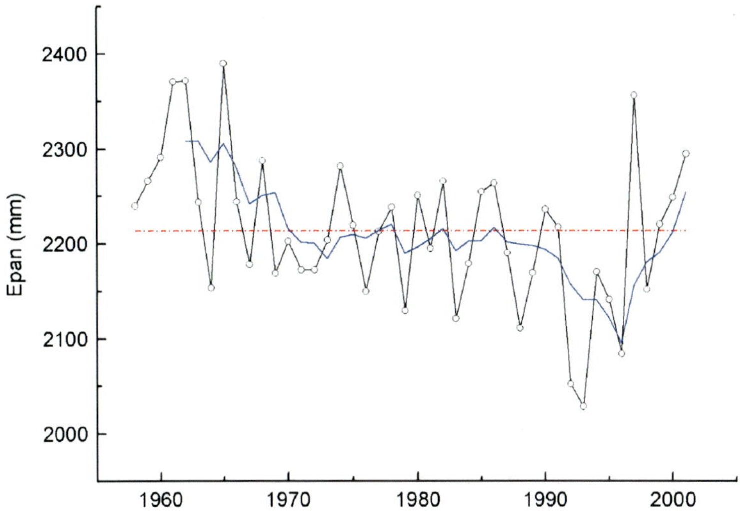

Fig. 7.2 Change of pan evaporation (E_{pan}) from 1958 to 2001

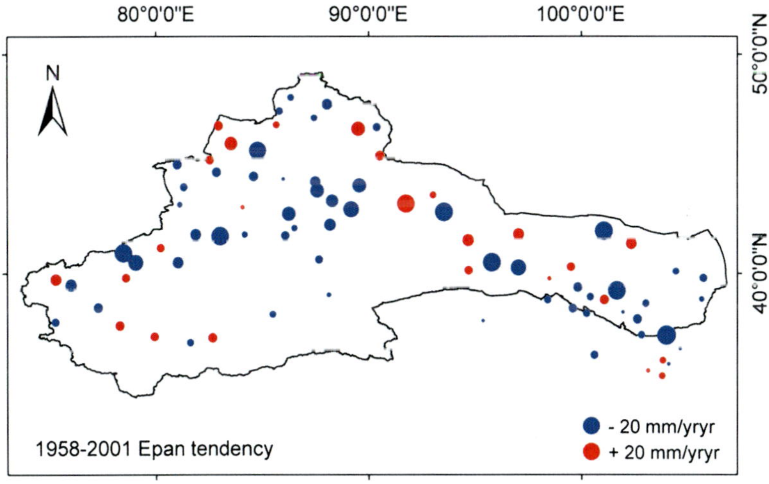

Fig. 7.3 Trends in annual E_{pan} for the meteorological stations in the Northwest China from 1958 to 2001

observation stations shown decreasing trends of pan evaporation at an average rate of -4.2 mm/yr^2 in the study area over the past 50 years (Fig. 7.3). Among the 81 stations, 56 stations had overall decreasing trends in pan evaporation for the study period. In the 25 increasing stations, 9 were in the northern Xinjiang, 6 were in the southern Xinjiang, and 10 in Hexi corridor. The decreasing rate of stations in Tianshan Mountains and Hexi corridor were larger than others.

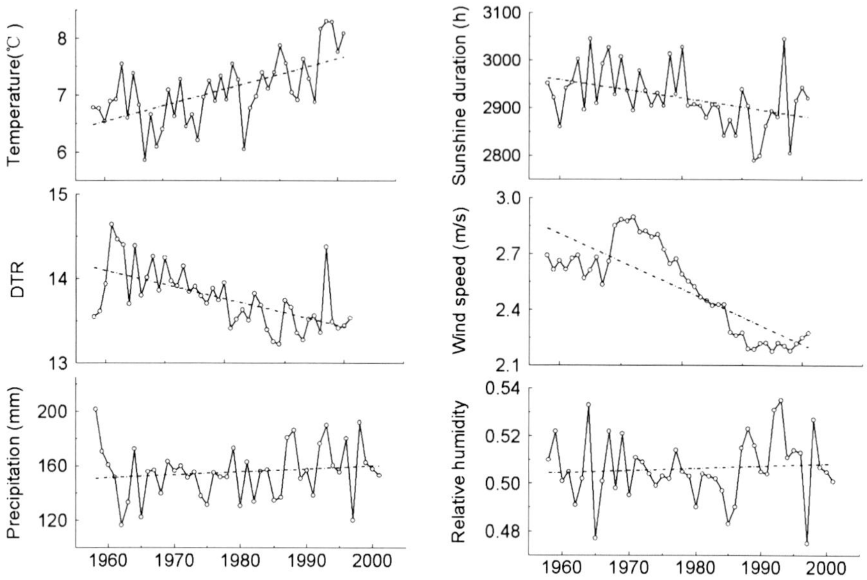

Fig. 7.4 Temporal changes of meteorological factors

7.1.2 Influencing Factors of Pan Evaporation

Pan evaporation measurements contain integrated information on atmospheric conditions. Changes in pan evaporation are often considered an indicator of climate change, which influenced by both radiation factors and atmospheric factors. Many factors can affect pan evaporation, including radiation, aerodynamic and land surface factors. The factors potentially affecting pan evaporation can be categorized into two types: wetness-related factors such as precipitation, relative humidity and low cloud cover and thermodynamic factors such as wind speed (WS), air temperature (T_a) and diurnal temperature range (DTR).

For aerially averaged pan evaporation across the full study area, the most closely related factor to decreasing pan evaporation was diurnal temperature range (DTR), which decreased with a similar trend as pan evaporation over the study period (Fig. 7.4). The correlation coefficient between DTR and E_{pan} was 0.78, passing the F-test at a significance level of 0.01. The second closest related factor was wind speed, which had a statistically significant decreasing trend as well. The relative humidity increase suggested that vapor pressure of near surface air had increased, which might be related to the expansion of irrigated areas. Although the physical mechanism linking DTR and pan evaporation was unclear, statistically the two had corresponding trends. The decrease in DTR was mainly caused by a faster increase in minimum temperature than maximum temperature, indicating a warming climate. This decrease in DTR might be mainly contributed from the increase in vapor and aerosols in the air, which reduced the daytime incoming solar radiation and also the nighttime outgoing longwave radiation from the land surface, resulting in a higher minimum temperature.

7.2 Change of Potential Evapotranspiration

7.2.1 Temporal and Spatial Distribution of ET_0

7.2.1.1 ET_0 Estimation

(1) Penman–Monteith method

Penman–Monteith equation (Allen et al. 1998) could be developed to better account for the specific local condition. This research adopts the Penman–Monteith method to estimate daily mean potential evaporation (mm d^{-1}).

$$ET_0 = \frac{0.408\Delta(R_n - G) + \gamma \dfrac{900}{T_{mean} + 273} U_2(\mathrm{vp_s} - \mathrm{vp})}{\Delta + \gamma(1+0.34U_2)} \tag{7.1}$$

where ET_0 is potential evaporation, R_n is the net radiation at the crop surface (MJ m^{-2} d^{-1}), G is the soil heat flux density (MJ m^{-2} d^{-1}), T_{mean} is the air temperature at 2 m height (°C), u_2 is the wind speed at 2 m height (m s^{-1}), vp_s is the saturation vapor pressure (kPa), vp is the actual vapor pressure (kPa), γ is the psychrometric constant (kPa°C^{-1}) and Δ is the slope vapor pressure curve (kPa°C^{-1}).

$$\Delta = \frac{4098 \times \left[0.6108 \exp\left(\dfrac{17.27T}{T + 237.3}\right)\right]}{(T + 237.3)^2} \tag{7.2}$$

where T is the mean temperature (°C).

We use the following equation transform the mean wind speed at 10–2 m above the ground:

$$u_2 = u_{10} \frac{4.87}{\ln(67.8z - 5.42)} \quad z = 10\,\mathrm{m} \tag{7.3}$$

For the calculation of R_n:

$$Rn = (1 - a')\left(a + b\frac{n}{N}\right)R_a - Q\left(\frac{T^4{}_{max,k} - T^4{}_{min,k}}{2}\right)(0.56 - 0.08\sqrt{\mathrm{vp}})\left(0.1 + 0.9\frac{n}{N}\right) \tag{7.4}$$

$$Ra = \frac{24(60)}{\pi} G_{sc} d_r [\omega_s \sin(\varphi)\sin(\delta) + \cos(\varphi)\cos(\delta)\sin(\omega_s)] \tag{7.5}$$

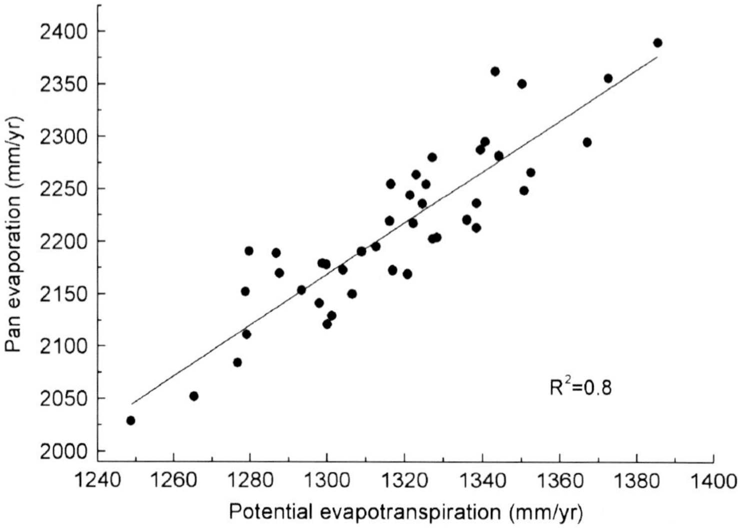

Fig. 7.5 Comparison of annual pan evaporation and potential evapotranspiration from 1958 to 2001

where R_a is solar radiation of the atmosphere (MJ m^{-2} d^{-1}), n is sunshine hours (h), N is maximum sunshine hours (h), Q is Boltzmann constant (4.903×10^{-9} MJ K^{-4} m^{-2} d^{-1}), a' is surface reflectivity (=0.23), a=0.25, b=0.5 recommended by FAO; G_{sc}: 0.082 MJ m^{-2}, d_r is the distance between the earth and sun, φ is latitude, ω_s is solar elevation at sunset: $\omega_s = \arccos[-\tan(\varphi)\tan(\delta)]$.

(2) Mann–Kendall non-parametric test

In this study, we used the Mann–Kendall test to detect trends in ET_0 and related meteorological factors. The Mann–Kendall test is often used to detect whether a time series has a significant trend or not (Gan 1998).

7.2.1.2 Trends of Annual ET_0

We used the Penman–Monteith equation to calculate the daily potential evapotranspiration in each site using complete observations.

The comparison of observed pan evaporation (1958–2001) and calculated potential evapotranspiration (1958–2001) trends were equivalent, with the $R^2=0.8$ (Fig. 7.5; Li et al. 2013).

Trends in the potential evapotranspiration anomaly were calculated relative to the average value for the period 1958–2010, in which the average ET_0 was 1,202 mm in the study area. As shown in Fig. 7.6 (Li et al. 2013), the annual ET_0 of the study area exhibited an obvious decreasing trend from 1958 to early 1990s, since the early 1990s, the trend has inversed to increase. The turning point is around 1993.

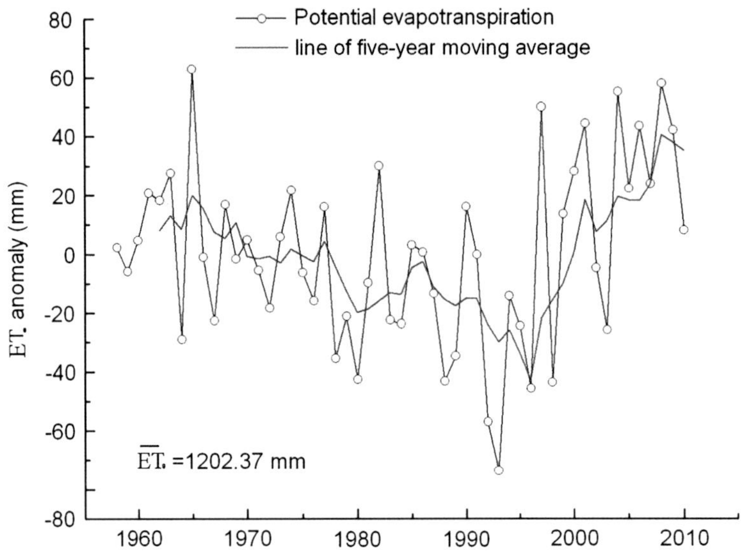

Fig. 7.6 Change in ET_0 in the arid region of Northwest China from 1958 to 2010

7.2.1.3 Trends of Seasonal ET_0

The potential evapotranspiration trends were also analyzed seasonally (Fig. 7.7; Li et al. 2013): winter (DJF), spring (MAM), summer (JJA) and autumn (SON). We found that the spring and summer have the most similar trends with the annual trend, autumn and winter were relatively less. This phenomenon suggested that changes in annual ET_0 mainly stem from changes in summer and autumn. The reversion trends of ET_0 since the early 1990s were all happened obviously in the four seasons and annual scales (Li et al. 2013).

7.2.1.4 Spatial Distribution of ET_0

The rank based non-parametric Mann–Kendall statistical test has been commonly used for trend detection (Li et al. 2013). Assuming normal distribution at the significant level of $P=0.05$, a positive Mann–Kendal statistics Z larger than +1.96 indicates a significant increasing trend, while a negative Z lower than −1.96 indicates a significant decreasing trend. Figure 7.8 (Li et al. 2013) shows the results of the trend analysis on ET_0 for each of the 81 ground stations over two periods: 1958–1993 and 1994–2010 using the Mann–Kendall test. As shown in Fig. 7.8a, during 1958–1993, annual ET_0 in most stations (about 60 %) had decreasing trends, and the exceptions were largely located in the Hexi Corridor in the east part of the study area, the northern Xinjiang had 8 slightly increasing stations. Figure 7.8b, on the other hand, during 1994–2010, about 76 % stations had increasing trends, with a cluster of

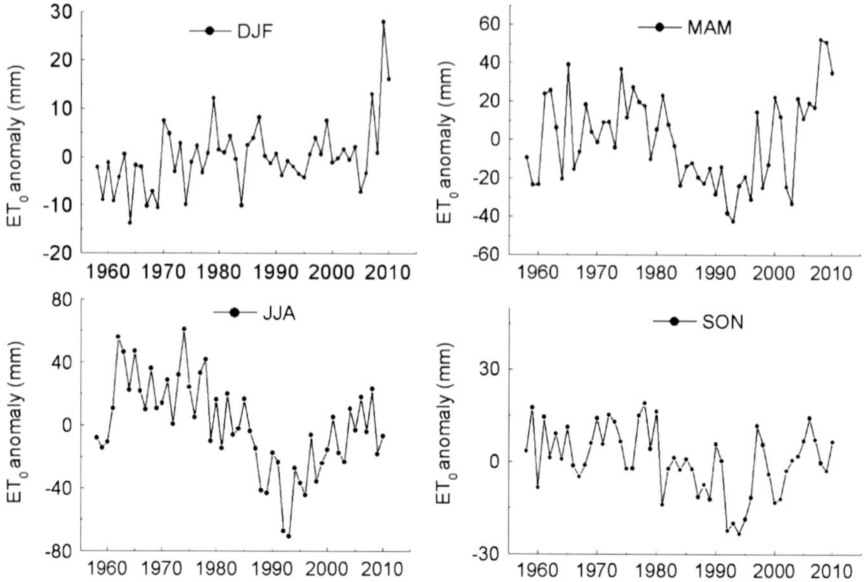

Fig. 7.7 Trends in ET_0 in four seasons (DJF, MAM, JJA, SON) from 1958 to 2010

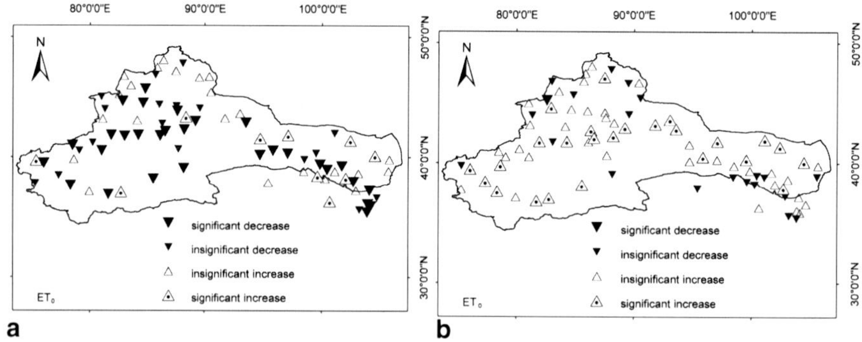

Fig. 7.8 Trends significance in annual ET_0 for the meteorological stations in the Northwest China during the periods (**a**) 1958–1993, and (**b**) 1994–2010

exceptions (insignificant decrease) in the Qilian Mountains, also in the eastern part of the study area; another cluster seems to had occurred in the Altay Mountains. During 1994–2010, there was only one station having a significant decreasing trend. The station was located in North Xinjiang, near the Northwestern Tianshan Mountains. The stations in the South Xinjiang had the most significant increase trend.

As shown in Fig. 7.9a and b (Li et al. 2013), annual ET_0 at most stations (almost 62 %) decreased during 1958–1993 at an average rate of -0.8 mm yr^{-2}, and with the largest decreased rate of -12.3 mm yr^{-2}. Another cluster stations were in the North Xinjiang and Hexi corridor. From 1994 to 2010, more than 80 % stations increased at an average rate of 5.7 mm yr^{-2}. Among the 16 decreasing stations

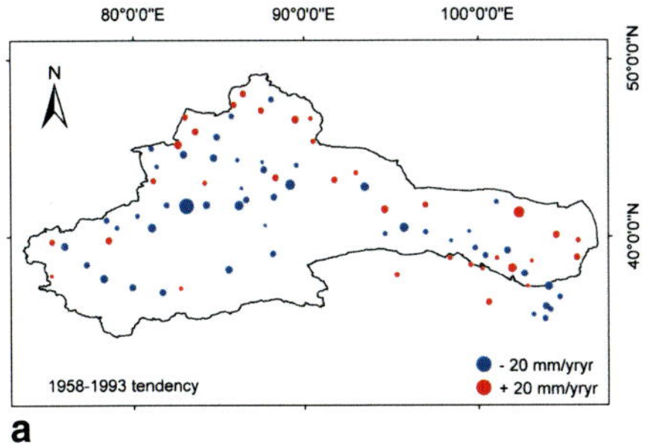

Fig. 7.9 Trends in annual ET_0 for the meteorological stations in the Northwest China during the periods (**a**) 1958–1993, and (**b**) 1994–2010

during 1994–2010, 6 were in the northern Xinjiang, 5 were in the southern Xinjiang, and 5 were in the Hexi Corridor.

In a word, the annual pan evaporation and potential evapotranspiration of the study area exhibited an obvious decreasing trend from 1958 to early 1990s, since the early 1990s, the trend has inversed to increase.

7.3 Attributing Factors to ET_0

7.3.1 Methodology

(1) Correlation analysis

A non-dimensional complete correlation analysis is conducted to identify main factors influencing potential evapotranspiration (ET_0). ET_0 is on a significant change trend over the past 50 years. Therefore, a potential attributing factor should show both a significant trend and a high correlation with ET_0 during the corresponding period (Liu et al. 2006). The complete correlation coefficient R can be calculated for this purpose:

$$R = |r_e \cdot r_t| \tag{7.6}$$

where r_e is the correlation coefficient between the climate factor and the potential evapotranspiration, r_t is the correlation coefficient between the climate factor and the time series.

7.3.2 Impact of Meteorological Variables

Factors that may affect the potential evapotranspiration can be classified into three groups (Liu et al. 2010a): dynamic factor such as wind speed (WS); thermodynamic factors such as air temperature (T_a), sunshine duration (SD), and diurnal

Table 7.1 Climate trends of meteorological factors (/10a)

Factors	T_a	SD	DTR	P	RH	WS
The period: 1958 to the early 1990s						
Tendency	0.0164	−2.949	−0.022	0.189	0.0002	−0.014
The period: The early 1990s–2010						
Tendency	0.042	0.654	−0.038	0.984	−0.001	0.014

temperature range (DTR); and water factors such as precipitation (P) and relative humidity (RH).

We took 1993 as the inflection point to divide time series into two parts: 1958–1993 and 1994–2010 (Li et al. 2013). Firstly, we conducted trend analysis on annual T_a, DTR, P, RH, WS and SD, using the Mann–Kendall test, based on the data from each of the 81 ground stations in the Northwest China (Table 7.1). Table 7.1 indicated that air temperature and precipitation increased sustained, while diurnal temperature range decreased continuously over the past 50 years. Relative humidity with minor changes was considered unchanging in the two periods. However, sunshine duration and wind speed had different trends in the two periods, which turn the significant decreasing trend to the upward trend since early 1990s.

Figure 7.10 (Li et al. 2013) shown the spatial distributions of the identified trends in annual T_a, DTR, P and RH, since SD and WS had decreasing tendency before early 1990s but increase after then, for these two variables we conducted the trend analysis for the periods before and after early 1990s separately (Fig. 7.11).

Figure 7.10 demonstrated that almost all stations had significant increasing trend in temperature except one site in the south slope of Tianshan Mountains had decreased trend. About 84 % stations had decreased in annual DTR, which main due to the increasing minimum temperature. And about 87 % stations in precipitation had increased trend, the decreasing stations main concentrate in the eastern of Qilian Mountains. Most stations in arid western China shown detectable increases in precipitation, Chen et al. (2006) also reported that precipitation in Xinjiang portrays an increasing trend. These trends indicated that the climate in the hyperarid regions is changing to warmer and wetter conditions. For the relative humidity (RH), about 48 % stations have decreasing trend, which main distribute in Tianshan Mountains and Hexi Corridor, and about 52 % stations have increasing trend, which main concentrate in South Xinjiang, the relative humidity increase suggested that vapor pressure of near surface air had increased, which might be related to the expansion of irrigated areas. The decreasing potential evapotranspiration during 1958–1993 may related to the increasing precipitation, decreasing wind speed and decreasing diurnal temperature range during 1958–1993.

It was noteworthy that wind speed (WS) at most of the 81 stations had opposite trends for the periods before and after early 1990s (Fig. 7.11; Li et al. 2013). During 1958–1993, annual WS in most stations (about 77 %) had decreasing trends, while about 68 % stations had increasing trends. The exceptional areas were the southern slope of Qilian Mountains and North Xinjiang, where most stations kept decreasing trends for the entire period of 1958–2010. In contrast, the temporal variation of sunshine duration is much less dramatic. The continuance decrease sunshine

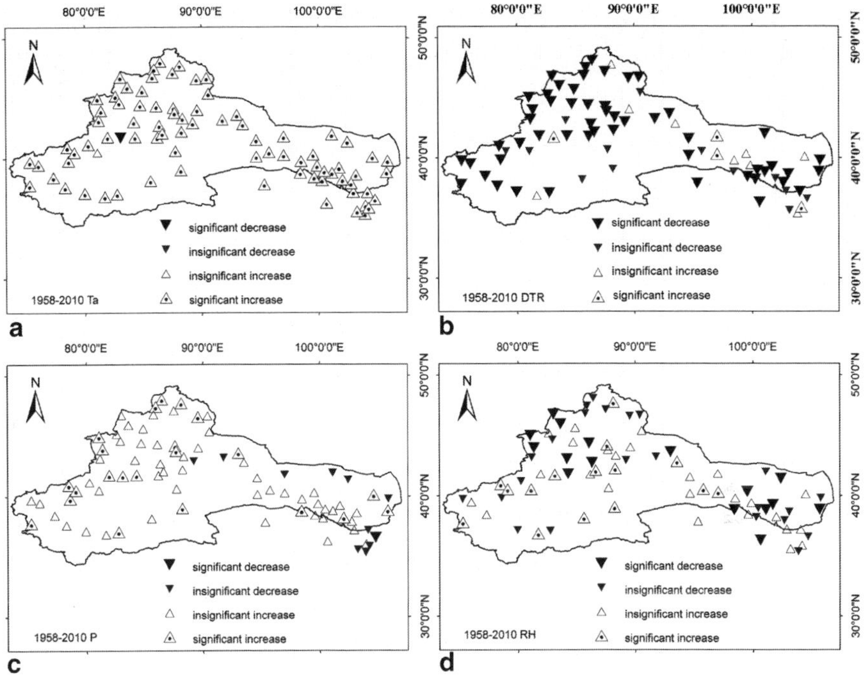

Fig. 7.10 Spatial distribution of climate factors (Ta, DTR, P and RH) trends identified by Mann–Kendall test in northwest China in 1958–2010. **a** Trend of T_a in 1958–2010, **b** Trend of DTR in 1958–2010, **c** Trend of P in 1958–2010, **d** Trend of RH in 1958–2010

duration in Gansu Province may be related to the air pollution of the provincial capital Lanzhou, which is the industrial base of China and it is the most polluted areas. There was good correspondence in the spatial distributions of the stations with potential evapotranspiration and wind speed during 1958–2010.

We calculated the complete correlation coefficients between each meteorological factor and potential evapotranspiration (Table 7.2). During 1958–1993, among all tested meteorological factors, DTR (diurnal temperature range) had the highest complete correlation coefficient with ET_0, the correlation coefficient between DTR and ET_0 was 0.41, passing the F-test at a significance level of 0.01, which shown that DTR had both a significant trend and a high correlation with ET_0 during the corresponding period. SD (sunshine duration) and WS (wind speed) also had high complete correlations with ET_0, with the correlation coefficient 0.27 and 0.29 respectively. During 1994–2010, WS (wind speed) had the highest complete correlation with ET_0, the correlation coefficient between WS and ET_0 was 0.47. And also T_a (temperature) and RH (relative humidity) also had high complete correlations with ET_0, with the correlation coefficient 0.27 and 0.28 respectively (Li et al. 2013).

Wind speed had being the most prominent factor. The present study suggests that, like in Australia, the decrease in wind speed in arid northwestern China is one of the major reasons for the decline in potential evapotranspiration. The decrease in wind speed probably relates to changes in large-scale circulation associated with

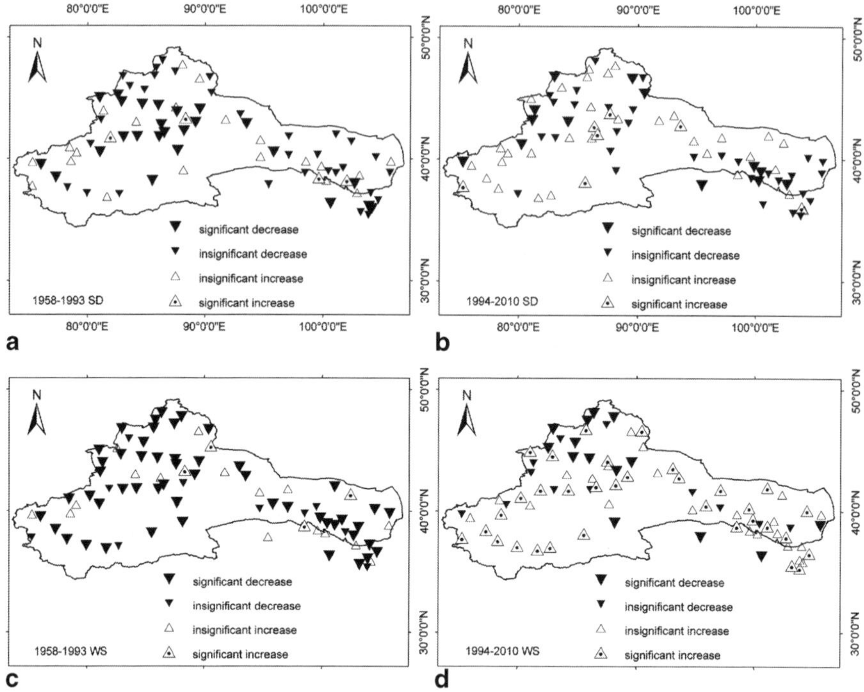

Fig. 7.11 Spatial distribution of climate factors (SD and WS) trends identified by Mann–Kendall test in northwest China for the two periods: 1958–1993, 1994–2010. **a** Trend of SD in 1958–1993, **b** Trend of SD in 1994–2010, **c** Trend of WS in 1958–1993, **d** Trend of WS in 1994–2010

climate warming. Wang et al. (2004) reported that the decrease in wind speed in China can be attributed to the weakening of the East Asian Monsoon over the last 50 years. However, an increase in land surface roughness due to reforestation and urban development near observation sites (which were usually located at the fringe of cities or towns when they was established) can also result in observations of decreasing wind speed. Changes in atmospheric circulation, air humidity, irrigation and atmospheric conditions are all have resulted in the change of wind speed, it is difficult to quantitatively pinpoint the cause of the converse changing wind speed. Further modeling and monitoring studies are needed to address this question.

7.4 Implications of the Changes in ET_0

The present study suggests that the decrease in wind speed in arid northwestern China is one of the major reasons for the decline in potential evapotranspiration before 1990s. The decrease in wind speed probably relates to changes in large-scale circulation associated with climate warming. Wang et al. (2004) reported that the decrease in wind speed in China can be attributed to the weakening of the East

Table 7.2 Annual complete correlation coefficients between ET_0 and climate factors

Factors	T_a	SD	DTR	WS	P	RH
The period: 1958 to the early 1990s						
r_e(with ET_a)	0.14	0.52**	0.62**	0.41*	−0.56**	−0.61**
r_t(with time series)	0.37*	−0.51**	−0.66**	−0.67**	0.10	0.14
R	0.05	0.27	*0.41*	0.29	0.06	0.08
The period: The early 1990s to 2010						
r_e(with ET_a)	0.56*	0.74**	0.18	0.53*	−0.59*	−0.79**
r_t(with time series)	0.48	0.06	−0.57*	0.88**	0.24	−0.36
R	0.27	0.04	0.10	*0.47*	0.14	0.28

Superscripts in this table indicate significance
** Significant at 99 % confidence level; * Significant at 95 % confidence level

Asian Monsoon over the last 50 years. An increase in land surface roughness due to reforestation and urban development near observation sites (which were usually located at the fringe of cities or towns when they was established) can also result in observations of decreasing wind speed. However, the increase potential evapotranspiration in recent years also showed a good consistence with the observations of increasing wind speed. McVicar (2012) assessing the relative contribution of the four primary meteorological variables on atmospheric evaporative demand (AED) dynamics, he has found that wind speed trends are often indentified as being important.

7.4.1 *Underlined Enhancement in Hydrological Cycles*

In the hyperarid inland basins, the land surface system behaves as energy abundant and water limited type. The relationship between precipitation and actual evapotranspiration is an important factor to determine the water resources and hydrological environment in a region. The potential evapotranspiration could reach to 3000 mm/y in this arid region, while the annual precipitation is less than 50 mm in plain area and about 500 mm in mountains. However, due to the uneven distribution of precipitation, there is runoff generated during the precipitation event and accumulated in the river networks to form water resources.

The complementary relationship between actual and potential evapotranspiration was firstly proposed by Bouchet (1963) and then formulated by Morton (1983) as follows:

$$E_a = 2 \times E_w - E_p \tag{7.7}$$

where, E_a and E_p are actual and potential evapotranspiration, respectively, E_w is evapotranspiration from wet surface. Figure 7.12 shows the concept of the complementary relationship.

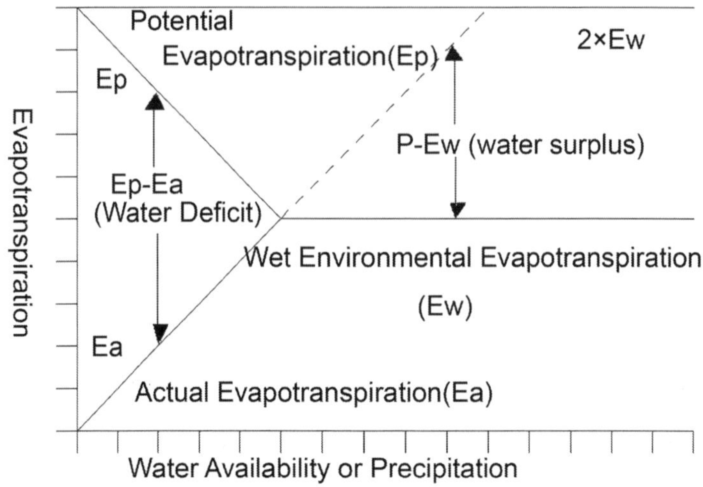

Fig. 7.12 Concept of the complementary relationship

In arid region, the actual evapotranspiration is far less then potential evapotranspiration, and falls into the lower left corner in Fig. 7.12. The actual evapotranspiration over the region is subject to increasing trend when potential evapotranspiration decrease, according to the complementary relationship. The effectiveness of the complementary relationship in non-humid regions has been testified by Sun (2007), they found actual evapotranspiration has increased over the past five decades in such regions.

At the same time, most sites in this region show detectable increases in precipitation and air temperature. Chen et al. (2006) also reported that precipitation in Xinjiang portrays has an increasing trend. These trends indicate that the climate in the hyperarid regions is changing to warmer and wetter conditions. In most parts of the study area, the aridity index (the ratio between annual potential evaporation and precipitation) has decreased over the study period (Shen et al. 2010). This suggests that the climate of most of the study area has become slightly less arid during the past half century. Chen et al. (2006) and Xu et al. (2009) have also reported an increasing trend in runoff in the past five decades. It can be concluded that the major components of the hydrological cycle, i.e. precipitation, runoff and evapotranspiration, are all undergoing increasing trends. This phenomenon indicates an enhanced hydrological cycle in the arid regions over the period of monitoring. However, it is necessary to clarify to what extent water can be renewed in the enhanced hydrological cycle or how much water/vapor may join the local cycle in the arid regions. Detailed research on water cycle processes and tracing technologies can help elucidate mechanisms of change in these hydrological processes at both the local and basin scales.

7.4.2 *Water Balance and Water Consumption Management*

The increase in actual evapotranspiration is mainly because the enlargement of irrigated land. It is reported that the irrigated farmland area in northwestern arid region of China was about 2.7 millions ha in the end of 1990s, but dramatically increased to 4.6 millions ha in 2010. Evapotranspiration from farmlands therefore increased significantly. Shen et al. (2013) estimated that the water withdrawal for irrigation has amounted to 44.2 billion m^2 only for 5 major crops in 2010. This number showed a 41.2% increase comparing with the number in the end of 1990s. As a matter of excessive exploitation, plenty of lakes and wetlands dried up and disappeared, the river way in low reaches even runs drying. The unplanned reclamation and water resources exploitation has caused serious ecological degradations and threatening the sustainability in this region. It is necessary to planning the water resources according to the water balance of the basins by fully considering the water yielding and the consumptions for ecosystems and humans.

The water balance equation at basin scale on annual basis is expressed as

$$d_w = P - R - E \tag{7.8}$$

where P is precipitation, R runoff, E evapotranspiration, d_W change in water storage in soil layer, river channel, and lake/reservoirs. In inland basins, such as the Tarim River Basin, the mountainous region functions for yielding the water resources and hydrological processes of precipitation and runoff almost happen there. We can simply divide this kind of basin into two parts: water yielding part and water consumption part. The former is the mountainous region and the latter is the arid plain region where the precipitation is rare and unable to form a water resource.

Thus, for the water yielding part, the mean annual water balance is

$$R_m - P_m - E_m \tag{7.9}$$

where m means mountainous part, and R_m can be considered as the total water resource generated from the mountain regions. It is the upper limit of available water resource (Q) for the water consumption part of the basin in plain area ($Q = R_m$). The mean annual runoff in inland basins that runs into its destination lakes finally evaporates into atmosphere. The water balance for plain area is then

$$Q = ET_{ag} + ET_{veg} + E_w \tag{7.10}$$

where ET_{ag}, ET_{veg}, and E_w are the evapotranspiration from agricultural area, native vegetation, and water surface respectively.

As mentioned earlier, the potential ET is far larger than the actual ET in the arid region, $ET_p >> ET$, and, if there is enlargement in the agricultural land, the excessive water should be withdrawn for crop consumption from the river channel. Given that there is no climate change and Q keeps at a stable level at mean annual scale, the water discharged into its destination lake then decreases and E_w is reduced. If the irrigated reclamation continued, the destination lake would disappear firstly and the n the native vegetation in the lower reaches would suffer from deterioration. This situation has been experienced in Tarim and Hei River Basins in past 30–40 years.

Under a climate change situation, Q in Eq. (7.10) has changed. If the water yields in mountainous region increase, there will be a good opportunity for humans to restore ecology. If the situation is reverse, i.e. water yields decrease, then ecosystems will be facing great risk of deterioration. One of the scientific questions is what is the critical value or threshold to balancing the water for agricultural and ecological use for different hydrological years. To restore the degraded ecosystems in these basins, people have to save water from the agricultural use by reducing the irrigated area or practicing water-saving agriculture. To study the potential of agricultural water saving, the ecological water requirement of native vegetation, appropriate groundwater level for surviving the native vegetation, etc. are the urgent issues for ecological restoration in these basins (Chen et al. 2010; Hao et al. 2010). On the other hand, the interactive processes between vegetation and groundwater associated with both drying and rewatering are most important scientific questions for understanding the ecohydrological processes in the arid environment.

In arid basins, water determines the greenness; this means that in arid basins the greenness (i.e. vegetation cover) is determined by the limited water. If the water is used for reclamation of natural land for agricultural use, then native vegetation somewhere will be affected and will suffer degradation. In the Tarim River Basin, for instance, the reclamation of barren land for grain production increased significantly between 1950 and 2000. As a result, a large part of the native vegetation at the fringes of oases and in the riparian lower reach died. Furthermore, croplands in the oases are exposed to the threats of sandstorms and desertification. Many of the traditional groundwater use facilities, such as Karez or Khanat, have dried up and have been destroyed. The local society and ecosystems are largely influenced then. Similar problems have occurred in the Amu Darya and Syr Darya Basins, where the increase in irrigated land in past decades caused significant shrinking of the water surface of the Aral Sea, and the fishery in the Aral Sea has almost been destroyed by now.

Water management in arid basins is a complicated issue because hydrology, ecology, and society are tightly linked by the limited water, and changes in one part of the system can result in large impacts on other parts. It is necessary for water management to consider this inter-relationship among mass and nutrient transfer and cycling at basin scale, ecological services, and their different functions in upper and lower reaches, and ecological conservation and economic development or social prosperity. An ecohydrological approach needs to be incorporated in an integrated watershed management framework for implementing the harmony of humans and nature in arid basins.

Summary Based on the daily data from 81 meteorological stations in the arid region of Northwest China, we got the summaries as follows:

[1] Because of the observational data of pan evaporation afford by meteorological administration ended in 2001, we analyzed the trends in the pan evaporation relative to the average value for the period 1958–2001, in which the average annual pan evaporation was 2,213 mm in the arid region of Northwest China. Pan evaporation in the study area has an obvious decreasing trend from 1958 to 2001, with an average rate of $-4.2\ mm/yr^2$.

[2] During the period 1958–2010, the potential evapotranspiration (ET_0) in this region have a turning point in early 1990s, the annual ET_0 of the study area exhibited an obvious decreasing trend from 1958 to early 1990s, since the early 1990s, the trend has inversed to increase. The turning point is around 1993.

[3] Overall, decreases in wind speed and diurnal temperature range had strong correlation to the decrease of ET_0 during 1958–1993, after early 1990s, much of the trend in ET_0 was due to the changes in the wind speed, increasing in wind speed was strongly associated with the increase of ET_0.

[4] From the implications of the changes in ET_0, considered the hydrological cycles and water balance, it is necessary for water management to consider this interrelationship among mass and nutrient transfer and cycling at basin scale, ecological services, and their different functions in upper and lower reaches, and ecological conservation and economic development or social prosperity

References

Allen RG, Pereira LS, Raes D, Smith M (1998) Crop evapotranspiration- guidelines for computing crop water requirements—FAO irrigation and drainage paper 56, U.N. Rome, pp 1–15

Bouchet RJ (1963) Evapotranspiration reele et potentielle, signification climatique. in General Assembly of Berkeley, Red Book, 62: 134–142, IAHS, Gentbrugge, Belgium

Brutsaert W, Parlange MB (1998) Hydrologic cycle explains the evaporation paradox. Nature 396:30. doi:10.1038/23845

Burn DH, Hesch NM (2007) Trends in evaporation for the Canadian prairies. J Hydrol 336(1–2):61–73. doi:10.1016/j.jhydrol.2006.12.011

Cannarozzo M, Noto LV, Viola F (2006) Spatial distribution of rainfall trends in Sicily (1921–2000). Phys Chem Earth 31(18):1201–1211. doi:org/10.1016/j.pce.2006.03.022

Chen YN, Kuniyoshi T, Xu CC, Chen YP, Xu ZX (2006) Regional climate change and its effects on river runoff in the Tarim basin, China. Hydrol Process 20(10):2207–2216. doi:10.1002/hyp.6200

Chen YN, Chen YP, Xu CC et al (2010) Effects of ecological water conveyance on groundwater dynamics and riparian vegetation in the lower reaches of Tarim river, China. Hydrol Process 24(2):170–177. doi:10.1002/hyp.7429

Gan TY (1998) Hydroclimatic trends and possible climatic warming in the Canadian prairies. Water Resour Res 34(11):3009–3015. doi:10.1029/98WR01265

Gavilan P, Castillo-Llanque F (2009) Estimating reference evapotranspiration with atmometers in a semiarid environment. Agric Water Manage 96(3):465–472. doi:org/10.1016/j.agwat.2008.09.011

Gong LB, Xu CY, Chen DL, Halldin S, Chen YD (2006) Sensitivity of the Penman–Monteith reference evapotranspiration variables in Changjiang (Yangtze river) Basin. J Hydrol 329:620–629. doi:10.1016/j.jhydrol.2006.03.027

Han SJ, Hu HP, Yang DW, Liu QC (2009) Differences in changes of potential evaporation in the mountainous and oasis regions of the Tarim basin, northwest China. Sci China Ser E 52(7):1981–1989

Han SJ, Wang SL, Yang DW (2010) Agricultural influences on evaporation paradox in China. Trans CSAE 26(10):1–8. doi:10.3969/j.issn.1002-6819.2010.10.001

Hao XM, Li WH, Huang X et al (2010) Assessment of the groundwater threshold of desert riparian forest vegetation along the middle and lower reaches of the Tarim river, China. Hydrol Process 24(2):178–186. doi:10.1002/hyp.7432

Hupet F, Vanclooster M (2001) Effect of the sampling frequency of meteorological variables on the estimation of the reference evapotranspiration. J Hydrol 243(3–4):192–204. doi:10.1016/S0022-1694(00)00413-3

IPCC (2007) Climate change 2007: the physical science basis, contribution of working group I to the fourth assessment report of the intergovernmental panel on climate change. Cambridge University Press, Cambridge

Li Z, Chen YN, Yang J et al (2013) Potential evapotranspiration and its attribution over the past 50 years in the arid region of northwest China. Hydrol Process. doi:10.1002/hyp.9643

Liu CM, Zhang D (2011) Temporal and spatial change analysis of the sensitivity of potential evapotranspiration to meteorological influencing factors in China. Acta Geogr Sin 66(5):579–588

Liu BH, Xu M, Henderson M, Gong WG (2004) A spatial analysis of pan evaporation trends in China, 1955–2000. J Geophys Res 109:D15102. doi:10.1029/2004JD004511

Liu B, Ma ZG, Ding YG (2006) Characteristics of the changes in pan evaporation over Northern China during the past 45 years and the relations to environment factors. Plateau Meteorol 25(5):840–848

Liu Q, Yang ZF, Cui BS (2008) Spatial and temporal variability of annual precipitation during 1961–2006 in Yellow river basin, China. J Hydrol 361(3–4):330–338. doi:org/10.1016/j.jhydrol.2008.08.002

Liu XM, Zheng HX, Liu CM, Cao YJ (2009) Sensitivity of the potential evapotranspiration to key climatic variables in the Haihe river basin. Resour Sci 31(9):1470–1476

Liu CM, Chen YN, Xu ZX (2010a) Eco-hydrology and sustainable development in the arid regions of China Preface. Hydrol Process 24(2):127–128. doi:10.1002/hyp.7481

Liu M, Shen YJ, Zeng Y, Liu CM (2010b) Trend in pan evaporation and its attribution over the past 50 years in China. Science China Press 20(4):557–568. doi:10.1007/s11442.010-0557-3

Ma JZ, Wang XS, Edmunds WM (2005) The characteristics of groundwater resources and their changes under the impacts of human activity in the arid northwest China-a case study of the Shiyang River Basin. J Arid Environ 61(2):277–295. doi:10.1016/j.jaridenv.2004.07.014

McVicar TR, Roderick ML, Donohue RJ, Niel TGV (2012) Ecohydrology bearings—invited commentary less bluster ahead? Ecohydrological implications of global trends of terrestrial near-surface wind speeds. Ecohydrology 5:381–388. doi:10.1002/eco.1298

Moonen AC, Ercoli L, Mariotti M, Masoni A (2002) Climate change in Italy indicated by agrometeorological indices over 122 years. Agric Forest Meteorol 111:13–27. doi:10.1016/S0168-1923(02)00012-6

Morton FI (1983) Operational estimates of areal evapotranspiration and their significance to the science and practice of hydrology. J Hydrol 66:1–76. doi:10.1016/0022-1694(83)90177-4

Oguntunde PG, Friesen J, Giesen N, Savenije HHG (2006) Hydroclimatology of Volta river basin in West Africa: trends and variability from 1901 to 2002. Phys Chem Earth 31(18):1180–1188. doi:org/10.1016/j.pce.2006.02.062

Peterson T, Golubev V, Groisman P (1995) Evaporation losing its strength. Nature 393:687–688. doi:10.1038/377687b0

Roderick ML, Farquhar GD (2002) The cause of decreased pan evaporation over the past 50 years. Science 298:1410–1411. doi:10.1126/science.1075390

Roderick ML, Hobbins MT, Farquhar GD (2009a) Pan evaporation trends and the terrestrial water balance. I. Principles and observations. Geogr Compass 3/2:746–760. doi:10.1111/j.1749-8198.2008.00213.x

Roderick ML, Hobbins MT, Farquhar GD (2009b) Pan evaporation trends and the terrestrial water balance. II. Energy balance and interpretation. Geogr Compass 3/2:746–760. doi:10.1111/j.1749- 8198.2008.00214.x

Shen YJ, Li S, Chen YN, Li ZQ, Zhang SW (2013) Estimation of regional irrigation water requirement and water supply risk in the arid region of Northwestern China 1989–2010. Agric Water Manage, 128:55–64. doi:10.1016/j.agwat.2013.06.014

Shen YJ, Liu CM, Liu M, Zeng Y, Tian CY (2010) Change in pan evaporation over the past 50 years in the arid region of China. Hydrol Process 24:225–231. doi:10.1002/hyp.7435

Sun FB (2007) Study on watershed evapotranspiration based on the Budyko hypothesis. Tsinghua University, June, pp 34

Wang ZY, Ding YH, He JH (2004) An updating analysis of the climate change in China in recent 50 years. Acta Meteorologica Sinica, 62(2):228–236. (in Chinese)

Xu ZX, Pang JP, Liu CM, Li JY (2009) Assessment of runoff and sediment yield in the Miyun reservoir catchment by using SWAT model. Hydrol Process 23(25):3619–3610. doi:10.1002/hyp.7475

Zeng Y, Qiu X, Liu CM (2007) Changes of pan evaporation in China in 1960–2000. Adv Water Sci 18(3):311–318

Zhang Y, Liu C, Tang Y, Yang Y (2007) Trends in pan evaporation and reference and actual evapotranspiration across the Tibetan plateau. J Geophys Res 112:D12110. doi:10.1029/2006JD008161

Chapter 8
The Nonlinear Hydro-climatic Process: A Case Study of the Tarim Headwaters, NW China

Jianhua Xu, Yaning Chen and Weihong Li

Abstract Some researchers have suggested that the hydro-climatic process is a complex system, with nonlinearity as its basic characteristic. But there is still a lack of effective means available to thoroughly discover the dynamics of hydro-climatic process at different time scales. Therefore, more studies are required to explore the nonlinearity of hydro-climatic process from different perspectives and using different methods. Based on the hydrologic and meteorological data in the areas of the Tarim headwaters, this chapter investigated the nonlinear hydro-climatic process by a comprehensive method including correlation dimension, R/S analysis, wavelet analysis, regression and artificial neural network modeling. The main findings are as follows:

(1) The hydro-climatic process in the Tarim headwaters presented periodic, nonlinear, chaotic dynamics, and long-memory characteristics.
(2) The correlation dimensions of the attractor derived from the AR time series for the Hotan, Yarkand, Aksu and Kaidu rivers were all greater than 3.0 and non-integral, implying that all four headwaters are dynamic chaotic systems that are sensitive to initial conditions, and that the dynamic modeling of hydro-climatic process requires at least four independent variables.
(3) The Hurst exponents indicate that a long-term memory characteristic exists in hydro-climatic process. However, there were some differences observed, with

J. Xu (✉)
The Research Center for East-West Cooperation in China, The Key Lab of GIScience of the Education Ministry PRC, East China Normal University, 200241 Shanghai, China
e-mail: jhxu@geo.ecnu.edu.cn

Y. Chen · W. Li
State Key Laboratory of Desert and Oasis Ecology, Xinjiang Institute of Ecology and Geography, Chinese Academy of Sciences, 830011 Urumqi, Xinjiang, China
e-mail: chenyn@ms.xjb.ac.cn

W. Li
e-mail: liwh@ms.xjb.ac.cn

Y. Chen (ed.), *Water Resources Research in Northwest China*,
DOI 10.1007/978-94-017-8017-9_8,

the Aksu, Yarkand and Kaidu rivers demonstrating a persistent trait, and the Hotan River exhibiting an anti-persistent feature.

(4) The variation pattern of runoff, temperature and precipitation was scale-dependent with time. Annual runoff (AR), annual average temperature (AAT) and annual precipitation (AP) at five time scales resulted in five variation patterns respectively.

(5) The nonlinear variation of runoff is resulted from regional climatic change. The variation periodicity of AR is close with that of AAT and AP. The multiple linear regression (MLR) and back-propagation artificial neural network (BPANN) based on wavelet analysis reveal the correlations between annual runoff (AR) with annual precipitation (AP), annual average temperature (AAT) at different time scales.

Keywords Hydro-climatic process · Nonlinearity · Periodic · Nonlinear and chaotic dynamics · Tarim headwaters

8.1 Introduction

Theoretically, hydro-climatic process can be evaluated to determine if they comprise an ordered, deterministic system, an unordered, random system, or a chaotic, dynamic system, and whether change patterns of periodicity or quasi-periodicity exist. However, it is difficult to achieve a thorough understanding of the nonlinear hydro-climatic process (Cannon and McKendry 2002; Xu et al. 2010).

In the last 20 years, many studies have been conducted to evaluate climatic change and hydrological processes in the arid and semi-arid regions in northwestern China (Chen and Xu 2005; Wang et al. 2010; Xu et al. 2011a, b; Zhang et al. 2010). A number of studies have indicated that there was a visible transition in the hydro-climatic processes in the past half-century (Chen and Xu 2005; Chen et al. 2006; Shi et al. 2007; Wang et al. 2010). This transition was characterized by a continual increase in temperature and precipitation, added river runoff volumes, increased lake water surface elevation and area, and elevated groundwater levels. This transition may inquiries a series questions if these changes represent a localized transition to a warm and wet climate type in response to global warming, or merely reflect a centennial periodicity in hydrological dynamics. To date, these questions have not received satisfactory answers; therefore, more studies are required to explore the nonlinear characteristics of hydro-climatic process from different perspectives and using different methods (Xu et al. 2012).

Because of the above reasons, this chapter investigated the nonlinear hydro-climatic process in the Tarim headwaters by an integrated approach including correlation dimension, R/S analysis, wavelet decomposition, regression analysis and artificial neural network.

8.1.1 *Materials and Methods*

8.1.1.1 Study Area

The Tarim River basin covers an area of 1.02×10^6 km^2 and is the largest continental river basin in China. The basin covers the entire southern part of Xinjiang in western China and is characterized by an abundance of rich natural resources and a fragile environment. This region has an extreme desert climate with an annual average temperature of 10.6 ~ 11.5 °C. In addition, the monthly mean temperature ranges from 20 to 30 °C in July and − 10 to − 20 °C in January and the highest and lowest temperatures are +43.6 °C and −27.5 °C, respectively. The accumulative temperature > 10 °C ranges from 4,100 to 4,300 °C. The average annual precipitation is approximately 116.8 mm for the entire area, with an uneven distribution of 200–500 mm in the mountainous area, 50 ~ 80 mm on the edges of the basin, and only 17.4–25.0 mm in the central portion of the basin. There is great temporal unevenness in precipitation within each year as well. More than 80 % of the total annual precipitation falls between May and September in the high-flow season, and less than 20 % of the total precipitation occurs between November and April.

The main channel of the Tarim River is 1,321 km in length. Naturally and historically the Tarim River basin consists of 114 rivers from nine drainage systems, which include the Aksu, Hotan, Yarkand, Qarqan, Keriya, Dina, Kaxgar, and Kaidu-Konqi rivers. The basin contains 20.44×10^6 ha of arable lands and has a human population of 8.26×10^6. The mean annual natural surface runoff is 3.98×10^{10} m^3, most of which originates from glaciers, snowmelt and precipitation in the surrounding mountains.

Intensive disturbances caused by human activities, particularly in response to excessive water resources exploitation, have brought about marked changes during the past 50 years. The drainage systems gradually disintegrated when the Weigan, Kaxgar, Dina, Keriya, and Qarqan rivers stopped flowing into the mainstream and were eventually disconnected from it. Today, there are only three drainage systems connected to the mainstream of the Tarim River, the Aksu River, Yarkand River and Hotan River. The Aksu River has two main tributaries (the Tongshigan and Kumalak) and originates from the Tianshan Mountains in the northwest portion of the basin. The Hotan River also has two main tributaries (the Kalaksh and Yulongkash) and originates in the Kunlun Mountains and flows through the southwestern portion of the basin. The Yarkand River originates from the Pamir Plateau and lies between the Aksu River and the Yarkand River (Fig. 8.1).

As mentioned above, glaciers, snowmelt and precipitation in the surrounding mountains are the major sources of runoff in the Tarim River. Specifically, glacier melt and snowmelt comprise 48.2 % of the total runoff of the river. Interannual runoff variability is small, with a coefficient of variation ranging from 0.15 to 0.25 and maximum and minimum modular coefficients of 1.36 and 0.79, respectively. Additionally, seasonal runoff is unevenly distributed, with runoff during the flood

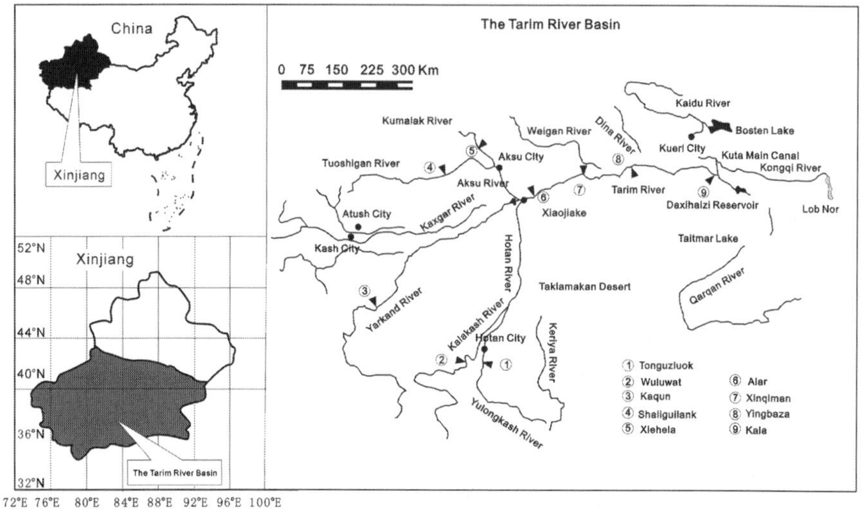

Fig. 8.1 Location of study area

season (June–August) accounting for 60–80 % of the total annual runoff (Chen and Xu 2005).

8.1.1.2 Data

To investigate hydro-climatic process of the Tarim headwaters, this study used the time series of annual runoff (AR) in the period of 1957~2008 from Xiehela and Shaliguilank hydrologic stations for the Aksu River, from the Kaqun hydrologic station for the Yarkand River, from the Tonguzluok and Wuluwat hydrologic stations for the Hotan River, and from the Dashankou hydrologic station for the Kaidu River. We also used annual average temperature (AAT) and annual precipitation (AP) from 28 meteorological stations in the Tarim River Basin for the same study period.

8.1.1.3 Methodology

(1) Correlation dimension

The correlation dimension method is usually applied to analyze a series and determine if the process exhibits a chaotic dynamic characteristic (Sivakumar 2007; Xu et al. 2009a). Consider $x(t)$, the time series of annual runoff, and suppose it is generated by a nonlinear dynamic system with m degrees of freedom. To restore the dynamic characteristic of the original system, it is necessary to construct an appropriate series of state vectors, $X^{(m)}(t)$, with delay coordinates in the m-dimensional phase space according to the basic ideas initiated by Grassberger and Procaccia (1983):

$$X^{(m)}(t) = \{x(t), x(t+\tau), \ \ldots, \ x(t+(m-1)\tau)\} \tag{8.1}$$

where m is the embedding dimension and τ is an appropriate time delay.

The trajectory in the phase space is defined as a sequence of m dimensional vectors. If the dynamics of the system can be reduced to a set of deterministic laws, the trajectories of the system converge toward a subset of the phase space, which is called an "attractor". Many natural systems do not conform with time to a cyclic trajectory. Some nonlinear dissipative dynamic systems tend to shift toward the attractors for which the motion is chaotic, i.e. not periodic and unpredictable over long times. The attractors of such systems are called strange attractors. For the set of points on the attractor, using the G-P method (Grassberger and Procaccia 1983), the correlation-integrals are defined to distinguish between stochastic and chaotic behaviors.

The correlation-integrals can be defined as follows:

$$C(r) = \frac{1}{N_R^2} \sum_{j=1}^{N_R} \sum_{i=1}^{N_R} \Theta(r - |X_i - X_j|) \tag{8.2}$$

where r is the surveyor's rod for distance, N_R is the number of reference points taken from N, and N is the number of points, $X^{(m)}(t)$. The relationship between N and N_R is $N_R = N-(m-1)\tau$. $\Theta(x)$ is the Heaviside function, which is defined as:

$$\Theta(x) = \begin{cases} 0 & x \le 0 \\ 1 & x > 0 \end{cases} \tag{8.3}$$

The expression counts the number of points in the dataset that are closer than the radius, r, within a hypersphere of the radius, r, and then divides this value by the square of the total number of points (because of normalization). As $r \to 0$, the correlation exponent, d, is defined as:

$$C(r) \propto r^d \tag{8.4}$$

It is apparent that the correlation exponent, d, is given by the slope coefficient of ln $C(r)$ versus ln r. According to (ln r, ln $C(r)$), d can be obtained by the least squares method (LSM) using a log-log grid.

To detect the chaotic behavior of the system, the correlation exponent has to be plotted as a function of the embedding dimension (as shown as Fig. 3.2 in Sect. 3.1). If the system is purely random (e.g. white noise) the correlation exponent increases as the embedding dimension increases, without reaching the saturation value.

If there are deterministic dynamics in the system, the correlation exponent reaches the saturation value, which means that it remains approximately constant as the embedding dimension increases. The saturated correlation exponent is called the correlation dimension of the attractor. The correlation dimension belongs to the invariants of the motion on the attractor. It is generally assumed that the correlation dimension equals the number of degrees of freedom of the system, and higher embedding dimensions are therefore redundant. For example, to describe the position of the point on the plane (two-dimensional system), the third dimension is not necessary because it is redundant. In addition, the correlation dimension is often fractal and represented as a non-integral dimension, which is typical for chaotic dynamical systems that are very sensitive to initial conditions.

(2) R/S analysis

R/S analysis, which is also called rescaled range analysis, is usually applied to analyze long-term correlation characteristics of a time series (Xu et al. 2008a; Li et al. 2008). The principle of R/S analysis is briefly introduced as follows (Mandelbrot and Wallis 1969; Turcotte 1997):

Considering a time series $x(t)$, such as the annual runoff sequence of a certain river, for any positive integer $\tau \geq 1$, the mean value series is defined as:

$$\langle x \rangle_\tau = \frac{1}{\tau}\sum_{t=1}^{\tau} x(t) \quad \tau = 1, 2, \ldots \tag{8.5}$$

The accumulative deviation is:

$$X(t,\tau) = \sum_{u=1}^{t}\left(x(u) - \langle x \rangle_\tau\right) \quad 1 \leq t \leq \tau \tag{8.6}$$

The extreme deviation is:

$$R(\tau) = \max_{1\leq t\leq \tau} X(t,\tau) - \min_{1\leq t\leq \tau} X(t,\tau) \quad \tau = 1, 2, \ldots \tag{8.7}$$

The standard deviation is:

$$S(\tau) = \left[\frac{1}{\tau}\sum_{t=1}^{\tau}\left(x(t) - \langle x \rangle_\tau\right)^2\right]^{\frac{1}{2}} \quad \tau = 1, 2, \ldots \tag{8.8}$$

When analyzing the statistic rule of $R(\tau)/S(\tau) \triangleq R/S$, H E Hurst discovered a relational expression,

$$R/S \propto \left(\frac{\tau}{2}\right)^H \tag{8.9}$$

which can be used to identify the Hurst phenomenon in the time series, where H is known as the Hurst exponent. It is evident that H is given by the slope coefficient of R/S versus $\tau/2$. According to (ln $\tau/2$, ln (R/S)), H can be obtained by the least squares method (LSM) in a log-log grid.

Hurst et al. (1965) once demonstrated that if $x(t)$ is an independently random series with limited variance, the exponent, $H=0.5$, and H $(0<H<1)$ is dependent on a correlation function $C(t)$:

$$C(t) = 2^{2H-1} - 1 \tag{8.10}$$

When $H>0.5$ and $C(t)>0$, the process has a long, enduring characteristic, and the future trend of the time series will be consistent with the past. In other words, if the past showed an increasing trend, the future will also show an increasing trend. When $H<0.5$ and $C(t)<0$, the process has an anti-persistence characteristic, and the future trend of the time series will be opposite from the past. In other words, if the past showed an increasing trend, the future will assume the reducing trend. When $H=0.5$ and $C(t)=0$, the process is stochastic; in other words, there is no correlation or only a short-range correlation in the process (Ai and Li 1993).

(3) Wavelet analysis

Wavelet transformation has been shown to be a powerful technique for characterization of the frequency, intensity, time position, and duration of variations in climate and hydrological time series (Torrence and Compo 1998; Smith et al. 1998; Chou 2007; Xu et al. 2009b). Wavelet analysis can also reveal the localized time and frequency information without requiring the time series to be stationary, as required by the Fourier transform and other spectral methods.

A continuous wavelet function $\Psi(\eta)$ that depends on a nondimensional time parameter η can be written as (Labat 2005):

$$\Psi(\eta) = \Psi(a,b) = |a|^{-1/2} \Psi\left(\frac{t-b}{a}\right) \tag{8.11}$$

where, t denotes time, a is the scale parameter and b is the translation parameter. $\Psi(\eta)$ must have a zero mean and be localized in both time and Fourier space (Farge 1992). The continuous wavelet transform (CWT) of a discrete signal, $x(t)$, such as the time series of runoff, temperature, or precipitation, is expressed by the convolution of $x(t)$ with a scaled and translated $\Psi(\eta)$,

$$W_x(a,b) = |a|^{-1/2} \int_{-\infty}^{+\infty} x(t)\Psi^*\left(\frac{t-b}{a}\right)dt \tag{8.12}$$

where, * indicates the complex conjugate, and $W_x(a, b)$ denotes the wavelet coefficient. Thus, the concept of frequency is replaced by that of scale, which can characterize the variation in the signal, $x(t)$, at a given time scale.

Selecting a proper wavelet function is a prerequisite for time series analysis. The actual criteria for wavelet selection include self-similarity, compactness, and smoothness (Ramsey 1999). For the present study, symlet 8 was chosen as the wavelet function according to these criteria.

The nonlinear trend of a time series, $x(t)$, can be analyzed at multiple scales through wavelet decomposition on the basis of the discrete wavelet transform (DWT). The DWT is defined taking discrete values of a and b (Banakar and Azeem 2008). The full DWT for signal, $x(t)$, can be represented as (Mallat 1989):

$$x(t) = \sum_k \mu_{j_0,k}\phi_{j_0,k}(t) + \sum_{j=1}^{j_0}\sum_k \omega_{j,k}\psi_{j,k}(t) \tag{8.13}$$

where $\phi_{j_0,k}(t)$ and $\psi_{j,k}(t)$ are the flexing and parallel shift of the basic scaling function, $\phi(t)$, and the mother wavelet function, $\psi(t)$, and $\mu_{j_0,k}\left(j < j_0\right)$ and $\omega_{j,k}$ are the scaling coefficients and the wavelet coefficients, respectively. Generally, scales and positions are based on powers of 2, which is the dyadic DWT (Sun et al. 2006).

Once a mother wavelet is selected, the wavelet transform can be used to decompose a signal according to scale, allowing separation of the fine-scale behavior (detail) from the large-scale behavior (approximation) of the signal (Bruce et al. 2002). The relationship between scale and signal behavior is designated as follows: low scale corresponds to compressed wavelet as well as rapidly changing details, namely high frequency; whereas high scale corresponds to stretched wavelet and slowly changing coarse features, namely low frequency. Signal decomposition is typically conducted in an iterative fashion using a series of scales such as a=2, 4, 8, ……, 2^L, with successive approximations being split in turn so that one signal is broken down into many lower resolution components.

(4) Regression analysis based on wavelet decomposition

For understanding the relationship between annual runoff with its related climatic factors at different time scales, we employed Multiple linear regression (MLR) based on wavelet decomposition. This method fits multiple linear regression equation (MLRE) between AR with AAT and AP by using multiple linear regression (MLR) based on the results of wavelet approximation (Xu et al. 2008b).

The multiple linear regression model is

$$y = a_0 + a_1x_1 + a_2x_2 + \ldots + a_kx_k \tag{8.14}$$

where, y is dependent variable, x_i the independent variables; a_i is the regression coefficient, which is generally calculated by method of least squares (Xu 2002). In this study, the dependent variable is the annual runoff (AR) and the independent variables are related climatic factors, such as the annual average temperature (AAT) and annual precipitation (AP), etc.

(5) BPANN based on wavelet decomposition

In order to disclose the relationship between the annual runoff with its related climatic factors at different time scales, this study employed the back propagation artificial neural network (BPANN) based on the results of wavelet decomposition. We first approximated the variation patterns of runoff and its related climate factors, such as AR, ATT and AP using wavelet decomposition on the basis of the discrete wavelet transform (DWT) at different time scales. Then the relationship between AR with AAT and AP were revealed by using BPANN based on the wavelet approximation.

In back-propagation networks, a number of smaller processing elements (PEs) are arranged in layers: an input layer, one or more hidden layers, and an output layer (Hsu et al. 1995). The input from each PE in the previous layer (x_i) is multiplied by a connection weight (w_{ji}). These connection weights are adjustable and may be likened to the coefficients in statistical models. At each PE, the weighted input signals are summed and a threshold value (θ_j) is added. This combined input (I_j) is then passed through a non-linear transfer function (f(·)) to produce the output of the PE (y_j). The output of one PE provides the input to the PEs in the next layer. This process can be summarized in equations as follows (Maier and Dandy 1998):

$$I_j = \sum w_{ji} x_i + \theta_j \tag{8.15}$$

$$y_i = f(I_j) \tag{8.16}$$

Our ANN model is a three-tier structure: an input X with two variables (i.e. AAT and AP) is linearly mapped to intermediate variables (called hidden neurons) H, which are then nonlinearly mapped to the output y (i.e. AR).

By comparing the advantages and disadvantages of artificial neural network transfer functions (Dorofki et al. 2012), we selected the activation function as hyperbolic tangent sigmoid transfer function as follows:

$$f(\mathrm{I}) = \frac{(1-e^{-I})}{(1+e^{-I})} \tag{8.17}$$

where f(I) represents transfer function, and I represents input.

The purpose of the model is to capture the relationship between a historical set of model inputs and corresponding outputs. As mentioned above, this is achieved by repeatedly presenting examples of the input/output relationship to the model and adjusting the model coefficients (i.e. the connection weights) in an attempt to minimize an error function between the historical outputs and the outputs predicted by the model. This calibration process is generally referred to as 'training'. The aim of the training procedure is to adjust the connection weights until the global minimum in the error surface has been reached.

The back-propagation process is commenced by presenting the first example of the desired relationship to the network. The input signal flows through the network, producing an output signal, which is a function of the values of the connection weights, the transfer function and the network geometry. The output signal produced is then compared with the desired (historical) output signal with the aid of an error (cost) function.

The model parameters are optimized by minimizing the mean square error given by the cost function:

$$E = <\| y - y^{obs} \|^2> \tag{8.18}$$

Where y^{obs} is the observed data, $<\cdot>$ denotes a sample or time mean.

Because it can train any network as long as its weight, net input, and transfer functions have derivative functions (Kermani et al. 2005), we selected Levenberg–Marquardt (trainlm) as the training function in the computing environment of MATLAB.

(6) Coefficient of determination and Akaike information criterion

In order to identify the uncertainty of the estimated model for a given timescale, the coefficient of determination was calculated as follows:

$$\mathrm{CD} = 1 - \frac{RSS}{TSS} = 1 - \frac{\sum_{i=1}^{n}(y_i - \hat{y}_i)^2}{\sum_{i=1}^{n}(y_i - \overline{y})^2} \tag{8.19}$$

where CD is the coefficient of determination; $\hat{y}_i$ and y_i are the simulate value and actual data of runoff respectively; $\overline{y}$ is the mean of $y_i (i = 1, 2, \ldots, n)$; $RSS = \sum_{i=1}^{n}(y_i - \hat{y}_i)^2$ is the residual sum of squares; $TSS = \sum_{i=1}^{n}(y_i - \overline{y})^2$ is the total sum of squares. The coefficient of determination is a measure of how well the simulate results represent the actual data. A bigger coefficient of determination indicates a higher certainty and lower uncertainty of the estimates (Xu 2002).

To compare the relative goodness between the ANN and multiple linear regression (MLR) fit for a given timescale, we also used the measure of Akaike information criterion (AIC) (Anderson et al. 2000). The formula of AIC is as follows:

$$AIC = 2k + n\ln(RSS / n) \tag{8.20}$$

where k is the number of parameters estimated in the model; n is the number of samples; RSS is the same as in formula (15). A smaller AIC indicates a better model.

For small sample sizes (i.e., n/K≤40), the second-order Akaike Information Criterion (AIC_c) should be used instead

$$AIC_c = AIC + \frac{2k(k+1)}{n-k-1} \tag{8.21}$$

where n is the sample size. As the sample size increases, the last term of the AIC_c approaches zero, and the AIC_c tends to yield the same conclusions as the AIC (Burnham and Anderson 2002).

8.2 The Chaotic Dynamics of Runoff Process

This study employed the correlation dimension to demonstrate the dynamic characteristics of the runoff time series of four headwaters. The AR time series of Aksu River was first used to reconstruct the phase space, while the correlation dimension

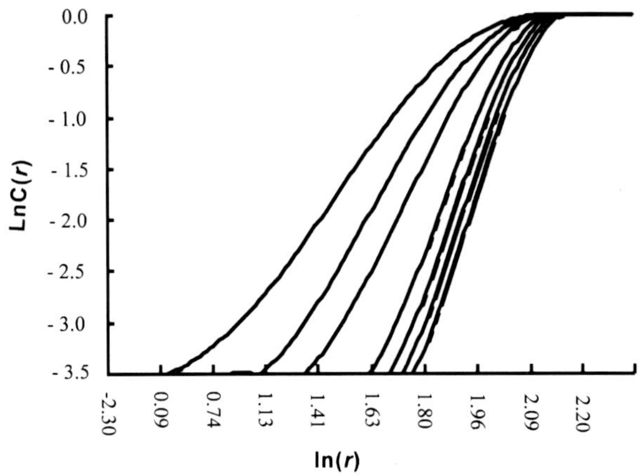

Fig. 8.2 A plot of ln C(r) versus ln (r)

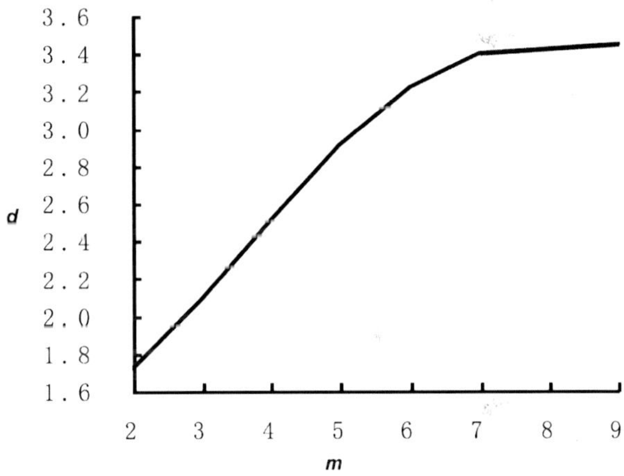

Fig. 8.3 The correlation exponent (d) versus embedding dimension (m)

of the attractor was calculated. Different values for the radius, r, were first selected to compute the values of the correlation-integrals, $C(r)$, which were used to plot the curves within a dual logarithmic coordinate system (Fig. 8.2). This diagram shows the relationship between ln $C(r)$ and ln $C(r)$ for the annual runoff with a number of different embedding dimensions, m. The slope coefficient of ln $C(r)$ versus ln r, i.e. the correlation exponent, d, which was used to embed dimension $m = 1, 2, \ldots$, was calculated using the least square method (LSM).

The diagram in Fig. 8.3 shows the gradual saturation process of the correlation exponent. It is evident that the correlation exponent increases with embedding dimension, m, and a saturated correlation exponent, the correlation dimension of attractor (D), was obtained when $m \geq 7$.

The same procedure was used to derive the correlation dimensions of the attractors for other headwaters, i.e. the Kaidu, Yarkand and the Hotan River. The index values of all four headwaters are shown in Table 8.1.

Table 8.1 The correlation dimensions for AR in the Tarim headwaters

Rivers	Kaidu River	Aksu River	Yarkand River	Hotan River
Embedding dimension (m)	7	7	7	7
Attractor dimension (D)	3.1104	3.2092	3.2118	3.2227

Table 8.2 Hurst exponents for AR in the Tarim headwaters

River Name	Kaidu River	Aksu River	Yarkand River	Hotan River
H	0.7897	0.6991	0.5113	0.4367
C(t)	0.4942	0.3178	0.0158	−0.0840

The fact that none of the correlation dimensions is an integer indicates that the annual runoff time series of all four headwaters are chaotic dynamic systems that are very sensitive to the initial conditions. Because the value of the index is above 3 for all headwaters, at least four independent variables are needed to describe the dynamics of the annual runoff process in each river.

8.3 The Long-memory Characteristic of Runoff Process

All statistical methods used for time series analysis are based on the assumption that all data from a given time series are independent (i.e. conforming to the Gauss distribution); hence, the series is stochastic. When Hurst (1951) and Hurst et al. (1965) analyzed the water level of the Nile River, he found that time-series variables such as the river water level did not exhibit the stochastic characteristic, but instead showed the characteristic of durability. Based on the empirical findings of Hurst et al., Mandelbrot and Wallis (1968) led to a breakthrough regarding fundamental theories of traditional statistical methods. Specifically, they found that many time series no longer presented a random Brownian movement unrelated to the past, but instead showed a characteristic of long-term correlation (Comte and Renault 1996), which he called "fractal".

This study employed the rescaled range (R/S) analysis method to characterize the fractal of annual runoff processes in the four headwaters of the Tarim River. Using the R/S analysis method, the Hurst exponent (H) and the correlation function ($C(t)$) were computed for the annual runoff time series of each headwater during the period of 1957~2008 (Table 8.2).

The results shown in Table 8.2 suggest that the annual runoff time series for each of the Tarim's headwaters possess the characteristic of long memory. The only difference among the headwaters is that the Kaidu, Aksu and Yarkand rivers demonstrate the persistent trait, whereas Hotan shows the anti-persistent feature. The H values for the Kaidu, Aksu and Yarkand River are greater than 0.5 and the C(t) values are greater than 0, which indicates that the future tendency of the annual runoff associated with these systems is consistent with the past runoff.

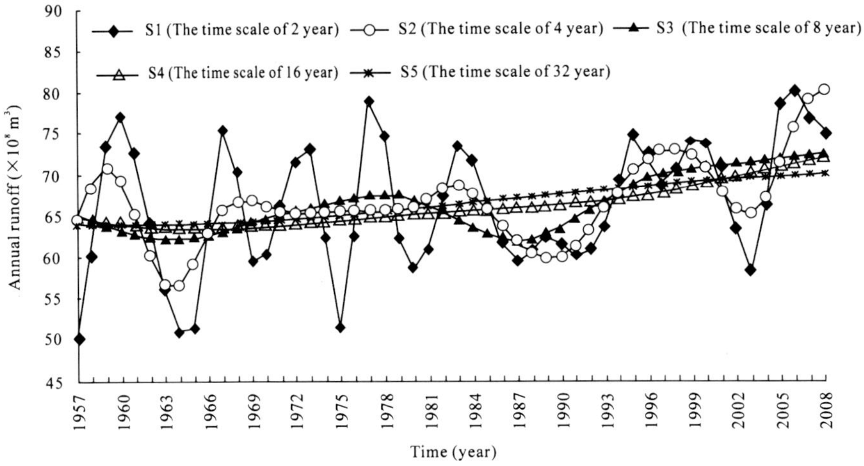

Fig. 8.4 The nonlinear variation patterns for AR in the Yarkand River at the different time scales

However, the H value is less than 0.5 and the C(t) value is less than 0 for the Hotan River, which implies that the future tendency of annual runoff is opposite to that of the past.

8.4 Variation Patterns of AR and its Related Regional Climate Factors

Our previous study indicated that (Xu et al. 2008b), the annual average temperature (AAT) and annual precipitation (AP) are the most important factors that related with the annual runoff (AR). The result was also supported by the other studies for the headwaters of the Tarim River Basin (Hao et al. 2008; Chen et al. 2009; Ling et al. 2012).

The raw data of AR, AAT and AP showed in fluctuating. It is difficult to identify any patterns simply based on the raw data. In order to show the scale-dependent with time for the hydro-climatic process, the wavelet analysis was used.

The nonlinear variation for AR in the Yarkand River and its related climate factors were analyzed at multiple-year scales through wavelet decomposition on the basis of the discrete wavelet transform (DWT).

The wavelet decomposition for the time series of AR in Yarkand River at five time scales resulted in five variants of nonlinear variations (Fig. 8.4). The S1 curve retains a large amount of residual noise from the raw data, and drastic fluctuations along the entire time span. These characteristics indicate that, although AR varied greatly throughout the study period, there was a hidden increasing trend. The S2 curve still retains a considerable amount of residual noise, as indicated by the

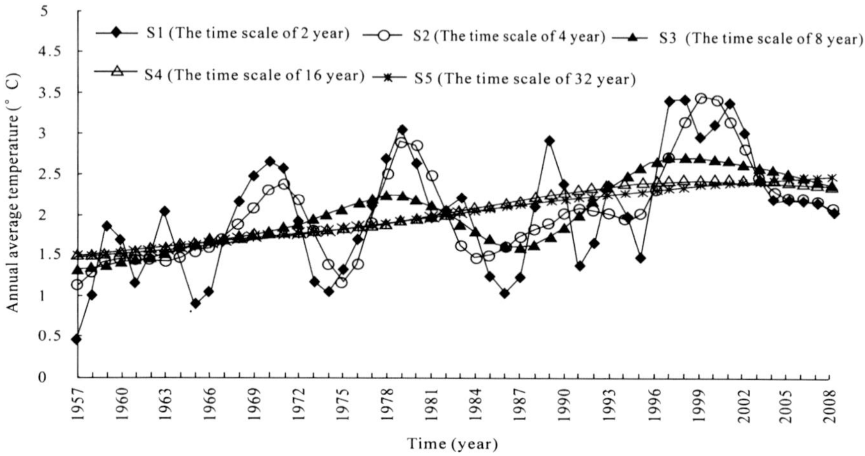

Fig. 8.5 The nonlinear variation patterns for AAT in the Yarkand River Basin at the different time scales

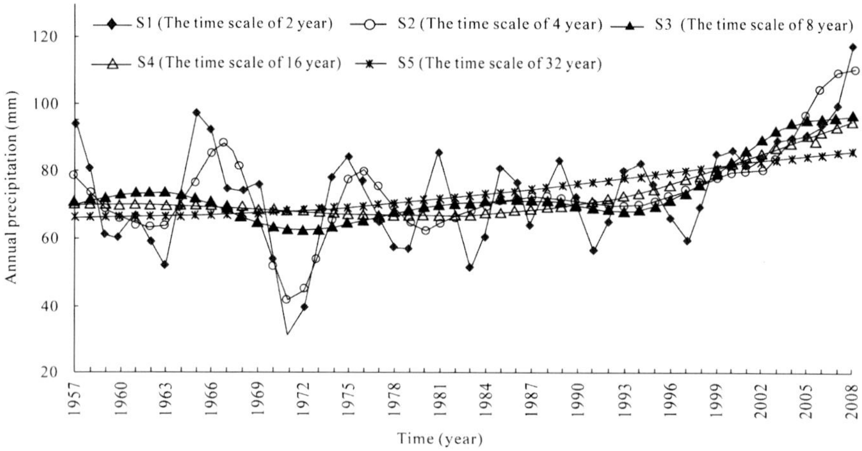

Fig. 8.6 The nonlinear variation patterns for AP in the Yarkand River Basin at the different time scales

presence of 4 peaks and 4 valleys. However, the S2 curve is much smoother than the S1 curve, which allows the hidden increasing trend to be more apparent. The S3 curve retained much less residual noise, as indicated by the presence of 2 peaks and 2 valleys. Compared to S2, the increase in runoff over time was more apparent in S3. Finally, the S5 curve presents an ascending tendency, whereas the increasing trend is obvious in the S4 curve.

Accordingly, Figs. 8.5 and 8.6 provide us a comparison for the nonlinear variations of AAT and AP at different time scales. The wavelet decomposition for the time series of AAT and AP in the Yarkand River Basin at five time scales resulted in five nonlinear variations respectively. These five time scales are also designated as S1 to S5. The curves present an ascending tendency although drastic fluctuations in S1 and S2. Then, the curves are getting much smoother and the increasing trend becomes even more obvious as the scale level increases.

The upper analysis showed that the hydro-climatic process in Yarkand River was dependent on time scales.

Figure 8.7 showed the nonlinear variation patterns of AAT, AP and AR in the Kaidu River at different time scales, which also indicated that the hydro-climatic process was scale-dependent in time.

By using the same approach, we also came to the similar results for the Aksu and Hotan River.

8.5 Simulation for Streamflow with Regional Climate Change

8.5.1 Simulation by BPANN Based on Wavelet Decomposition

For the purpose of understanding hydro-climatic process, based on the results of wavelet decomposition at different time scales, the back-propagation artificial neural network (BPANN) was employed to simulate the relationship between AR with AAT and AP.

Our study considered a three-layer BPANN, i.e. input layer, output layer and hidden layer, to simulate the nonlinear relationship between AR with AAT, and AP at each time scales. The input layer contains two variables, i.e. AAT and AP, the output layer contains one variable i.e. AR, and the neuron number of hidden layer is 4. Table 8.3 shows the optimized parameters for BPANN in the Kaidu River.

The numerical work was carried out using MATLAB. We selected tansig as the transfer function, and trainlm as the training function to train network.

Based on wavelet decomposition results of AR, AAT and AP from 1957 to 2008, we randomly extracted 80, 10 and 10% of the data as training, validation and testing samples, respectively. The results show that, at the time scales of S1, S2, S3, S4 and S5 (i.e. 2-year, 4-year, 8-year, 16-year, and 32-year), all network models have reached the expected error target (0.001) with learning rate of 0.01. The optimized parameters of the BPANN for the hydro-climatic process at different time scales are showed in Table 8.3.

Table 8.3 reveals that, as the time scale increased from S1 to S5, the estimated error decreases. The average absolute error and relative error for the simulation of AR at time scale of S1 are 2.3163×10^8 m^3 and 6.65% respectively, but those at time scale of S5 only are 0.0617×10^8 m^3 and 0.18% respectively.

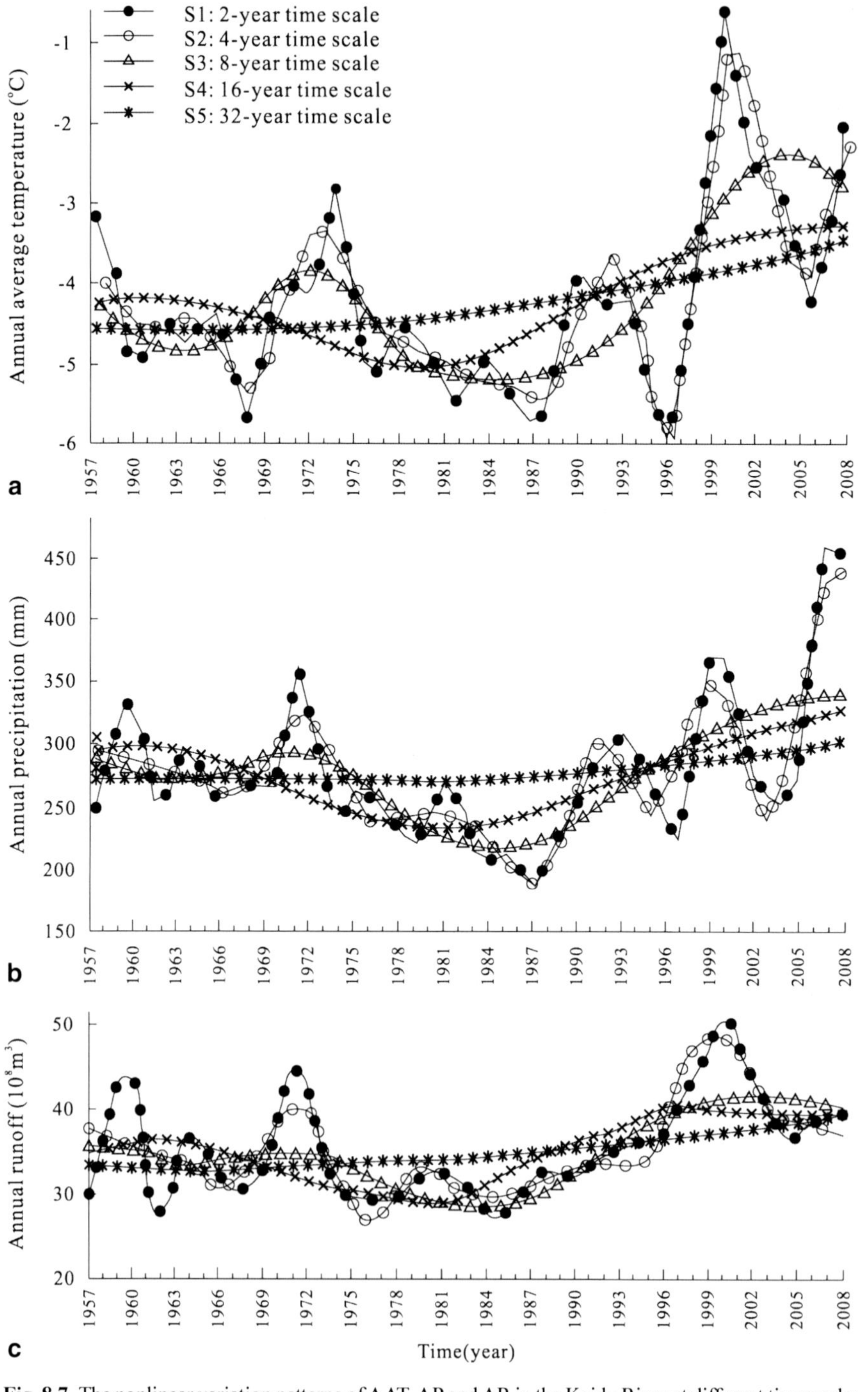

Fig. 8.7 The nonlinear variation patterns of AAT, AP and AR in the Kaidu River at different time scales

Table 8.3 Basic parameters of the BPANN for hydro-climatic process in the Kaidu River at different time scales

Time scale	Neuron number of the hidden layer	Input variable	Output variable	Transfer function	Train function	Best epoch	Average absolute error	Average relative error (%)
S1	4	AAT, AP	AR	tansig	trainlm	4	2.3163	6.65
S2	4	AAT, AP	AR	tansig	trainlm	5	1.4335	4.25
S3	4	AAT, AP	AR	tansig	trainlm	6	0.9905	2.85
S4	4	AAT, AP	AR	tansig	trainlm	6	0.1938	0.57
S5	4	AAT, AP	AR	tansig	trainlm	10	0.0617	0.18

S1, S2, S3, S4 and S5 represent 2-year, 4-year, 8-year, 16-year and 32-year time scales, respectively. "tansig" is hyperbolic tangent sigmoid transfer function, and "trainlm" is a network training function that updates weight and bias values according to Levenberg–Marquardt optimization
AR annual runoff, *AAT* annual average temperature, *AP* annual precipitation

Table 8.4 MLREs for hydro-climatic process in the Kaidu River at different time scales

Time scale	Regression equation	R^2	F	Average absolute error	Average relative error (%)
S1	$AR=2.251AAT+0.036AP+34.420$	0.541	28.932	2.9997	8.41
S2	$AR=4.098AAT-0.002AP+52.921$	0.745	71.584	1.9987	5.85
S3	$AR=2.922AAT+0.048AP+34.263$	0.917	269.877	0.9714	2.69
S4	$AR=5.644AAT+0.010AP+56.036$	0.978	1,066.793	0.4501	1.26
S5	$AR-5.216AAT-0.009AP+59.490$	0.999	30,814.085	0.0779	0.22

Significance level $\alpha=0.01$. S1, S2, S3, S4 and S5 represent 2-year, 4-year, 8-year, 16-year and 32-year time scales, respectively
AR annual runoff, *AAT* annual average temperature, and *AP* annual precipitation

8.5.2 Simulation by MLR Based on Wavelet Decomposition

To compare the simulated results from BPANN with those from the regression model, a group of multiple linear regression equations (MLREs) were fitted based on the results of wavelet decomposition at different time scales (Table 8.4).

Table 8.4 shows that on the five time scales, AR has positive correlations with both AAT and AP at a high significant level of 0.01. In another words, although the runoff, temperature and precipitation displayed nonlinear variations, the runoff presented a linear correlation with the temperature and precipitation. In addition, both of the statistic F and coefficient of determination R^2 for each regression equation showed an increasing trend with the time scale. This pattern indicated that the impact of AAT and AP on AR is more significant at a larger time scale than at a smaller time scale.

Table 8.4 also shows that, the simulated error of MLR is large at the time scale of S1 and S2 (i.e. 2-year and 4-year scale), moderate at the time scale of S3 (i.e. 8-year scale), and small at the time scale of S4 and S5 (i.e. 4-year and 5-year scale).

8.5.3 Comparison Between BPANN and MLR on Wavelet Decomposition

Figure 8.8 reveals the original data of AR in the Kaidu River and the simulated values by MLR and BPANN on wavelet decomposition at different time scales respectively.

Table 8.5 showed the coefficient of determination (i.e. CD) as well as the AIC value for BPANN and MLRE at each time scale.

A higher coefficient of determination and a lower AIC value indicate a better model. Overall, comparing the two modeling methods above, i.e. the multiple linear regression (MLR) and back-propagation artificial neural network (BPANN) based on wavelet analysis, we conclude that the BPANN is better than the MLR at each time scale. In other words, both MLR and BPANN successfully simulated the hydro-climatic process based on wavelet analysis, but the effect from BPANN is better than that from MLR.

By using the same approach, we also came to the similar results for other headwaters of the Tarim River, i.e. the Aksu, Yarkand and Hotan River.

Summary Based on the hydrologic and meteorological data in the areas of the Tarim headwaters, this chapter investigated the nonlinear hydro-climatic process by an integrated method including correlation dimension, R/S analysis, wavelet decomposition, regression analysis and artificial neural network modeling. The main findings are as follows:

[1] The results of this study showed that the hydro-climatic process in the Tarim headwaters presented periodic, nonlinear, chaotic dynamics, and long-memory characteristics.

[2] The correlation dimensions of the attractor for the time series of annual runoff in the four Tarim headwaters were all greater than 3.0 and non-integral, implying that all four headwaters are dynamic chaotic systems that are sensitive to initial conditions, and that the dynamic modeling requires at least four independent variables.

[3] The Hurst exponents indicate that a long-term memory characteristic exists in hydrological process in the Tarim headwaters. However, there were some differences among the headwaters. The Kaidu, Aksu, and Yarkand River presented a persistent trait, But the Hotan River exhibited an anti-persistent feature.

[4] The hydro-climatic process at a large time scale (e.g. 32-year scale) is basically a linear process, but at a small time scale (e.g. 2-year or 4-year scale) it is essentially a nonlinear process with complicated causations. Therefore, the estimated precision is high at a large time scale (e.g. 32-year scale) because the time series of runoff are monotonically related to long-term climatic changes. However, it is difficult to accurately predict a nonlinear hydro-climatic process at a small time scale (e.g. 2-year or 4-year scale).

[5] Our integrated approach conducted by this study can be used to explore the hydro-climatic process of other inland rivers in northwest China.

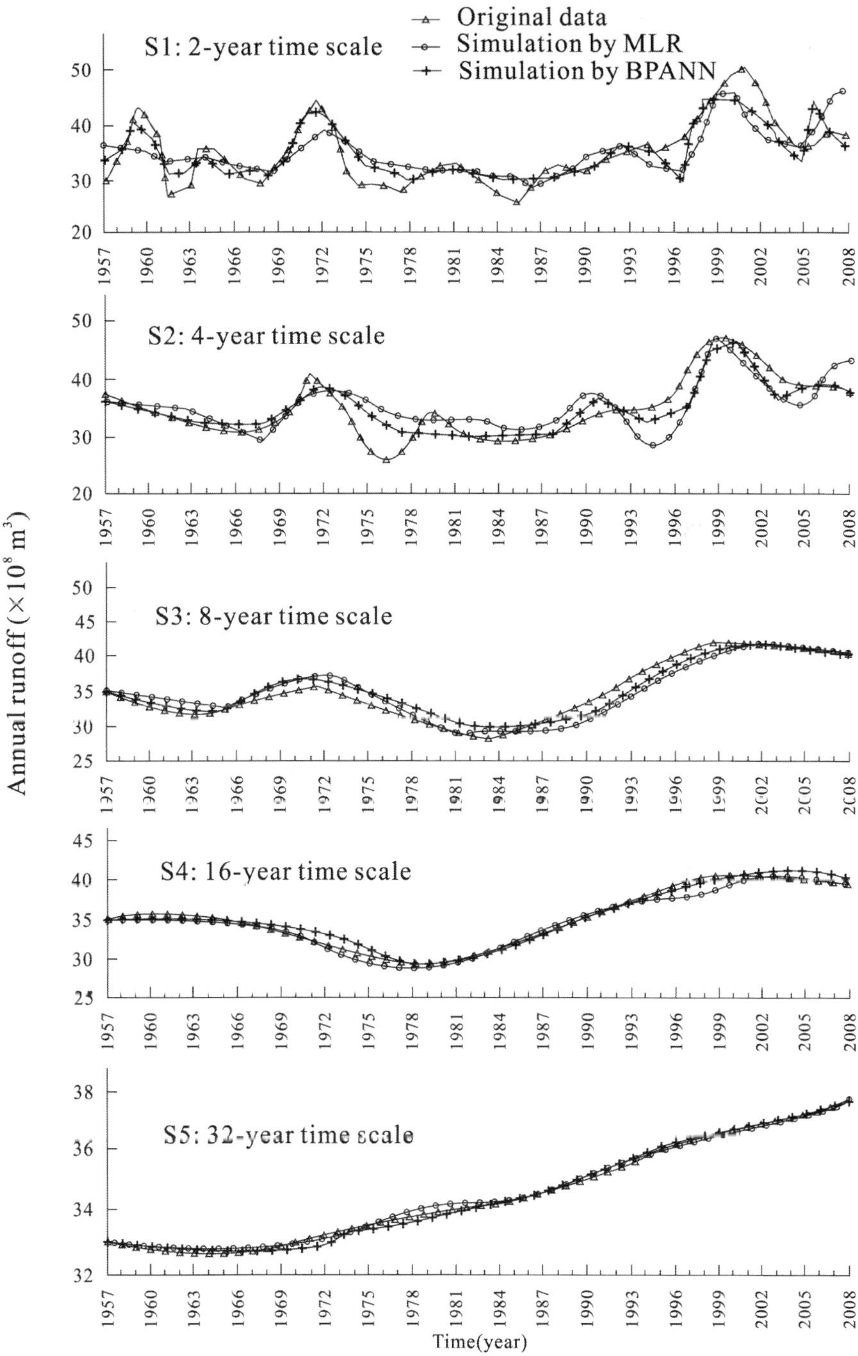

Fig. 8.8 Simulated results for AR by BPANN and MLR at the different time scales

Table 8.5 Comparison between BPANN and MLR model in the Kaidu River at different time scales

Time scale		S1	S2	S3	S4	S5
MLRE	CD	0.541	0.556	0.916	0.977	0.978
	AIC	143.347	102.296	24.869	−63.774	−248.01
BPANN	CD	0.715	0.856	0.920	0.998	0.999
	AIC	118.4882	72.50646	22.28173	−151.494	−259.208

CD is the coefficient of determination; AIC is Akaike information criterion

Acknowledgments This work was supported by National Basic Research Program of China (973 Program; No: 2010CB951003).

References

Ai NS, Li HQ (1993) The nonlinear science methods for quaternary studies. Quatern Res 2:109–120. (in Chinese)

Anderson DR, Burnham KP, Thompson WL (2000) Null hypothesis testing: problems, prevalence, and an alternative. J Wildl Manage 64(4):912–923

Banakar A, Azeem MF (2008) Artificial wavelet neuro-fuzzy model based on parallel wavelet network and neural network. Soft Comput 12:789–808

Bruce LM, Koger CH, Li J (2002) Dimensionality reduction of hyperspectral data using discrete wavelet transform feature extraction. IEEE T Geosci Remote 40(10):2331–2338

Burnham KP, Anderson DR (2002) Model selection and multimodel inference: a practical information-theoretic approach (2nd ed). New York: Springer-Verlag, pp 49–97

Cannon AJ, McKendry IG (2002) A graphical sensitivity analysis for statistical climate models: application to Indian monsoon rainfall prediction by artificial ne6ural networks and multiple linear regression models. Int J Climatol 22:1687–1708

Chen YN, Xu ZX (2005) Plausible impact of global climate change on water resources in the Tarim River Basin. Sci China (D) 48(1):65–73

Chen YN, Takeuchi K, Xu CC, Chen YP, Xu ZX (2006) Regional climate change and its effects on river runoff in the Tarim Basin, China. Hydrol Process 20:2207–2216

Chen YN, Xu CC, Hao XM, Li WH, Chen YP, Zhu CG, Ye ZX (2009) Fifty-year climate change and its effect on annual runoff in the Tarim River Basin, China. Quatern Int 208:53–61

Chou CM (2007) Efficient nonlinear modeling of rainfall-runoff process using wavelet compression. J Hydrol 332:442–455

Comte F, Renault E (1996) Long memory continuous time models. Journal of Econometrics 73(1):101–149

Dorofki M, Elshafie AH, Jaafar O, Karim OA, Mastura S (2012) Comparison of artificial neural network transfer functions abilities to simulate extreme runoff data. Int Proc Chem Biol Environ Eng 33:39–44

Farge M (1992) Wavelet transforms and their applications to turbulence. Annu Rev Fluid Mech 24:395–457

Grassberger P, Procaccia I (1983) Characterization of strange attractor. Phys Rev Lett 50(5):346–349

Hao XM, Chen YN, Xu CC, Li WH (2008) Impacts of climate change and human activities on the surface runoff in the Tarim river basin over the last fifty years. Water Resour Manage 22(9):1159–1171

Hsu K, Gupta HV, Sorooshian S (1995) Artificial neural network modeling of the rainfall-runoff process. Water Resour Res 31(10):2517–2530

Hurst HE (1951) Long-term storage capacity of reservoirs. Trans Am Soc Civ Eng 116:770–808
Hurst HE, Black RP, Simaike YM (1965) Long-term storage: an experimental study. Constable, London, pp 1–155
Kermani BG, Schiffman SS, Nagle HG (2005) Performance of the Levenberg–Marquardt neural network training method in electronic nose applications. Sensors Actuat B: Chem 110(1):13–22
Labat D (2005) Recent advances in wavelet analyses: part 1. A review of concepts. J Hydrol 314:275–288
Li ZL, Xu ZX, Li JY, Li ZJ (2008) Shift trend and step changes for runoff time series in the Shiyang River basin, northwest China. Hydrol Process 22:4639–4646
Ling HB, Xu HL, Fu JY (2012) Temporal and spatial variation in regional climate and its impact on runoff in Xinjiang, China. Water Resour Manag. doi:10.1007/s11269-012-0192-0
Maier HR, Dandy GC (1998) The effect of internal parameters and geometry on the performance of back-propagation neural networks: an empirical study. Environ Modell Softw 13:193–209
Mallat SG (1989) A theory for multiresolution signal decomposition: the wavelet representation. IEEE Trans Pattern Anal Mach Intell 11(7):674–693
Mandelbrot BB, Wallis JR (1968) Noah, Joseph and operational hydrology. Water Resource Res 4(5):909–918
Mandelbrot BB, Wallis JR (1969) Robustness of the rescaled range R/S in the measurement of noncyclic long-run statistical dependence. Water Resource Res 5(5):967–988
Ramsey JB (1999) Regression over timescale decompositions: a sampling analysis of distributional properties. Econ Systems Res 11(2):163–183
Shi YF, Shen YP, Kang ES, Li DL, Ding YJ, Zhang GW, Hu RJ (2007) Recent and future climate change in northwest china. Climatic Change 80:379–393
Sivakumar B (2007) Nonlinear determinism in river flow: prediction as a possible indicator. Earth Surf Process Landf 32(7):969–979
Smith LC, Turcotte DL, Isacks BL (1998) Streamflow characterization and feature detection using a discrete wavelet transform. Hydrol Process 12:233–249
Sun GM, Dong XY, Xu GD (2006) Tumor tissue identification based on gene expression data using DWT feature extraction and PNN classifier. Neurocomputing 69:387–402. doi:10.1016/j.neucom.2005.04.005
Torrence C, Compo GP (1998) A practical guide to wavelet analysis. Bull Am Meteorol Soc 79(1):61–78
Turcotte DL (1997) Fractals and chaos in geology and geophysics (2nd ed). New York: Cambridge University Press, pp 158–162
Wang J, Li H, Hao X (2010) Responses of snowmelt runoff to climatic change in an inland river basin, northwestern China, over the past 50 years. Hydrol Earth Syst Sci 14(10):1979–1987
Xu JH (2002) Mathematical methods in contemporary geography. Higher Education Press, Beijing, pp 37–105. (in Chinese)
Xu JH, Chen YN, Li WH, Dong S (2008a) Long-term trend and fractal of annual runoff process in mainstream of Tarim river. Chinese Geogr Sci 18(1):77–84
Xu JH, Chen YN, Ji MH, Lu F (2008b) Climate change and its effects on runoff of Kaidu river, Xinjiang, China: a multiple time-scale analysis. Chinese Geogr Sci 18(4):331–339
Xu JH, Chen YN, Li WH, Ji MH, Dong S (2009a) The complex nonlinear systems with fractal as well as chaotic dynamics of annual runoff processes in the three headwaters of the Tarim River. J Geog Sci 19(1):25–35
Xu JH, Chen YN, Li WH, Ji MH, Dong S, Hong YL (2009b) Wavelet analysis and nonparametric test for climate change in Tarim river basin of Xinjiang during 1959–2006. Chinese Geogr Sci 19(4):306–313
Xu JH, Li WH, Ji M, Lu FH, Dong S (2010) A comprehensive approach to characterization of the nonlinearity of runoff in the headwaters of the Tarim river, western China. Hydrol Process 24(2):136–146
Xu JH, Chen YN, Lu F, Li WH, Zhang LJ, Hong YL (2011a) The nonlinear trend of runoff and its response to climate change in the Aksu river, western China. Int J Climatol 31(5):687–695

Xu JH, Chen YN, Li WH, Yang Y, Hong YL (2011b) An integrated statistical approach to identify the nonlinear trend of runoff in the Hotan river and its relation with climatic factors. Stoch Env Res Risk A 25(2):223–233

Xu JH, Chen YN, Li WH, Nie Q, Hong YL, Yang Y (2012) The nonlinear hydro-climatic process in the Yarkand river, northwestern China. Stoch Env Res Risk A. doi:10.1007/s00477-012-0606-9

Zhang Q, Xu CY, Tao H, Jiang T, Chen D (2010) Climate changes and their impacts on water resources in the arid regions: a case study of the Tarim river basin, China. Stoch Env Res Risk A 24(3):349–358

Chapter 9
Climate Change Scenarios and the Impact on Runoff

Zhaofei Liu and Zongxue Xu

Abstract

1. The performance of General Circulation Models (GCMs) in simulating near-surface air temperature and precipitation is assessed in the Tarim River Basin (TRB). The criteria included skills on reproducing the observed statistics of two variables, such as long-term mean and standard deviation, seasonal variation, temporal and spatial distributions, and probability density functions (PDFs). The results show that some GCMs performed relatively better than others, usually simulating temperature better than precipitation.
2. Two statistical downscaling models, the Non-Homogeneous hidden Markov Model (NHMM) and the Statistical Down-Scaling Model (SDSM), which have been widely applied in the world and were proved skillful in term of downscaling purpose, were evaluated based on observed daily precipitation in the Tarim River Basin (TRB). The evaluated merits included relative errors of residual functions for annual mean and several percentile values, correlation analysis of annual cycle and spatial distributions, and two skill scores based on probability density functions (PDF). Results indicated that NHMM had a better model performance on daily precipitation than SDSM in the study area, and the former was a more stable model than the later. The main reason is that NHMM could capture precipitation spatial distribution pattern, while SDSM is a single-site model, which downscales each individual station independently. SDSM also lost its skill on modeling extreme values of precipitation amount.
3. The NHMM and SDSM were applied to generate future scenarios of both mean and extremes in the Tarim River Basin, which were based on nine combined scenarios including three GCMs (CSIRO30, ECHAM5, GFDL21) predictor sets and three special report on emission scenarios (SRES A1B, SRES A2,

---

Z. Xu (✉) · Z. Liu
Key Laboratory of Water and Sediment Sciences, Ministry of Education,
College of Water Sciences, Beijing Normal University, 100875 Beijing, China
e-mail: zxxu@bnu.edu.cn

Z. Liu
Institute of Geographic Sciences and Natural Resources Research,
Chinese Academy of Sciences, 100101 Beijing, China
e-mail: zfliu@igsnrr.ac.cn

Y. Chen (ed.), *Water Resources Research in Northwest China*,
DOI 10.1007/978-94-017-8017-9_9,

SRES B1). Results showed that trends magnitude projected by statistical downscaling models were the greatest under SRES A2 scenario, and was the smallest under B1 scenario, with A1B scenario in–between. In general, the trends magnitude was greater in the period of 2081–2100 than that in the period of 2046–2065.

4. Driven by these climate change scenarios, a distributed macro-scale hydrological model (Variable Infiltration Capacity model) was applied to assess the impact of climate change on hydrological processes in the Headwater Catchment (HC) of the TRB. Results showed that it tends to be warmer and drier for the HC under the combined climate change scenarios. It is different from air temperature that the magnitude of changes for extreme values of precipitation is obviously greater than that for mean values. It tends to show a decreasing trend for runoff in the HC, driven by the combined climate change scenarios. But it showed an increasing trend for winter runoff. It showed an inconsistent intra-annual distribution for the changes of precipitation and runoff in the HC, which might be explained by the increasing snowmelt runoff resulted from the increasing air temperature. It was concluded that uncertainties from different GCMs outputs are more significant than emission scenarios in the assessment on the impact of climate change.

Keywords General Circulation Models (GCMs) · Non-Homogeneous hidden Markov Model (NHMM) · Statistical Down-Scaling Model (SDSM) · Variable Infiltration Capacity model · Climate change scenarios · Rank score · Hydrological process

9.1 Introduction

General Circulation Models (GCMs) are the primary tool for understanding the attribution of past climate variations and for the projection of future climate at both global and regional scales (Murphy et al. 2004). However, GCM climate projections have large uncertainties, particularly at regional scales with their coarse horizontal resolution (Trenberth 1997; Murphy et al. 2004), a major shortfall for assessing regional impact of climate change (McCarthy et al. 2001). Uncertainties stem from a hierarchy of sources (Giorgi and Francisco 2000): (1) uncertainty related to the forcing scenarios; (2) uncertainty between different GCMs; (3) uncertainty from different realizations for a given scenario and GCM; and (4) uncertainty related to sub-grid scale forcing and processes. Of these, inter-model variability between different GCMs represents the largest uncertainty in the prediction of regional climate change (Giorgi and Francisco 2000; Fu and Charles 2007).

Inter-model variability as a source of uncertainty has been assessed by comparing different GCMs at regional scales (Kittel et al. 1998; Giorgi and Mearns 2002). Several global model intercomparison projects have been carried out to gain insight into model behavior by comparing model results among themselves and with observations (Lambert and Boer 2001). These projects include the Atmospheric Model Intercomparison Project (AMIP), Paleoclimatic Model Intercomparison Project

(PMIP), Seasonal Forecast Model Intercomparison Project (SMIP) and the Coupled Model Intercomparison Project (CMIP). However, the assessed performance of models depends heavily on the assessment criteria used.

Besides that, the spatial resolutions of GCMs are usually too coarse (100 km) to be used directly for impact studies (Schubert and Sellers 1997). Generating information below the grid scale of GCMs, which is referred to as downscaling, is needed in assessments for the impact of climate change studies. There are two main approaches for downscaling: dynamical and statistical (Fowler et al. 2007; Christensen et al. 2007). Fowler et al. (2007) reviewed downscaling techniques and concluded that dynamical downscaling methods provide little advantage over statistical techniques, at least for present day climates. Given the advantages of being computationally inexpensive, able to access finer scales than dynamical methods and relatively easily applied to different GCMs, parameters, and regions (Timbal et al. 2003; Wilby et al. 2004; Wood et al. 2004), statistical downscaling (SD) methods were used in this study.

SD methods use an empirical statistical technique to establish relationships between observed large-scale and regional/local climate (predictors and predictands) for a baseline period (Karl et al. 1990; Christensen et al. 2007). These relationships are then applied to downscale future climate scenarios using GCM outputs. SD methods can be basically classified into three types (Wilby et al. 2004): regression models (transfer functions), weather generators, and weather classification. Regression models, based on linear or nonlinear methods, represent relationships between predictands and the large scale atmospheric forcing. Weather generators, based on unconditional or conditional, replicate the statistical attributes of a local climate variable (such as mean and variance) but not observed sequences of events (Wilks and Wilby 1999). Weather classification methods group days into finite number of discrete weather types or "states" according to their synoptic similarity. In general, SD methods which combine these techniques are most effective (Christensen et al. 2007).

The SDSM has both of deterministic transfer functions and stochastic components (Wilby et al. 2002). It has been widely applied in the statistical downscaling studies for both climate variables and air quality variables, and has been recommended by the Canadian Climate Impacts and Scenarios (CCIS) project (http://www.cics.uvic.ca).

The NHMM, which was originally developed by Hughes and Guttorp (1994), is used to simulate multi-site patterns of daily precipitation occurrence conditioned on a finite number of weather states which referenced as 'hidden' states. Therefore, the NHMM has weather classification and stochastic components. The NHMM has also had many successful applications of downscaling daily precipitation. These studies have documented that NHMM is a promising method to reproduce both daily precipitation occurrence and amount.

Therefore, NHMM and SDSM were selected to generate future climate change scenarios for precipitation and air temperature. In the evaluation of each model performance, indices of discrete mean or percentile values were easy to make a misleading because these statistics only explained model performance on limited values. Therefore, distributions of variables, which could capture entire statistic characters of variables, were also used for model evaluation. The focus in most of previous works for downscaling has been on reproducing the mean behavior of climate variables and few has been done to downscale variables with emphasis on reproducing extreme values. However, it is the extremes of these variables that have severe consequences and there is a growing need for reproducing and being able to reasonably project future scenarios of extreme values for impact studies. This work is, therefore, aimed at applying two different statistical downscaling methods in the generation of future scenarios for both mean and extremes values.

Although there were a few studies to use empirical methods (e.g. Zeng et al. 2011; Teegavarapu 2010), hydrological models had advantages for assessing the impacts of future climate change on local hydrology (Gleick 1986). There were many hydrological models had been used, including Variable Infiltration Capacity (VIC) model. It is important to identify whether simple models or complex ones should be used for selecting a suitable model (Jiang et al. 2007). There are many factors for the choice of a model for a particular study (Gleick 1986), while the dominant ones include the purpose of study and model and data availability (Xu 1999). Based on the main purpose of assessing hydrological impacts of climate change in a macro-scale basin, a macro-scale hydrological model, the VIC model (Liang et al. 1996) was used in this study. The VIC model had been widely used at large scale, including the application in the arid regions of China.

Overall, the method which uses climate change scenarios downscaled from GCMs outputs to driven hydrological models is the most useful in assessing the impacts of future climate change on regional hydrology. Besides that, based on multi-GCMs and multi-emission scenarios, climate projections are able to represent an uncertainty envelope for the projected runoff. These methods had been widely applied over different countries and regions recently. In general, there are large differences in future runoff changes under different climate change projections at different regions. All of these studies projected an increasing trend for air temperature, and most of them predicted an increasing trend for winter runoff, which was mainly attributed to more snowmelt runoff caused by warmer winter. In China, there were also several studies for estimating impacts of climate change on local hydrology at some regions, including the Yellow River basin (Liu et al. 2011), Yangtze River basin (Zeng et al. 2011), Songhua River basin (Meng and Mo 2012), Tarim River basin (Liu et al. 2010), and the Poyang Lake region (Ye et al. 2011). However, those studies did not use hydrological models, or did not use downscaling methods, or used only one GCM. There is few studies using multi-GCMs outputs, downscaling methods, and hydrological models to assess the impact of future climate change on hydrology at the basin scale in China.

In this study, the performance of IPCC AR4 General Circulation Models (GCMs) was assessed by a score-based method in the Tarim River Basin (TRB). Then, multi-GCMs and multi-emission scenarios were downscaled for regional climate projections, which could represent an uncertainty envelope for the impact assessment. Finally, Multi-climate change scenarios were used to drive the VIC model to assess the impact of future climate change on hydrological processes in the HC of TRB. Runoff at the arid regions is more sensitive to climate changes compared with the humid regions. Therefore, assessment for the impact of climate change on water availability at arid regions is of great scientific and practical merits for water resources management in the arid regions of the world.

9.2 Assessment on the Performance of AR4 GCMs in the Tarim River Basin

9.2.1 Data Description

(1) Observed data Observed monthly mean air temperature was from 35 National Meteorological Observatory (NMO) stations in and around the TRB for the period of 1960–2000. This data set passed NMO data quality control. These station (point) data were interpolated on a 2.5° × 2.5° grid using a inverse distance weighting method for data from at least three nearest stations. The geographical locations and spatial distribution of these stations are shown in Fig. 9.1.

(2) National Centers for Environmental Prediction (NCEP) reanalysis data The NCEP reanalysis precipitation data over the TRB was used due to the lack of observed data in the mountainous regions which contribute most of the basin precipitation. The maximum and minimum values of the NCEP annual precipitation over the TRB is 716.3–9.7 mm, for mountainous and desert regions, respectively. This range is consistence with historical observations as Tang (1990), who showed that the maximum long-term mean annual precipitation was 831 mm in the mountainous regions. We also compared the observed and NCEP reanalysis annual time series over the entire TRB. The NCEP reanalysis precipitation captures the annual variability and long-term trend of the observed series, although absolute values differ each other due to the lack of stations in the mountainous regions (not shown).

(3) GCM data Simulations from 25 GCMs (Table 9.1) were obtained from the IPCC Data Distribution Centre at http://ipcc-ddc.cru.uea.ac.uk for the period of 1961–2000/1999. The outputs of all GCMs were interpolated to a 2.5° × 2.5° common grid.

9.2.2 Method Description

A rank score (RS) was computed for each assessment statistic to assign a score value of as:

$$RS_i = \frac{x_i - x_{\min}}{x_{\max} - x_{\min}} * 10 \tag{9.1}$$

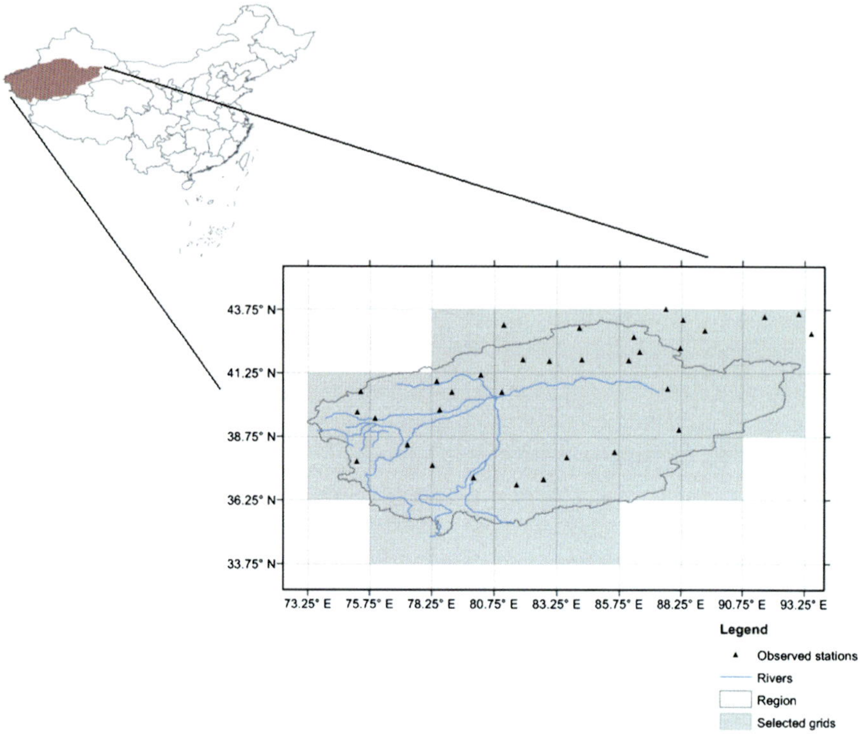

Fig. 9.1 Location of the Tarim River Basin in China, meteorological stations and the selected NCEP Reanalysis grids used in this study

where x_i is the absolute value of relative error between GCM output and observed value for the *i*th statistics. Low errors lead to low rank scores. The total RS for each GCM for a specific climate variable was obtained by summing all RS for all assessment statistics used. All assessment statistics used have equal weight in this summation except trend analysis, trend magnitude, the first two leading modes of each EOF, and two PDF criteria, which are weighted 0.5 each. This total RS was then used to rank the GCMs for three climate variables used: monthly mean air temperature, precipitation, and sea level pressure.

Ten assessment statistics were used to characterize long-term mean and standard deviation, seasonal variation, temporal and spatial distributions, and probability density functions (PDFs) of climate variables. Firstly, three statistics examined the similarity between the observed monthly climate variables and the GCMs outputs, including monthly mean and standard deviation. The relative error (RE) quantifies the similarity between modeled and observed mean values, defined by:

$$RE = \frac{1}{n}\sum_{i=1}^{n}\frac{X_{mi} - X_{oi}}{X_{oi}} \tag{9.2}$$

Table 9.1 Climate models data description

GCMs	ID	Originating group(s)	Country	Resolution	Selected period
BCC:CM1	1	Beijing climatic center	China	1.9° × 1.9°	1961–2000
BCCR:BCM20	2	Bjerknes centre for climatic research	Norway	1.9° × 1.9°	1961–2000
CCSM3	3	National center for atmospheric research	USA	1.4° × 1.4°	1961–2000
CGCM2.3.2	4	Meteorological research institute	Japan	2.8° × 2.8°	1961–2000
CGCM3.1_T47	5	Canadian centre for climatic modelling & analysis	Canada	2.8° × 2.8°	1961–2000
CGCM3.1_T63	6			1.9° × 1.9°	
CNRM:CM3	7	Météo-France/Centre National de Recherches Météorologiques	France	1.9° × 1.9°	1961–1999
CSIRO:MK30	8	Commonwealth Scientific and Industrial Research Organisation (CSIRO) atmospheric research	Australia	1.9° × 1.9°	1961–2000
CSIRO:MK35	9			1.9° × 1.9°	
ECHAM4	10	Climatic Research Centre	Germany	2.8° × 2.8°	1961–2000
ECHAM5	11	Max Planck Institute for Meteorology	Germany	1.9° × 1.9°	1961–2000
ECHO_G	12	Meteorological Institute of the University of Bonn, Meteorological Research Institute of KMA, and Model and Data group	Germany/ Korea	3.9° × 3.9°	1961–2000
FGOALS:g10	13	National Key Laboratory of Numerical Modelling for Atmospheric Sciences and Geophysical Fluid Dynamics (LASG)/Institute of Atmospheric Physics	China	2.8° × 2.8°	1961–1999
GFDL:CM20	14	US Department of Commerce/ National Oceanic and Atmospheric Administration (NOAA)/Geophysical Fluid Dynamics Laboratory (GFDL)	USA	2.0° × 2.5°	1961–2000
GFDL:CM21	15			2.0° × 2.5°	
GISS:AOM	16	National Aeronautics and Space Administration (NASA)/ Goddard Institute for Space Studies (GISS)	USA	3° × 4°	1961–2000
GISS:EH	17			4° × 5°	
GISS:ER	18			4° × 5°	
HadCM3	19	Hadley Centre for Climatic Prediction and Research/Met Office	UK	2.5° × 3.75°	1961–1999
HadGEM1	20			1.3° × 1.9°	
INM:CM30	21	Institute for Numerical Mathematics	Russia	4° × 5°	1961–2000
IPSL:CM4	22	Institut Pierre Simon Laplace	France	2.5° × 3.75°	1961–2000
MIROC3.2_hires	23	Center for Climatic System Research (University of Tokyo), National Institute for Environmental Studies, and Frontier Research Center for Global Change (JAMSTEC)	Japan	1.1° × 1.1°	1961–2000
MIROC3.2_ medres	24			2.8° × 2.8°	
PCM	25	National Center for Atmospheric Research	USA	2.8° × 2.8°	1961–1999

where X_{mi} and X_{oi} are the modeled and observed *i*th values of time series, *n* is the sample length (480 months in this study). Using the standard deviations of observed and GCM climate variables in Eq. (9.2) give the relative bias for standard deviation (RBSD), quantifying the difference in variance between GCM and observed series. The normalized root mean square error (NRMSE), defined as root mean square error dividing by corresponding standard deviation of the observed field (Randall et al. 2007), is an index considering both mean value and standard deviation:

$$NRMSE = \frac{\sqrt{\frac{1}{n}\sum_{i=1}^{n}\left(X_{mi} - X_{oi}\right)^2}}{\sqrt{\frac{1}{n-1}\sum_{i=1}^{n}\left(X_{oi} - \overline{X_o}\right)^2}} \tag{9.3}$$

The correlation coefficient was used to evaluate the "goodness-of-fit" of the annual cycle and spatial distribution of monthly climate variables. For annual cycle assessment, the correlation coefficient was calculated with observed and modeled long-term monthly mean values (i.e. a sample size of 12). For spatial distribution assessment, the correlation coefficient was computed with observed and modeled long-term means for each individual grid cell. The sample size is thus 25, i.e. the number of grid cells in the study region.

The rank-based nonparametric Mann–Kendall test and trend magnitude method were applied to detect the long-term monotonic trends and their magnitudes. The annual time series was used in this case, instead of monthly ones.

EOF analysis was used to compare the space-time variability of observed and modeled data. One advantage of using EOFs is the ability to identify and quantify the spatial structures of correlated variability. The first two leading modes of each EOF (EOF1 and EOF2), accounting for the majority of the total variance, were compared each other.

Finally, two skill scores, the Brier score (*BS*) and the skill score (S_{score}), based on probability density functions (PDFs) were used to measure how well each GCM captures the observed PDF of climate variables, for monthly time series:

$$BS = \frac{1}{n}\sum_{i=1}^{n}(P_{mi}, P_{oi})^2 \tag{9.4}$$

$$S_{score} = \sum_{i=1}^{n} Minimum\left(P_{mi}, P_{oi}\right) \tag{9.5}$$

where P_{mi} and P_{oi} are the modeled and observed *i*th probability values of each bin and *n* is the number of bins. All datasets were binned based on their data ranges, with bin sizes varying but the number of bins fixed at 100. The *BS* is a mean squared error measure for probability forecasts (Brier 1950), and the S_{score} calculates the cumulative minimum value of observed and modeled distributions for each bin, measuring the overlap between two PDFs (Perkins et al. 2007; Fig. 9.2).

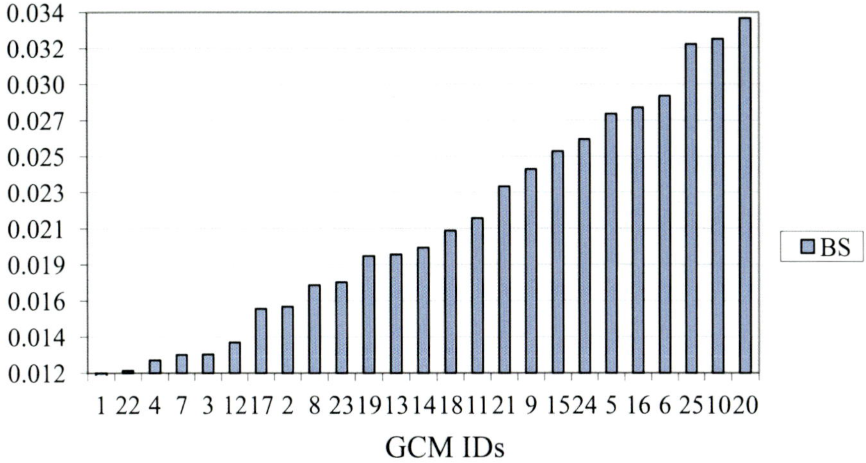

Fig. 9.2 Example of ranking 25 GCMs based on BS statistics (GCM IDs could be found in Table 9.1)

9.2.3 Results and Discussion

9.2.3.1 Monthly and Annual Mean Air Temperature

Table 9.2 presents the mean, standard deviation, NRMSE, correlation coefficients of annual cycle (monthly distribution) and spatial distribution, Kendall Z values and magnitudes of trend, EOF, and two PDF scores, for the 25 GCM data sets for monthly air temperature. A total score, in which weights of trend statistics and magnitude, EOF 1 and 2 and two PDF statistics (BS and S_{score}) were 0.5 while others were 1, is given zed in the last column to represent overall performance (Table 9.2). The results indicate that the HadCM3 and ECHAM4 models simulate monthly air temperature better than other 23 GCMs.

The long-term mean observed annual air temperature for TRB was 5.6 °C while the GCM values vary from −0.9 to 7.8 °C (Table 9.2) with median and mean values of 2.3 and 3.0 °C, respectively. This implies that using the mean or median of all available GCMs may not be appropriate for climate change applications. The standard deviations of GCM values vary from 3.6 to 14.7 °C while the observed value was 11.5 °C. The NRMSE of the GCM values ranges from 0.4 to 0.94, where a smaller value indicates a better fit between observed and simulated values. The NRMSE provides additional information because it measures the differences of every pair of observed and GCM values. For example, two datasets with the same mean and standard deviation could have a large NRMSE if the spatial patterns are very different.

All GCMs can reproduce the annual cycle (monthly distribution) of air temperature accurately, with a minimum correlation coefficient of 0.906. This is not the case for the spatial distribution, particularly where air temperature has a strong relationship with elevation, i.e. it is relatively high in the valley areas and decreases with

Table 9.2 Model performance for monthly mean air temperature

GCMs	Mean value (°C)	Stdev	NRMSE	CC_i	CC_s	Mann–Kendal		EOF		PDF		Total score
						Zc	Slope (°C/10a)	EOF1	EOF2	BS	S_{score}	
0	5.6	11.5				2.67	0.20	77.5	9.8			
1	1.9	3.6	0.94	0.906	0.70	3.90	0.09	83.8	8.4	0.08	0.31	45
2	1.3	10.3	0.52	0.993	0.84	0.90	0.04	73.2	11.0	0.05	0.86	16
3	3.2	10.0	0.50	0.978	0.82	2.97	0.12	74.7	11.1	0.04	0.76	11
4	3.0	9.8	0.65	0.989	0.54	5.84	0.24	89.9	5.1	0.04	0.79	28
5	0.9	11.4	0.63	0.991	0.68	6.05	0.36	93.1	3.8	0.04	0.83	28
6	1.6	11.1	0.54	0.990	0.84	5.98	0.47	82.7	7.7	0.03	0.86	17
7	2.3	8.6	0.53	0.990	0.85	1.09	0.16	79.2	8.0	0.05	0.68	15
8	2.0	11.1	0.48	0.993	0.85	5.21	0.25	84.0	6.9	0.05	0.83	15
9	7.0	10.7	0.47	0.992	0.85	3.53	0.24	73.1	13.8	0.04	0.71	11
10	6.4	11.6	0.43	0.986	0.89	3.79	0.29	72.1	13.6	0.04	0.71	8
11	7.4	10.3	0.43	0.996	0.88	3.18	0.13	85.1	4.7	0.05	0.70	14
12	−0.6	8.2	0.79	0.975	0.41	4.53	0.21	86.3	6.6	0.05	0.65	35
13	1.8	10.6	0.68	0.981	0.53	1.40	0.14	76.4	12.7	0.05	0.74	21
14	1.1	9.7	0.54	0.992	0.84	1.27	0.08	81.0	11.6	0.05	0.81	16
15	2.6	9.8	0.57	0.990	0.67	2.55	0.13	79.3	11.8	0.04	0.79	11
16	7.8	8.6	0.52	0.989	0.84	4.86	0.21	82.3	9.5	0.04	0.66	13
17	6.1	9.6	0.58	0.991	0.72	2.35	0.21	83.1	7.5	0.04	0.76	18
18	5.0	9.7	0.55	0.993	0.72	3.88	0.17	86.6	4.7	0.03	0.78	18
19	5.1	10.5	0.40	0.997	0.89	2.01	0.20	69.9	8.2	0.02	0.89	4
20	−0.7	14.7	0.74	0.996	0.78	1.94	0.21	53.9	16.1	0.04	0.70	29
21	0.8	11.3	0.65	0.992	0.63	1.83	0.18	76.3	11.9	0.04	0.89	16
22	−0.9	8.3	0.74	0.980	0.69	3.20	0.32	84.9	7.5	0.05	0.69	31
23	5.7	11.0	0.45	0.988	0.89	3.62	0.21	62.3	16.0	0.03	0.85	11
24	2.9	11.4	0.66	0.972	0.53	5.74	0.32	83.7	7.6	0.03	0.77	22
25	0.9	10.3	0.66	0.982	0.66	3.19	0.19	81.7	8.7	0.04	0.86	14

‘0’ in the column of Climatic models means observed values

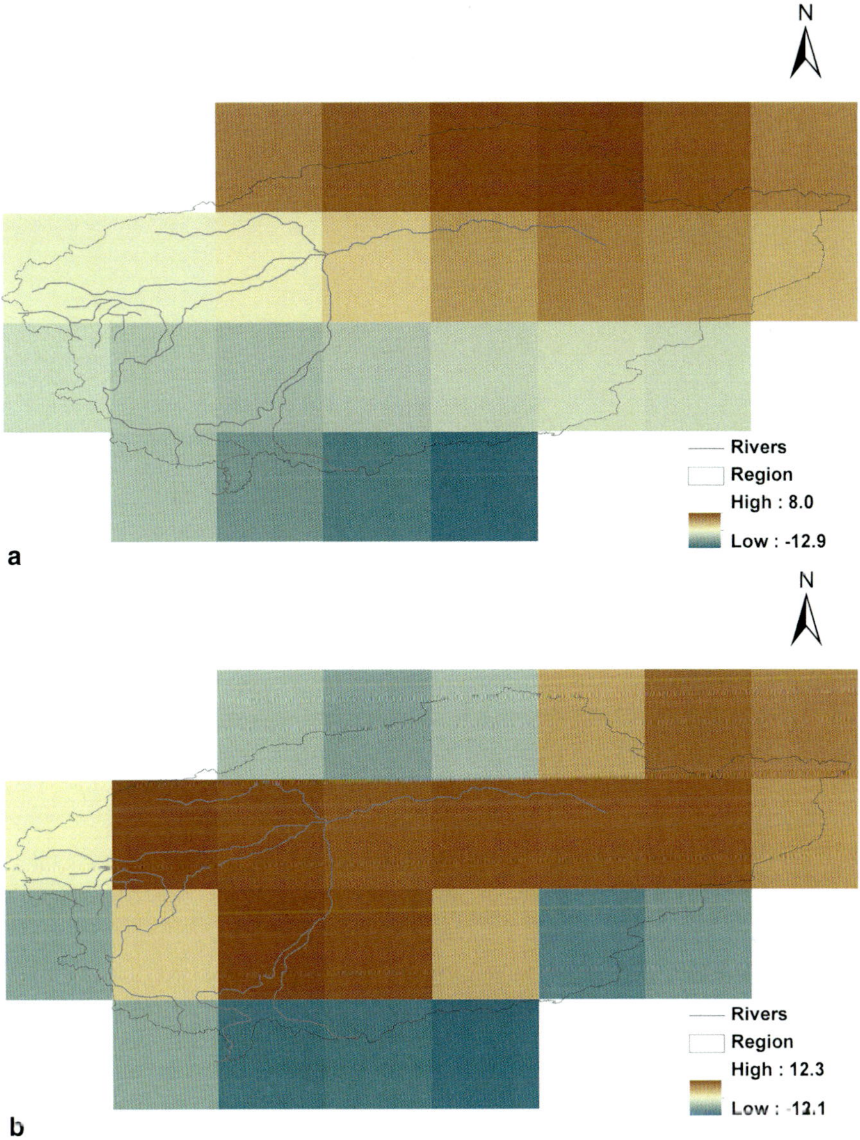

Fig. 9.3 Spatial distributions of annual temperature of the TRB, both **a** observed; and **b** simulated by ECHO_G

elevation for both the northern and southern TRB, while the GCMs usually simulate a north–south temperature gradient (Fig. 9.3). This results in much smaller correlation coefficients for spatial distribution (Table 9.2).

All GCMs simulate an increasing trend of annual air temperature for the period of 1961–1999/2000. This is consistent with observed data, which has a Kendall *Z* statistic of 2.67. However, the range of GCM Z statistics varies from 0.9 to 6.5.

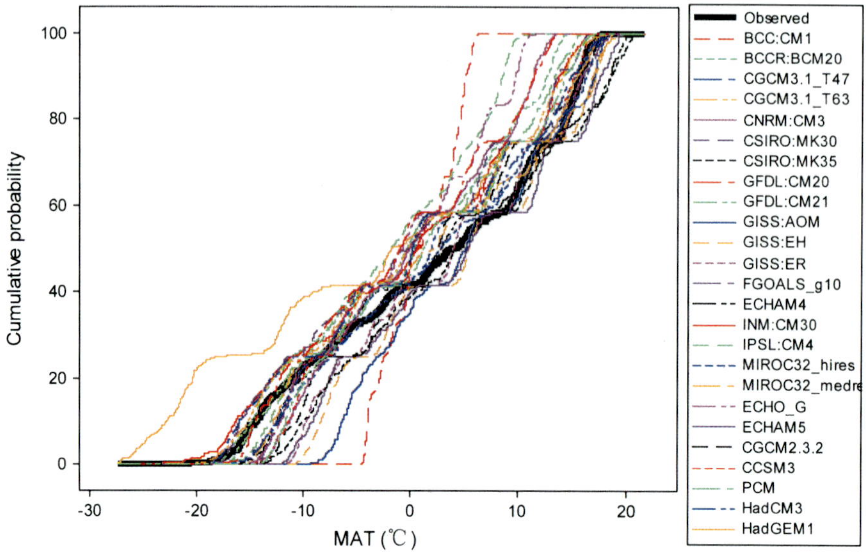

Fig. 9.4 Empirical cumulative probabilities of observed and modeled monthly mean temperature for the TRB

This implies that some GCMs overestimate and some underestimate the temperature trend for the period of 1961–1999/2000: 16 GCMs simulated a larger Z statistic and 9 less than observed. Both median and mean Z statistic values, 3.2 and 3.4, respectively, overestimated the observed temperature trend. The magnitude of the observed annual air temperature trend was 0.20 °C/decade, while the GCMs range between +0.04 to +0.47 °C/decade. However, both the median and mean GCM trends magnitudes were +0.21 °C/decade, close to the observed value. The Kendall trend testing indicated that the observed annual air temperature trend was statistically significant at the level of $\alpha=0.05$, and 19 out of 25 GCMs were also significant at the level of $\alpha=0.05$.

The spatial and temporal variability of monthly air temperature was characterized by EOF analysis. The first two EOFs for monthly mean air temperature account for 77.5 and 9.8 % of the total variance. In general, the GCMs can simulate this variability well: the mean and median values for first two EOFs being 79.1 and 9.4 %, and 81.7 and 8.4 %, respectively (Table 9.2). This suggests that the physical processes dominating air temperature variability are captured by the GCMs. However, they generally have larger explained variance for the first EOF than observed. This implies that GCMs simulate a more uniform spatial distribution. This is consistent with the spatial correlation coefficients showing that GCMs usually simulate a too smooth north–south air temperature gradient (Fig. 9.3).

Figure 9.4 shows empirical cumulative probability distributions for observed and GCM monthly mean air temperature for the TRB. Overall, the GCMs simulate probability distributions close to observed except for the NCAR:PCM and BCC:CM1 models. The results of the S_{score} and the *BS* from each GCM for each

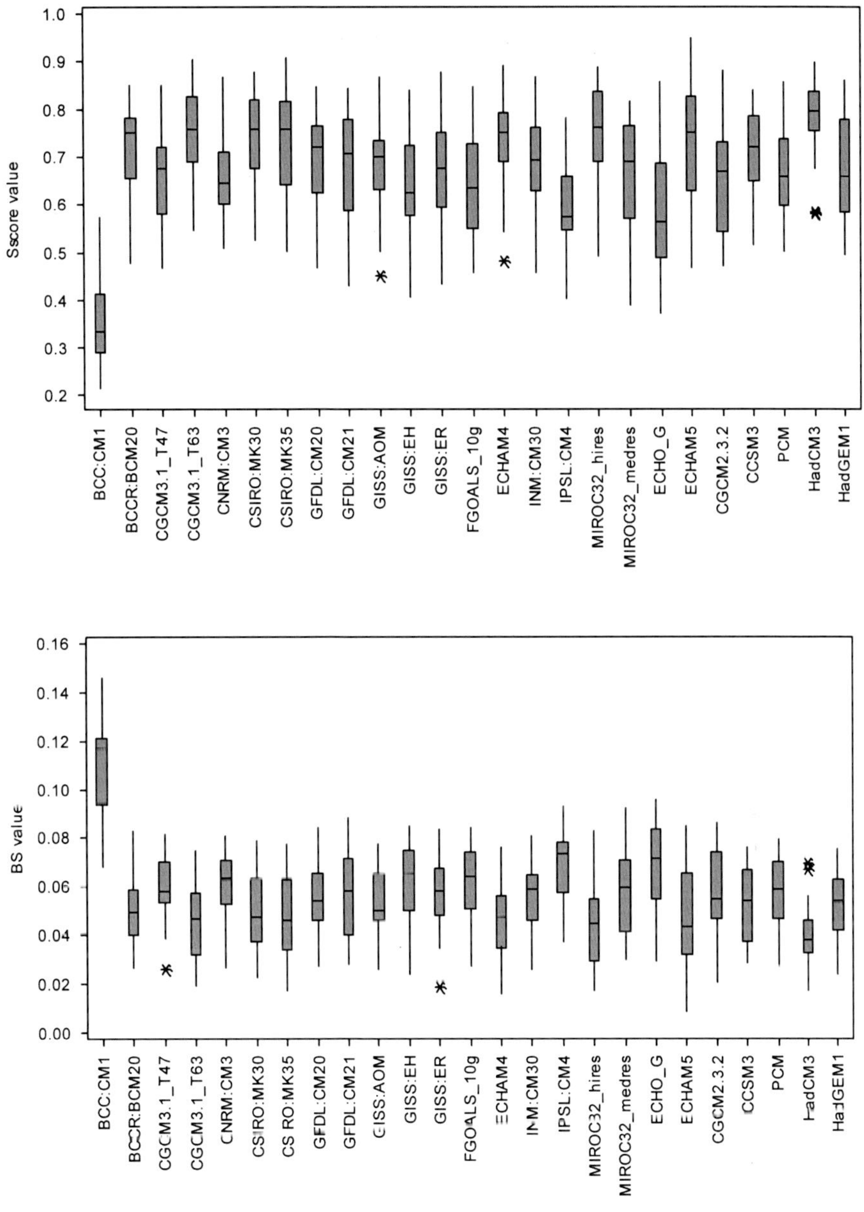

Fig. 9.5 Boxplots of ensemble PDF-based skill scores of monthly mean air temperature over all grids in the study area

grid over the TRB are presented in Fig. 9.5. The S_{score} of area averaged means for 20 of the 25 models exceeded 0.7, with the a few being close to 0.9. This is consistent with the empirical cumulative probability plots. The relatively poorer performance of the BCC:CM1 model is confirmed by these two indices. The variation of scores implies spatial differences, i.e., monthly mean air temperature probability distribution was close to the observed one in some grid cells but not in others.

9.2.3.2 Monthly and Annual Precipitation

Table 9.3 presents mean, standard deviation, NRMSE, annual cycle (monthly distribution) and spatial distribution correlation coefficients, Kendall Z-statistics and magnitudes, EOFs, and two PDF scores for monthly and annual precipitation. The total score is then summarized in the last column to represent overall performance (Table 9.3).

The long-term mean observed annual precipitation for TRB was 250 mm, whereas the GCM values vary from 250 to 736 mm (Table 9.3). The median and mean values of the 25 GCMs were 160 and 167 %, respectively, of the annual precipitation. This implies that using the mean or median of all available GCMs results may not provide an appropriate precipitation projection for climate change studies. GCM precipitation standard deviation values vary between 11.2 and 36.7 mm compared to the observed value of 21.6 mm. The NRMSE values range from 1.36 to 4.07, much greater than that obtained for monthly mean temperature (0.40–0.94) and MSLP (0.33–1.42). This implies that GCM simulation of precipitation is not as accurate as that for air temperature or MSLP.

The annual cycle of precipitation is also not simulated as accurately as that of air temperature and MSLP. The worse performing GCM has a correlation of only 0.17 with a peak rainy season in March–May, whereas the observed rain peaks is in May–July (Fig. 9.6). The spatial distribution of precipitation results were better than that for MSLP, as there were no negative correlation coefficients, but poorer (correlation coefficients of 0.17–0.72) than that for air temperature (0.41–0.89). In general, most GCMs simulate the correct pattern of precipitation variation with elevation across the TRB.

Observed annual precipitation has a statistically insignificant increasing trend for the period of 1961–1999/2000 with a Kendall Z-statistic of 1.34, corresponding to a p-value of 0.1802. GCMs simulate both increasing and decreasing trends: the range of Z-statistics ranging from −1.95 to 3.04. The fact that GCM precipitation trends have different signs makes the use of GCM precipitation more uncertain and challenging for decision makers. Addressing this issue, various downscaling techniques have been developed to predict regional precipitation conditional on projected variables such as MSLP, geopotential height etc. (Charles et al. 2007). The observed annual precipitation trend was +5.5 mm/decade, while the GCM values ranged from −5.3 to +11.1 mm/decade. Only a few GCMs (CSIRO:MK30, ECHAM5, GFDL:CM21 and HadCM3) could simulate a trend of similar magnitude to observed ones.

The first two EOFs for monthly precipitation explain 52.7 and 9.5 %, respectively, of the total variance. In general, GCMs simulate this spatial and temporal variability with mean and median values of 47.9 and 10.3 %, and 48.0 and 10.1 %, respectively (Table 9.3). This implies that the physical processes dominating monthly precipitation variability result from large scale circulation features. However, the range is

Table 9.3 Model performance for precipitation

GCMs	Mean value (mm/a)	Stdev	NRMSE	CC_i	CC_s	Man–Kendal		EOF		PDF		Total score
						Zc	Slope (mm/10)	EOF1	EOF2	BS	S_{score}	
0	249.9	21.6				1.34	5.5	52.7	9.5			
1	279.9	14.9	1.46	0.25	0.33	−0.27	−0.2	28.1	19.2	0.02	0.60	34
2	504.0	30.1	2.77	0.74	0.17	0.62	3.2	30.0	14.5	0.02	0.57	35
3	435.9	13.8	1.76	0.53	0.50	−0.73	−2.4	42.5	9.8	0.03	0.38	29
4	276.2	13.2	1.45	0.56	0.69	0.83	2.1	44.1	13.7	0.02	0.68	13
5	260.8	15.6	1.64	0.81	0.44	0.48	0.9	61.4	6.4	0.02	0.70	15
6	354.4	21.0	1.68	0.74	0.22	−0.27	−0.6	49.7	8.8	0.01	0.78	17
7	699.8	36.7	3.38	0.71	0.40	−0.34	−1.2	30.5	11.1	0.03	0.33	46
8	263.7	16.3	1.48	0.79	0.22	1.60	3.3	34.7	11.9	0.01	0.78	14
9	321.5	22.9	1.81	0.84	0.25	−0.57	−3.6	56.8	7.5	0.01	0.75	18
10	249.6	14.6	1.57	0.49	0.25	−0.13	−0.5	49.2	7.7	0.01	0.80	22
11	257.0	11.2	1.36	0.62	0.26	1.69	4.2	48.0	11.2	0.02	0.69	19
12	430.4	15.2	1.94	0.77	0.61	−1.95	−3.6	47.5	11.1	0.03	0.45	24
13	673.7	29.6	2.93	0.17	0.56	−0.17	−2.1	66.1	7.5	0.03	0.34	43
14	400.3	19.7	1.76	0.67	0.38	0.64	2.7	52.2	11.1	0.02	0.63	18
15	316.8	15.0	1.44	0.44	0.47	1.27	5.6	55.1	8.4	0.02	0.73	13
16	508.8	20.6	2.00	0.42	0.47	2.34	9.8	52.5	8.0	0.03	0.44	25
17	476.9	20.6	2.86	0.50	0.44	0.51	1.6	45.2	11.9	0.03	0.43	27
18	522.7	18.9	2.43	0.50	0.62	2.11	6.8	52.7	9.4	0.03	0.28	23
19	324.1	21.5	1.73	0.69	0.23	1.21	8.4	60.0	6.5	0.01	0.78	15
20	311.7	19.8	1.57	0.83	0.30	0.46	1.5	43.6	10.1	0.01	0.75	13
21	554.3	30.4	3.00	0.64	0.72	−0.55	−5.3	57.6	9.4	0.02	0.51	30
22	736.4	26.1	4.07	0.80	0.48	2.02	11.1	44.6	12.6	0.03	0.20	39
23	348.2	19.0	2.00	0.75	0.28	3.04	11.1	39.0	12.3	0.02	0.70	22
24	459.9	19.3	2.23	0.74	0.56	0.94	2.9	45.4	12.0	0.02	0.51	18
25	447.4	28.7	2.07	0.58	0.64	1.60	9.8	62.1	5.4	0.02	0.67	19

'0' in the column of climatic models means NCEP reanalysis data

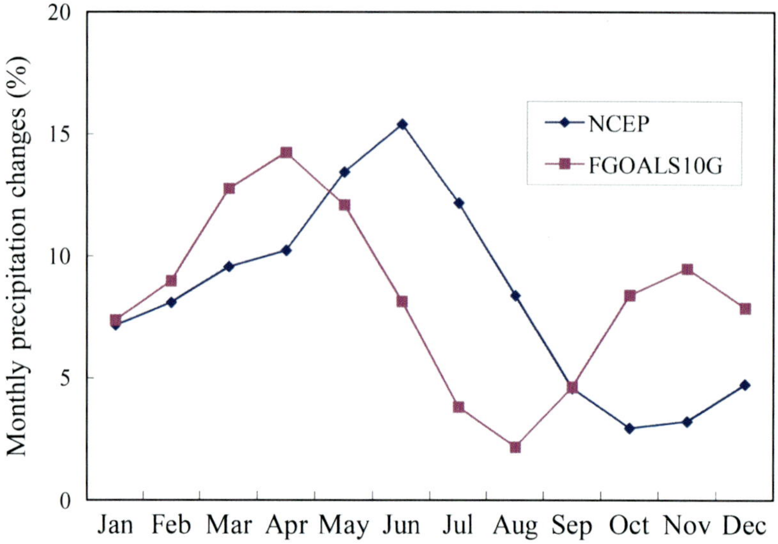

Fig. 9.6 NCEP and GCM monthly precipitation distribution over TRB

much greater than that of air temperature with EOF1 explaining 28.1–66.1 % and EOF2 explaining 5.4–19.2 %. This indicates that some GCMs simulate a more uniform while others a more varied spatial distribution. Several GCMs (GISS:AOM and GISS:ER) simulate EOFs that explain similar amounts to observed ones.

Figure 9.7 shows empirical cumulative probability distributions for observed and modeled monthly precipitation. Overall, the GCMs do not simulate the probability distributions well, especially for the lower tails of the distributions. Almost all GCMs overestimate monthly precipitation amounts, consistent with the mean annual bias shown in Table 9.3. The results of S_{score} and *BS* for each GCM and grid are presented in Fig. 9.8. The S_{score} of precipitation was generally smaller than that of mean monthly air temperature and MSLP for the majority of GCMs. The "outliers" in the *BS* plot indicate inconsistency in spatial distribution between GCM and observed precipitation, corresponding to those GCMs with poor results for the correlation coefficients of spatial distribution and EOF analysis.

9.2.3.3 Overall Model Performance

An overall assessment for all three variables is summarized in Table 9.4, leading to several conclusions:

(1) In general, models from CSIRO (Australia), ECHAM (Germany), GFDL (USA), Hadley Centre (UK), and NCAR (USA) perform relatively better than other models. This may be due, in part, to their relatively higher resolution in the ocean and sea-ice components compared to other coupled climate models. A higher

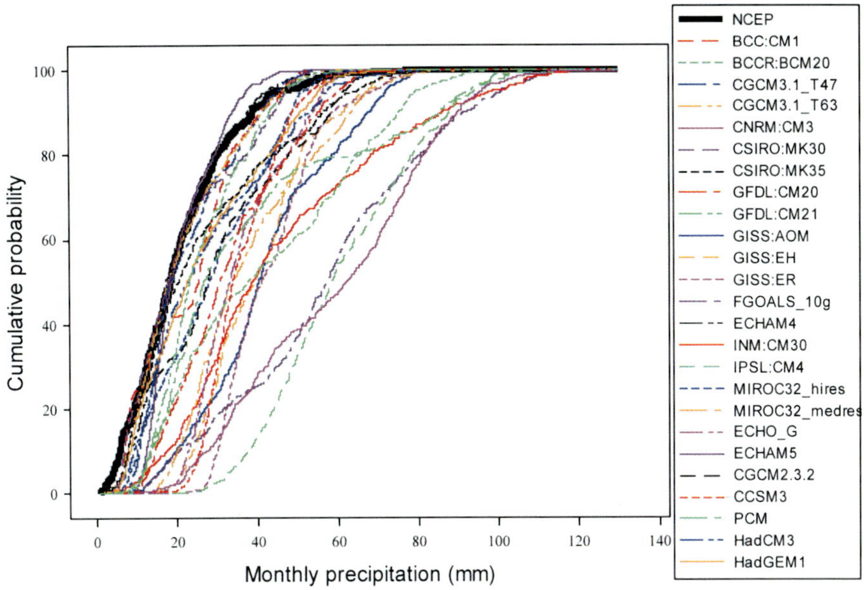

Fig. 9.7 Cumulative empirical probabilities of observed and modeled monthly precipitation

resolution usually leads to better simulation of regional changes and often relatively smaller climate sensitivity and corresponding transient climate response.

(2) Of the variables assessed, GCMs generally simulate monthly air temperature the best and precipitation the worst. The poor performance of GCM precipitation is the rationale behind statistical downscaling. The relative performance in simulating three variables is consistent with the results of a global scale study of AR4 models (Randall et al. 2007). Most of GCMs were able to capture the probability distribution of monthly air temperature and MSLP, but not for precipitation. All GCMs, except ECHAM4, overestimated precipitation. This is consistent with the results of Giorgi and Mearns (2002), in which precipitation was greatly overestimated over the Tibet region including the TRB. This overestimation of precipitation occurred over the mountainous and desert regions in the middle and eastern parts of the TRB.

(3) For GCMs developed by same organizations, such as the Hadley Centre models, the MIROC32 models and the GFDL, model performance was often better for a higher resolution or a later version model, most likely due to improved physical parameterization and numerical improvements. The GFDL:CM21 model was better performing than the GFDL:CM20 model for all three variables, with global scale studies confirming a reduction in model biases. HadGEM1 performs better than HadCM3 for both precipitation and MSLP, which could be a result of increased horizontal and vertical resolution, new dynamics, and improvements in physical parameterizations. However, monthly air temperatures did not improve. Similar differences are seen between the results for CSIRO MK30 and MK35 models.

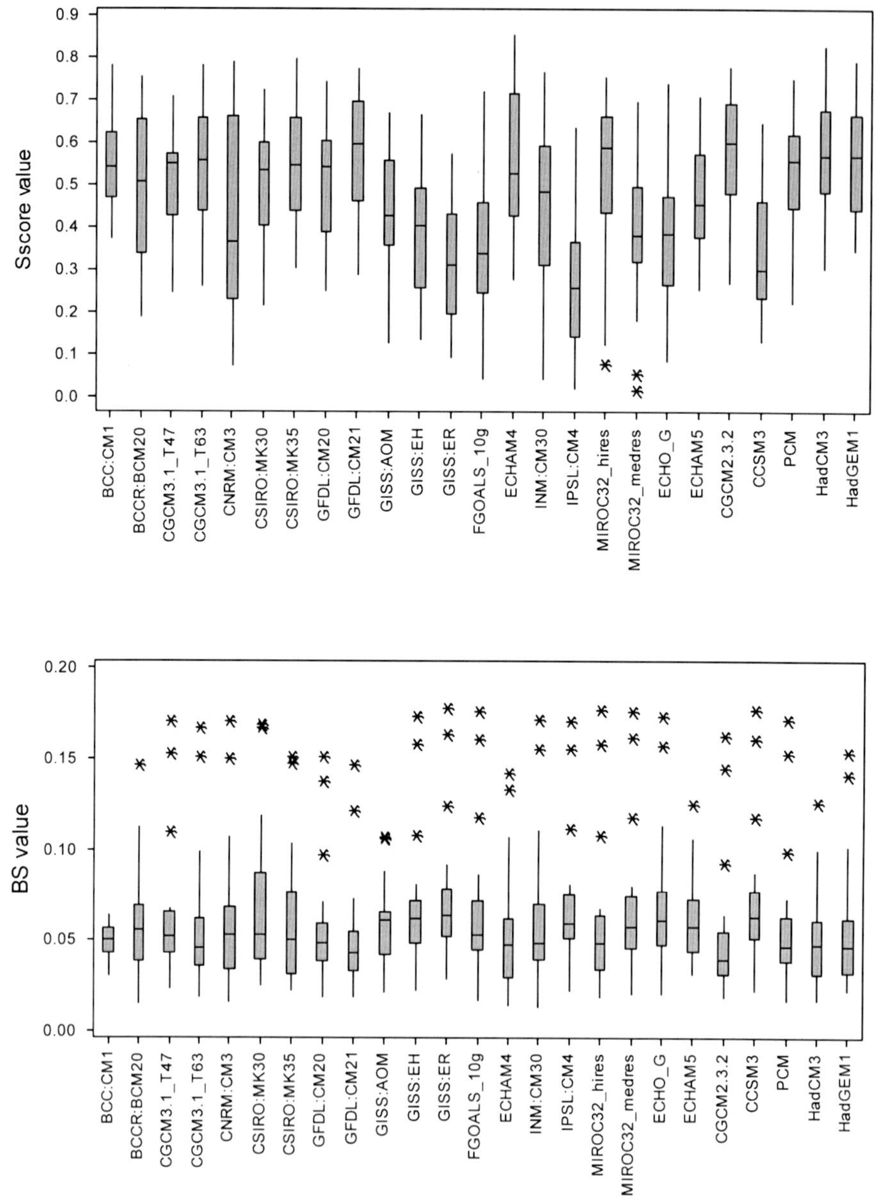

Fig. 9.8 Boxplots of ensemble PDF-based skill scores of monthly mean precipitation over all grids of study region

(4) Depending on the intended application of GCM ranking results, not all climate variables or criteria need necessarily be assessed. For example, for statistical downscaling, MSLP as a primary variable would be taken into account but precipitation could be ignored. For flood applications, more focus could be put on the upper-tail of the relevant probability distributions.

Table 9.4 Comparison of Scores for monthly MSLP, mean temperature and precipitation

Climatic model	MSLP	Temperature	Precipitation	Total
BCC:CM1	32	22	29	83
BCCR:BCM20	15	8	32	54
CCSM3	23	5	29	57
CGCM2.3.2	17	11	17	45
CGCM3.1_T47	9	13	16	38
CGCM3.1_T63	10	9	18	36
CNRM:CM3	25	7	44	75
CSIRO:MK30	13	7	16	35
CSIRO:MK35	11	4	17	32
ECHAM4	9	4	22	35
ECHAM5	14	6	17	37
ECHO_G	17	17	26	59
FGOALS:g10	12	8	41	61
GFDL:CM20	20	8	16	44
GFDL:CM21	14	6	15	35
GISS:AOM	15	7	23	45
GISS:EH	8	4	24	36
GISS:ER	9	5	23	36
HadCM3	16	2	14	32
HadGEM1	5	14	13	31
INM:CM30	7	6	32	45
IPSL:CM4	22	13	34	69
MIROC3.2_hires	23	4	21	47
MIROC3.2_medres	19	9	20	47
PCM	6	7	21	34

9.2.4 *Conclusions*

A score-based multi-criteria method was developed to assess and rank the performance of 25 AR4 GCMs. For the TRB region, GCM performance in simulating monthly temperature was better than for precipitation, with MSLP in between. This conclusion is consistent with results of previous global scale studies. The method is easily applied to different study regions, with assessment results suitable to guide the selection of GCMs for use in regional climate change impact studies.

It is not necessary to use all assessment criteria presented, dependent on the objective of the assessment. For example, flood studies are interested in the high-tails of precipitation PDFs.

A GCM that does not simulate current climate (1961–2000) accurately does not necessarily imply that it cannot produce accurate projections. However, the critical question is how we would assess such projections. All results and conclusions were based on GCM 20c3m runs, but other sources of uncertainty due to model limitations in sub-grid scale forcing and processes also exist. A research priority in the future is to quantify the full range of uncertainty.

Most of climate change impact studies require daily climate data, whereas this study concentrated on monthly data because of data availability and processing time. The assumption is that if a GCM simulates monthly climate variable accurately, then it has high

Table 9.5 NCEP/NCAR candidate predictors

ID	Full name	Classification
air	Air temperature	Pressure variables (Three pressures 400, 500 and 600 hPa)
hgt	Geopotential height	
shum	Sensible humidity	
uwnd	East–west velocity	
vwnd	North–south velocity	
slp	Air pressere at sea level	Surface variables
lhtfl	Latent heat net flux at surface	
shtfl	Sensible heat net flux at surface	
nlwrs	Net longwave radiation flux at surface	
nswrs	Net shortwave radiation flux at surface	

probability of simulating daily climate variable accurately; if a GCM cannot simulate monthly climate variables accurately, then it is impossible to simulate daily climate variable accurately. The first part of this assumption will be tested in future studies.

9.3 Evaluation on the Performance of Two Statistical Downscaling Models to Generate Daily Precipitation Over the Tarim River Basin, China

9.3.1 Dataset Description

(1) Observed data Observed daily precipitation was used as the predictand for downscaling in the TRB. There are 27 National Meteorological Observatory stations in or around the TRB, in which 25 stations with continuous data series of 1960–2000 were selected in this study. All observed data series had been passed the data quality control.

(2) NCEP reanalysis data The NCEP reanalysis datasets could be found in Sect. 9.2. Both pressure and surface variables were selected as candidate predictors (Table 9.5), which include five pressure variables at three pressures (400, 500, and 600 hPa) and five surface ones. All reanalysis data has been standardized before downscaling. Standardized predictors are routinely employed in order that the same models could be applied to future climate scenario generation using standardized GCM output for future studies.

9.3.2 Methods Description

9.3.2.1 Choice of Predictors and Predictor Domains

Choice of predictor variables should be given high attention in statistical downscaling methods. Circulation related predictors are usually selected as candidate predictors, because observations with relatively long series are available and

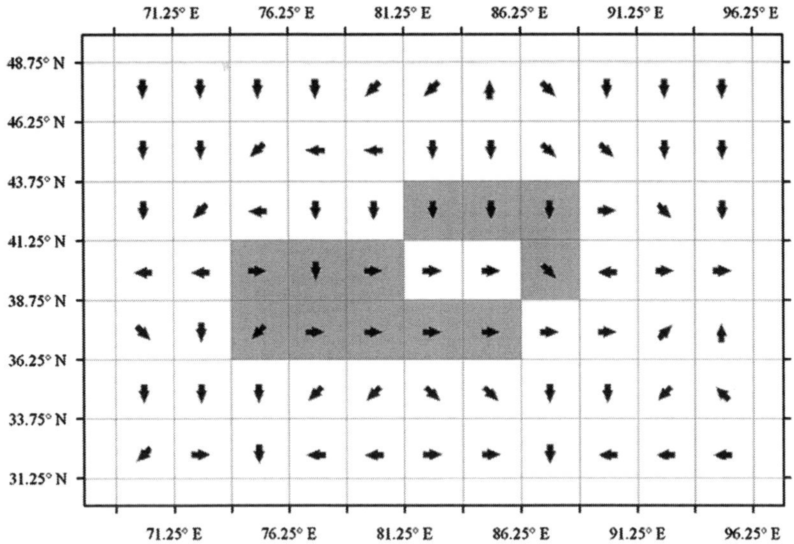

Fig. 9.9 MSLP gradients over the Tarim River basin (*gray* grids mean covering observed stations)

GCMs simulate these with some skills. However, it is increasingly acknowledged that circulation variables alone are not sufficient, as they usually fail to capture key precipitation mechanisms based on thermodynamics and moisture content. Thus humidity has increasingly been used to downscale precipitation (Karl et al. 1990), especially as it may be an important predictor under a changed climate. It is suggested that the optimal predictors must be strongly correlated with the predictand, be physically sensible and capture multiyear variability (Wilby et al. 2004). In this study, 20 NCEP reanalysis large-scale variables (Table 9.5), which have potential physical relationships with precipitation, were selected as candidate predictors. These predictors include not only circulation variables (i.e. geopotential and wind component), but also air temperature, radiation and moisture variables (specific humidity). Exploratory and partial correlation analyses were used to determine which candidate predictors had the strongest relationship with precipitation over the TRB.

The selection of predictor domain, which is generally ignored, has been found a critical factor affecting the downscaled results, especially for downscaling precipitation, because only one grid predictor used for downscaling is not necessarily related solely to a location. Therefore, multi-grid predictor domains were for each observed station in the application of SDSM. Potential grids include one grid nearest to the observed station and eight grids neighboring it. The predictor domains were selected from these candidate potential grids based on mean sea level pressure gradients (Fig. 9.9). As for each station, the centralized grid in which a station is located and its neighboring grids which have their mean sea level pressure gradient point to the centralized grid were selected as grids related with this station.

9.3.2.2 SDSM

As mentioned above, SDSM is a hybrid between stochastic weather generator and multilinear regression method. That is because regional circulation patterns and atmospheric moisture variables are used to condition local precipitation at each station. The regression component represents the deterministic part of the model, while the stochastic component is a random element of the model (Prudhomme and Davies 2009). The multilinear regression of the model is used to derive a statistical relationship between predictors and precipitation, which includes some transform functions in order to obtain secondary data series of precipitation and predictors that have stronger correlations than the original data series. It allows the prediction of local weather conditions from large-scale predictor variables simulated by GCMs under baseline and changed climatic scenarios. Precipitation is then modeled by stochastic weather generator conditioned on predictors. The stochastic component was added to replicate a variance closer to the observed variability because the chaotic nature of local weather conditions cannot be fully explained by the variations in predictor variables (Prudhomme and Davies 2009). Therefore, SDSM enables to generate multiple simulations with slightly different time series attributes, but the same overall statistical properties. The detailed technical information of SDSM can be found from Wilby et al. (2002).

Generally, the application of SDSM contains five steps (Wilby et al. 2002): (1) selection of predictors; (2) model parameters calibration; (3) weather generator; (4) model validation; and (5) generation of future series of the predictand. Since the purpose of this study is to evaluate the performance of two downscaling methods for downscaling observed precipitation, step 5 was not included in this study. Model calibration is based on multiple linear regression equations, given daily precipitation and large scale atmospheric variables. Then, daily precipitation was then simulated by weather generator in the model, which is based on NCEP re-analysis predictors. Evaluation of the model is based on comparison between the simulated and observed daily precipitation series.

9.3.2.3 NHMM

The NHMM can be defined by a state transition probability matrix and a precipitation occurrence probability distribution (Charles et al. 1999). It defines stochastic conditional relationships between multi-site daily precipitation occurrence patterns and a discrete set of weather states, which are referenced as hidden states because they are not directly observable. Then, the transition probabilities between these hidden states, which condition on a set of predictors but are not held fixed as in the homogeneous HMM (so called nonhomogeneous HMM), are defined by a first-order Markov chain. Finally, daily precipitation amounts for each station are simulated based on Gamma distribution.

There are two assumptions in the NHMM. The first one is that precipitation occurrence is only conditioned on a small number of given hidden states. By conditioning precipitation occurrence on hidden states, rather than atmospheric circulation patterns, NHMM captures much of the spatial and temporal variability of daily multi-site precipitation occurrence records (Charles et al. 1999). The other assumption describes that the hidden state on a given day only depends on the previous day's state or predictors, which corresponds to the Markov property.

Application of NHMM contains five parts: (1) choice of predictors; (2) selection for hidden states number; (3) generation of precipitation occurrence; (4) simulation for precipitation amount; (5) model validation. Predictors are selected in Sect. 3.1. The most appropriate hidden states number is estimated based on maximum likelihood, which is evaluated by the Bayes Information Criterion (BIC). Once hidden states number has been estimated, the most likely daily sequence of hidden states, along with dry states (zero precipitation) can be determined by the Viterbi algorithm, a dynamic programming scheme. Finally, daily precipitation amount in wet days at observed stations is modeled for each state based on Gamma distribution. In all cases, the NHMM was run 20 times from random seeds in the simulation. Detailed description of NHMM could be found from Kirshner (2005).

9.3.2.4 Model Settings and Evaluation Merits

(1) Model settings The observed data series (1961–2000) were split into two periods, 1961–1990 and 1991–2000, used for model calibration and validation, respectively. Although cross-validation may be more effective for shorter records, it is not necessary when long observed records are available (Wilby et al. 2004) as the case of this study. The NHMM was fitted on a seasonal basis with the wet season (April–September) and the dry season (October–March), respectively, in which precipitation take up 84.3 and 15.7 % of annual precipitation. The SDSM was based on monthly type, in which model parameters are derived for each month.

(2) Evaluation merits Several objective functions were selected for model evaluation in this study. Firstly, residual functions were used to compare the differences of statistical properties between observed and modeled precipitation series. The model bias was evaluated by relative error (RE) of long-term annual precipitation at each individual station. Extreme events, which are of interest in climate impact assessments, were evaluated by the percentiles of 50, 90, 95 and 99 (referenced as 50th, 90th, 95th, 99th, respectively) of the wet-day precipitation amount (wpa), wet-spell length (wsl) and dry-spell length (dsl). In addition, the mean and the maximum values of simulated wpa, wsl and dsl were also evaluated.

Secondly, the correlation coefficient, a widely used indicator to evaluate the "goodness-of-fit", was then used to assess the inner-annual cycle and spatial distribution of precipitation. For inner-annual cycle assessment, the correlation coefficient was calculated with observed and modeled long-term monthly mean values. The sample size is only 12 as it is the number of months. For spatial distribution assessment, the correlation coefficient was computed with observed and modeled long-term mean values for each individual grid cell. The sample size is then the number of stations in the study area, which is 25 in this study.

Finally, Brier score (BS) and skill score (S_{score}), which could be found in Sect. 9.2, were used to measure how well each model can capture the PDFs of wpa, wls and dsl distributions.

All objective functions were evaluated both for model calibration (1961–1990) and validation (1991–2000).

9.3.3 Results Analysis

9.3.3.1 Selected Predictors

It is well known that there should be weak correlation between selected predictors, which were confirmed in this study because the model performance was generally improved while different types of variables were included. If we only focused on explained variance of precipitation, then the same type variables (e.g. shum400, shum500 and shum600) with high inter-correlations each other were more likely to be all selected. Therefore, both explained variance and partial correlation analysis were used for the choice of predictors in this study to avoid this problem. Explained variance describes the correlation between precipitation and candidate predictors, while partial correlation analysis expresses the correlation among candidate predictors themselves.

The shum400, vwnd500, shtfl, lhtfl, nlwrs, slp and air500 were selected as predictors for the NHMM, based on explained variances and partial correlation analysis. It was supported with the dynamic mechanism of precipitation in the TRB, which is mainly controlled by westerly water vapor transport. Predictor sets were varied for each individual station for the SDSM. Overall, the selected predictor number for each station was between five and seven: shum and vwnd were the most selected predictors, air, shtfl, lhtfl, nlwrs, and slp followed, while other variables were rarely selected.

9.3.3.2 Model Calibration and Validation

(1) SDSM model calibration and validation Three parameters of the SDSM, event threshold, bias correction and variance inflation, were adjusted to get the best statistical agreement between observed and simulated daily precipitation for model calibration. Results showed that event threshold was best set to 0.1 mm/day, which indicated that dry days were defined when daily precipitation less than 0.1 mm/day, while wet days were defined when it was greater than 0.1 mm/day.

(2) NHMM model calibration and validation The choice of the appropriate number of hidden states in the NHMM was based on BIC indices for both wet and dry season models, respectively. Results showed that the BIC reached their minimum values when hidden states numbers were six and three for wet and dry seasons, respectively (Fig. 9.10). Therefore, they were selected as the appropriate number of hidden states for the NHMM.

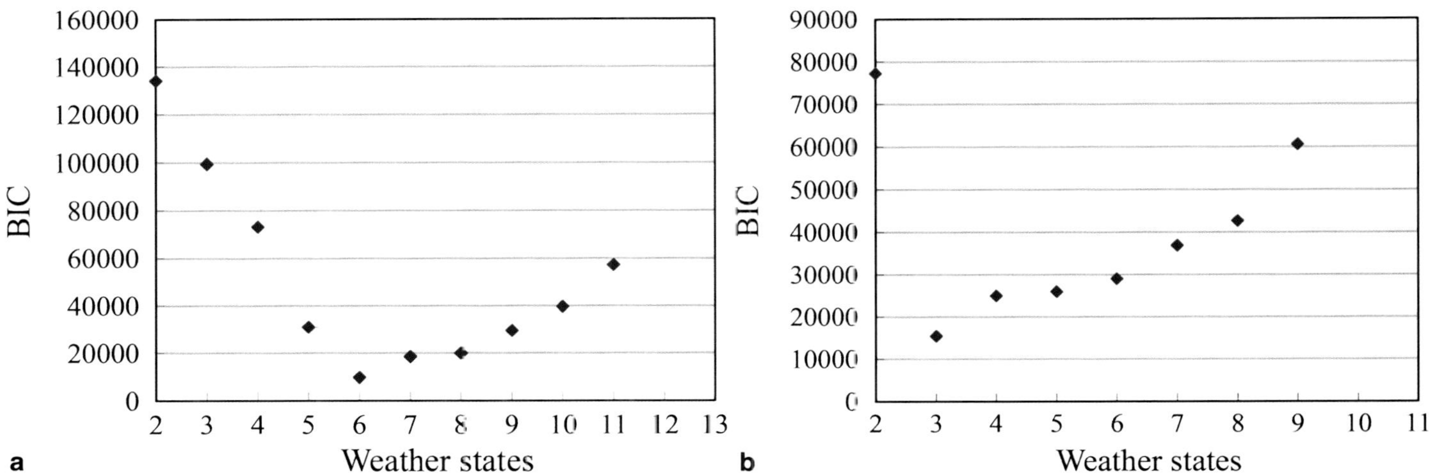

Fig. 9.10 BIC with different numbers of hidden states (**a** is for the wet season model; **b** is for the dry season model)

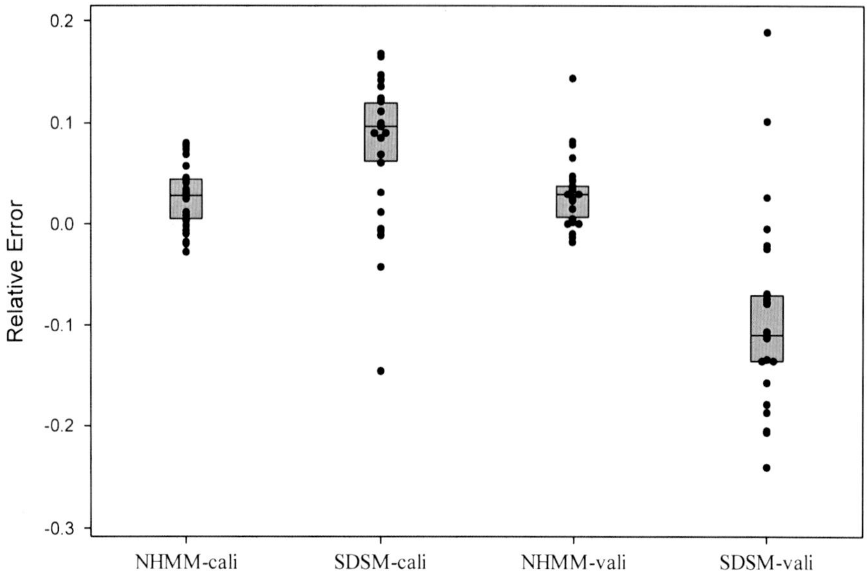

Fig. 9.11 Relative errors between modeled and observed annual precipitation for each station in the model calibration and validation ('*cali*' and '*vali*' mean in the calibration and validation; *gray box* indicate median values at the level of 0.95)

Several event thresholds for wet and dry days were calibrated and a similar result as that of SDSM was obtained: the best model performance was achieved with an event threshold of 0.1 mm/day.

9.3.3.3 Comparison Between Two Downscaling Models

(1) Residual functions Model biases of long-term mean annual precipitation for SDSM and NHMM were shown in Fig. 9.11. In general, both SDSM and NHMM performed well in simulating mean annual precipitation with relative errors at most stations being less than 20 %. Model performance was generally a little better in calibration period than that in validation period, but difference was not very large. Overall, NHMM showed a little better model performance for the mean annual precipitation than SDSM.

Model performance on monthly precipitation was shown in Fig. 9.12. In general, NHMM simulated monthly precipitation better than SDSM. It may be attributed to the fact that NHMM was fitted separately for wet and dry season models, which made more meanings in term of precipitation physical processes. In both model calibration and validation periods, NHMM was able to simulate monthly precipitation well for all dry months, while SDSM could only have satisfied performance from January to March, but have very poor performance from October to December. Model performance of NHMM changed little from model calibration to validation periods, while SDSM exhibited a big difference between model calibration and

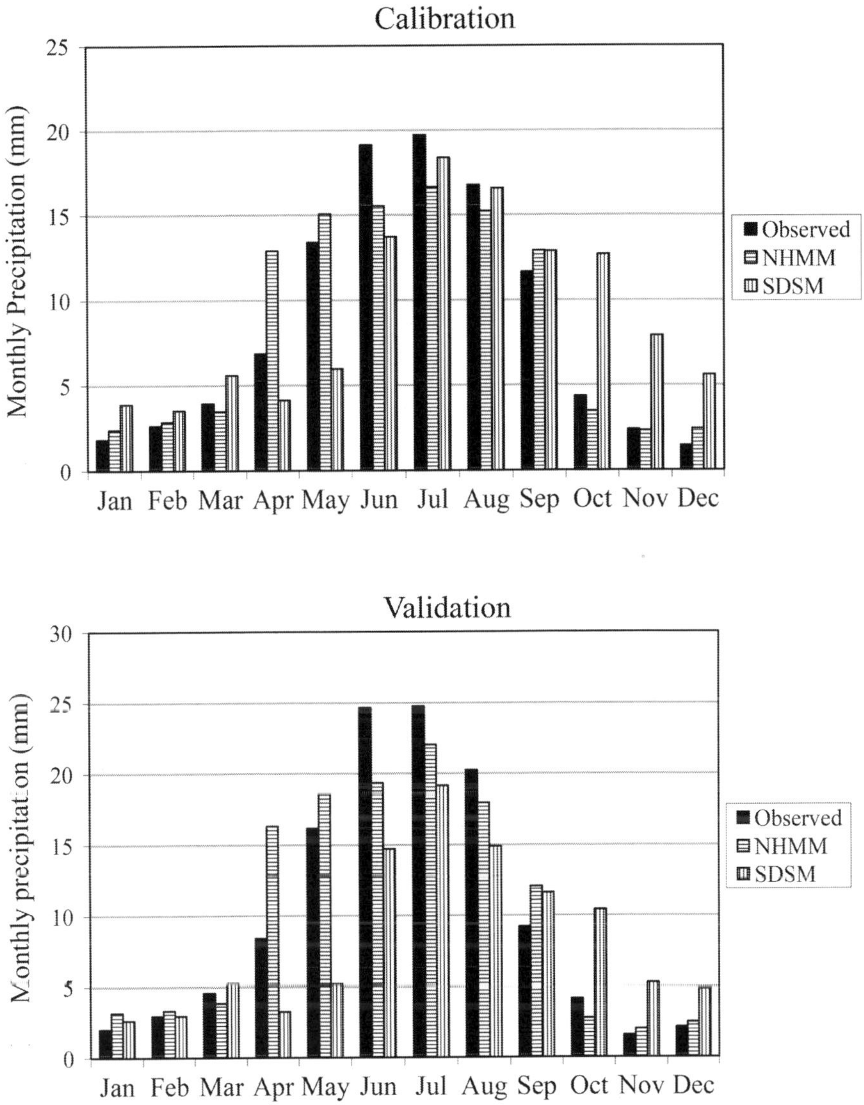

Fig. 9.12 Modeled and observed monthly precipitation for both model *calibration* and *validation*

validation periods. This fact indicated that NHMM had a better model stability than SDSM in simulating monthly precipitation.

Table 9.6 showed statistics for mean and percentile values of dsl, wsl and wpa in calibration and validation periods, respectively. Both NHMM and SDSM underestimated almost all statistics of dsl, wsl and wpa in calibration and validation periods. However, the situation changed a lot when statistics distributions were considered (Fig. 9.13). It showed that observed wsl was only 1 day for about 50 % of wsl in calibration and validation periods. Both NHMM and SDSM could capture

Table 9.6 Statistics of mean and percentile values for dry spell length, wet spell length and wet-day precipitation amount in the calibration and validation periods

Statistic		Calibration						Validation					
		Mean	50th	90th	95th	99th	Max	Mean	50th	90th	95th	99th	Max
Dsl (day)	Observed	8.3	4	20	33	76	120	7.9	3	18	35	71	109
	NHMM	5.3	3	12	18	33	58	5.3	3	13	18	33	50
	SDSM	5.9	3	14	20	38	66	6.2	3	15	23	48	64
Wsl (day)	Observed	2.4	2	5	6	11	20	2.4	2	5	6	11	13
	NHMM	1.6	1	3	4	6	9	1.5	1	3	4	6	8
	SDSM	1.6	1	3	4	6	11	1.6	1	3	4	7	9
Wpa (mm/day)	Observed	1.07	0.27	2.50	4.85	13.60	42.59	1.42	0.40	3.62	6.65	16.52	33.98
	NHMM	0.96	0.31	2.37	4.05	9.92	31.91	1.40	0.52	3.49	5.75	13.79	29.30
	SDSM	0.75	0.31	2.13	2.87	4.51	7.82	0.90	0.44	2.38	3.11	4.81	7.15

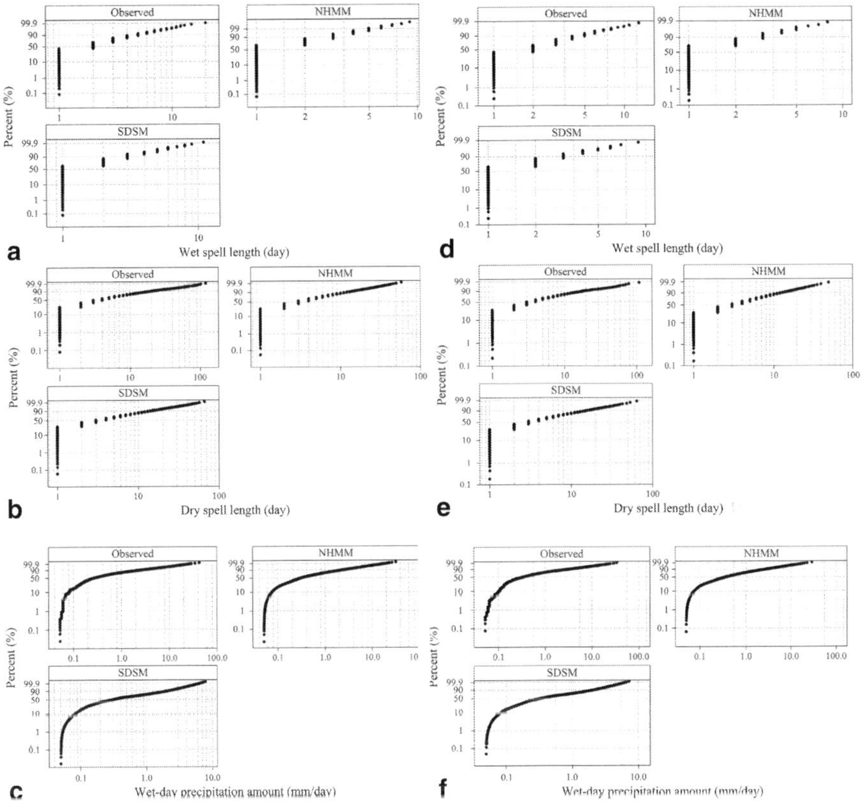

Fig. 9.13 Distributions of modeled and observed wet spell length, dry spell length and wet-day precipitation amount in the model calibration and validation (The distributions were fit based on the Gamma distribution; **a**, **c**, **e** describe wsl, dsl and wpa in calibration respectively; while **b**, **d**, **f** was same as that but for validation)

that well. In general, these two models were able to capture most characteristics of the wsl distribution in calibration and validation periods except underestimation of extreme values. Similar results were also shown for dsl. Nearly 90% of dsl was less than 10 days, which was well simulated by both of NHMM and SDSM, except for the underestimated extreme values as the case of wsl. For the simulation of wpa, NHMM and SDSM performed differently. NHMM had a better model performance on mean wpa than SDSM. Simulated wpa by NHMM were closed to observed values for all statistics. There was little model performance difference between calibration and validation periods. While SDSM could only model it well for percentile values less than 90th, but had a very poor performance on larger values. For example, the modeled maximum wpa is only 7.82 and 7.15 mm/day in model calibration and validation, respectively, which were much lower than the observed ones of 42.6 and 34.0 mm/day (Table 9.6). Distribution results (Fig. 9.13) showed that NHMM performed pretty well in simulating wpa for both the shape of distribution and extreme values, but SDSM had a poor performance on extreme values.

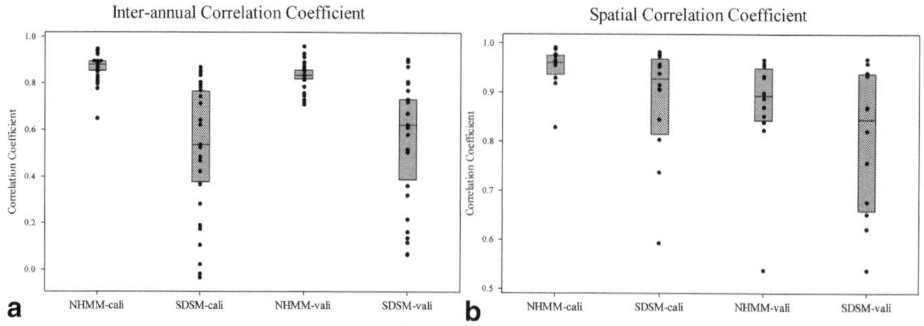

Fig. 9.14 Box plots of both *inter-annual correlation coefficients* for each observed station and *spatial correlation coefficients* for each month in the model calibration and validation ('*cali*' and '*vali*' mean in the calibration and validation; *gray box* indicate median values at the level of 0.95)

It was very difficult to improve this result since any attempt to increase the model variance inflation which controls the extreme values will affect the mean values which were originally well calibrated in SDSM.

(2) Correlation analysis of inner-cycle and spatial distribution Results for correlation analysis of inner-cycle and spatial distribution were shown in Fig. 9.14. NHMM modeled inner-annual data series with high correlation coefficients with observed ones for almost all observed stations in both calibration and validation periods, with a median correlation coefficient being larger than 0.8. Although SDSM could also simulate long-term annual mean precipitation well (above Fig. 9.11), it performed relatively poor on inner-annual cycle characteristics. In model calibration, inner-annual cycle correlation coefficients were less than 0.5 for 11 stations, 2 stations even with negative correlation coefficients. Overall, NHMM had a much better model performance on inner-annual cycle than SDSM.

In general, both models could capture well of the spatial distribution characteristics of observed precipitation, as 11 months had a spatial correlation coefficient being larger than 0.9 and 0.8, respectively for NHMM and SDSM in model calibration. It could also be noted that NHMM performed a little better than SDSM. The median spatial correlation coefficient was larger than 0.95 for NHMM model calibration period, and NHMM also performed well for validation period with median values nearly to 0.9. Although performed not as well as NHMM, SDSM was also able to capture spatial distribution characteristics of precipitation, with a median value of spatial correlation coefficient nearly to 0.9 for both calibration and validation periods.

(3) Skill scores based on PDFs The results of two PDFs based skill scores S_{score} and BS indicated that there was little difference between NHMM and SDSM while simulating dsl and wsl, and both models showed good skills with high stabilities (Fig. 9.15). However, skills from two models showed a large difference while simulating wpa, and NHMM was more powerful on simulating wpa than SDSM. Overall, NHMM had a very good skill in modeling wpa, dsl and wsl, especially for wpa with S_{score} being greater than 0.9 for both calibration and validation periods. SDSM also had a good skill in modeling dsl and wsl with S_{score} being greater than 0.75 at

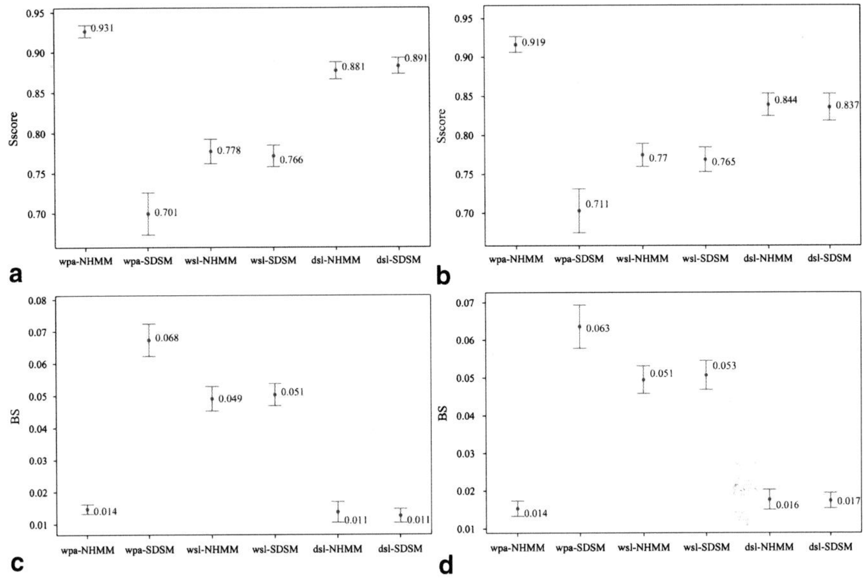

Fig. 9.15 Interval plots of two skill scores for wet spell length, dry spell length and wet-day precipitation amount of each station in the model calibration and validation (**a** and **b** were for Sscore in the calibration and validation respectively; while **c** and **d** were the same as that but for BS)

all stations used in this study. However, it did not have a good skill for wpa, which maybe because of its poor performance on extreme values.

9.3.4 Discussion and Conclusion

This study evaluated two widely used statistical downscaling models, NHMM and SDSM, on simulating daily precipitation over the TRB. Overall, NHMM, based on weather states which made it more physical meanings, seems to be more skillful for downscaling daily precipitation than SDSM in the TRB. The main reasons might be that NHMM downscale daily precipitation independently for wet and dry seasons that could capture the physical precipitation processes, and that the spatial distribution pattern of precipitation is taken into account. More specifically, the following conclusions are made:

The selection of predictors and predictor domains was an important step in the application of statistical downscaling techniques because it would have a great impact on the downscaling results. In this study, multi-grids, which were selected based on mean sea level pressure gradients, were used as predictor domains to capture more circulation conditions related to precipitation mechanism. It was found that moisture and wind variables (shum and vwnd) had the strongest relationship with precipitation, which was consistent with the dynamic mechanism of precipitation in the TRB. Other candidate predictors such as air temperature, mean sea level pressure, and radiation variables, which had strong statistical relationships with precipitation, were also selected for downscaling models because they may contribute to local precipitation.

Based on monthly precipitation amount and inner-annual correlation analysis, NHMM showed a better model performance on simulating monthly precipitation than SDSM: the former was able to model precipitation well for all months while the later only well for several months. NHMM also exhibited a better model stability than SDSM with little model performance difference between calibration and validation periods. Fitting the wet and dry season models independently in the NHMM made it more helpful to capture the physical precipitation processes.

Discrete mean or percentile values could be easy to make a misleading conclusion for model evaluation because these statistics only explained model performance on limited values. Distributions or PDFs based measures, which could capture entire statistical characteristics of climate variables, were used in this study. On the bias of these measures, it was found that NHMM could have a very well performance on modeling dsl, wsl and wpa, especially for wpa. It was able to simulate wpa pretty well for both the shape of distribution and extreme values. However, SDSM had a poor performance on extreme values of wpa.

Both NHMM and SDSM could capture well of the spatial distribution characteristics of precipitation in the TRB, and the former performed a little better. It may be attributed to the facts that NHMM includes spatial correlations for modeling multi-site precipitation, which is missing in SDSM as it models each individual station separately.

9.4 Climate Change Scenarios Generated by Statistical Downscaling

9.4.1 Dataset Description

Observed station data and NCEP reanalysis data were same as that in Sect. 9.3.

Model output from three different GCM datasets, which include CSIRO30, ECHAM5, and GFDL21, were used to assess TRB precipitation and air temperature changes in the twenty-first century. All GCMs runs used in this study were run1. Three emission scenarios of the Intergovernmental Panel on Climate Change (IPCC) Special Report on Emissions Scenarios (SRES), SRES A1B, A2, and B1 simulations from each of the GCMs were selected. Data series include both present control period (1960–2000) and future period (2046–2065 and 2081–2100).

9.4.2 Methodology Description

Long observed data series (1961–2000) were split into two periods: 1981–2000 and 1961–1980, and used for model calibration and validation, respectively. The period of 1981–2000 was selected for model calibration because both observed and NCEP

Table 9.7 Explained variances of each predictor and SDSM and NHMM selected predictors

Variables	T_{max}		T_{min}		Prec	Variables	T_{max}		T_{min}		Prec
	Explained variance	NM	Explained variance	NM			Explained variance	NM	Explained variance	NM	
air400	0.500		0.797			shum600	0.363	25	0.692	25	★
air500	0.496	2	0.778			uwnd400	0.032		0.083		
air600	0.513	23	0.825	25		uwnd500	0.036		0.064		
hgt400	0.303	25	0.435	25	★	uwnd600	0.025		0.055	2	
hgt500	0.124		0.077			vwnd400	0.015		0.030		
hgt600	0.051		0.015			vwnd500	0.016		0.017		
shum400	0.273		0.583		★	vwnd600	0.047	2	0.120	2	★
shum500	0.279		0.565		★	slp	0.440	25	0.650	25	★

"*NM*" is the number of stations selected as predictor by SDSM, ★ indicates selected by NHMM, "T_{max}", "T_{min}" and "*Prec*" describe maximum air temperature, minimum air temperature, and precipitation, respectively

data have more quality data than previous periods. The NHMM was fit on a seasonal basis with the wet season (April–September) and the dry season (October–March), respectively, in which precipitation takes up 84.3 and 15.7 % of annual precipitation. The SDSM was based on monthly type, in which model parameters are derived for each month. Models calibrated and validated using observed predictors derived from reanalysis data were then driven by predictors derived from large–scale predictor variables simulated by GCMs in the control period representing present climate (1981–2000) and future periods (2046–2065 and 2081–2100). Finally, plausible future changes of daily precipitation and air temperature were estimated as the difference of downscaled values between those using predictors derived from GCMs outputs corresponding to the future period under the IPCC SRES A1B, A2, and B1 emission scenarios and the corresponding values run for the control period.

Several model performance criteria could be found from Sect. 9.2. All objective functions were evaluated for both calibration and validation periods.

Selection of predictor variables and domain were based on the method in Sect. 9.2.

9.4.3 Results Analysis

9.4.3.1 Predictor Selection

Table 9.7 presents the predictors selected for T_{max} and T_{min} in SDSM, and for precipitation in NHMM, respectively. The shum400, shum500, shum600, vwnd600, hgt400, slp and air600 are selected as predictors for the NHMM, based on explained variances. These predictors are consistent with the dynamic mechanism of precipitation in the TRB, which is mainly controlled by westerly water vapor transport (Dai et al. 2007; Qian and Qin 2008). Predictor sets are varied for each individual station for the SDSM. Overall, the selected predictor number for each station is five and six: air600, shum600, hgt400 and slp are the most selected predictors, air500 and vwnd600 are followed, while other variables were not selected.

9.4.3.2 Model Performance

Model biases of air temperature and precipitation for both mean and percentile values were shown in Fig. 9.16. Mean value of model biases at all stations was about 0.5 °C for the simulated T_{max} and T_{min} during both calibration and validation periods. It could be found in Fig. 9.16 that NHMM tends to underestimate three statistics of precipitation, including dsl, wsl and wpa. Underestimation of dsl and wsl means that the simulated wet or dry spell lengths are shorter than observed ones. However, there is no significant difference between simulated and observed precipitation, with relative errors of three statistics of precipitation smaller than 15 %. Results of two PDFs based skill scores S_{score} and BS for air temperature and precipitation were shown in Fig. 9.17. It could be seen from Fig. 9.17 that NHMM showed better skills on simulating dsl than that on wsl in the TRB, with greater S_{score} values and smaller BS values for dsl than that for wsl. This might be resulted from the fact that values of dsl are much higher than wsl in the arid region, and the statistical downscaling model shows better ability on greater values of time series. That is consistent with results from Liu et al. (2011). NHMM had skills in modeling wpa with S_{score} being greater than 0.9 in both calibration and validation periods. SDSM also showed good skills on downscaling T_{max} and T_{min}, with S_{score} values at most stations greater than 0.8 in both calibration and validation periods. In general, SDSM and NHMM showed abilities to downscaling daily precipitation, T_{max} and T_{min} over the TRB.

It could be seen from Figs. 9.16 and 9.17 that there was little difference for model performance of NHMM and SDSM between calibration and validation periods, which indicated that both models had abilities on the stability in downscaling precipitation and air temperature over the TRB. Besides that, when NCEP predictors were replaced by GCM (CSIRO30, ECHAM5, and GFDL21) predictors in downscaling precipitation, T_{max} and T_{min} in calibration period, there were no significant differences for model performance with little changes on the values of S_{score} and BS. It indicated that the relationship between predictors and predictands calibrated by NCEP predictors could also be used for these three GCMs predictors.

9.4.3.3 Local Climate Change Scenarios

Figure 9.18 described the changes for mean values of monthly, seasonal and annual precipitation, T_{max} and T_{min} generated by statistical downscaling models. Changes of annual precipitation projected by statistical downscaling model were not obviously, with values lower than 15 % under all combined scenarios. The magnitude of annual precipitation changes in the period of 2081–2100 was smaller than that in the period of 2046–2065. In both periods, CSIRO30 tended to project increasing trends for annual precipitation, while ECHAM5 and GFDL21 projected decreasing trends for that under all three combined scenarios. As for monthly precipitation, it showed much greater changes than annual precipitation based on the statistical downscaling model, with changes ranged from − 80 to 60 % approximately. It should be pointed out that monthly precipitation showed increasing trends only in March and July under most of combined scenarios for the period of 2046–2065. It exhibited increasing trends in April, May, October and November, while showed decreasing trends in

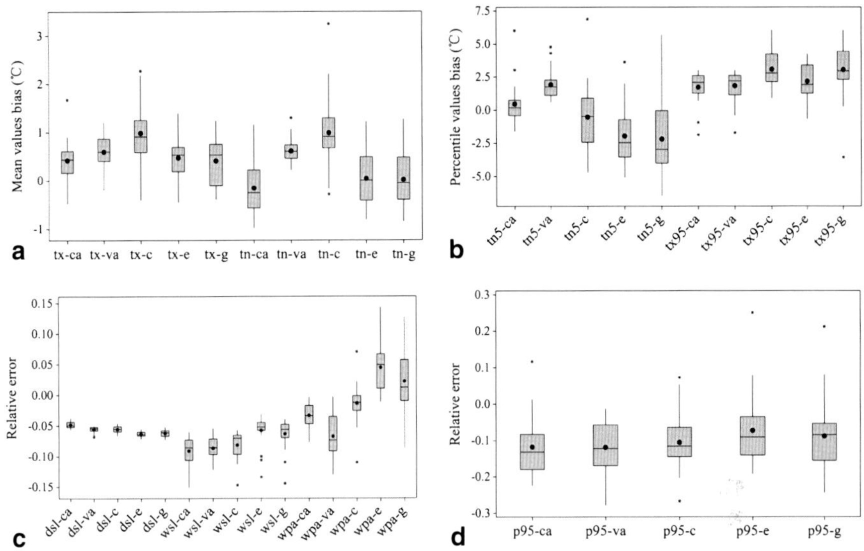

Fig. 9.16 Model bias for precipitation and temperature (**a**, **b** describe bias of mean and percentile values for temperature, respectively; **c**, **d** are the same as that but for precipitation. "*–ca*" and "*–va*" mean calibration and validation respectively; "*–c*", "*–e*", "*–g*" mean predictors of CSIRO30, ECHAM5 and GFDL21 respectively; "tx", "tn" mean maximum and minimum air temperatures respectively; "*tn5*", "*tx95*", "*p95*" mean 5th value of minimum air temperature, 95th value of maximum air temperature and wpa, respectively)

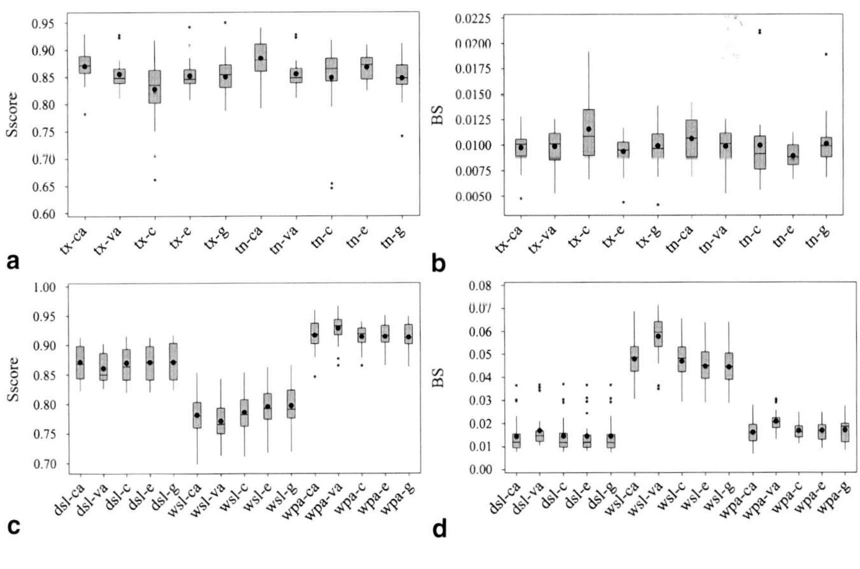

Fig. 9.17 Boxplots for precipitation, maximum and minimum air temperatures based on skill scores (**a**, **b** describe air temperature; **c**, **d** describe precipitation. "*–ca*" and "*–va*" mean calibration and validation respectively; "*–c*", "*–e*", "*–g*" mean predictors of CSIRO30, ECHAM5 and GFDL21 respectively; "tx", "tn" mean maximum and minimum air temperatures respectively; "tn5", "tx95", "p95" mean 5th value of minimum air temperature, 95th value of maximum air temperature and wpa, respectively)

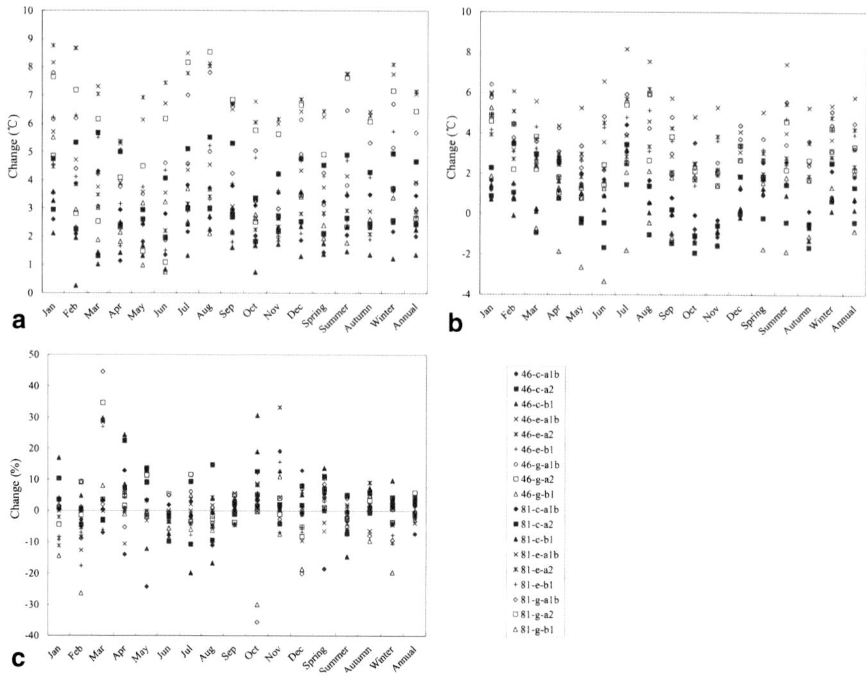

Fig. 9.18 Changes of mean values for precipitation, maximum and minimum air temperature projected by statistical downscaling models (**a**, **b**, **c** describe changes of maximum and minimum air temperatures, precipitation, respectively; "46" and "81" represent periods of 2046–2065 and 2081–2100, respectively; "-c", "-e", "-g" mean predictors of CSIRO30, ECHAM5 and GFDL21, respectively; "a1b", "a2" and "b1" represent SRES A1B, A2, and B1, respectively. The following is the same)

July and August under all combined scenarios for the period of 2081–2100. Changes of monthly precipitation in period of 2081–2100 were also smaller than that in the period of 2046–2065.

It could be found from Fig. 9.18a that T_{max} exhibited increasing trends under all combined scenarios. The magnitude of increasing trends in the period of 2081–2100 was greater than that in the period of 2046–2065, with the highest change larger than 7.0 °C. The magnitude of the increasing trend was the greatest under A2 scenario, and was the smallest under B1 scenario, with A1B scenario in-between. In both periods, the increasing trend of T_{max} projected by CSIRO30 was smaller than that by ECHAM5 and GFDL21. The difference between changes of monthly and annual T_{max} was not as obviously as that for precipitation.

It could be found from Fig. 9.18b that T_{min} exhibited increasing trends under most of combined scenarios, except for CSIRO30 related scenarios and the combined GFDL21-B1 scenario. Changes of annual T_{min} ranged from −0.4 to 3.3 °C in the period of 2046–2065 and ranged from −0.9 to 5.9 °C in the period of 2081–2100. The difference between changes of monthly and annual T_{min} was not as obvious as that for precipitation. It was worthy to point out that winter T_{min} showed increasing trends under all combined scenarios for both periods. That might result in more snowmelt runoff in the HC.

9.4.4 *Conclusions and Discussion*

Regional climate change scenarios, statistical downscaled from three IPCC (Intergovernmental Panel on Climate Change) Special Report on Emission Scenarios (SRES A1B, A2 and B1) and three General Circulation Models (GCMs) outputs, were used to assess the impact of climate change on hydrological processes in the Headwater Catchment (HC) of the TRB.

It showed an obvious increasing trend for minimum and maximum air temperatures in TRB under nine combined scenarios, in which the magnitude of the increasing trend was the greatest under A2 scenario, and was the smallest under B1 scenario, with A1B scenario in-between. This was consistent with the changes of air temperature under different emission scenarios over the world. It tends to exhibit decreasing trend for precipitation at the HC, with the magnitude of decreasing trend being smaller than 10 % at all stations, and most of stations were smaller than 5 %. The magnitude of increasing air temperature in the period of 2081–2100 was greater than that in the period of 2046–2065, while the magnitude of changes for annual precipitation in the period of 2081–2100 was smaller than that in the period of 2046–2065. It was different from air temperature that the magnitude of changes for extreme values of precipitation is obviously greater than that for mean values.

9.5 Assessment on the Hydrological Impact of Climate Change in Headwater Catchment of the Tarim River Basin

9.5.1 *Study Area Description*

The headwater catchment of the TRB, as shown in Fig. 9.19, can be regarded as a single catchment with relatively independent hydrological units. It was, therefore, selected as the study area. The catchment has its own characteristics of runoff generation. Annual precipitation in the plain region is usually lower than 100 mm, so it is not able to generate runoff. Runoff is only generated in the high mountainous region, but is consumed in the low plain region by evaporation or irrigation.

9.5.2 *Dataset Description*

(1) Observed stations data Observed daily data series from 14 National Meteorological Observatory (NMO) stations located in or around the TRB were selected, which included precipitation, maximum and minimum air temperature (T_{max} and T_{min}), and mean wind speed over a period of 1961–2010. These data were provided by the National Climatic Centre of China. Location of the stations is shown in Fig. 9.1. Another ten precipitation stations and five hydrological stations with daily data series

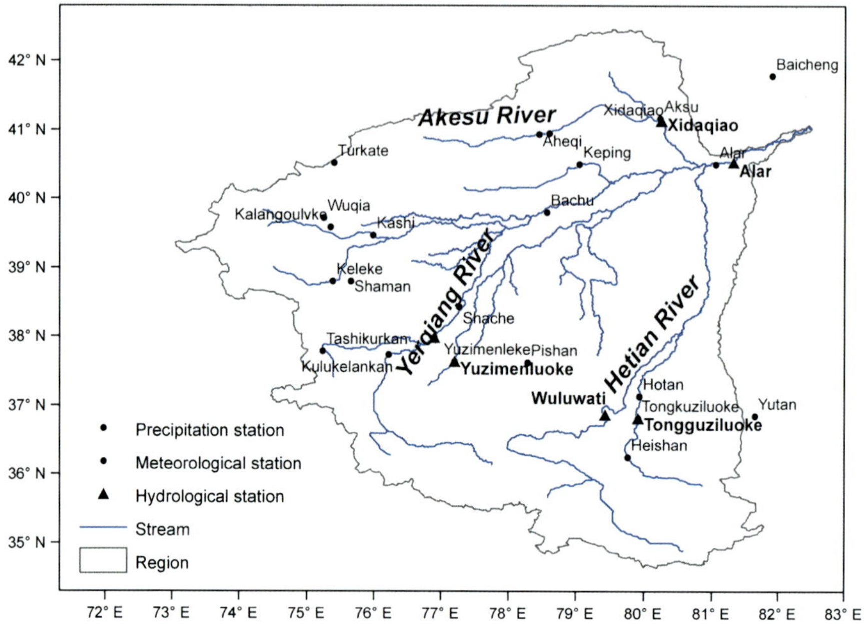

Fig. 9.19 The headwater catchment of the TRB

of precipitation and monthly streamflow, which is provided by the Bureau of Hydrology, China Ministry of Water Resources, were also selected. The sum streamflow from five hydrological stations (Fig. 9.1) were considered as the total discharge of the HC instead of the streamflow at Alar station, which was the control station for the catchment. Because of that the streamflow at Alar station had been greatly changed by human activities, such as operation of reservoirs and irrigations. Observed streamflow data was only available for several non-consecutive series, which include monthly series for 1964, 1971, 1976–1979, 1981, 1983–1984, 1986–1987, and 2001–2005.

(2) DEM, soil and land cover data Digital Elevation Map (DEM) data: The Shuttle Radar Topographic Mission (SRTM) DEM data for the TRB was obtained from the SRTM website (http://srtm.csi.cgiar.org).

Soil data: The spatial soil data, with a resolution of 30 by 30 min, was derived from the ISRIC-WISE soil database (http://www.isric.org). Twenty-two soil variables, identified as being useful for agro-ecological zoning, land evaluation, crop growth simulation, modeling of soil gaseous emissions and analyses of global environmental change, were considered in the soil database. Soil texture classes were reclassified into 12 texture classes according to the United States Department of Agriculture (USDA) soil classes.

Land cover data: The University of Maryland's 1 km global land cover data was available from: http://www.geog.umd.edu/landcover/1km-map.html. The 1 km spatial resolution land cover classification data was from the Advanced Very High Resolution Radiometer (AVHRR). All of pixels were used within a hierarchical tree structure to classify the AVHRR data into 12 classes.

9.5.3 Method Description

9.5.3.1 Hydrological Model

The grid network version of the three-layer VIC (VIC-3 L) model (Liang and Xie 2001), a macro-scale hydrological based land surface model considering both infiltration and saturation excess runoff, was used in this study. Vertically, the soil column is divided into three soil layers in the VIC-3 L model, including a upper layer, a lower layer and a thin layer on the top of the upper layer (Liang et al. 1996). The upper layer is designed to represent the dynamic behavior of the soil column that responds to rainfall events, and the lower layer is used to characterize the slowly varying between-storm soil moisture behavior, and the thin top soil layer allows for quick bare soil evaporation following small summer rainfall events. Horizontally, each grid cell can have partial surface coverage by a number of land cover classes.

The simulated runoff at each grid was routed to the outlet of the catchment (Alar station) with the unit hydrograph method for overland flow (include surface flow and subsurface flow) and the linear Saint–Venant method for channel flow. The area fraction of each grid was applied in the routing model to deal with the area dismatch between the digital boundary based on grids and the actual boundary.

9.5.3.2 Model Settings

The long observed data series (1961–2000) were split into two periods: 1981–2000 and 1961–1980, used for model calibration and validation, respectively. The period of 1981–2000 was selected for model calibration because both observed and NCEP data have more quality data than previous period. Then, the models calibrated and validated using observed predictors derived from reanalysis data were later driven by predictors derived from large-scale predictor variables simulated by GCMs in the control period representing the control period (1981–2000) and the future periods (2046–2065 and 2081–2100). Finally, local climate (daily precipitation and air temperature) change scenarios, which include multi-GCMs and multi-emission scenarios, were estimated by the difference of downscaled variables between the future periods and the control period, under the IPCC SRES A1B, A2, and B1 emission scenarios. The VIC-3 L model was applied in the HC of the TRB with the resolution of $0.5° \times 0.5°$. Either full energy balance and water balance or only water balance were developed in the VIC-3 L model. In this study, full energy balance and water balance were adopted to simulate evapotranspiration, runoff and soil moisture. The running step of the model was 24 h. The vegetation data set, soil data set and meteorological forcing data set at each grid cell were applied to the VIC-3 L model as model inputs for simulations in the catchment from 1963 to 2005, in which data series of 1963–1964 were used for model initialization. Meteorological forcing data which included daily precipitation, maximum air temperature, minimum air temperature and mean wind speed from 24 stations in the catchment were interpolated to 203 grids based on the Thiessen polygons method. The inconsecutive observed streamflow data, which included 1960s, 1970s, and 1980s, was used to

Table 9.8 Model performance on monthly runoff in calibration and validation periods

Period	Nash efficiency coefficient	Relative error (%)	Correlation coefficient
Calibration	0.821	6.3	0.894
Validation	0.782	8.1	0.877

model calibration. The period from 2001 to 2005, which was with relative long and consecutive observed streamflow data series, was selected for model validation.

9.5.4 Results Analysis

9.5.4.1 Model Performance

Comparisons between simulated and observed flows for calibration and validation are shown in Table 9.8 and Fig. 9.20. In generally, both simulated annual runoff volumes and hydrograph shape were consistence with observed values, for example, Nash and Sutcliffe (1970) efficiency coefficient for the model was 0.821 and 0.782 during calibration and validation periods, respectively. Streamflow was overestimated during all simulated periods. Total runoff was well simulated in the catchment with relative error smaller than 10 %. The correlation coefficient between simulated and observed flows was greater than 0.87 for all simulated periods. Overall, the calibrated VIC-3 L model showed satisfied model performance in simulating streamflow in the HC.

9.5.4.2 Impact of Climate Change on Monthly Runoff

Daily meteorological data at each station, generated by downscaling models, were inputted into the calibrated VIC-3 L model to simulate hydrological processes under climate change scenarios in the study area.

Monthly, seasonal and annual changes for runoff at the HC, which is controlled by the Alar station, were shown in Fig. 9.21. Driven by different combined climate change scenarios, annual runoff for the HC changed from −20.7 to 4.9 % in the period of 2046–2065, and changed from 0.5 to 7.2 % in the period of 2081–2100. In generally, it did not show obvious change for annual runoff, with the magnitude of these changes smaller than 8 % at most of combined climate change scenarios. It showed obviously greater changes under SRES A2 scenario than that under A1B and B1 scenarios.

Although there was little change for annual runoff driven by climate change scenarios at the HC, it showed obvious changes for monthly and seasonal runoff, with the greatest magnitude was close to 70 %. In the period of 2046–2065, it showed the greatest change magnitude under A2 scenario, while exhibited the smallest change magnitude under B1 scenario; it showed the greatest and smallest changes in ECHAM5 and CSIRO30 models, respectively; it tended to exhibit

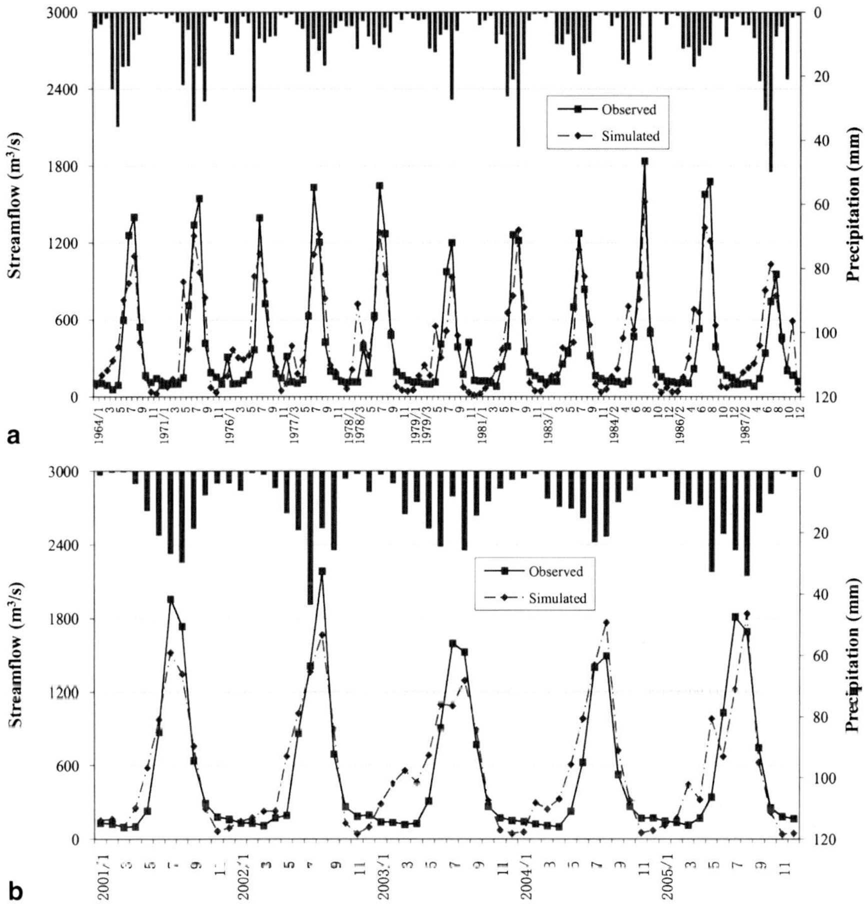

Fig. 9.20 Mean monthly hydrographs for calibration and validation periods in headwater catchment of the TRB. ("**a**" and "**b**" describe hydrographs for calibration and validation periods, respectively)

decreasing trends from June to August (Summer), while showed increasing trends from October to February of the next year (include Winter). Monthly and seasonal runoff changes in the period of 2081–2100 were different with that in the period of 2046–2065. In the period of 2081–2100, it showed increasing trends for seasonal runoff except for summer runoff, in which the winter runoff increased greatest with the magnitude of larger than 20 %; monthly runoff changed less than 10 % under most of combined climate change scenarios except that from November to January of the next year, in which monthly runoff changes were larger than 20 %; although it showed relative large change magnitude for runoff from November to January of the next year, the increased water volume was not so great because it was in low flow seasons. In both periods, it showed an obviously increasing trend for runoff in April, which might be beneficial to agricultural irrigation in spring; monthly and

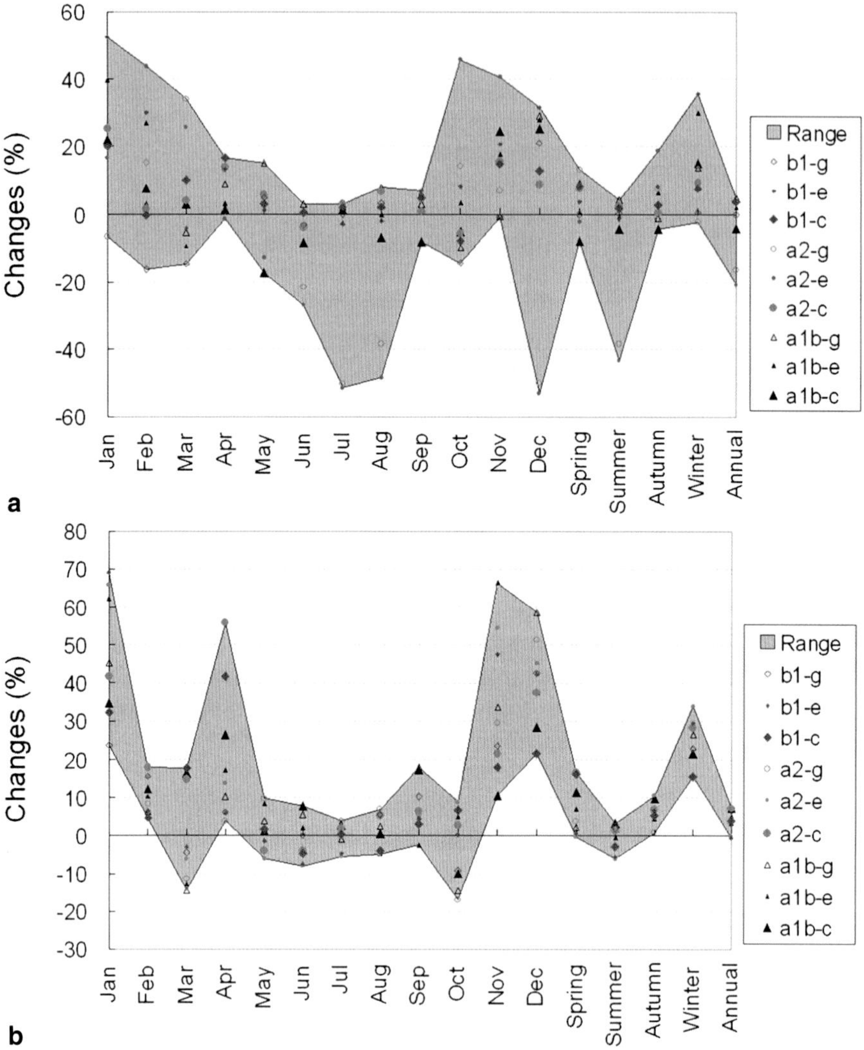

Fig. 9.21 Monthly, seasonal and annual runoff changes under climate change scenarios in the future (**a** and **b** describe periods of 2046–2065 and 2081–2100, respectively)

seasonal runoff changed greater under A2 and A1B scenarios than that under B1 scenario, while there was little difference among three GCMs.

Although it showed decreasing trends for precipitation from June to August (summer), runoff in these months was tend to increase in the period of 2081–2100 (Fig. 9.21). Besides that, it showed decreasing trends for precipitation from November to January of the next year (Fig. 9.21), in which runoff exhibited obviously increasing trends. For example, it could be seen from Fig. 9.21 that runoff increased by more than 20 % in January, while corresponded precipitation showed

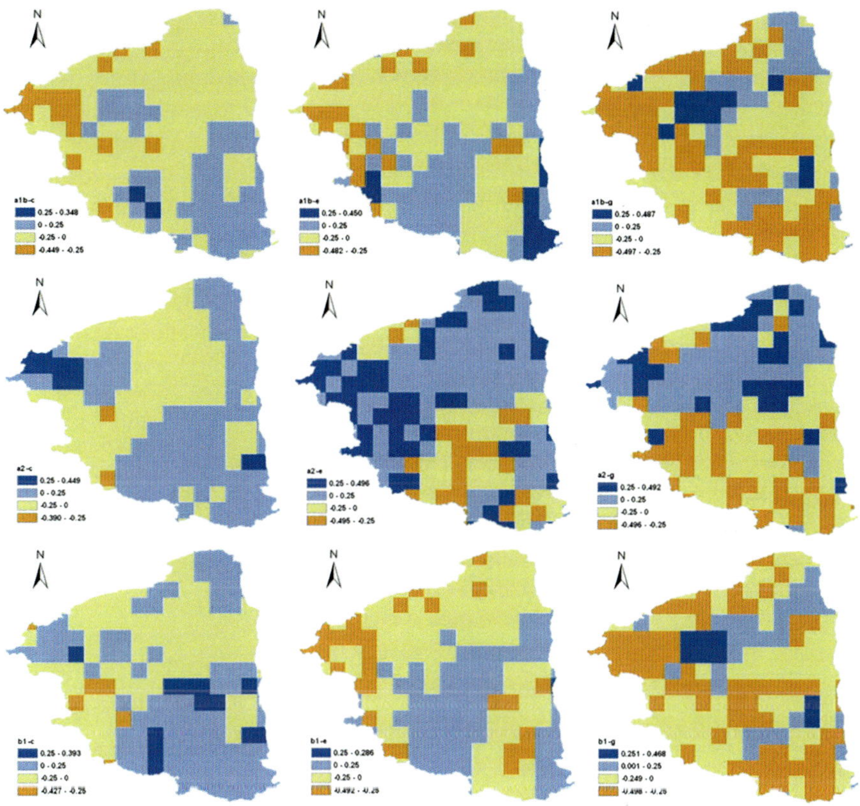

Fig. 9.22 Spatial distribution for runoff changes in the HC during the period of 2046–2065

little changes with the magnitude of smaller than 3 % and even decreased by more than 10 % under some combined climate change scenarios. That might be due to the increased snowmelt runoff resulted from the increased air temperature.

9.5.4.3 Impact of Climate Change on Spatial Distribution of Runoff

Spatial distributions of runoff changes in the HC during the period of 2046–2065, which were compared with runoff in the benchmark period (1981–2000), were shown in Fig. 9.22. Driven by climate change scenarios, runoff in the period of 2046–2065 will change from − 50 to 50 % when compared with that in the benchmark period. Runoff showed obviously decreasing trend under A1B scenario, especially under the combined GFDL21-A1B scenario with the decreasing trend area exceeded 4/5 of the entire study area, in which the area with the magnitude of decreasing trend larger than 25 % took up 1/3 of the entire region. Runoff tends to increase under A2 scenario, especially under the combined ECHAM5-A2 scenario with the increasing trend area that is close to

3/4 of the entire study area. It showed obviously increasing trend for runoff at the upstream regions of the Akesu River and the Yerqiang River, with the magnitude of increasing trend at most of regions being larger than 25 %. That was beneficial to total runoff for the HC, because runoff at the upstream regions takes up most of total runoff. Although it showed little trend for runoff under the combined CSIRO30-B1 scenario, runoff exhibited obviously decreasing trend under the combined ECHAM5-B1 and GFDL21-B1scenarios, in which spatial distributions were similar with that under A1B scenario. Overall, runoff in the HC during the period of 2046–2065 tends to decrease when compared with that in the benchmark period, because it showed more combined scenarios with decreasing trend than that for increasing trend. It was more consistent for spatial distributions of runoff under the same GCM than that under the same emission scenario. In other words, impact of different GCMs on spatial distribution of runoff might be more significant than emission scenarios. Runoff at the Akesu River catchment exhibited decreasing trend under most of climate change scenarios except for A2 scenario, while runoff at the Yerqiang River and Hetian River catchments showed increasing trend under some combined scenarios and exhibited decreasing trends under other combined scenarios.

9.5.4.4 Impact of Climate Change on Spatial Distribution of Evapotranspiration

Figure 9.23 describes spatial distribution of evapotranspiration changes in the HC during the period of 2046–2065, which were compared with runoff in the benchmark period. It showed obvious difference for evapotranspiration changes at spatial distribution. The maximum magnitude of increasing and decreasing trends was larger than 80 and 50 %, respectively, which was under the combined GFDL21-A2 scenario. The minimum magnitude of evapotranspiration changes was under the combined CSIRO-A2 scenario, in which the value changes from −2 to 18.3 %. In generally, evapotranspiration at most regions of the HC changed from −20 to 40 % under combined climate change scenarios in the period of 2046–2065, in which the area with increasing trend was larger than that with decreasing trend. It was also more consistent for spatial distributions of evapotranspiration under a GCM than that under an emission scenario. For example, when evapotranspiration was driven by climate change scenarios of GFDL21, it always showed the maximum increasing evapotranspiration at the northern region with the magnitude of increasing trend larger than 36 %, and it always showed the maximum decreasing trend at the upstream regions of the Yerqiang River and the Hetian River. It also showed the maximum increasing trend at the northern region when evapotranspiration was driven by climate change scenarios of ECHAM5, but it exhibited maximum decreasing trend at the downstream region of the Akesu River and the Yerqiang River, with the magnitude of these trends smaller than that of GFDL21. The increasing trend of evapotranspiration at the northwestern region is greater than that at other regions.

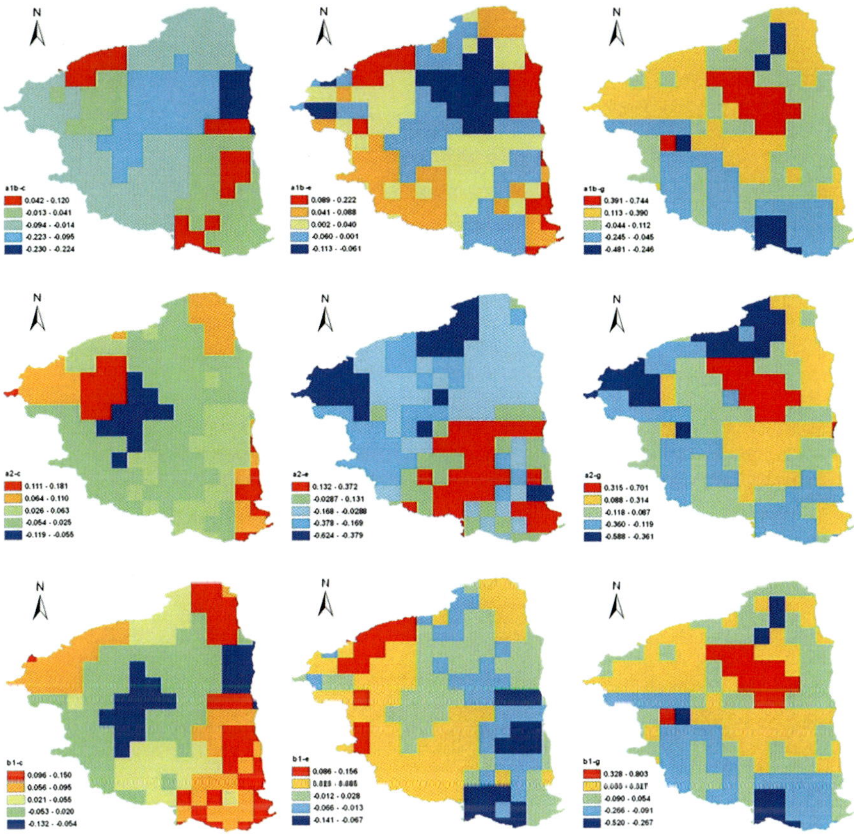

Fig. 9.23 Spatial distribution for evapotranspiration changes in the HC during the period of 2046–2065

9.5.5 *Discussion and Conclusions*

Runoff in the HC tended to exhibit a decreasing trend under climate change in the future. But it showed an increasing trend for winter runoff, which was mainly attributed to more snowmelt runoff resulted from warmer winter. Both precipitation and runoff for the HC showed obviously increasing trend in April. This might be beneficial to the spring irrigation in the study area. The spatial distribution of runoff under different GCMs was more consistent with that under different emission scenarios. It includes that the impact of different GCMs outputs on spatial distribution of runoff was more significant than emission scenarios, which was consistent with previous studies and show that uncertainties from different GCMs outputs are the most important factor in the assessment on the impact of climate change. It shows an inconsistent intra-annual distribution for the changes of precipitation and runoff in the HC, which might be explained by the increasing snowmelt runoff resulted from the increasing air temperature.

References

Brier GW (1950) Verification of forecasts expressed in terms of probability. Mon Wea Rev 78:1–3

Charles SP, Bates BC, Hughes JP (1999) A spatiotemporal model for downscaling precipitation occurrence and amounts. J Geophys Res 104(D24):31657–31669

Charles SP, Bari MA, Kitsios A, Bates BC (2007) Effect of GCM bias on downscaled precipitation and runoff projections for the serpentine catchment, Western Australia. Int J Climatol 27:1673–1690

Christensen JH, Hewitson B, Busuioc A, Chen A, Gao X, Held I, Jones R, Kolli RK, Kwon W-T, Laprise R, Magaña Rueda V, Mearns L, Menéndez CG, Räisänen J, Rinke A, Sarr A, Whetton P (2007) Regional climate projections. In: Solomon S, Qin D, Manning M, Chen Z, Marquis M, Averyt KB, Tignor M, Miller HL (eds) Climate change 2007: the physical science basis. Contribution of working group I to the fourth assessment report of the Intergovernmental Panel on Climate Change. Cambridge University Press, Cambridge

Dai XG, Li WH, Ma ZG, Wang P (2007) Water-vapor source shift of Xinjiang region during the recent twenty years. Prog Nat Sci 17(5):569–575

Fowler HJ, Blenkinsop S, Tebaldi C (2007) Linking climate modeling to impacts studies: recent advances in downscaling techniques for hydrological modeling. Int J Climatol 27:1547–1578

Fu GB, Charles SP (2007) Multi-model climate change scenarios for southwest Western Australia and potential impacts on streamflow. In: Eva B et al (eds) Quantification and reduction of predictive uncertainty for sustainable water resources management. IAHS Press, Oxfordshire: UK pp 16–24

Giorgi F, Francisco R (2000) Evaluating uncertainties in the prediction of regional climate change. Geophys Res Lett 27(9):1295–1298

Giorgi F, Mearns LO (2002) Calculation of average, uncertainty range, and reliability of regional climate changes from AOGCM simulations via the "reliability ensemble averaging" (REA) method. J Climate 15(10):1141–1158

Gleick PH (1986) Methods for evaluating the regional hydrologic impacts of global climatic change. J Hydrol 88:97–116

Hughes JP, Guttorp P (1994) Incorporating spatial dependence and atmospheric data in a model of precipitation. J Appl Meteorol Climatol 33(12):1503–1515

Jiang T, Chen YQ, Xu CY, Chen XH, Chen X, Singh VP (2007) Comparison of hydrological impacts of climate change simulated by six hydrological models in the Dongjiang Basin, South China. J Hydrol 336:316–333

Karl TR, Wang WC, Schlesinger ME, Knight RW, Portman D (1990) A method of relating general circulation model simulated climate to observed local climate. Part I: seasonal statistics. J Climate 3:1053–1079

Kirshner S (2005, March) Modeling of multivariate time series using hidden Markov models. PhD thesis. University of California, Irvine

Kittel TGF, Giorgi F, Meehl GA (1998) Intercomparison of regional biases and doubled CO2-sensitivity of coupled atmosphere-ocean general circulation model experiments. Clim Dynam 14:1–15

Lambert SJ, Boer GJ (2001) CMIP1 evaluation and intercomparison of coupled climate models. Clim Dynam 17:83–106

Liang X, Lettenmaier DP, Wood EF (1996) One-dimensional statistical dynamic representation of subgrid spatial variability of precipitation in the two-layer variable infiltration capacity model. J Geophys Res 101:21403–21422

Liang X, Xie ZH (2001) A new surface runoff parameterization with subgrid-scale soil heterogeneity for land surface models. Adv Water Resour 24:1173–1193

Liu ZF, Xu ZX, Huang JX, Charles SP, Fu GB (2010) Impacts of climate change on hydrological processes in headwater catchment of the Tarim River Basin. Hydrol Process 24:196–208

Liu LL, Liu ZF, Ren XY, Fischer T, Xu Y (2011) Hydrological impacts of climate change in the Yellow River Basin for the 21st century using hydrological model and statistical downscaling model. Quatern Int 244(2):211–220

McCarthy JJ, Canziani OF, Leary NA, Dokken DJ, White DS (2001) Climate change 2001: impacts, adaptation, and vulnerability. Cambridge University Press, Cambridge: UK p 155

Meng DJ, Mo XG (2012) Assessing the effect of climate change on mean annual runoff in the Songhua River basin, China. Hydrol Process 26(7):1050–1061

Murphy JM, Sexton DMH, Barnett DN, Jones GS, Webb MJ, Collins M, Stainforth DA (2004) Quantification of modelling uncertainties in a large ensemble of climate change simulations. Nature 430:768–772

Nash JE, Sutcliffe JV (1970) River flow forecasting through conceptual models part 1: a discussion of principles. J Hydrol 10:282–290

Perkins SE, Pitman AJ, Holbrook NJ, Mcaneney J (2007) Evaluation of the AR4 climate models' simulated daily maximum temperature, minimum temperature, and precipitation over Australia using probability density functions. J Climate 20:4356–4376

Prudhomme C, Davies H (2009) Assessing uncertainties in climate change impact analyses on the river flow regimes in the UK. Part 1: baseline climate. Clim Change 93:177–195

Qian WH, Qin A (2008) Precipitation division and climate shift in China from 1960 to 2000. Theor Appl Climatol 93:1–17

Randall DA, Wood RA, Bony S, Colman R, Fichefet T, Fyfe J, Kattsov V, Pitman A, Shukla J, Srinivasan J, Stouffer RJ, Sumi A, Taylor KE (2007) Cilmate models and their evaluation. In: Solomon S, Qin D, Manning M, Chen Z, Marquis M, Averyt KB, Tignor M, Miller HL (eds) Climate change 2007: the physical science basis. Contribution of working group I to the fourth assessment report of the Intergovernmental Panel on Climate Change. Cambridge University Press, Cambridge

Schubert S, Sellers AH (1997) A statistical model to downscale local daily temperature extremes from synoptic-scale atmospheric circulation patterns in the Australian region. Clim Dynam 13:223–234

Tang QC (1990) Formation and transformation of runoff in arid regions of china. J Nat Resour 5(1):1–10

Teegavarapu RSV (2010) Modeling climate change uncertainties in water resources management models. Environ Modell Softw 25:1261–1265

Timbal B, Dufour A, McAvaney B (2003) An estimate of future climate change for western France using a statistical downscaling technique. Clim Dynam 20:807–823

Trenberth KE (1997) The use and abuse of climate models. Nature 386(6621):131–133

Wilby RL, Dawson CW, Barrow EM (2002) SDSM-A decision support tool for the assessment of regional climate change impacts. Environ Modell Softw 17:147–159

Wilby RL, Charles SP, Zorita E, Timbal B, Whetton P, Mearns LO (2004) Guidelines for use of climate scenarios developed from statistical downscaling methods. Supporting material of the Intergovernmental Panel on Climate Change (IPCC), prepared on behalf of Task Group on data and scenario support for Impacts and Climate Analysis (TGICA). http://www.ipcc-data.org/guidelines/dgm_no2_v1_09_2004.pdf

Wilks DS, Wilby RL (1999) The weather generation game: a review of stochastic weather models. Prog Phys Geogr 23:329–357

Wood AW, Leung LR, Sridhar V, Lettenmaier DP (2004) Hydrologic implications of dynamical and statistical approaches to downscaling climate model outputs. Clim Change 62:189–216

Xu CY (1999) From GCMs to river flow: a review of downscaling techniques and hydrologic modeling approaches. Prog Phys Geogr 23(2):229–249

Ye XC, Zhang Q, Bai L, Hu Q (2011) A modeling study of catchment discharge to Poyang lake under future climate in China. Quatern Int 244(2):221–229

Zeng XF, Kundzewicz ZW, Zhou JZ, Su BD (2011) Discharge projection in the Yangtze River basin under different emission scenarios based on the artificial neural networks. Quatern Int 282:113–121

Chapter 10
Changes in Extreme Hydrological Events

Huaijun Wang, Yaning Chen and Guili Sun

Abstract There has been a paucity of information on trends in hydrological extremes. Based on daily precipitation and temperature and streamflow records in the arid region, northwest of China, the temporal and spatial characteristics of hydrological extremes were analyzed during 1960–2010. The results showed that: (1) The temperature extremes show patterns consistent with warming, with a large proportion of stations showing statistically significant trends. Warming trends in indices derived from daily minimum temperature are of greater magnitudes than those from maximum temperature. The decreases in frequency for cold extremes mainly occur in summer and autumn, while warm extremes show significant increases in frequency in autumn and winter; (2) Most precipitation indices exhibited increasing trends across the region, and the precipitation increase in the arid region was due to the increase in both precipitation frequency and intensity; (3) There were increasing trends for frequency of extreme hydrological events from 1901 to 2010, particularly post 1970s. Extreme hydrological events occurred mainly during summer. Based on the results obtained in this study, the frequency, magnitude, and intensity of extreme hydrological events are expected to increase over time; (4) Frequency distribution exhibited a general symmetry along the Tianshan Mountains, distributed evenly in Junggar Basin and Tarim Basin. More extremes were observed in the north-west than that in the southeast. The maximum incidence was in west Tianshan Mountains and generally decreased south-eastward.

Keywords Temperature extremes · Precipitation extremes · Extreme hydrological events · Analysis of frequency · Magnitude · Intensity

---

Y. Chen (✉) · H. Wang
State Key Laboratory of Desert and Oasis Ecology, Xinjiang Institute of Ecology and Geography, Chinese Academy of Sciences, No. 818 South Beijing Road, 830011, Urumqi, Xinjiang, China
e-mail: chenyn@ms.xjb.ac.cn

H. Wang
e-mail: wanghj@ms.xjb.ac.cn

G. Sun
College of Forestry and Horticulture, Xinjiang Agricultural University, No. 311 East Nongda Road, 830052, Urumqi, China
e-mail: sxfgl@126.com

Y. Chen (ed.), *Water Resources Research in Northwest China*,
DOI 10.1007/978-94-017-8017-9_10,

The fourth assessment report of the Intergovernmental Panel on Climate Change (IPCC) (IPCC 2007) stated that, by the end of this century, the global mean temperature is expected to rise about 0.74 °C. In the context of global warming, damage caused by extreme natural hazards has increased worldwide. For example, of the total human deaths in China each year, 75 % are caused by extreme natural disasters, as are 40 % of total economic losses. In addition, variations and trends of extreme climate events have recently received increased attention, as such events are more sensitive to climate change than their mean values (Alexander et al. 2007; Tank and Konnen 2003; Williams et al. 2010).

Changes in extreme climatic events are among the most serious challenges to mankind today due to their significant broad-spectrum impacts (Manton 2010) on the economy (Linnenluecke and Griffiths 2010), human health (Rocklov and Forsberg 2009), animals (Welbergen et al. 2008), and the natural ecosystem (Mantua et al. 2010; Parker et al. 2008). Assessing changes in extreme climate events at the regional scale can identify indicators that cause environmental and other problems and help us gain positive information in order to formulate rational countermeasures. Extreme precipitation events can influence flood and soil erosion differently with changes in frequency versus intensity, while for temperature extremes, economic losses may be incurred. For example, the heat waves that occurred in the summer of 2003 resulted in over 22,000 heat-related deaths across Europe (Schar and Jendritzky 2004), and the unprecedented freezing in January in Southern China caused heavy losses in the national economy (Ding et al. 2008). Generally, in China, cold waves are common in northern regions in the winter, hot days and heat waves are common in southern areas in the summer, and the arid and semi-arid regions of Northwest China often suffer drought (Wang et al. 2011).

Temperature extremes have been studied recently in many countries and regions around the world, such as Italy (Toreti and Desiato 2008), Uruguay (Rusticucci and Renom 2008), South America (Vincent et al. 2005), Southern and west Africa (New et al. 2006), and Greece (Kioutsioukis et al. 2010). Global temperature extremes and changes have also been studied recently (Alexander et al. 2006; Frich et al. 2002). The studies concluded that widespread significant changes in temperature extremes were associated with warming, especially for those indices derived from daily minimum temperatures. Recent studies on changes in temperature extremes in China suggested decreasing trends in cold extremes but increasing trends in warm extremes (Wang et al. 2011).

Precipitation extremes have also been studied in South Asia (Endo et al. 2009), Emilia-Romagna (Pavan et al. 2008), India (Pal and Al-Tabbaa 2011), Greece (Kioutsioukis et al. 2010), South and West Africa (New et al. 2006), the Indo-Pacific (Caesar et al. 2011), South Portugal (Durao et al. 2010), northern central Italy (Di Baldassarre et al. 2006) and Germany (Tromel and Schonwiese 2007). As well, global changes in daily climate extremes have been analyzed (Alexander et al. 2006; Frich et al. 2002; Kiktev et al. 2003). These studies concluded that changes in precipitation extremes present a mixed pattern and regionally dependent variations. For example, significant downward trends in extreme precipitation events have been found in some areas of Germany but no such trend occurs in South and West

Africa. In China, Wang and Zhou (2005) found that annual extreme precipitation events decreased in central, north, and northeast China but increased in the east and the arid Northwest, with the latter region also experiencing significant increasing trends in extreme summertime precipitation events.

You et al. (2011) comprehensively analyzed the characteristics of daily precipitation indices at 303 meteorological stations in China. Their findings showed that regional annual total precipitation exhibited an increasing trend and that most precipitation indices were strongly correlated with annual total precipitation. Additional regional studies include those done in the Karsts region of Guizhou province (Zhang et al. 2010), the Zhujiang River Basin (Gemmer et al. 2011), the Yangtze River Basin (Su et al. 2006), the eastern and central Tibetan Plateau (You et al. 2008), and the Loess Plateau (Li et al. 2010). A precipitation-based index indicated a more mixed pattern of change and few stations revealed any significant trends.

Climate change appears to be altering the distribution of extreme hydrological events. Although occurring infrequently, these events are often sudden and have a major impact on human society and the natural environment (Chebana et al. 2010; Committee 2007). In the arid areas of Northwest China, such events include hailstorms, floods (from rainstorms, snowmelts, icicles, and glacier lake outbursts), landslides, debris flows (storm and glacial), and snowfalls (blizzards and avalanches) (Bell et al. 2004; Reiser and Kutiel 2010; Strom and Korup 2006). Of particular concern are areas such as northwestern Xinjiang, which have experienced severe damage due to extreme flooding and snow events (Sun et al. 2012, 2013).

It is commonly believed that extreme hydrological events have significantly increased, both in scope and frequency (Shi et al. 2007). Exploring changing characteristics of hydrological extreme events in the arid region of China is a prerequisite for a good assessment of climate change impacts on the region's ecological environment and agricultural development. In this section, extreme hydrological events are divided into two categories: climate extremes and hydrological hazards (including hailstorms, floods, landslides, snowfalls, etc.). The remainder of the paper is organized as follows. Section 10.2 presents changes in temperature and precipitation extremes, focusing mainly on research by Wang et al. (2012), Wang et al. (2013a), and Wang et al. (2013b). In Sect. 10.3, changes in hydrological extreme events are discussed, based primarily on Sun et al.'s (2012, 2013) studies. Finally, in Sect. 10.4, the connection between hydrological extremes and climate is investigated.

10.1 Changes in Climate Extremes

10.1.1 Study Area and Methods

The arid region of Northwest China is situated from 34° to 50° N and from 73° to 108° E, covering a vast area measuring 2.53 million km^2 (Fig. 10.1). Within this zone lie numerous basins west of the Helanshan Mountains in the Ningxia

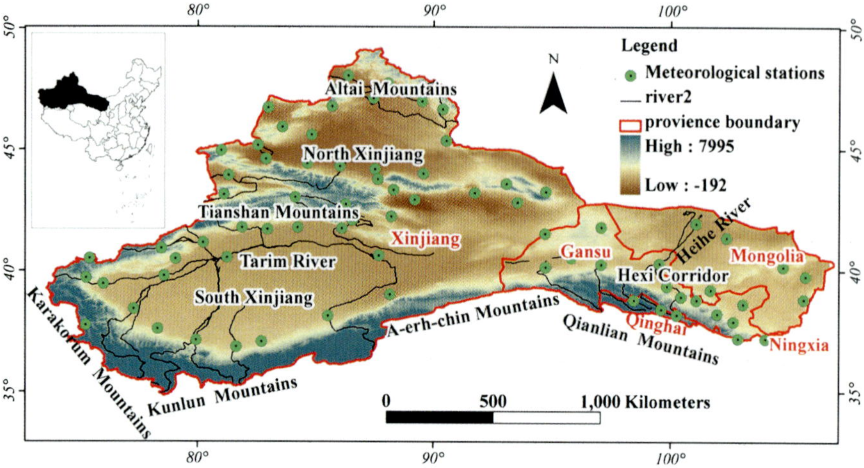

Fig. 10.1 Location of the arid region and meteorological stations

Hui Autonomous Region, and west of the Ushaoling Mountains in Gansu Province. Of these, there are many endorheic drainage basins including the Tarim Basin, the Tsaidam Basin, the Badanjilin Desert, and the Tengger Desert, as well as other endorheic drainage basins in the northern Xinjiang Uygur Autonomous Regions, the Alashangqi Desert in western Inner Mongolia, and the Hexi Corridor Gobi Desert in Gansu Province. The climate of the arid region is typical of inner-continental land masses, with a wide temperature range, low precipitation, and low humidity. The climate is dominated by continental arid conditions, with lesser effects from the East Asian monsoons (Liu et al. 2010a).

Data concerning daily, monthly and yearly precipitation and temperature of China's arid Northwest region, from January 1, 1960 to February 28, 2011, were provided by the National Climate Center (NCC) of the China Meteorological Administration (CMA). For this area, 76 stations passed the internal homogeneity check of the China National Meteorological Center (CNMC), including the moving t test (Peterson et al. 1998), standard normal homogeneity test (Alexandersson 1986), and departure accumulating method (Buishand 1982). Data were analyzed using the RclimDex package (soft and documentation available for download from http://cccma.seos.uvic.ca/ETCCDMI). Table 10.1 lists the indices used in this study. It should be noted that some of the indices recommended by ETCCDMI are not relevant for China's arid region. For example, R20 (days above 20 mm) is not considered here, since the index R10 (days above 10 mm) is deemed sufficient for our study. R99p (extremely wet days) was also not included in the analysis because there were too many zero readings for this index in the arid region. Based on the CMA classification, a 24-hour period is considered a rainy day if precipitation of at least 0.1 mm day^{-1} can be measured in it.

Data are analyzed using the RclimDex package (soft and documentation available for download from http://cccma.seos.uvic.ca/ETCCDMI, which can calculate

Table 10.1 Definitions of 15 temperature indices and 9 indices used in this study, all the indices are calculated by RClimDEX

ID	Indicator name	Definitions	UNITS
Cold extremes			
FD0	Frost days	Annual count when TN (daily minimum) < 0 °C	Days
ID0	Ice days	Annual count when TX (daily maximum) < 0 °C	Days
TN10p	Cool nights	Percentage of days when TN < 10th percentile	Days
TX10p	Cool days	Percentage of days when TX < 10th percentile	Days
TNn	Coldest nights	Monthly minimum value of daily minimum temp	°C
TXn	Coldest days	Monthly minimum value of daily maximum temp	°C
CSDI	Cold spell duration indicator	Annual count of days with at least 6 consecutive days when TN < 10th percentile	Days
Warm extremes			
TN90p	Warm nights	Percentage of days when TN > 90th percentile	Days
TX90p	Warm days	Percentage of days when TX > 90th percentile	Days
TNx	Warmest nights	Monthly maximum value of daily minimum temp	°C
TXx	Warmest days	Monthly maximum value of daily maximum temp	°C
SU25	Summer days	Annual count when TX(daily maximum) > 25 °C	Days
TR20	Tropical nights	Annual count when TN(daily minimum) >20 °C	Days
WSDI	Warm spell duration indicator	Annual count of days with at least 6 consecutive days when TX > 90th percentile	Days
Variability extremes			
DTR	Diurnal temperature range	Monthly mean difference between TX and TN	°C
Precipitation extremes			
RX1day	Max 1-day precipitation amount	Monthly maximum 1-day precipitation	Mm
Rx5day	Max 5-day precipitation amount	Monthly maximum consecutive 5-day precipitation	Mm
SDII	Simple daily intensity index	Annual total precipitation divided by the number of wet days (defined as PRCP ≥ 1.0 mm) in the year	Mm/day
R10	Number of heavy precipitation days	Annual count of days when PRCP ≥ 10 mm	Days
R0.1	Number of days above nn mm	Annual count of days when PRCP ≥ 0.1 mm	Days
CDD	Consecutive dry days	Maximum number of consecutive days with RR < 1 mm	Days
CWD	Consecutive wet days	Maximum number of consecutive days with RR ≥ 1 mm	Days
R95p	Very wet days	Annual total PRCP when RR > 95th percentile	Mm
PRCPTOT	Annual total wet-day precipitation	Annual total PRCP in wet days (RR ≥ 1 mm)	mm

16 temperature extremes and 11 precipitation indices based on daily maximum and minimum temperatures. In this paper, we select only 15 temperature indices and 9 precipitation indices for analysis (Table 10.1). This paper also analyses seasonal values (ANN for annual change, DJF for winter, MAM for spring, and JJA and SON for summer and autumn, respectively) of some indices, such as absolute indices (TNn, TNx, TXn, TXx), percentile temperature indices (TN10p, TN90p, TX10p, TX90p) and DTR (whenever possible). Regional averages were calculated as an arithmetic mean of values at all stations in the study.

In this section, we use a nonparametric Kendall's tau-based slope estimator (Sen 1968), but statistical significance for the trends in extreme climate indices is performed using the Mann–Kendall (M–K) test (Mann 1945; Kendall 1975). A trend was considered to be statistically significant if it is significant at the 5 % level. The results of the M–K test are heavily affected by serial correlation of the time serial correlation on M–K results. We adopt the Yue and Pilon method by using the R package "ZYP" to remove lag-1 autocorrelation (Yue et al. 2002).

10.1.2 Cold Extremes

The spatial distribution pattern of the temporal trends in cold extremes is shown in Figs. 10.2, 10.3 demonstrates the regional annual and seasonal trends for temperature indices from 1960 to 2010, and Table 10.2 shows the proportion of stations with negative, insignificant and positive trends for yearly temperature extremes.

For cold nights (TN10) and cold days (TX10), about 96 and 61.8 % of stations had significant decreasing trends. The regional trend magnitudes for these two indices were − 1.89 and − 0.89 days/decade, respectively. The temperature of the coldest nights (TNn) and coldest days (TXn) had also statistically significant increases at approximately 60 and 10 % of stations, respectively. Stations situated in northern and central Xinjiang (along the Tianshans) showed the largest changes, and regional trend magnitudes were 0.65 and 0.36 °C/decade, respectively. The number of frost days (FD0) had decreased, with 90 % of stations showing 0.05 significant levels. Stations with larger trend magnitudes were again distributed in central Xinjiang (along the Tianshans). For ice days (ID0), about 38 % of the stations showed significant decreased trends, and stations in the Qilian Mountains (southeastern arid areas) seemed to show more statistically significant changes. The number of cold spell duration indicators (CSDI) (not shown) decreased at a rate of − 0.66 days/decade, while only 20 % of stations showed significant changes.

Table 10.3 illustrates the proportion of stations where trends in indices are of a particular relative magnitude. About 83 % of stations show larger trend magnitudes in TNn than in TXn. For FD0 and ID0, 79 % of stations have greater trend magnitudes in FD0. About 96 % of stations show larger trend magnitudes in TN10p than TX10p. From these data, we can see that cold extremes derived from minimum temperatures have a larger change than those derived from maximum temperatures.

The seasonal variations of cold nights (TN10p) are displayed in Fig. 10.4. It is apparent that the greatest contribution to cold extremes was from SON (Fig. 10.3).

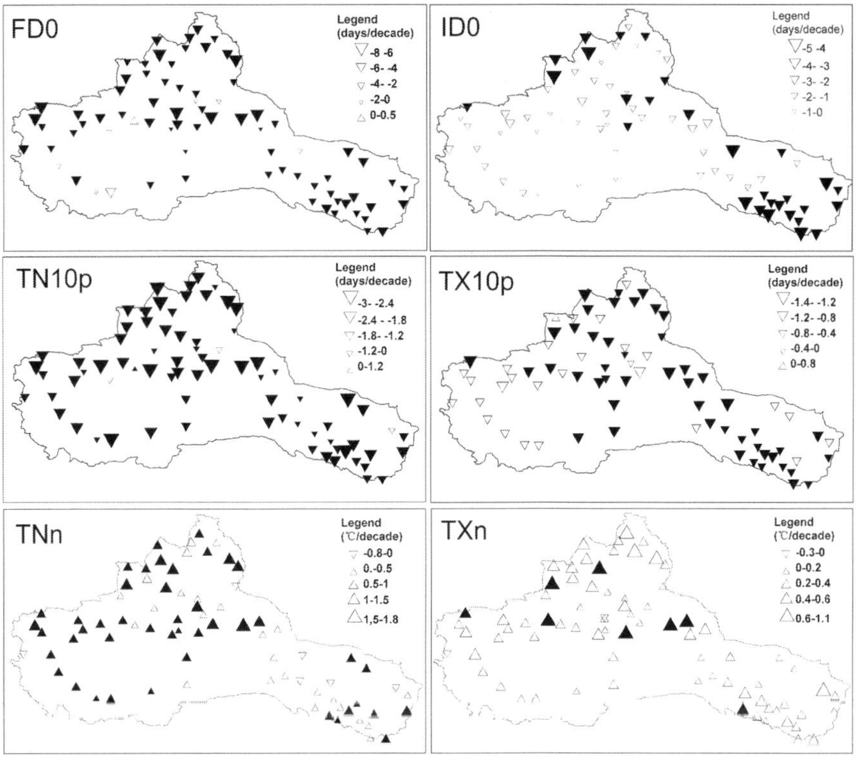

Fig. 10.2 Spatial patterns of trends per decade during 1960–2010 in the arid region of China of cold extremes (FD0, ID0, TN10p, TX10p, TNn and TXn); *Upward*-pointing (*downward* pointing) *triangles* indicate increasing (decreasing) trends

With coldest nights (TNn), for example, there was an increase of 0.83 days/decade in autumn for the arid region, far more than the increase in any other season. However, the number of stations showing a more statistically significant trend was likely to occur in JJA. For instance, cold nights (TN10p) (Fig. 10.4) in MAM, JJA, SON and DJF had 34, 84, 53 and 40 %, respectively, of stations showing significant decreasing trends. Seasonal cold extremes derived from minimum temperatures also showed larger magnitude trends than those derived from maximum temperatures (Table 10.3).

When comparing probability density functions (PDFs) within different periods, data from all stations were used. Of these, station indices of ID0, FD0, TX10p and TN10p were used to calculate PDFs (Fig. 10.5). Results showed that the distribution function of cold extremes have changed over the past 50 years, which was in accordance with the trend analysis; furthermore, cold extremes moved to the left, especially after 2000. For example, the number of frost day (FD0) in the 1960s, 1970s, 1980s and 1990s was 175d, 173d, 171d and 169d, respectively, with the reduced rate being relatively small. After 2000, the inter-annual slope reach the maximum absolute value and the number of frost days quickly dropped to 158d.

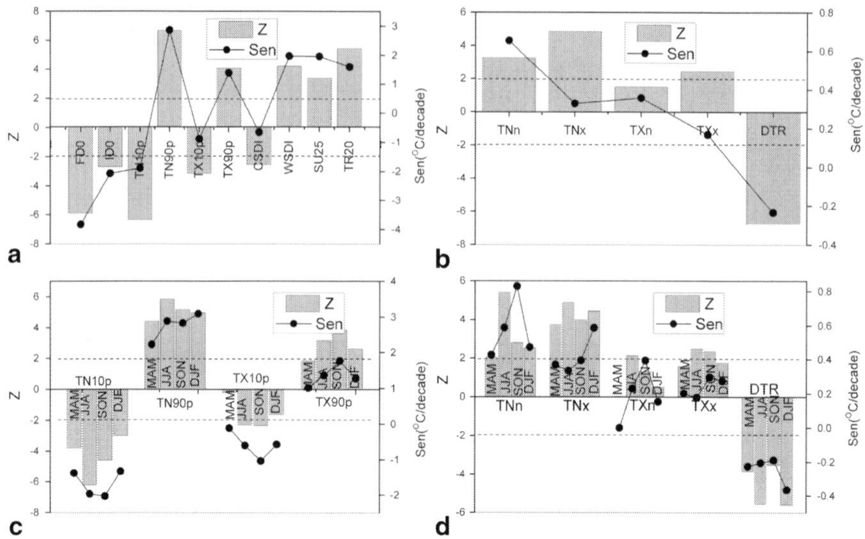

Fig. 10.3 Regional annual and seasonal trends for temperature indices; The dot line is the 95 % confidence level for Mann–Kendall test (*a*, *b* annual change, *c*, *d* seasonal change)

Power spectrum analysis was used to find a credible period for TX10p and ID0 by continuous wavelet transform (Fig. 10.6). The ID0 have quasi-three-year and quasi-two-year periods within the timeframes of 1965–1970 and 1973–1976, respectively. A quasi-eight-year period was also observed from 1985 to 2002, but this period was not significant. The inter-annual variance of ID0 was above the moving average during the periods 1961–1979, 1982–1984 and 1990–1996, which indicated that Tx10p was on the high side prior to 1996. The curve was below its moving average after 2000, which is consistent with the probability density distribution's weak decreasing trend. The inter-annual variance curve of the TX10p was relatively simple, showing a high side from 1961 to 1987 and a low side after 1987. Overall, the number of TX10p continued to decline, and a three distinct quasi-periods were observed for this index.

10.1.3 Warm Extremes

Figure 10.7 shows changes in warm extremes from 1960 to 2010, while Fig. 10.3 demonstrates regional trends for warm extremes. The number of stations with negative, insignificant, and positive trends is listed in Table 10.2. Of these stations, 96 and 92 % showed statistically significant trends for warm nights (TN90) and warm days (TX90), respectively, and regional magnitudes for these two indices were 2.85 and 1.37 days/decade, respectively. The warmest days (TXx) and warmest nights (TNx) also reveal an increasing trend, but the spatial patterns were less consistent, especially for the warmest days. About 62 % of stations had significant increasing

Table 10.2 Percentage of stations showing significant annual trends for each index

	TN10p	TN90p	TX10p	TX90p	TNn	TNx	TXn	TXx	DTR	FD0	ID0	CSDI	SU25	TR20	WSDI
SP	1.3	96.1	0	91.8	59.2	61.8	10.5	28.9	2.63	0	0	0	69.7	60.5	69.7
ISP	0	1.3	1.3	8.2	32.9	31.6	88.2	55.3	7.9	1.3	1.3	1.3	21.1	9.2	28.9
ISN	2.6	1.3	36.8	0	7.9	3.9	1.3	11.8	14.5	9.2	60.5	78.9	1.3	5.3	1.3
SN	96.1	1.3	61.8	0	0	2.6	0	2.63	75	89.5	38.2	19.7	2.6	2.6	0

SP significant positive trend, *ISP* insignificant positive trend, *ISN* insignificant negative trend, *SN* significant negative trend

Table 10.3 Proportion of individual stations where the trend in one index if of greater magnitudes than the trend in a second

Index	Comparison	Proportion				
		ANN	MAM	JJA	SON	DJF
TX90>TX10	abs	0.89	0.87	0.89	0.84	0.99
TN90>TN10	abs	0.89	0.87	0.79	0.66	0.96
TXn>TXx	rel	0.78	0.26	0.70	0.57	0.34
TNn>TNx	rel	0.74	0.58	0.86	0.91	0.36
TNx>TXx	rel	0.78	0.79	0.76	0.46	0.88
TNn>TXn	rel	0.83	0.91	0.86	0.83	0.79
FD0>ID0	abs	0.79				
TN90>TX90	abs	0.87	0.86	0.83	0.66	0.91
TN10>TX90	abs	0.71	0.63	0.67	0.55	0.51
TN10>TX10	abs	0.96	0.92	0.95	0.96	0.93
TN90>TX10	abs	0.97	0.96	0.95	0.89	0.99

abs indicates that the absolute magnitudes of trends are compared, *rel* indicates that the sings of trends are retained during comparison

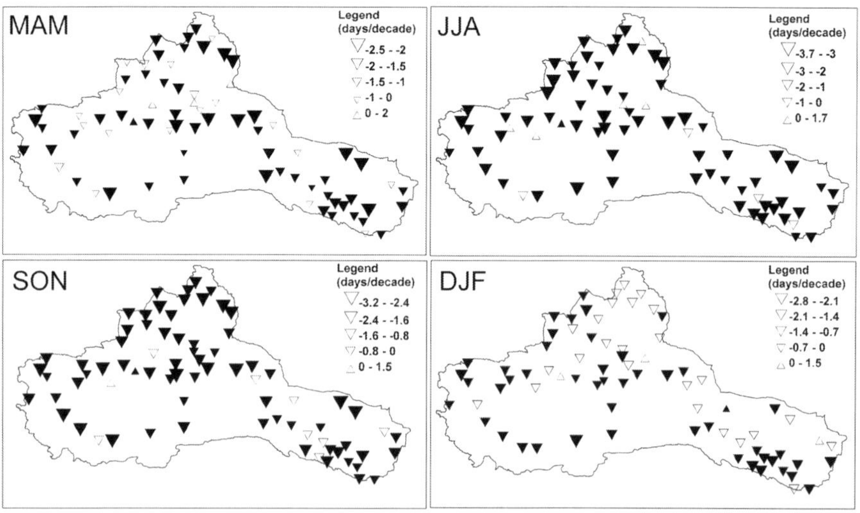

Fig. 10.4 Spatial patterns of trends per decade during 1960–2010 in the arid region of China of the seasonal occurrence of cold nights (TN10p)

trends for TNx, but only 29 % of stations showed statistically significant trends for TXx. For the summer days (SU25) and tropical nights (TR20) (not shown), about 53 and 46 % of the stations showed statistically significant increasing trends. The regional trend magnitudes SU25 and TR20 were 1.96 and 1.59 days/decade, respectively. The warm spell duration indicator (WSDI) (not shown) also increased, with a regional average increase of 1.97 days/decade. Stations showing significant trends were distributed in southern Xinjiang and in the western Hexi Corridor.

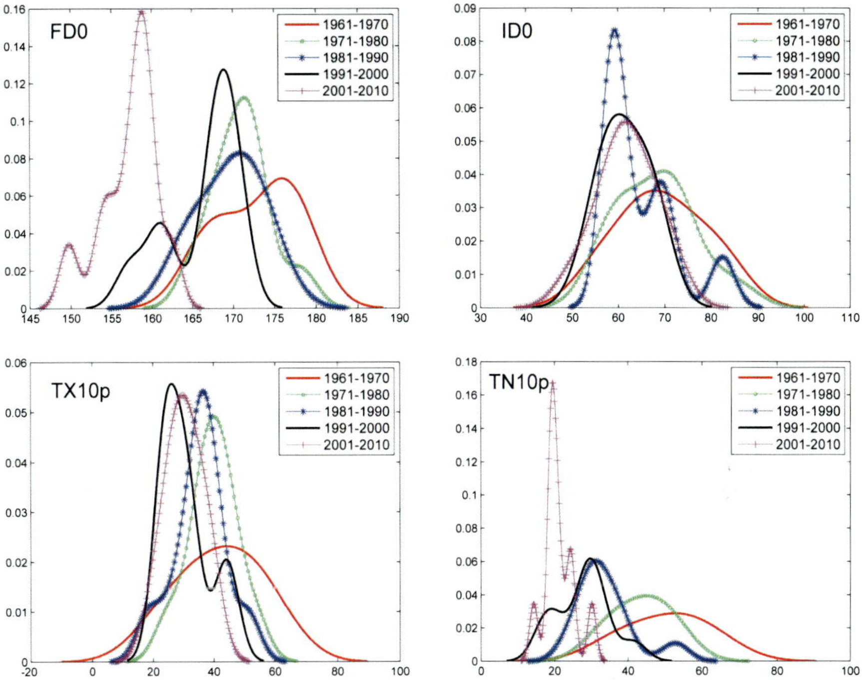

Fig. 10.5 The probability distribution of FD0, ID0, TX10p and TN10p, respectively

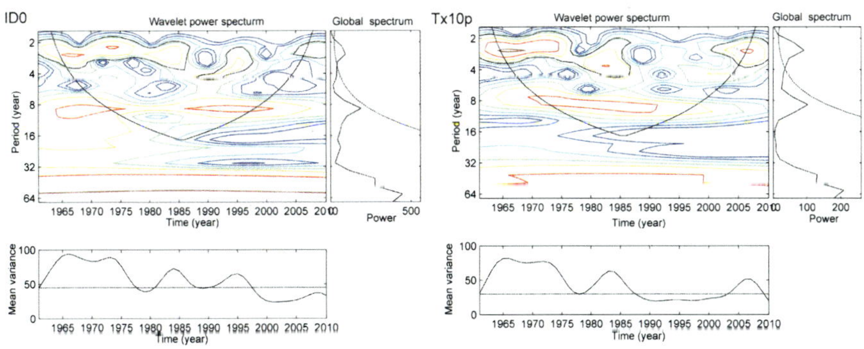

Fig. 10.6 Power spectrum of ID0 and TX10 (surrounded by the *black line* in Figure is significant at 0.05 level. The *red*-alignment is the high energy spectrum)

Figure 10.8 shows the spatial distribution of seasonal trend magnitudes of warm nights (TN90p). On the whole, seasonal warm extremes had increasing trends. For TN90p and TNx, the regional trends in DJF showed the largest trend magnitudes compared with other seasons. For TX10p and TXx, SON showed the largest regional trend magnitudes. The patterns of numbers of stations that were most statistically significant were similar to the region trend (not shown). For TN90p, TNx, TX90p

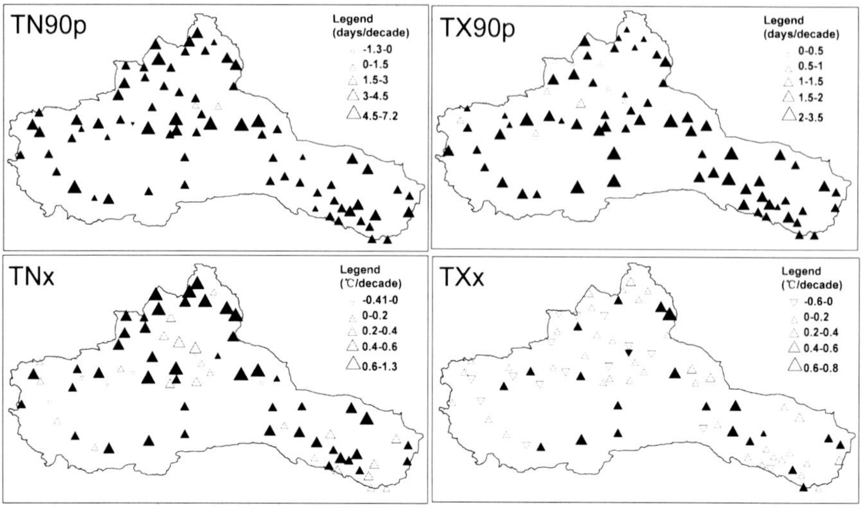

Fig. 10.7 Spatial patterns of trends per decade during 1960–2010 in the arid region of China of the warm extremes

and TXx, the largest proportion of stations with positive seasonal trends was 93 % (DJF), 76 % (DJF), 84 % (SON), and 46 % (SON), respectively. Like cold extremes, seasonal warm extremes derived from minimum flow had greater magnitudes compared with those from maximum flow (Table 10.3).

The PDFs of TX90p and TN90p wre calculated and the results shown in Fig. 10.9. Here, we can see that the PDFs of warm extremes moved to the right. For example, the minimum number of days of TX90p in the 1960s and 1970s was 31 and 30d. This number increased to 32 and 34d in the 1980s and1990s, and then, after 2000, increased rapidly to 55d. These changes indicate that climate was very different before and after 2000. The probability density distributions of these indices were significantly different during the five time periods, with notable shifts in the distribution. Overall, the temperature indices showed the most marked shifts toward fewer cold nights and more warm nights.

The continuous wavelet analysis of warm days (TX90p) was also discussed (not shown). From 1994 to 2004 and from 1990 to 2001, warm days (TX90p) have a 3–5 year period and a quasi-9-year period, although this latter period is not significant. The inter-annual variability of series variance showed more warm days on the high side during 1991–2005, while during 1961–1991 and 2007–2010, there were fewer warm days (TX90p). Overall, there were fewer than normal warm days before 1996, and this index significantly increased after 1996.

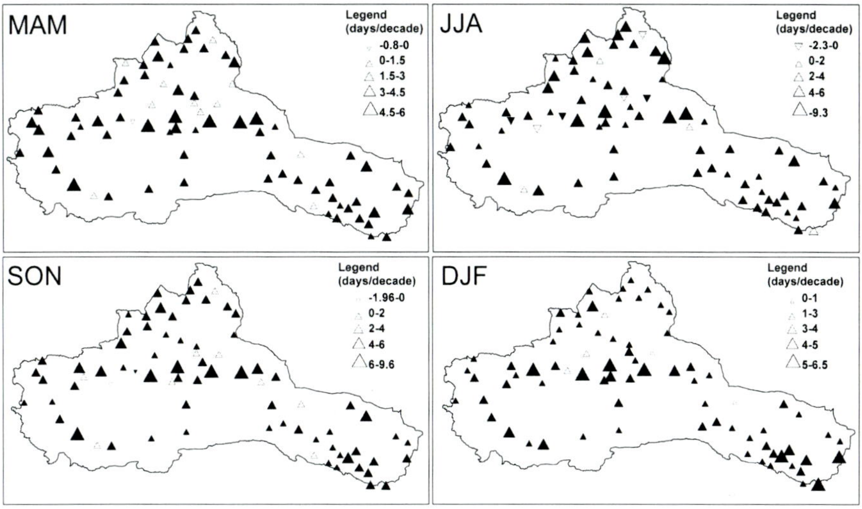

Fig. 10.8 Spatial patterns of trends per decade during 1960–2010 in the arid region of China of but for the seasonal TN90p

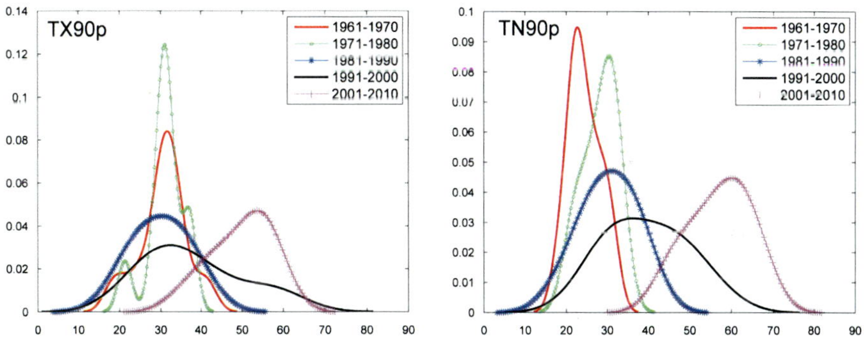

Fig. 10.9 Probability function of warm extremes (TX10p and TN90p, respectively)

10.1.4 DTR

The annual and seasonal magnitudes trends of DTR are shown in Fig. 10.10. Seventy-five percent of stations showed a significant decreasing trend at annual scale (Table 10.2), and the regional trend magnitude in DTR was −0.23 °C/decade. Stations with larger trend magnitudes were once again located along the Tianshan Mountains, especially for winter. DJF showed the most significant change, with a regional magnitude of −0.37 °C/decade. About 54, 61, 47 and 68 % of the stations in MAM, JJA, SON, and DJF showed significant decrease trends, respectively. The decrease in DTR results from the minimum temperature increasing faster than the

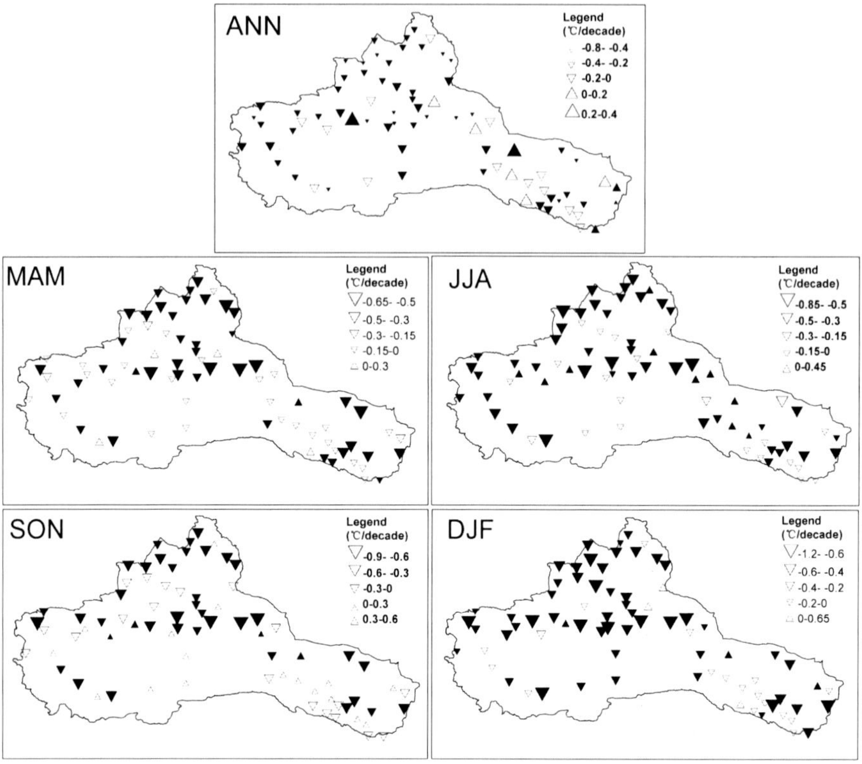

Fig. 10.10 Spatial patterns of trends per decade during 1960–2010 in the arid region of China of but for the annual and seasonal diurnal temperature range (DTR)

maximum temperature. The decrease in DTR might be mainly due to the increase in vapor and aerosols in the air, which reduces daytime incoming solar radiation and also nighttime outgoing longwave radiation from the land surface, resulting in a higher minimum temperature (Shen et al. 2010).

10.1.5 Comparison of the Temperature Extreme Indices

It is useful to compare trends in warm and cold indices, since these provide information on the relative changes in the tails of the daily temperature distributions. The results are shown in Table 10.3. For TX90 versus TX10, the regionally averaged trend in TX90 (1.37 days/decade) was of greater magnitude than that of TX10 (−0.89 days/decade). When looking at individual stations, a greater proportion (89 %) of the stations do have trends in TX90 that are greater than TX10. For TN90 versus TN10, the regional trend in TN90 (2.85 days/decade) was higher than that in TN10 (−1.89 days/decade), and roughly 89 % of stations show a larger trend in TN90. However, for TXx versus TXn, the regional trend in TXn (0.36 °C/decade) was much higher than in TXx (0.17 °C/decade), and about 78 % of stations showed larger trends in TXn. The magnitude of regional trends in TNn (0.65 °C/decade) was more than twice than that of TNx (0.33 °C/decade). This is also reflected at individual stations, where 74 % of them have larger magnitude trends in TNn.

Therefore, we can conclude that changes in percentile-based warm extremes (TN90 and TX90) seemed to be have larger trend magnitudes than those in percentile-based cold extremes (TN10 and TX10), while absolute-based warm extremes (TNx and TXx) seem to have a smaller trend than those in absolute-based cold extremes (TNn and TXn), which is in agreement with the results of eastern and central Tibetan (You et al. 2008). By the preceding analysis, we also can see that changes in extremes derived from daily minimum temperatures (TN10, TN90, TNn, TNx) were generally larger than those from daily maximum temperatures (TX10, TX90, TXn, TXx, respectively).

10.1.6 Precipitation Extremes

The spatial distribution of the trends and trend magnitudes of precipitation extremesis shown in Fig. 10.11. The seasonal spatial distribution of the trends and magnitudes for indices of RX1day is shown in Fig. 10.12. For PRCPTOT and R0.1, about 76–92 % of stations showed a positive trend, respectively, but only 14–21 % of stations showing statistically significant changes, respectively. As for SDII, the positive and negative trends were 78 % (14 % significant) and 22 % (1 % significant) of stations, respectively; stations with decreasing trend magnitudes were scattered throughout the study area and thus were less coherent. Indices reflecting heavy precipitation events (HPE), such as R10, R95, RX1day, Rx5day, were dominated by increasing trends. Significant decreasing trends can be observed at 26 % of stations for CDD, while only 8 % of stations showed significant changes , mainly in northern North Xinjiang, the middle Tianshan Mountains areas, and also northern and western South Xinjaing.

Figure 10.12 shows seasonal variability of RX1day. Indices related to intense precipitation generally showed the largest increasing trends in winter, while the RX1day index had a significant increase in DJF (27 stations or 36 %) and SON (16 stations or 21 %). In the annual and other seasons, most of the stations also experienced a positive trend, but only few stations were statistically significant. The annual and seasonal RX5day can be detected in areas that were similar to the spatial distribution of the RX1day.

The regional average series of precipitation extremes is shown in Fig. 10.13. Trend magnitudes of PRCPTOT and R0.1 show a significant positive trend, with a slope of 6.82 mm/decade and 1.26 days/decade, respectively, while for SDII, the tendency is weaker but nonetheless significant at 0.1. Results regarding heavy precipitation events (HPE) for R10, R95, RX1day, and RX5day are also dominated by positive trends. The total precipitation reflected by annual precipitation and PRCPTOT changed significantly, with changes in rain days (R0.1) and heavy precipitation

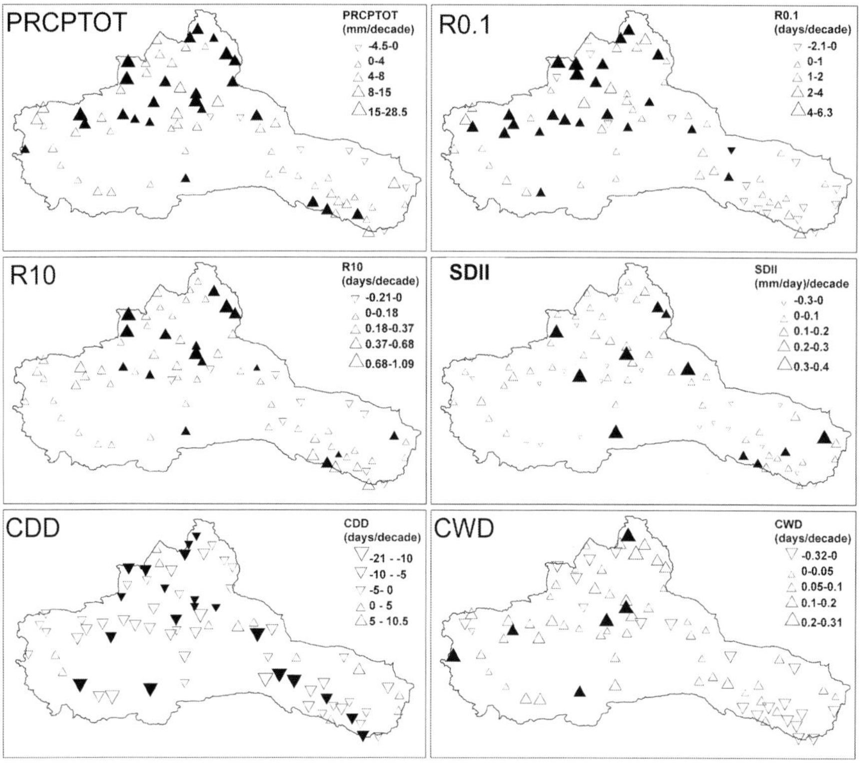

Fig. 10.11 Spatial distribution of trends (*left* column, a1–a4) and trend magnitudes (*right* column, b1–b4) of precipitation extremes

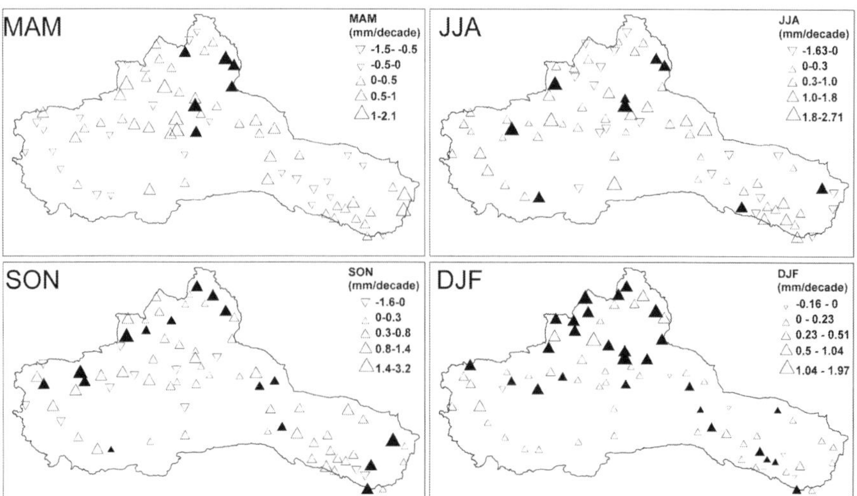

Fig. 10.12 Seasonal spatial trends (*left* column, a1–a2) and trend magnitudes for RX1day (*right* column, b1–a2)

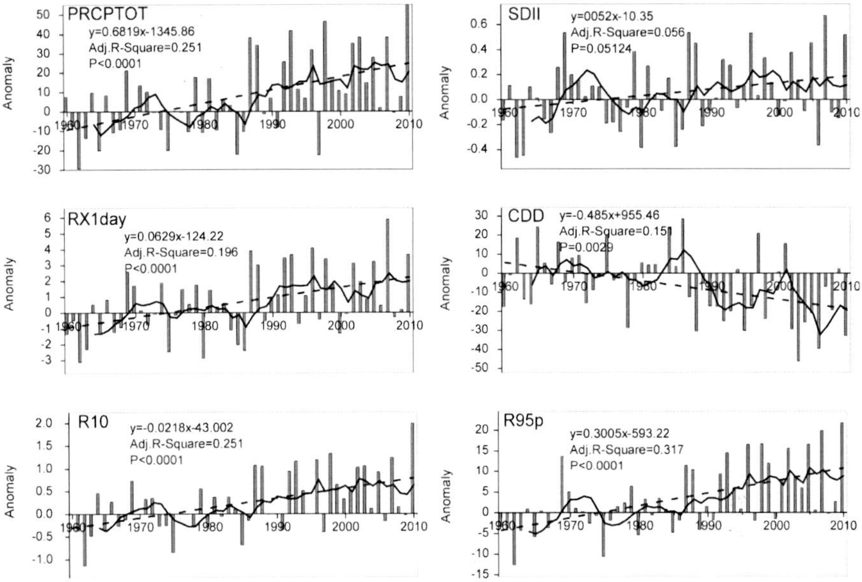

Fig. 10.13 Regional trends of precipitation extremes

events (R10, R95, RX1day, and RX5day) as well as changes in the mean simple daily intensity index (SDII). These results indicate that increasing precipitation resulted from a combination of increasing rainfall frequency and intensity. Regional CDD and CWD showed significant decreasing (−4.85 days/decade) and increasing trends (0.047days/decade).

10.1.7 Relationship Between Extremes, Averages, and Elevation

Relationships between extreme climate trends, trend magnitudes and elevation were analyzed at individual surface stations in China's arid Northwest region (Table 10.4). There were slight correlations between trend and elevation, but they were not significant. Interestingly, the relationship between elevation and cold extremes was negative, but and in warm extremes, this relationship was positive. The relationship between extreme precipitation trends and elevation showed a mixed pattern, with positive relations in PRCPTOT, R10 and RX1day but the opposite relationship in CDD, R95p and SDII. Hence, the relationship between climate extremes and elevation was not clear.

Whether high elevations around the world have been warming faster than lower elevations nearby (or faster than global averages) was an inconclusive question (Beniston et al. 1997; Pepin and Lundquist 2008; Seidel and Free 2003). Liu and Chen's (2000) research revealed greater warming at stations at high elevations compared with surrounding regions in the Tibetan Plateau. This elevation dependency

Table 10.4 Relationship between climate extremes (unit/decade), elevation (m) and mean climate (unit/decade) (the coefficient is the linear slope between two variables; slope significant are marked in bold)

Indices		Elevation		Mean	
		Trend	Magnitude	Trend	Magnitude
Cold extremes	FD0	−0.0003	−0.0005	0.08792	**0.01813**
	ID0	−0.0002	−0.0002	−0.04497	−0.02799
	TN10p	−0.0003	−0.0006	−0.00851	0.06286
	TNn	−0.00008	−0.0002	0.08781	0.01134
	Tx10p	0.00006	0.00008	0.01477	0.0037
	TXn	−0.0003	−0.0002	0.07136	0.02223
DTR	DTR	0.0003	−0.00005	0.03866	0.00879
Warm extremes	TN90p	0.0004	0.0002	**−0.13789**	−0.05536
	TNx	0.0003	0.00003	**−0.19313**	**−0.03201**
	TX90p	0.0002	0.00003	**−0.07756**	0.039381
	TXx	**0.0006**	**0.0001**	**−0.12585**	**−0.02323**
	CSDI	0.0002	0.0004	**−0.06113**	**−0.05603**
Precipitation extremes	CDD	0.00007	0.0036	−0.0008	−0.000017
	PRCPTOT	0.00007	0.0002	−0.00346	**−0.02212**
	R10	0.00006	0.00005	**−0.00184**	**−0.00076**
	R95p	0.00002	−0.001	**0.0009**	**−0.008**
	RX1day	0.00002	−0.0002	−0.00103	−0.00065
	SDII	−0.00001	0.0004	−0.00116	−0.00012

also has been reported by many other regional studies (Chen et al. 2003). However, many studies failed to find an elevation dependency in climate trends or a decreasing warming rate at higher elevations (Pepin and Lundquist 2008; You et al. 2010). Our study also showed that there were no simple linear relationships between climate extreme trends and elevation, and therefore suggesting that there is an element of elevation dependency in climate extremes in China's Northwest could be misleading.

Table 10.4 shows the relationships between trends (trend magnitudes) and mean climate. There were strong negative correlations between warm extremes and mean temperature ($p<0.05$). Faster warming trends occur at lower temperatures, which is in accordance with the study of Tibetan Plateau (You et al. 2010). For cold extremes, the positive relationships can be found in most indices, but this relationship is insignificant. Precipitation extremes show negative relationships (except for PRCPTOT), but only the PRCPTOT and R10 show significant relationship. Our study suggests that, in the arid region, there are significant negative correlations between warm extremes and mean temperature.

Compared with other regions in the world (Table 10.5), patterns in China's Northwest were broadly similar, but there were some differences. For instance, decreases in cold extremes and decreases in extreme minimum temperatures were greater than increases in extreme maximum temperatures, results which agreed with earlier global (Alexander et al. 2006) and regional (Kioutsioukis et al. 2010) studies.

The tendency of extreme indices also presented differences; for example, the rate of decline for DTR was much higher than the global rate or the rate of other regions (Table 10.5). Comparing the Tibetan Plateau of China with climate extremes, trends related to extreme cold events seemed to be decrease greatly and warm extreme events seemed to be less increase. For precipitation indices, only a few stations were statistically significant for any index. This is perhaps to be expected, as secular trends are difficult to detect against the larger inter-annual and decadal-scale variability of precipitation over regions (New et al. 2006).

Conclusion Using the RClimdex indices, we analyzed the results of the extreme indices in the daily climate data. In the arid Northwest zone, the trends of most temperature extremes were greater than the corresponding trends in the mean annual surface temperatures. Stations along the Tianshan Mountains showed larger trend magnitudes for temperature indices, indicating that the warming climate will affect the distribution of glaciers, which will then affect the distribution of water supplies throughout the Northwest. Extremes derived from daily minimum temperatures (TN10 and TN90) are generally larger than changes in extremes derived from daily maximum temperatures (TX10 and TX90), and changes in the minimum of daily minimum and maximum temperatures (TNn and TXn) generally have higher trend magnitudes than those in the maximum of daily minimum and maximum temperatures (TNx and TXx), indicating a long-term decrease in DTR. In general, the arid region of China is warming up, as demonstrated by decreases in cold extremes (FD0, ID0, TN10p, and TX10p) and increases in warm extremes (TN90p, TX90p, TXx, TNx). Moreover, decreases in extremes derived from minimum temperatures were greater than increases in extremes derived from maximum temperatures.

Seasonal warming was apparent in all seasons. For cold extremes, the greatest contribution to frequency was from SON; however, the number of stations showing enhanced statistical significance was for the season of JJA. For warm extremes, the regional trends of TN90p and TNx in DJF showed the largest magnitudes; for TX90p and TXx, SON was the season showing the largest regional trend magnitudes. The seasonal patterns for the number of stations showing the most statistical significance were for regional trends. The largest proportion of stations with positive seasonal trends was DJF for TN90p and TNx, and SON for TX90p and TXx, respectively.

Precipitation extremes have increased except in consecutive dry days, which showed the opposite trend. Changes in precipitation extremes also showed seasonal variability. On an annual basis, the number of rain days (R0.1) had significantly increased. Meanwhile, the precipitation intensity, as reflected by simple daily intensity index (SDII), the number of heavy precipitation days (R10), very wet days (R95p), max 1-day precipitation amount (RX1day) and max 5-day precipitation amount (RX5day), had also significantly increased. This suggested that the precipitation increase in the arid region was due to the increase in both precipitation frequency and intensity.

Table 10.5 Trends of temperature and precipitation extremes from this study and other works (Trends significant are marked in bold)

Index	This study	China	Global	Eastern and central Tibetan Plateau	Middle East	Central and south Asia	Southern and west Africa
Temperature							
FD0 (days/decade)	**3.24**	**−3.73**	–	**−4.32**	**−0.6**	–	–
ID0 (days/decade)	**2.75**	–	–	**−2.46**	–	–	–
TNn (°C/decade)	**0.85**	**0.63**	**0.71**	**0.69**	**0.28**	**0.73**	**0.27**
TXn (°C/decade)	**0.61**	**0.35**	**0.37**	**0.30**	**0.2**		**0.18**
TN10p (days/decade)	**−0.93**	**−2.06**	**−1.26**	**−2.38**	**−1.3**	**−5.70**	**−1.63**
TX10p (days/decade)	**−2.36**	**0.62**	**−0.62**	**−0.85**	**−0.4**	**−2.60**	−1.00
DTR (°C/decade)	**−0.24**	**−0.18**	**−0.08**	**−0.20**	**−0.12**	**−0.12**	**−0.01**
TNx (°C/decade)	**0.32**	**0.21**	**0.30**	**0.25**	**0.23**		**0.19**
TXx (°C/decade)	**0.17**	**0.07**	**0.21**	**0.28**	**0.07**	**0.17**	**0.16**
TN90p (days/decade)	**2.10**	**1.75**	**1.58**	**2.54**	**1.2**	**6.86**	**2.35**
TX90p (days/decade)	**1.25**	**0.62**	**0.89**	**1.26**	**0.66**	**4.72**	**2.24**
WSDI (days/decade)	**1.08**	–	–	–	–		**2.74**
Precipitation							
RX1day (mm/decade)	0.41	1.37	**0.85**	0.27	0	1.02	0.05
SDII ((mm/day)/decade)	0.04	0.06	0.05	0.03	−0.006	–	**0.088**
CDD (days/decade)	−2.52	−1.22	−0.55	**−4.64**	**−5.0**	–	3.57
PRCPTOT (mm/decade)	2.85	3.21	**10.59**	6.66	−0.3	6.87	−0.052
R10 (days/decade)	0.07	–	–	0.23	−0.03	0.11	−0.38
R95p (mm/decade)	1.64	**4.06**	**4.07**	1.28	−0.3	**6.46**	0.024
Sources		(Kang et al. 2011)	(Alexander et al. 2006)	(You et al. 2008)	(Zhang et al. 2005)	(Tank et al. 2006)	(New et al. 2006)

Table 10.6 The concentration degree of extreme hydrological events in Xinjiang

Areas	C
Xinjiang	0.55
Altay	0.27
Tacheng	0.4
Changji	0.51
Urumqi	0.72
Kuytun–Shihezi	0.71
Bozhou	0.64
Ili	0.42
Turpan	0.85
Hamilton	0.39
Kashi	0.59
Hotan	0.64
Kezhou	0.51
Bazhou	0.6
Aksu	0.72

10.2 Changes in Extreme Hydrological Events

10.2.1 Study Area and Methods

Xinjiang Province (Fig. 10.14) is located in Northwest China and is a typical inland arid region. Three mountain ranges—the Altai, the Tianshans, and the Kunluns—run north to south, with two major basins between these ranges, namely the Junggar Basin in North Xinjiang and the Tarim Basin in South Xinjiang. Xinjiang is divided into southern and northern areas by its natural landscape, resulting in two mountain-basin systems with different hydrology and heat conditions (Xu et al. 2004). As breathtaking as it is, this unique mountain-basin landscape is responsible for the frequent occurrence of natural disasters. Alpine areas are distributed widely in Xinjiang and possess a large number of glaciers and permanent snow, which increases the frequency of outburst floods. The middle area experiences the highest precipitation in Xinjiang (Gao et al. 2002), and water erosion is very intense, making floods and snowfalls the predominant extreme events (Hong and Adler 2008). In the low mountain areas, active fault zones increase the occurrence of floods, landslides, and debris flows during rainstorms and seasonal snowmelts.

Most statistical data were collected from the administrative divisions of Xinjiang, with information on the hydrological events coming from the historical natural disaster records of Xinjiang. Few documents or records were available for extreme hydrological events before 1949, with data mainly based on scattered information and records obtained from county annals and personal journals. Data on catastrophic natural disasters from 1949 to 1993 were collected from Research on Xinjiang Natural Hazards, 40 years of Disaster Reduction and Mitigation in Xinjiang, and A History of Calamities in Xinjiang. Data from 1994 to 2010 were obtained from the statistical yearbooks of Xinjiang, local county and city history documents, and

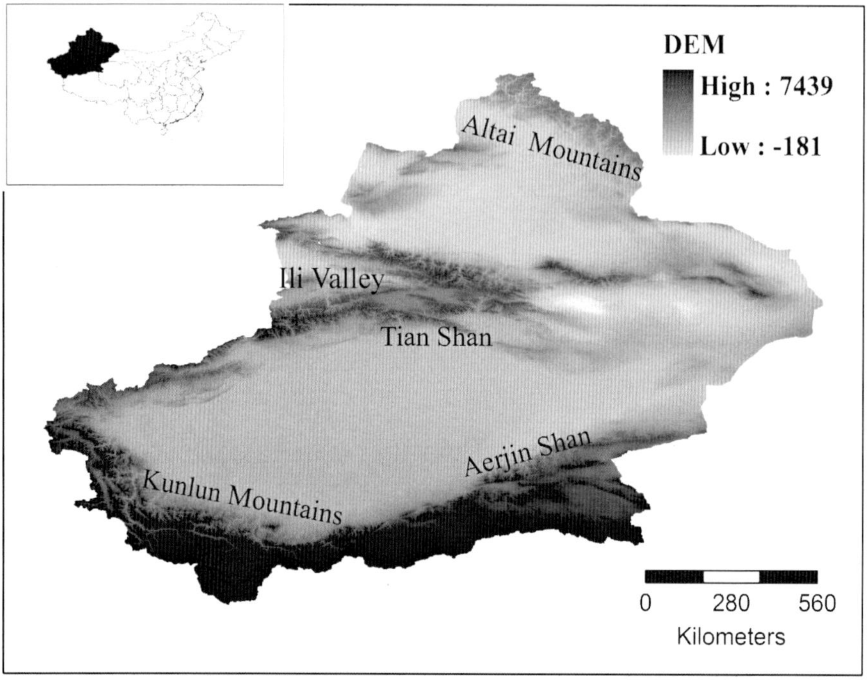

Fig. 10.14 Location and topography of the research area

civil blogs of Xinjiang. These comprehensive data were compiled from a number of meteorological stations by the Meteorological and Hydrological Hazards statistical institutions and departments. Data related to hydrological disasters are sparse in the Kunlun Mountains region in southern Xinjiang due to their being only a few hydrological stations and a relatively small population. For this reason, the Kunlun Mountains region is excluded from this paper.

We studied the mechanisms of extreme hydrological events in Xinjiang based on the intra-annual characteristics and decadal changes reflected by concentration degree and concentration index, as well as t test, Mann–Kendall tests, and correlation analysis (Xu et al. 2010b, 2004).

The vector resultant of durative extreme hydrological event (C) days can be expressed as

$$C_n = \frac{R}{w} = \frac{\sqrt{R_x^2 + R_y^2}}{\sum_{i=1}^{12} r_i} \tag{10.1}$$

$$R_x = \sum_{i=1}^{12} r_i \sin\theta_i, R_y = \sum_{i=1}^{12} r_i \cos\theta_i \quad (10.2)$$

where C is the concentration degree (C) of the corresponding durative extreme hydrological events, w is the total frequency of extreme hydrological events, R is the vector of the total frequency for a period of extreme hydrological events, r_i is the monthly extreme hydrological event amount, and i is the total duration time of the corresponding durative extreme hydrological event. The vector angle θ_i represents each month angle value of extreme hydrological events.

The formula for determining the concentration index can be expressed as the following, based on previous studies (Reiser and Kutiel 2010; Yao et al. 2010; Zhang et al. 2009):

$$I = \left({}^{r_n}\!/_{r_z} - {}^{n}\!/_{365} \right) * 100 \quad (10.3)$$

where I is the concentration index (I) of extreme hydrological events, r_n is the frequency of n-day extreme hydrological events, r_z is the amount of extreme hydrological events, and n is the length of research time. Thus, r_n/r_z is the ratio of the actual frequency of extreme hydrological events and the total annual frequency of extreme hydrological events in a period of time, and n/365 is the average daily frequency of hydrological disasters. $I>0$ or $I<0$ shows whether the actual hydrological disaster intensity is greater or less than the average distribution intensity for a period of time in a region.

On the basis of the statistical characteristics of this study, we applied universal kriging (Demarsily and Ahmed 1987; Kumar 2007) to calculate the past frequency of extreme hydrological events in each area of Xinjiang and to estimate the uncertainty of extreme hydrological events.

10.2.2 Intra-annual Changes in Extreme Hydrological Events

We obtained intra-annual spatial distribution and used statistical data (Fig. 10.15) to determine the reliability of concentration degree and concentration index.

The maximum frequency of extreme hydrological events occurred during the Xinjiang summer (June–August) (Fig. 10.15), mainly because rainstorm floods were the dominant hydrological disaster type in Xinjiang, and these happened mainly in summer. Generally, the incidence of rainstorm floods was greatly affected by both the weather systems and the terrain features. Xinjiang is impacted by west wind drift, and thus rainstorm floods accounted for more than 80 % of extreme hydrological events (Jiang et al. 2005). Xinjiang possesses numerous mountain ranges and intermountain basins. Tall mountains can intercept moisture by westerly circulation from mid-latitude circulation, and a huge disparity in terrain elevation make formation of rainstorms and subsequent rainstorm floods and debris flow easily.

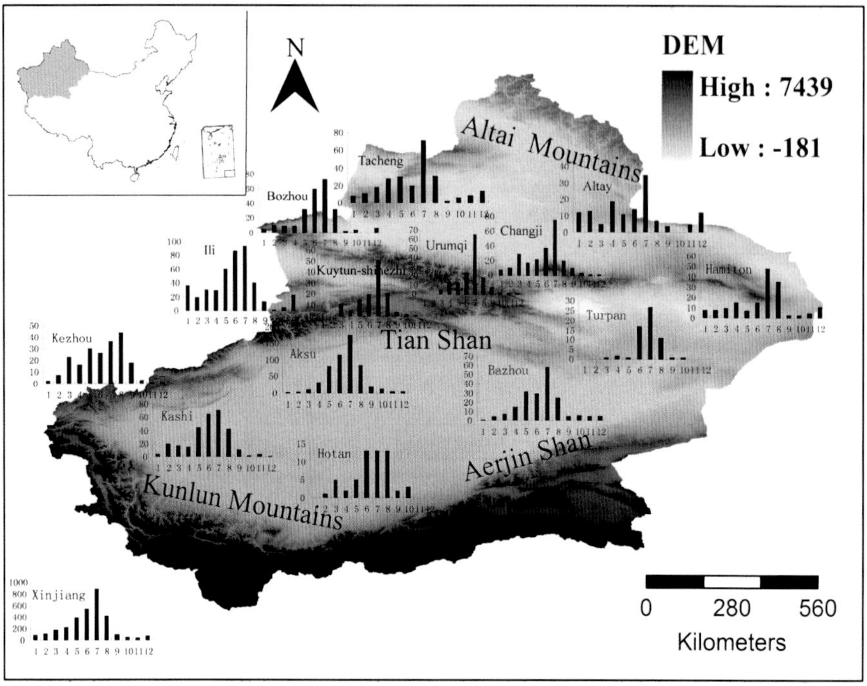

Fig. 10.15 Inter-annual spatial distribution of extreme hydrological events in Xinjiang (abscissa for the month/month; ordinate for the frequency of disaster/times)

We obtained the intra-annual spatial distribution of extreme hydrological events in Xinjiang using concentration degree (Table 10.6) and concentration index (Table 10.7). The overall concentration degree of Xinjiang was 0.55 but varied by region (Table 10.6). The highest concentration degree was obtained for Turpan (0.85), which indicates that the inter-annual distribution of extreme hydrological events was very concentrated, with most of these events occurring over a very short period of time during the year. The concentration degree in Urumqi, Kuytun–Shihezi, and Aksu area was the next highest, at a degree of 0.72, 0.71, and 0.72, respectively. The lowest occurred in northern Altay (0.27), which indicated that the intra-annual distribution of extreme hydrological events was scattered. These results were consistent with the statistical data (Fig. 10.15).

The concentration index in summer was highest overall for each region (Table 10.7). This indicates that extreme hydrological events mainly occurred in summer (June to August) in Xinjiang. The maximum concentration index in summer was observed in the Turpan area (65.51), followed by Urumqi, Kuytun–Shihezi, Bozhou, Hotan, and Aksu (46.12, 45.34, 43.24, 43.76, and 44.03, respectively). This indicates that extreme hydrological events in these areas were more concentrated in summer than in other areas. The minimum concentration index in summer was observed in Altay (16.22), with Tacheng, Ili and Kezhou demonstrating slightly higher concentration indices (24.54, 25.23, and 26.55, respectively).

Table 10.7 The concentration index of extreme hydrological events in Xinjiang

Areas	I_{win}	I_{spr}	I_{sum}	I_{aut}	$I_{sum\ y}$	$I_{win\ y}$
Xinjiang	−15.80	0.55	34.91	−18.29	59.16	−9.43
Altay	2.35	0.89	16.22	−18.09	41.77	8.92
Tacheng	−11.35	5.99	24.54	−17.8	55.18	−4.5
Changji	−16.05	4.44	30.67	−17.69	59.77	−9.08
Urumqi	−24.01	−3.88	46.12	−16.87	66.9	16.22
Kuytun–Shihezi	−24.66	−0.04	45.34	−19.27	69.96	19.27
Bozhou	−15.6	−4.08	43.24	−22.19	63.82	13.13
Ili	−6.68	2.76	25.23	−19.94	52.65	−1.96
Turpan	−24.66	−18.1	65.51	−21.38	22.88	21.38
Hamilton	−8.59	−5.61	33.68	−18.11	52.72	−2.04
Kashi	−16.21	1.02	35.48	−18.91	61.15	10.47
Hotan	−22.9	−3.6	43.76	−15.89	64.82	14.13
Kezhou	−19.83	8.68	26.55	−14.03	59.88	−9.2
Bazhou	−19.48	3.32	33.89	−16.37	61.87	11.19
Aksu	−22.82	−2.01	44.03	−17.84	66.69	−16

The concentration indices in winter were all negative (except for Altay). This is because extreme winter events, mainly in the form of snowfalls, were more common in Altay than in other areas. Altai is located in the northern-most region of Xinjiang; it experiences a longer cold season and features certain topographic conditions that lead to higher levels of precipitation and thus greater snowfalls. The concentration index in winter in Ili, Hamilton and Tacheng was the next highest (−6.68, −8.59 and −11.35, respectively), which were also mainly related to the occurrence of snow disasters.

Kezhou saw its maximum concentration index occur in spring, which was related to the greater frequency of snowfall in this area compared to the other southern Xinjiang areas. In addition, the temperature in spring was higher than in northern Xinjiang, leading to floods caused by snowmelt and rainstorms. The concentration indices for Tacheng and Changji in spring were the next highest, mainly due to their northern Xinjiang location. Negative values were observed in autumn and showed little difference. This indicated that the occurrence of extreme autumn events in all areas was low, which was consistent with the observed record.

Generally, the frequency and distribution of extreme hydrological events are strongly related to atmospheric circulation, while the percentage of such events is related to annual precipitation totals. For instance, in the western Tianshan Mountains (a region heavily affected by west wind drift), rainfall during extreme hydrological events contributes to more than half of the total annual precipitation (Shi et al. 2007). In addition, the distribution of extreme hydrological events is associated with topographical features. Western and northern areas of Xinjiang are particularly mountainous (Fig. 10.15), a feature which favors the occurrence of extreme events.

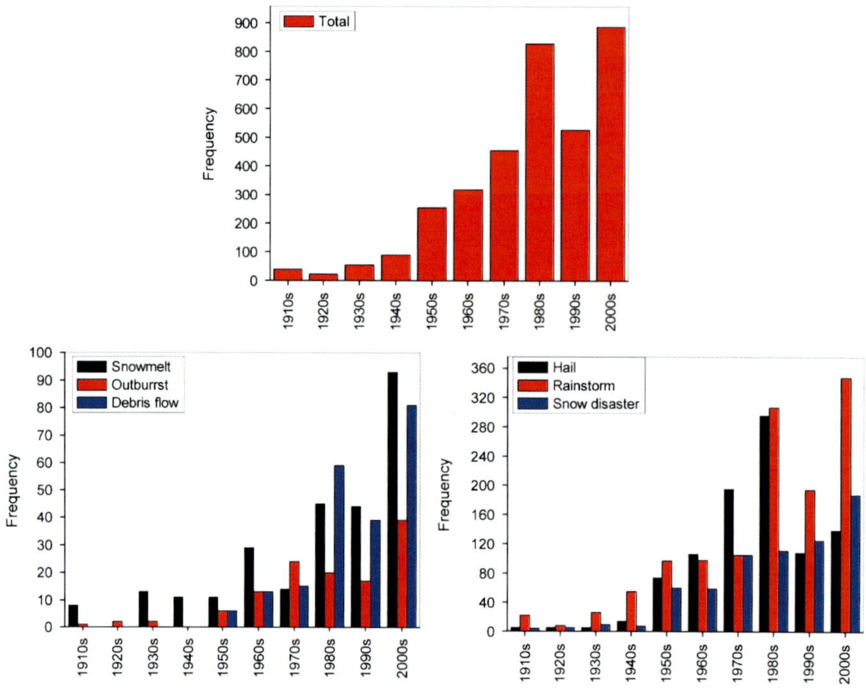

Fig. 10.16 The frequency change in various types of extreme hydrological events

10.2.3 Decadal Changes in Extreme Hydrological Events in All Areas

Figure 10.16 shows decadal changes in various types of extreme hydrological events from 1901 to 2010. Over the past 100 years, the frequencies of various extreme water-related events in Xinjiang have showed a generalized increasing trend, particularly in the past decade (Fig. 10.16).

Although the frequency of extreme hydrological events generally increased in all regions from 1901 to 2010, there were some exceptions (Fig. 10.17). The frequency of extreme hydrological events in the 1980s was about twice that of the 1990s for some regions (such as Tacheng, Kuytun–Shihezi, Bazhou, Kashi, Kezhou and Aksu). In the above areas, the frequency of extreme hydrological events was higher in the 1980s than in any other decade, with the sole exception of Tacheng, which experienced a higher rate of snow disasters and snowmelt floods. Similarly, extreme hydrological events increased from 1901 to 2010 in Changji, Urumqi, Turpan, and Hotan, which were dominated by rainstorm flood disasters. Snow-related disasters were common in Altay, with the increasing trend most evident after the 1980s and reaching a maximum between 2001 and 2010. Ili experienced frequent dominant hazard types, which in turn led to the steady increase in extreme events due to the occurrence of complementary multi-hazard types.

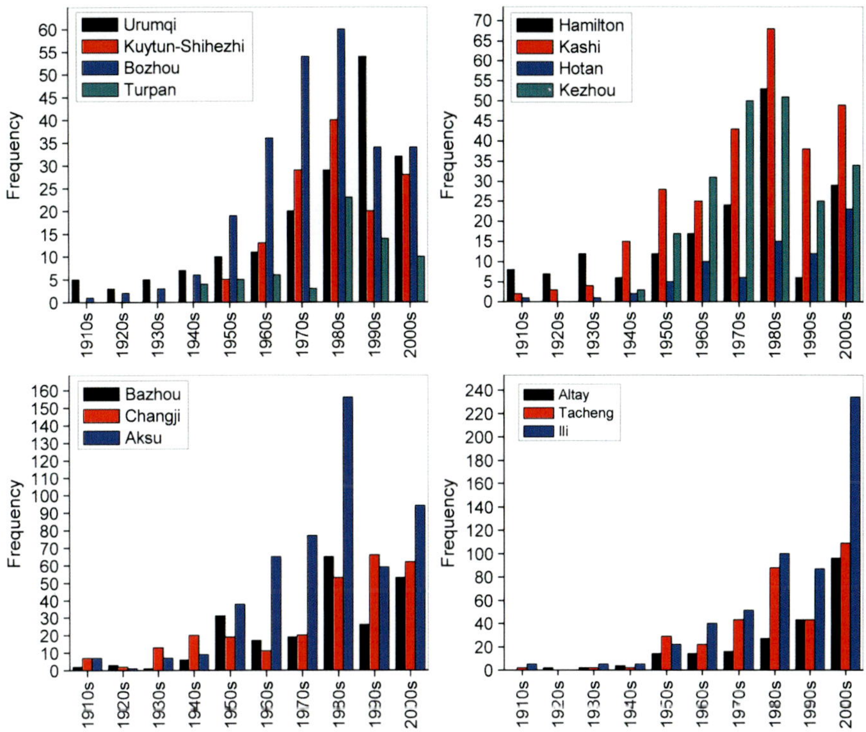

Fig. 10.17 Decadal changes of extreme hydrological events in Xinjiang

These profound changes were likely related to the climate changing from warm and dry to warm and wet, and the obviously rising temperatures after the 1980s in Xinjiang (Shi et al. 2007). Generally, the changing trend of extreme hydrological events is strongly related to climate change, and global warming has the potential to impact the frequency and intensity of extreme hydrological events (Shi et al. 2007). As a consequence of global warming and an enhanced water cycle, precipitation, glacial melt-water, river run-off and air temperature have all increased over the past few decades, making extreme hydrological events more likely to occur (Yao et al. 2010).

10.2.4 Trend Variations in Extreme Hydrological Events

The same increased occurrence of extreme hydrological events observed worldwide has also been observed in Xinjiang. This increased occurrence can be attributed to a variety of factors, the most obvious being climate change, which has resulted in temperature and precipitation increase (Arnell 1999). One of the most significant consequences of climate change is the spatial and temporal alternation of

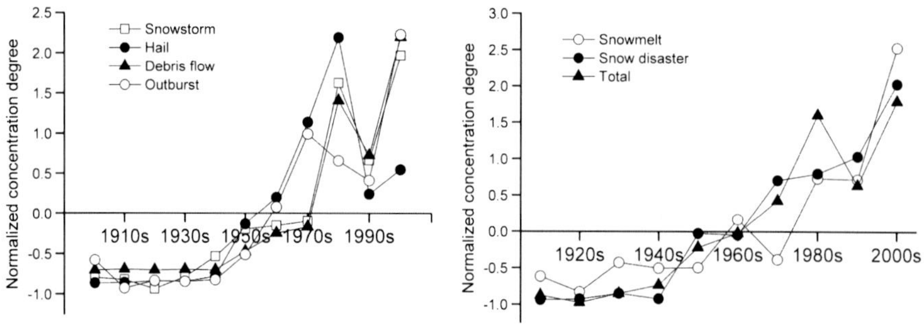

Fig. 10.18 Concentration degree decadal change from 1901 to 2010 in Xinjiang

regional hydrological cycles, resulting in an increase in the frequency and intensity of extreme hydrological events (Hirschi and Seneviratne 2010; Trenberth 1999). Over the past 100 years, the frequency of various extreme hydrological events has increased in Xinjiang, particularly since the 1970s and within the past ten years (Figs. 10.16 and 10.17).

Climate change has modified the pattern of precipitation and resulted in an alternation in both spatial and temporal regional hydrological cycles (Arnell 1999; Raziei et al. 2009). The changes in precipitation pattern have also resulted in an alternation of concentration degree and frequency of extreme hydrological events (Redano and Lorente 1993). Figure 10.18 shows the normalized values of concentration degree variations from type to type from 1901 to 2010. All concentration degree values exhibited large intra-annual variability, especially post-1970s, with the degree reaching a maximum value in the past ten years. The normalized values were almost all below zero prior to the 1970s, after which they became positive and continued to increase. Thus, it is reasonable to assume that extreme hydrological events will continue to strengthen in intensity according to the concentration degree and frequency trends.

10.2.5 Spatial Distribution of Extreme Hydrological Events

Incidences of extreme hydrological events were assessed in Xinjiang (Fig. 10.19). The frequency distribution of extreme hydrological events showed general symmetrical distribution along the entire length of the Tianshan Mountains, with even distribution in the Junggar and Tarim Basins. Greater frequency was observed in northwestern Xinjiang than in southeastern Xinjiang, with the northwest experiencing more than 80 % of the total extreme hydrological events. More extreme hydrological events occurred in the western Tianshan Mountains region, with moderately high values in the Altai Mountains, the central regions of the northern Tianshan Mountains, and the Karakorum Mountains. Extreme hydrological events also had a high incidence in the Hamilton Basin in the east region of Xinjiang. Low frequency

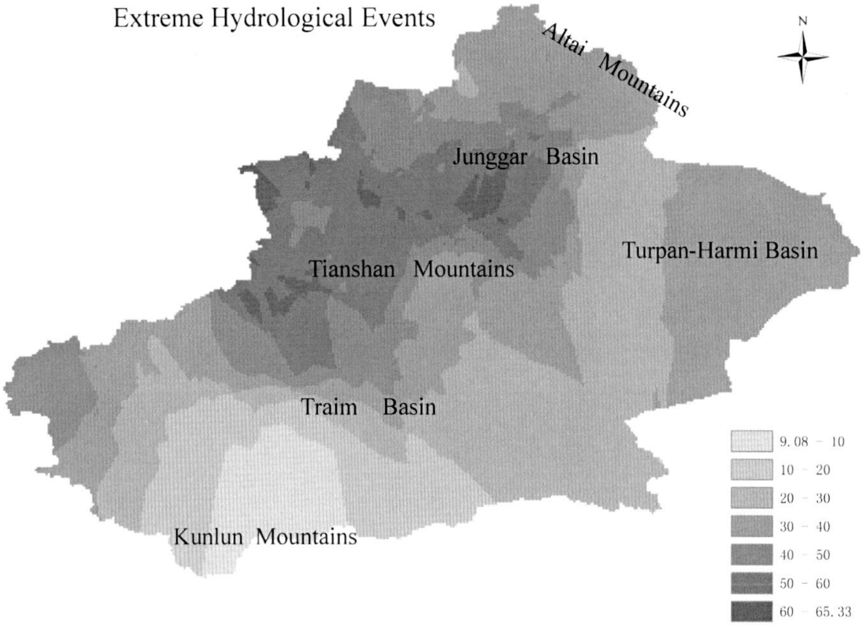

Fig. 10.19 The spatial distribution of extreme hydrological events

regions were observed in the Turpan Basin and Kunlun Mountains, which experienced less than 3 % of the total extreme hydrological events. These results may relate to the regional topography and sparse human population.

The reasons for the occurrence of hydrological disasters are complex and cannot be simply attributed to either macro-scale circulation patterns or geographic aspects. For instance, the frequency of extreme hydrological events could have changed due to an increase or decrease in precipitation and a rise or fall in temperature in various areas or a change in regional vulnerabilities to extreme precipitation and extreme temperature (Jiang et al. 2005). Earlier in this section, we analyzed the relationship between the frequency of extreme hydrological events and mean annual temperature and annual precipitation, ascertaining that the relationship was not significant. As extreme hydrological events tend to be anomalous hazards, we surmise that their occurrence is thus closely related to anomalous precipitation events or anomalous temperature events within a certain timeframe. However, because data related to daily minimum and maximum temperature and precipitation are limited, we could not do an in-depth analysis based on this supposition.

Type distribution of extreme hydrological events in Xinjiang was also analyzed, giving the following results (Fig. 10.20):

Snowmelt floods were mainly distributed in the vast northern areas of Xinjiang, and rainstorm floods were the dominant hydrological disaster type overall in Xinjiang. The distribution of these floods is shown in Fig. 10.20, which clearly indicates

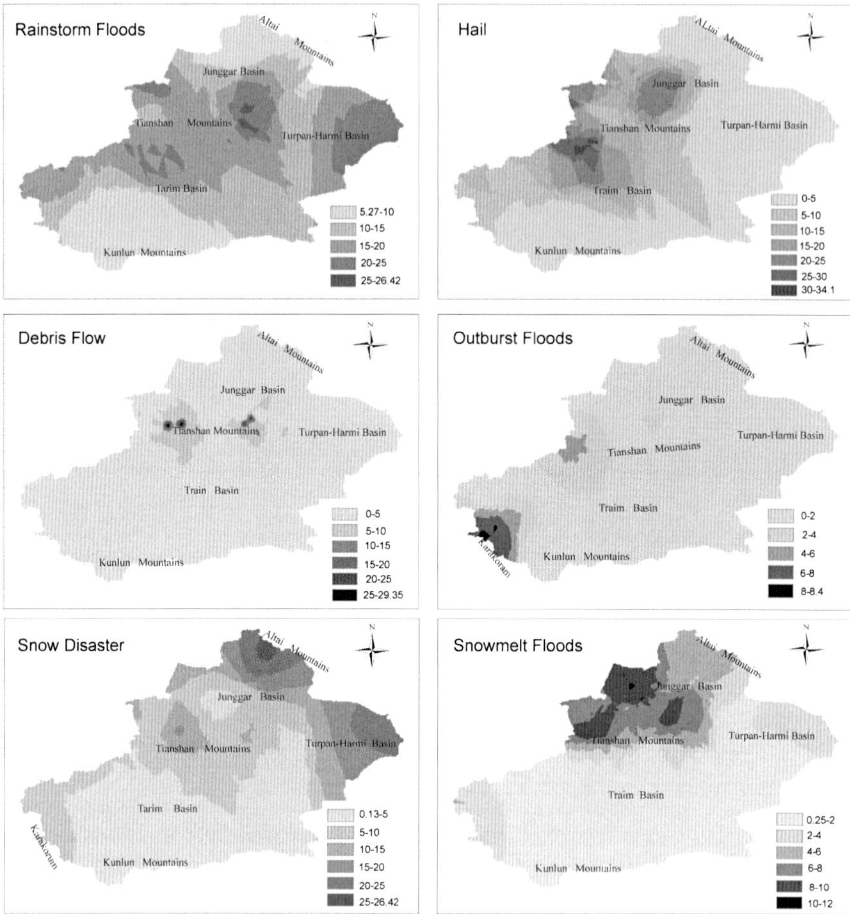

Fig. 10.20 The spatial distribution of extreme hydrological events in Xinjiang

that rainstorm floods were most frequent over vast central areas of Xinjiang, especially in the central and western Tianshan Mountains and the Tarim Basin. The incidence was also high in the Hamilton Basin but low in the Altai Mountains. Generally, the incidence of rainstorm floods was greatly affected by both the weather systems and the terrain features (Turkes et al. 2009; Turkes and Sumer 2004).

Rainstorm floods accounted for more than 80 % of extreme hydrological events in Xinjiang. In addition, the distribution of rainstorm floods was associated with topographical features. Xinjiang possesses numerous mountain ranges and inter-mountain basins. Tall mountains can intercept a significant amount of moisture flowing in via westerly circulation from mid-latitude circulation, and a huge disparity in terrain elevation is favorable for the formation of rainstorms and subsequent floods.

Hailstorms mainly occurred in the central Junggar Basin and northwest of the Tarim Basin. Specifically, hailstorms frequently occurred over the central Junggar Basin and up to 30 times per year in the northwestern edge of the Tarim Basin. The frequency of these storms generally decreased south-eastward, with the lowest values found in southeast Xinjiang (in the Taklimakan Desert and Turpan-Hamilton Basin). The frequency distribution of hail events in Xinjiang was strongly related to topography and precipitation, with a clear geographical distribution. As hail is usually accompanied by heavy rainfall, areas that experience substantial levels of precipitation also experience more hailstorms than areas receiving less precipitation (Yao et al. 2010). There were few or no hail events in extreme drought areas (such as the Taklimakan Desert) due to the dry conditions.

Debris flow at a particular location is traditionally expressed in terms of floods (Fig. 10.20). Flows occurred most frequently in the western and central Tianshan Mountains, with Ili Valley accounting for more than 85 % of such events. Ili Valley is particularly prone to debris flow and landslides due to its unique topographical features and high precipitation.

Outburst floods include glacier lake outburst floods and ice floods. In China's arid Northwest region, glacier lake outburst floods mainly occurred in southern Xinjiang and ice floods in northern Xinjiang. Outburst floods occurred most frequently in the Karakoram Mountains, the western Tianshan Mountains, and the southern edge of the Junggar Basin. Glacier lake outburst floods occur when water is suddenly and rapidly released from a glacier lake due to glacier damming, supra-glacial drainage or subglacial drainage. The rivers that most often generated glacier lake outburst floods were those interconnected with the Yarkant River, which originates in the northern slopes of Karakorum Mountain. This type of flood event is closely related to high-altitude temperatures.

Snow hazards generally occurred more often in northern Xinjiang than in southern Xinjiang, in western areas than in eastern areas, and in mountains than in plains. A maximum frequency was apparent in the northern Altai Mountains, with snow hazard events occurring up to 25 times per year. A secondary maximum was also apparent in the eastern Hamilton Basin, and a third maximum was observed in the central and western Tianshan Mountains, the northwest edge of the Junggar Basin, and the Karakorum Mountains. Xinjiang is affected by the mid-latitude westerlies of the northern hemisphere. Water vapor is mainly transported from the south-west, with most of it drifting on a westerly airflow from the Atlantic to Xinjiang. Significant amounts of precipitation (snow) occur once this vapor meets with favorable terrain. Westerly winds carrying water vapor can enter Xinjiang through any one of a number of western gaps, such as the Tacheng Basin, Ili Valley, and Erqis Valley. When intercepted by the northern Altai and central Tianshan Mountains, high levels of precipitation (snow) may occur. This is the main reason why snow hazards happen with greater frequency in northern, northwestern and western Xinjiang.

The frequency of snowmelt floods (Fig. 10.20) was five times higher in northern Xinjiang (especially in the northwest) than in southern Xinjiang. A maximum was apparent in the Tacheng Basin and Ili Valley, with snowmelt floods occurring up to ten times per year. A secondary maximum was also apparent in central Junggar

Basin and the Altai Mountains, and a third maximum was observed in the Hamilton Basin and Karakorum Mountains. The frequency of snowmelt flooding decreased south-eastward, with the lowest frequencies found in the southeast Taklimakan Desert. This may be due to the higher levels of precipitation and snowfalls in winter in northern Xinjiang. In addition, low temperatures in winter often result in heavy snowfalls. With increasing temperatures in spring and summer, this snow rapidly melts and can culminate in floods.

Overall, our data shows that rainstorm floods mainly occurred in the vast central areas of Xinjiang, while hailstorms mainly occurred in the central Junggar Basin and northwest of the Tarim Basin. Debris flows were most frequent in the western and central Tianshan Mountains, outburst floods were most common west of the Tianshan and Karakorum Mountains, and snow hazards primarily occurred in northern areas of Xinjiang, especially in the Altai Mountains and the Hamilton Basin.

Conclusion

(1) Most extreme hydrological events occurred in the warmer half of the year. Intra-annual distribution was concentrated in Turpan, followed by Urumqi, Kuytun–Shihezi, and Aksu, with Altay showing the lowest number. These results indicate that the methods used in this study were reliable and can be applied in future research.
(2) The frequency of various extreme hydrological events exhibited an obvious increasing trend over the past 100 years, particularly since the 1970s and within the past ten years. Variations in event type were observed from area to area.
(3) The frequency of extreme hydrological events is expected to increase over time. The concentration degree values increased from 1901 onwards, peaking in 2010. The intensity of extreme hydrological events will continue to strengthen in coming decades.
(4) Frequencies showed a coarse symmetry distribution trend along the Tianshan Mountains and a homogeneous distribution in the Junggar and Tarim Basins. The frequencies were higher in northwest than in southeast Xinjiang, with northwestern regions experiencing more than 80 % of the total extreme hydrological events. The frequency of extreme hydrological events was highest in the western portion of the Tianshan Mountains region, with moderately high values observed in the Altai Mountains, the central regions of the northern Tianshan Mountains, and the Karakorum Mountains.
(5) The types of extreme hydrological events showed an evident regional distribution. Rainstorm floods mainly occurred in the vast central areas of Xinjiang, and hailstorms in the central portion of the Junggar Basin and northwestern portion of the Tarim Basin. Debris flow was most prominent in the west and central Tianshan Mountains, while outburst floods occurred most frequently in the western Tianshan Mountains and the Karakorum Mountains. Snow hazard and snowmelt-related events primarily took place in the northern areas of Xinjiang.

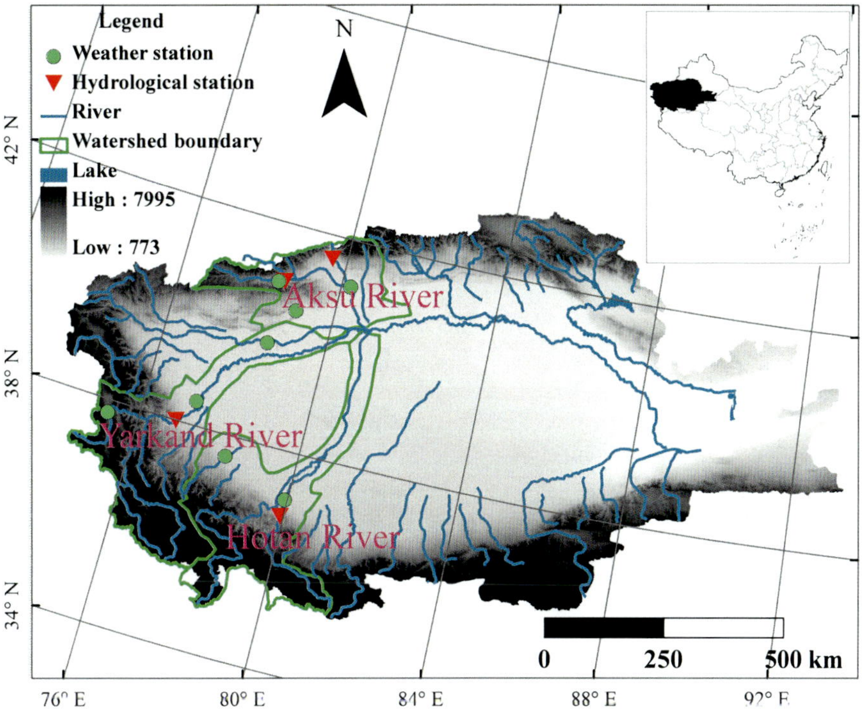

Fig. 10.21 Distribution of the river system in the Tarim River Basin

10.3 Relationships Between Extreme Hydrological Events and Climate Change

10.3.1 Study Area and Methods

The Tarim River Basin is China's largest inland river basin. It is situated between 73–97° E and 34–45° N, covers a total area of 1.02×10^6 km^2 (Fig. 10.21), and is composed of 114 rivers of 9 water systems. Today, only three water systems—the Aksu River, the Hotan River, and the Yarkand River—have a natural hydraulic relationship with the mainstream. Of these three flows, the Aksu River forms the mainstream of the Tarim River, accounting for 73.2 % of the total flow, while the Hotan and Yarkand account for 23.2 and 3.6 %, respectively (Chen et al. 2007).

The headwater of the Tarim River Basin, with an area of 36,200 km^2, combines typical topography of high mountains and low plains. The mean annual precipitation of the mountainous region exceeds 300 mm, with 70 % of it occurring from June to October. In the lower plains region, precipitation amounts range from 60 to 200 mm. The annual air temperature shows intra-annual variations and the mean is about 8.8 °C (Liu et al. 2010a, b). Run-off generated in the mountainous regions is consumed in the low plain region by evaporation or irrigation.

Daily hydrological data provided by the Xinjiang Water Bureau and covering the period from 1960 to 2007 were collected from four hydrological stations. Figure 10.21 shows the locations of the stations. Data for the Aksu River were obtained from the Xehela and Shaliguilank hydrological stations; Yarkand River data came from the Kaqun hydrological station, and Hotan River data from the Tonguzluok station. Because all of the stations are located in the source areas of the rivers, water amounts were less affected by human activities. Missing data are interpolated using regressive relations between neighboring stations.

In our study, we select a 1-day maximum flow and 1-day minimum flow to represent hydrologic extremes in the Tarim River. The homogeneity of the daily streamflow of the Tarim was assessed by using the Penalized Maximal F tests (Wang 2008a, b) through the RH-test software package (Version 3, (Wang et al. 2010)). Results show that all stream-flow time series satisfy the homogeneity.

The nonparametric Mann–Kendall (M–K) method (Mann 1945; Kendall 1975) was used in our study to detect possible trends in hydrological extremes. The results of the M–K test were heavily impacted by the serial correlation of the time serial correlation, so we adopted the Yue and Pilon method (Yue et al. 2002) to remove the serial correlation.

As a wavelet transform can expand time series into time frequency space, it can therefore find localized intermittent periodicities. Wavelet analysis can also reveal localized time and frequency information without requiring the time series to be stationary (Xu et al. 2010a).

The cross wavelet transform (XWT) of two series x_n and y_n is defined as

$$W^{XY} = W^{X} W^{Y*} \tag{10.4}$$

where * denotes complex conjugation. The cross wavelet power as $|W^{XY}|$. The complex argument arg(W^{XY}) can be interpreted as the local relative phase between x_n and y_n in time frequency space. The circular mean of the phase over regions with higher than 5% statistical significance that are outside the COI quantify the phase relationship. The circular mean of a set of *angles* $(a_i, i = 1,...,n)$ is defined as:

$$a_m = \arg(X,Y) \quad \text{with} \quad X = \sum_{i=1}^{n} \cos(a_i) \quad \text{and} \quad Y = \sum_{i=1}^{n} \sin(a_i) \tag{10.5}$$

The XWT phase angle within the 5% significant regions and outside the COI has the mean phase $-176 \pm 12^{\circ}$, where ± designates the circular standard deviation. The circular standard deviation is defined as $s = \sqrt{-2\mathrm{Ln}(R/n)}$, where $R = \sqrt{X^2 + Y^2}$.

Cross wavelet power reveals areas with high common power. Another useful measure is how coherent the cross wavelet transform is in time frequency space. Following Torrence and Compo (1998) and Torrence and Webster (1998), we define the wavelet coherence of two time series as

$$R_R^2(s) = \frac{\left|s\left(s^{-1}W_n^{XY}(s)\right)\right|}{s\left(s^{-1}W_n^X(s)\right) \cdot s\left(s^{-1}W_n^Y(s)\right)} \tag{10.6}$$

where S is a smoothing operator. The smoothing operator S as

$$s(W) = S_{scale}\left(S_{time}\left(W_n(s)\right)\right) \tag{10.7}$$

The statistical significance level of the wavelet coherence is estimated using Monte Carlo methods. The Monte Carlo estimation of the significance level requires in the order of 1,000 surrogate data set pairs. The number of scales per octave should be high enough to capture the rectangle shape of the scale smoothing operator while minimizing computing time. Empirically, and to be satisfactory, we set 12 scales per octave.

10.3.2 Relationship Between Extreme Hydrological Events and Climate

Temperature, precipitation and extreme hydrological events from 1950 to 2010 were basically on the rise (Fig. 10.22). During this timeframe, average temperature, precipitation and extreme hydrological events were 9.58°C, 135.6 mm and 57.5 times /10a, respectively. Air temperature and precipitation were below average and the frequency of extreme hydrological events was also relatively small during the period from 1950 to 1979 (40 times/10yr). However, from 1980 onwards, the occurrence of extreme events increased (to 78 times/10yr), as did changes in temperature and precipitation (Fig. 10.22). Likewise, the frequency of storm floods in Xinjiang experienced an increasing trend during the same time periods. The t-test and non-parametric Mann–Kendall test also found similar results (Table 10.8). These findings indicate that the increasing temperatures and precipitation intensified the hydrological cycle, which affected the speed and intensity of extreme hydrological events.

10.3.3 Relationship with Basin-wide Climate

A significant increasing trend was observed in the temperatures of all three basins, which was in accordance with other studies (Chen et al. 2007), whereas precipitation showed only an insignificant increasing trend (not shown for distribution). Furthermore, the effects of temperature fluctuations on stream-flow were minimal when the temperature was less than 15 °C (Fig. 10.23), but the impact of even small fluctuations in temperature (more than 15 °C) were often amplified by a factor of two or more in the headwater of the Tarim River Basin. The scatter between temperature and hydrological extremes was best fit by exponential distribution, which also

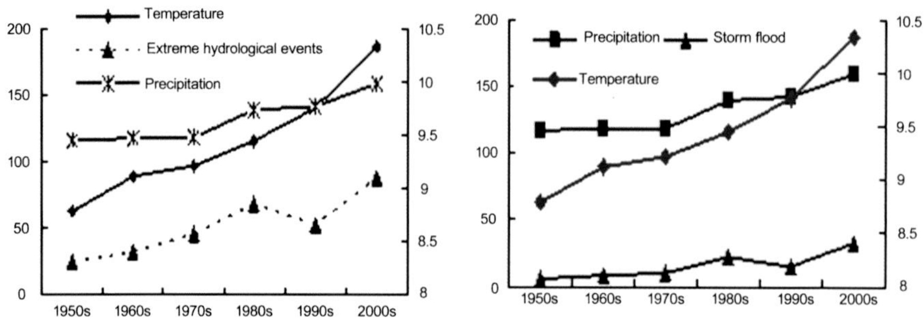

Fig. 10.22 Change relation of temperature, precipitation and extreme hydrological events in Xinjiang

Table 10.8 Monotonic trend test for temperature, precipitation and extreme hydrological events in Xinjiang

	t-test				Mann–Kendall test		
	β_0	β_1	T_C	H_0	Z_0	β	H_0
Temperature	6.313	0.0233	15.38	R	3.428	0.0252	R
Precipitation	141.22	0.6706	23.61	R	1.702	0.6883	A
Extreme events	126.57	0.4835	20.64	R	2.381	0.4950	R

R reject, *A* accept, α 0.05

indicated that there were only minor effects on stream-flow when the temperature was less than the threshold. As expected, correlations were positive for precipitation, i.e., an increase in precipitation led to an increase in hydrological extremes. The linear fitting between hydrological extremes and precipitation showed a significant positive correlation (*F*-test) between each other, while the Adj.R-Square was relatively slow. Interestingly, a higher correlation was found between hydrological extreme events and temperature than between such events and precipitation. The Adj.R-Square between climate variables and 1-day maximum extremes was larger than that of 1-day minimum flow, which indicted that the effects of climate change were severe for 1-day maximum extremes.[1]

Wavelet coherence between mean monthly hydrological extremes and monthly world-wide precipitation and temperature for the Aksu River were shown in Fig. 10.24. The coherence between hydrological extremes and temperature appeared as the most important factor throughout the spectra at the 0.5–1.7 year period. Phases between temperature and hydrological extremes were almost in synch with each other in all of the significant areas. In the Aksu River and Yarkand River regions, higher than 0.8 coherence coefficients were found across the spectrum for precipitation at the 0.7–1.5 year period. Significant coherence between precipitation

[1] From "Huaijun Wang et al. Hydrological extreme variability in the headwater of the Tarim River: links with atmospheric teleconnection and regional climate. Stochastic environmental research and risk assessment (submitted)."

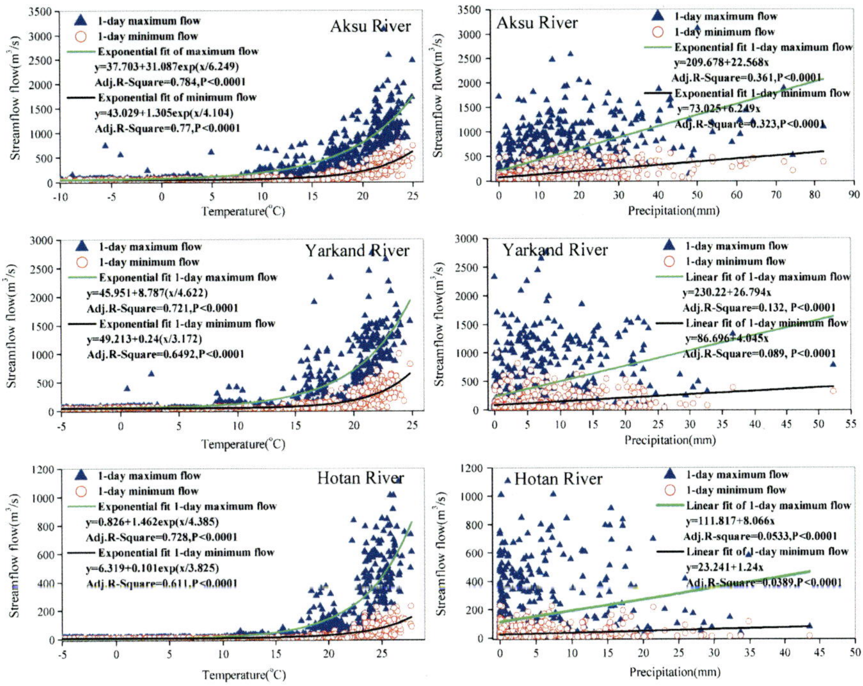

Fig. 10.23 Relationship between hydrological extremes and regional climate

variability and hydrological extremes was also found at the 0.7–1.5 year scale, but this relationship was not consistent throughout the entire time period. From this, we can surmise that the effects of temperature were greater than the effects of precipitation, which is in accordance with regression analysis results.[2]

Rising temperatures speed up the water cycle, causing the frequent occurrence of extreme weather events and intensifying the instability of the hydrological element. While it is generally accepted that heavy precipitation and extreme temperatures will result in run-off situations, extreme events related to run-off are a complex process whose formation process is difficult to simulate with conventional statistical tools. Therefore, we selected the indices of warm nights and strong precipitation to calculate the probability density function during the period before and after 1986.

The main factor impacting extreme runoff events can be determined according to shape similarities. Specifically, the number of short-term peaks in the Aksu River increased significantly from 1988 to 2007 (Fig. 10.25). Notable increases were also observed in warm nights and heavy precipitation, especially for the latter, which indicated that run-off events were the joint result of extreme temperatures and precipitation. Large increases in peak frequency contributed to large increases

[2] From "Huaijun Wang et al. Hydrological extreme variability in the headwater of the Tarim River: links with atmospheric teleconnection and regional climate. Stochastic environmental research and risk assessment (submitted)."

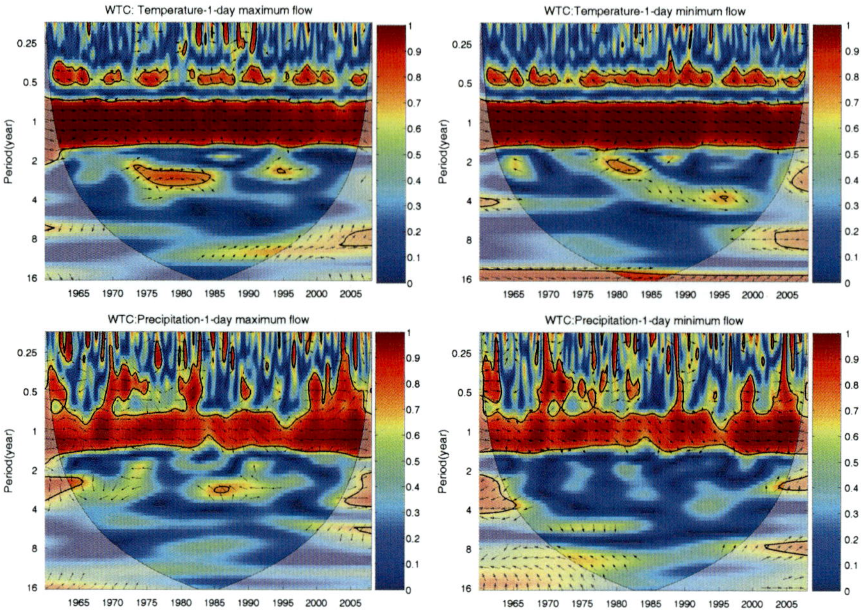

Fig. 10.24 Wavelet coherence of the hydrological extremes and regional climate

in the number of extreme temperatures in the Yarkand River. In contrast, insignificant changes in flood peaks in the Hotan River were due to insignificant changes in temperature and precipitation. The Kaidu River also experienced only minor changes, as the increase in heavy precipitation was offset by a decrease in warm temperatures. The above analysis shows that extreme changes in precipitation result in run-off extremes and that extreme temperature changes play a leading role in the Yarkand River.

10.3.4 Typical Extreme Hydrological Events in Response to Climate Change

(1) Hail and storm floods Storm floods are a common disaster in Xinjiang, especially in South Xinjiang where the main source of storm water vapor is the Indian peninsula and the Bay of Bengal. The moisture vapor flows northward to the south-center Qinghai-Tibet Plateau and then onwards into Xinjiang. Storm flooding mainly occurs in the low mountainous region, whose weather system is predominantly sub-tropical (80 %), followed by a trough from West Siberia. The center path of Xinjiang rainstorms also correspond to this weather system.

The systems which impact the Aksu River are sub-synoptic scale systems superimposed small and medium-scale systems. The sub-synoptic scale systems can be divided into three categories, namely the slot in Central Asia, the north branch

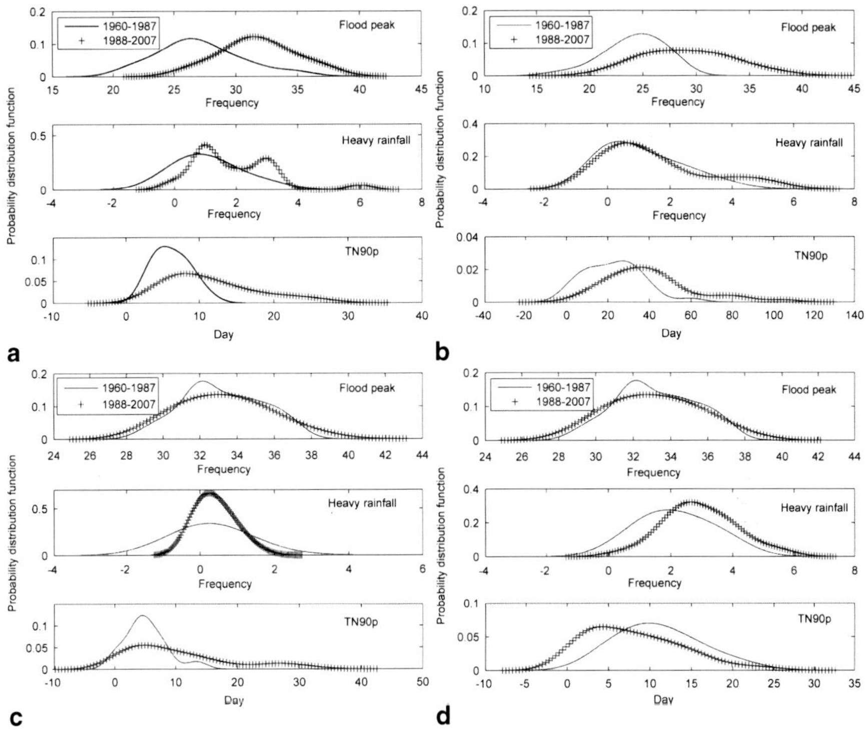

Fig. 10.25 Distribution probability of Tarim River (*a* Aksu River, *b* Yarkand River, *c* Hotan River, *d* Kaidu River)

slot, and the front small groove. The small and medium-scale systems in the lower troposphere are mesoscale shear line, mesoscale low pressure and convective cell.

Aksu Prefecture experiences the most severe hail storms due to it being controlled by the southern border vortex and zonal weather. The prefecture also has numerous leeward mountainsides and many flared valleys extending from west to east. This type of terrain results in large-scale cold air acceleration which speeds up and leads to kinetic energy increases and subsequent increases in hail opportunities. The hail in the Kashi region is often affected by vortex weather in southern Xinjiang and concentrates in Jiashi, Bachu, Shufu and Yuepuhu Counties, while hail in Bazhou and Kezhou is affected by zonal weather and vortex weather, respectively.

(2) Glacier-dammed lake outburst floods Glacier-dammed lake outburst floods (GLOFs) mainly occur in the Chogori glacier area in the Karakorum Mountains, the Tuomuer glacier area in the Tianshans, the northern slope of the Kunlun Mountains, and the Pamir glacier area. Of these areas, the highest frequency of GLOFs is found in the Yarkand River, which is located on the north slope of the Karakorum Mountains. The Yarkand River has widely developed glaciers, which is an important factor in the formation of GLOFs. The glaciers in the Karakorum Mountains show a

Table 10.9 Two-viable Wilcoxon testing the correlation among temperature, precipitation and runoff in Yarkant River

Test	Precipitation	Summer precipitation	Temperature	Summer Temperature
Z_W	−0.631	−0.754	−0.852	−1.059
Sig. (2-tailed)	0.517	0.423	0.332	0.238
Judgment	Accept H_0	Accept H_0	Accept H_0	Accept H_0
Conclusion	Peak flow is associated with precipitation and temperature			

Table 10.10 The change of the 0 °C level high of the flood peak discharge above 4,000 m^3/s

Year	Peak flow (m^3/s)	0 °C layer height difference (gpm)	
		0:00	12:00
1961	6,270	526	785
1971	4,570	835	689
1978	4,700	−486	−472
1984	4,570	633	798
1997	4,040	664	505
1999	6,070	543	496

generally retreating trend with the impact of climate warming, which triggered a significant increase in GLOF events.

The Yarkand River is a typical glacier recharge river whose peak flow is highly sensitive to climate change. Many meteorological factors, such as temperature and mountain precipitation, will affect peak flow. Relationships between mountain temperature and peak flow were tested by Two-viable Wilcoxon test (H0: the peak flow associated with temperature or precipitation), with the results (Table 10.9) showing that peak flow has a strong relationship with climate variables.

An analysis of data from the Kashi radiosonde station from 1961 to 2007 (daily readings at 0:00 and 12:00) found the region's high-altitude air temperature was related to the 0 °C layer height. The high-altitude air temperature and 0 °C layer height were lower from spring to early summer, which is beneficial for high mountain snow accumulation. The Kashi altitude temperature and 0 °C layer height rose rapidly and remained elevated for a few days from the end of July to early August, leading to the production of glacial lake outburst flooding conditions from August to September. The flood years showing peak flows of more than 4,000 m^3/s reveal similar changes at 0 °C heights (Table 10.10). Generally, differences in the 0 °C layer height eight days before and after flood events were larger than 500gpm when the GLOF occurred.

(3) Snowstorms and melt water Snowstorms and snowmelt water events mainly occur in North Xinjiang, especially in the areas around Altay, Tacheng, Ili and Changji. Snow days and precipitation from November to January showed a significant increasing trend since the 1980s, which is in accordance with overall warming trends since 1979. From this, it can be inferred that increased blizzard and winter precipitation activity in Xinjiang is linked to winter warming.

Table 10.11 Winter precipitation and cumulative depth of snowpack anomaly before snowmelt occurred (Unit: %)

Year		Tacheng	Altay	Ili	Shihezi	Changji	Urumqi
1985	Winter precipitation anomalies	35	52	14	9	−16	18
	Snow depth of ended February	26	35	16	12	−20	30
1988	Winter precipitation anomalies	31	50	76	68	31	32
	Snow depth of ended February	37	56	42	52	63	5
2005	Winter precipitation anomalies	−3.5	41	62	67	16	21
	Snow depth of ended February	−57	27	50	47	14	35
2010	Winter precipitation anomalies	110	130	87	65	81	96
	Snow depth of ended February	52	60	45	41	26	39

Winter precipitation is an important factor affecting snowmelt floods. When winter precipitation is above normal, the snow is deeper than normal. The correlation coefficient between winter precipitation and snowmelt floods is above 0.6 or more (Table 10.11), so winter precipitation can be considered a predictor of snowmelt floods. Moreover, winter precipitation anomalies and snow depth in the plains regions is generally greater in years that experience severe spring floods. The winter precipitation and snow depth anomalies were 30–70 % and 20–50, respectively (1985, 1988, 2005, 2010), which is more than 50–60 % for the historical average snow depth.

In addition to high winter precipitation levels and deep snow, low temperatures during the winter and spring warming periods will also affect snowmelt floods. In years that experience a wide range of snowmelt floods, winter temperatures are significantly lower and severe temperature changes are also observed before and after the spring. February temperatures are likewise relatively lower, and thus the temperature rise is comparatively larger (Table 10.12).

Conclusion

(1) Both precipitation and temperature have significant impacts on hydrological extremes, and exponential distribution can best fit the relationship between extremes and temperature. Moreover, in-phase relationships observed between extremes and climate variables show a significant coherence at 0.7–1.5 year period intervals.
(2) Large increases were observed in data measuring warm nights and heavy precipitation. In particular, the rise in heavy precipitation levels indicates that run-off extremes may be the joint result of extreme temperatures and precipitation. Extreme precipitation prompted significant changes in run-off extremes in the Aksu River, while extreme temperatures played a leading role in the Yarkand River.
(3) Increased snowstorm events appear to be related to increased winter precipitation and winter warming. The dominant factor for glacier-dammed lake outburst flooding is temperature, with frequency fluctuations corresponding to fluctuations in temperature.

Table 10.12 Winter temperature anomaly and spring temperature extreme change before snowmelt occurred (°C, °C/d, respectively)

Year		Tacheng	Altay	Ili	Shihezi	Changji	Urumqi
1985	Winter temperature anomalies	−2.6	−2.3	−2.1	−1.2	−1.1	−0.6
	Maximum warming in winter	2.8	2.3	1.2	1.8	2	1.5
1988	Winter temperature anomalies	−4.6	−4.8	−3.5	−2.6	−2.7	−1.7
	Maximum warming in winter	1.9	1.8	2.9	1.7	1.8	1.4
2005	Winter temperature anomalies	−3	−2.9	−2.2	−2.1	−2.1	−1.2
	Maximum warming in winter	2.1	2	1.1	2.2	2.6	2.1
2010	Winter temperature anomalies	−3.4	−3.8	−2.1	−3.6	−1.8	−1.6
	Maximum warming in winter	2.2	1.9	2.1	0.8	1.0	0.7

Summary

(1) Little information is currently available regading trends in daily climate and climate extremes in China's arid Northwest region. Based on daily precipitation and temperature records at 76 meteorological stations from 1960 to 2010, the temporal and spatial characteristics of climate extremes in the arid region were analyzed. Within this framework, 15 indices of extreme temperature and 9 indices of extreme precipitation were examined. Temperature extremes showed patterns consistent with warming, with a large number of stations showing statistically significant trends. Warming trends in indices derived from daily minimum temperatures were of greater magnitude than those from maximum temperatures. Most precipitation indices exhibited increasing trends across the region in relation to both frequency and intensity.

(2) Intra-annual characteristics and decadal changes in extreme hydrological events were examined in Xinjiang from 1901 to 2010 using concentration degree, concentration index, and Mann–Kendall tests. The results showed that extreme hydrological events occurred mainly during the summer. Intra-annual distribution was mostly concentrated in Turpan and Urumqi, followed by Kuytun–Shihezi and Aksu, while the smallest degree of concentration was observed in Altay. There were increasing trends for frequency of extreme hydrological events from 1901 to 2010, particularly post-1970s. Based on the results obtained in this study, the frequency, magnitude, and intensity of extreme hydrological events are expected to increase into the future.

(3) The spatial distribution of extreme hydrological events was also examined in Xinjiang using district data from 1901 to 2010. Frequency distribution exhibited a general symmetry along the Tianshan Mountains, distributed evenly in the Junggar and Tarim Basins. Greater extremes were observed in the northwest than in the southeast. Specifically, the maximum incidence was in the western portion of the Tianshans, with a general south-eastward decrease. There were significant regional variations in type distribution (e.g., rainstorm floods were more common in central Xinjiang). The incidence of extreme hydrological events was greatly affected by weather systems and terrain features, which made certain hazards occur in some regions but not in others.

(4) Extreme hydrological events in Xinjiang increased from 1980 onwards. The increased frequency of these events appears to be related to increases in temperature and precipitation. Moreover, extreme snow-related events are linked to winter warming, and 0 °C level height rising sharply are the main factors determining the snowmelt floods. In glacial lake outburst floods, the main control factor is temperature, with precipitation increasing the likelihood of occurrence at certain temperature conditions.

References

Alexander LV, Zhang X, Peterson TC et al (2006) Global observed changes in daily climate extremes of temperature and precipitation. J Geophys Res-Atmos 111. doi:10.1029/2005JD006290

Alexander LV, Hope P, Collins D, Trewin B, Lynch A, Nicholls N (2007) Trends in Australia's climate means and extremes: a global context. Aust Meteorol Mag 56:1–18

Alexandersson H (1986) A homogeneity test applied to precipitation data. J Climatol 6:661–675

Arnell NW (1999) Climate change and global water resources. Global Environ Change 9:31–49

Bell JL, Sloan LC, Snyder MA (2004) Regional changes in extreme climatic events: a future climate scenario. J Climate 17:81–87

Beniston M, Diaz HF, Bradley RS (1997) Climatic change at high elevation sites: an overview. Clim Change 36:233–251

Buishand TA (1982) Some methods for testing the homogeneity of rainfall records. J Hydrol 58:11–27

Caesar J, Alexander LV, Trewin B et al (2011) Changes in temperature and precipitation extremes over the Indo-Pacific region from 1971 to 2005. Int J Climatol 31:791–801

Chebana F, El Adlouni S, Bobee B (2010) Mixed estimation methods for Halphen distributions with applications in extreme hydrologic events. Stoch Environ Res Risk Assess 24:359–376

Chen B, Chao WC, Liu X (2003) Enhanced climatic warming in the Tibetan Plateau due to doubling CO2: a model study. Clim Dynam 20:401–413

Chen YN, Li WH, Xu CC, Hao XM (2007) Effects of climate change on water resources in Tarim river basin, northwest China. J Environ Sci (China) 19:488–493

Committee NS (2007) International symposium with NATO support on the problem of extreme hydrologic events. Water Resour 34:350–351

Demarsily G, Ahmed S (1987) Application of Kriging techniques in groundwater hydrology. J Geol Soc India 29:57–82

Di Baldassarre G, Castellarin A, Brath A (2006) Relationships between statistics of rainfall extremes and mean annual precipitation: an application for design-storm estimation in northern central Italy. Hydrol Earth Syst Sci 10:589–601

Ding YH, Wang ZY, Song YF, Zhang J (2008) The unprecedented freezing disaster in January 2008 in southern China and its possible association with the global warming. Acta Meteorologica Sinica 22:538–558

Durao RM, Pereira MJ, Costa AC et al (2010) Spatial-temporal dynamics of precipitation extremes in southern Portugal: a geostatistical assessment study. Int J Climatol 30:1526–1537

Endo N, Matsumoto J, Lwin T (2009) Trends in precipitation extremes over southeast Asia. Sola 5:168–171

Frich P, Alexander LV, Della-Marta P et al (2002) Observed coherent changes in climatic extremes during the second half of the twentieth century. Clim Res 19:193–212

Gao XJ, Zhao ZC, Giorgi F (2002) Changes of extreme events in regional climate simulations over east Asia. Adv Atmos Sci 19:927–942

Gemmer M, Fischer T, Jiang T, Su BD, Liu LL (2011) Trends in precipitation extremes in the Zhujiang river basin, south China. J Climate 24:750–761

Hirschi M, Seneviratne SI (2010) Intra-annual link of spring and autumn precipitation over France. Clim Dynam 35:1207–1218

Hong Y, Adler RF (2008) Predicting global landslide spatiotemporal distribution: Integrating landslide susceptibility zoning techniques and real-time satellite rainfall estimates. Int J Sediment Res 23:249–257

IPCC (2007) Intergovernmental panel on climate change. Climate change 2007: synthesis report [EB/OL]. [2007]. http://www.ipcc.ch

Jiang A, Du Y, Xie Z, Ding Y (2005) Climatic trends of heavy precipitation spatial and temporal concentration in China. Acta Geographica Sinica 60:1007–1014

Kendall MG (1975) Rank-correlation measures. Charles Griffin, London, p 202

Kiktev D, Sexton DMH, Alexander L, Folland CK (2003) Comparison of modeled and observed trends in indices of daily climate extremes. J Climate 16:3560–3571

Kioutsioukis I, Melas D, Zerefos C (2010) Statistical assessment of changes in climate extremes over Greece (1955–2002). Int J Climatol 30:1723–1737

Kumar V (2007) Optimal contour mapping of groundwater levels using universal kriging—a case study. Hydrolog Sci J 52:1038–1050

Li Z, Zheng FL, Liu WZ, Flanagan DC (2010) Spatial distribution and temporal trends of extreme temperature and precipitation events on the Loess Plateau of China during 1961–2007. Quatern Int 226:92–100

Linnenluecke M, Griffiths A (2010) Beyond adaptation: resilience for business in light of climate change and weather extremes. Bus Soc 49:477–511

Liu XD, Chen BD (2000) Climatic warming in the Tibetan plateau during recent decades. Int J Climatol 20:1729–1742

Liu CM, Chen YN, Xu ZX (2010a) Eco-hydrology and sustainable development in the arid regions of China preface. Hydrol Process 24:127–128

Liu Z, Xu Z, Huang J, Charles SP, Fu G (2010b) Impacts of climate change on hydrological processes in the headwater catchment of the Tarim River basin, China. Hydrol Process 24:196–208

Mann HB (1945) Non-parametric tests against trend. Econometrica 13:345–351

Manton MJ (2010) Trends in climate extremes affecting human settlements. Curr Opin Environ Sustain 2:151–155

Mantua N, Tohver I, Hamlet A (2010) Climate change impacts on streamflow extremes and summertime stream temperature and their possible consequences for freshwater salmon habitat in Washington State. Clim Change 102:187–223

New M, Hewitson B, Stephenson DB et al (2006) Evidence of trends in daily climate extremes over southern and west Africa. J Geophys Res-Atmos, 111. doi:10.1029/2005JD006289

Pal I, Al-Tabbaa A (2011) Assessing seasonal precipitation trends in India using parametric and non-parametric statistical techniques. Theor Appl Climatol 103:1–11

Parker BR, Vinebrooke RD, Schindler DW (2008) Recent climate extremes alter alpine lake ecosystems. Proc Natl Acad Sci U S A 105:12927–12931

Pavan V, Tomozeiu R, Cacciamani C, Di Lorenzo M (2008) Daily precipitation observations over Emilia-Romagna: mean values and extremes. Int J Climatol 28:2065–2079

Pepin NC, Lundquist JD (2008) Temperature trends at high elevations: patterns across the globe. Geophys Res Lett 35. doi:10.1029/2008GL034026

Peterson TC, Easterling DR, Karl TR et al (1998) Homogeneity adjustments of in situ atmospheric climate data: a review. Int J Climatol 18:1493–1517

Raziei T, Saghafian B, Paulo AA, Pereira LS, Bordi I (2009) Spatial patterns and temporal variability of drought in western Iran. Water Resour Manage 23:439–455

Redano A, Lorente J (1993) Modeling the spatial and temporal distribution of rainfall intensity at local scale. Theor Appl Climatol 47:25–32

Reiser H, Kutiel H (2010) Rainfall uncertainty in the Mediterranean: intraseasonal rainfall distribution. Theor Appl Climatol 100:105–121

Rocklov J, Forsberg B (2009) Comparing approaches for studying the effects of climate extremes—a case study of hospital admissions in Sweden during an extremely warm summer. Glob Health Action 2:28–38

Rusticucci M, Renom M (2008) Variability and trends in indices of quality-controlled daily temperature extremes in Uruguay. Int J Climatol 28:1083–1095
Schar C, Jendritzky G (2004) Climate change: hot news from summer 2003. Nature 432:559–560
Seidel DJ, Free M (2003) Comparison of lower-tropospheric temperature climatologies and trends at low and high elevation radiosonde sites. Clim Change 59:53–74
Shen YJ, Liu CM, Liu M, Zeng Y, Tian CY (2010) Change in pan evaporation over the past 50 years in the arid region of China. Hydrol Process 24:225–231
Shi YF, Shen YP, Kang E, Li DL, Ding YJ, Zhang GW, Hu RJ (2007) Recent and future climate change in northwest China. Clim Change 80:379–393
Strom AL, Korup O (2006) Extremely large rockslides and rock avalanches in the Tien Shan Mountains, Kyrgyzstan. Landslides 3:125–136
Su BD, Jiang T, Jin WB (2006) Recent trends in observed temperature and precipitation extremes in the Yangtze river basin, China. Theor Appl Climatol 83:139–151
Sun G, Chen Y, Li W, Pan C, Li J, Yang Y (2012) Intra-annual distribution and decadal change in extreme hydrological events in Xinjiang, northwestern China. Nat Hazard. doi:10.1007/s11069-012-0242-1
Sun G, Chen Y, Li W, Pan C, Li J, Yang Y (2013) Spatial distribution of the extreme hydrological events in Xinjiang, north-west of China. Nat Hazard. doi:10.1007/s11069-013-0574-5
Tank A, Konnen GP (2003) Trends in indices of daily temperature and precipitation extremes in Europe, 1946–99. J Climate 16:3665–3680
Toreti A, Desiato F (2008) Changes in temperature extremes over Italy in the last 44 years. Int J Climatol 28:733–745
Torrence C, Compo GP (1998) A practical guide to wavelet analysis. Bull Am Meteorol Soc 79:61–78
Trenberth KE (1999) Conceptual framework for changes of extremes of the hydrological cycle with climate change. Clim Change 42:327–339
Tromel S, Schonwiese CD (2007) Probability change of extreme precipitation observed from 1901 to 2000 in Germany. Theor Appl Climatol 87:29–39
Turkes M, Sumer UM (2004) Spatial and temporal patterns of trends and variability in diurnal temperature ranges of Turkey. Theor Appl Climatol 77:195–227
Turkes M, Koc T, Saris F (2009) Spatiotemporal variability of precipitation total series over Turkey. Int J Climatol 29:1056–1074
Vincent LA, Peterson TC, Barros VR et al (2005) Observed trends in indices of daily temperature extremes in South America 1960–2000. J Climate 18:5011–5023
Wang XL (2008a) Accounting for autocorrelation in detecting mean shifts in climate data series using the penalized maximal t or F test. J Appl Meteorol Climatol 47:2423–2444
Wang XL (2008b) Penalized maximal F test for detecting undocumented mean shift without trend change. J Atmos Oceanic Technol 25:368–384
Wang YQ, Zhou L (2005) Observed trends in extreme precipitation events in China during 1961–2001 and the associated changes in large-scale circulation. Geophys Res Lett 32:L09707
Wang XL, Chen H, Wu Y, Feng Y, Pu Q (2010) New techniques for the detection and adjustment of shifts in daily precipitation data series. J Appl Meteor Climatol 49:2416–2436
Wang Y, Xu X, Du YG, Tang JP (2011) Variations of temperature and precipitation extremes in recent two decades over China. Atmos Res 101:143–154
Wang HJ, Chen YN, Chen ZS (2012) Spatial distribution and temporal trends of mean precipitation and extremes in the arid region, northwest of China, during 1960–2010. Hydrol Process. doi:10.1002/hyp.9339
Wang HJ, Chen YN, Chen ZS, Li W (2013a) Changes in annual and seasonal temperature extremes in the arid region of China, 1960–2010. Nat Hazards 65:1913–1930
Wang HJ, Chen Y, Xun S, Lai D, Fan Y, Li Z (2013b) Changes in daily climate extremes in the arid area of northwestern China. Theor Appl Climatol 112:15–28
Welbergen JA, Klose SM, Markus N, Eby P (2008) Climate change and the effects of temperature extremes on Australian flying-foxes. Proc R Soc B-Biol Sci 275:419–425

Williams CJR, Kniveton DR, Layberry R (2010) Assessment of a climate model to reproduce rainfall variability and extremes over Southern Africa. Theor Appl Climatol 99:9–27

Xu ZX, Chen YN, Li JY (2004) Impact of climate change on water resources in the Tarim river basin. Water Resour Manage 18:439–458

Xu J, Li W, Ji M, Lu F, Dong S (2010a) A comprehensive approach to characterization of the nonlinearity of runoff in the headwaters of the Tarim river, western China. Hydrol Process 24:136–146

Xu Y-P, Booij MJ, Tong Y-B (2010b) Uncertainty analysis in statistical modeling of extreme hydrological events. Stoch Env Res Risk A 24:567–578

Yao C, Qian W, Yang S, Lin Z (2010) Regional features of precipitation over Asia and summer extreme precipitation over southeast Asia and their associations with atmospheric-oceanic conditions. Meteorol Atmos Phys 106:57–73

You QL, Kang SC, Aguilar E, Yan YP (2008) Changes in daily climate extremes in the eastern and central Tibetan Plateau during 1961–2005. J Geophys Res-Atmos 113. doi:10.1029/2007JD009389

You Q, Kang S, Pepin N, Fluegel W-A, Yan Y, Behrawan H, Huang J (2010) Relationship between temperature trend magnitude, elevation and mean temperature in the Tibetan Plateau from homogenized surface stations and reanalysis data. Global Planet Change 71:124–133

You QL, Kang SC, Aguilar E, Pepin N, Flugel WA, Yan YP, Xu YW, Zhang YJ, Huang J (2011) Changes in daily climate extremes in China and their connection to the large scale atmospheric circulation during 1961–2003. Clim Dynam 36:2399–2417

Yue S, Pilon P, Phinney B, Cavadias G (2002) The influence of autocorrelation on the ability to detect trend in hydrological series. Hydrol Process 16:1807–1829

Zhang Q, Xu C-y, Gemmer M, Chen YD, Liu C (2009) Changing properties of precipitation concentration in the Pearl river basin, China. Stoch Environ Res Risk Assess 23:377–385

Zhang QA, Xu CY, Zhang ZX, Chen X, Han ZQ (2010) Precipitation extremes in a karst region: a case study in the Guizhou province, southwest China. Theor Appl Climatol 101:53–65

Chapter 11
Water Resource Management

Yaning Chen, Yanjun Shen and Weihong Li

Abstract China's arid Northwest region has a unique inland water cycle that maintains a fragile balance. Water resources in this area come from high mountain glaciers, seasonal snowmelt, middle mountain precipitation, and water emanating from fractured bedrock in the lower mountains. The region's main problems with water resource development involve water shortages arising either from natural causes (i.e., climate and weather anomalies) or from engineering projects, including industrial and domestic wastage. Besides, we analyzed the spatial and temporal variations of irrigation water demand as well as crop water requirement, the results suggested that the demand of irrigation water in the arid region of Northwestern China showed increasing trend during the past two decades, which mainly caused by fast increase in cotton cultivation areas, because irrigation water requirement for cotton was much larger than the other crops. To resolve these issues, the following water resource management countermeasures are suggested: the construction of water management systems and mechanisms; the reformation of the current water resource management system; changes in water management properties; the unified management of surface and groundwater; the reduction of use; and the broadening of water sources.

Keywords Water resources · Irrigation water demand · Crop water requirement · Surface and groundwater · Water resource management

Y. Chen (✉) · W. Li
State Key Laboratory of Desert and Oasis Ecology,
Xinjiang Institute of Ecology and Geography, Chinese Academy of Sciences,
No. 818 South Beijing Road, 830011 Urumqi, Xinjiang, China
e-mail: chenyn@ms.xjb.ac.cn

W. Li
e-mail: liwh@ms.xjb.ac.cn

Y. Shen
Key Laboratory of Agricultural Water Resources,
Center for Agricultural Resources Research, Chinese Academy of Sciences,
No. 2 Huaizhong Road, 050021 Shijiazhuang, China
e-mail: yjshen@sjziam.ac.cn

Y. Chen (ed.), *Water Resources Research in Northwest China*,
DOI 10.1007/978-94-017-8017-9_11,

China's current total water resources, measuring 2.8×10^{12} m^3, are ranked sixth in the world. However, the country's water resources per capita are far less (only 2,300 m^3), which is about 1/4 of the world's average value. In this regard, China is one of the poorest countries globally (ranking 109th) while at the same time being the world's largest consumer of water. Recent estimates calculate China's freshwater consumption at $6{,}500 \times 10^8$ m^3, which is 14 % of the world's present consumption and 1.4 times that of the US.

The arid region of northwest China is located in the hinterland of the Eurasian continent and stretches from 35 to 50°N and 73 to 106°E. It lies west of the Helan and Wushao Mountains and north of the Kunlun Mountains. Geopolitically, the area incorporates the Xinjiang Uyghur Autonomous Region (Xinjiang for short), the Hexi Corridor in Gansu Province, and the western portion of the Helan Mountains in Inner Mongolia. The land mass comprises approximately 2.5 million km^2 and accounts for one quarter of the total land area of China.

Situated far from the sea, the relatively secluded Northwest has a typical continental climate marked by extremely low precipitation. Like arid regions in Africa, West Asia and Australia, China's Northwest has classic topography and geomorphology patterns such as the Altai, Tianshan, Karakoram, Kunlun, Altun and Qilian Mountains. These high-altitude mountains intercept the transmission of water vapor, resulting in enhanced precipitation in the immediate area and the provision of a steady stream of water for the formation and development of desert oases in the lowlands.

Due to the barrier of the towering mountain peaks, precipitation in mountainous areas is relatively abundant and serves as the formation zone for runoff. In contrast, precipitation in the plains areas is scant and serves as the runoff's dissipative zone. Average annual precipitation in the plains is less than 100 mm, while in the Tarim Basin in Xinjiang, annual precipitation is less than 50 mm. Such moderate amounts are of little hydrological significance and rarely produce surface runoff. Combined with the dry climate and strong evaporation, water scarcity is much more prevalent here than elsewhere in China. Hence, the natural shortage of water is the "bottleneck" constraining the region's sustainable economic and social development.

In addition to being a foundational resource for development in northwest China, water is also important for the realization of China's grand western development strategy goals. Average annual precipitation in the Northwest is about 230 mm, but the evaporation capacity is 8–10 times the precipitation average. The area's total water resources measure approximately 197.9 billion m^3, accounting for 5.84 % of the nation's total water resources. This is immensely disproportionate to the Northwest's land area in relation to that of the rest of the country (the Northwest covers one-quarter of China's land area). In both the Northwest and the rest of China, 87.35 % of the water resources are surface water and only 12.65 % are groundwater.

Arable land in the Northwest accounts for 5.6 % of the nation's total, while the population in only 2.1 %. The per capita possession of water resources is approximately 80 %, which gives people a false sense that there is no water shortage. However, in measuring the relationship between land area and water resources, we see that the per square kilometer water resource is only 44×10^3 m^3/km^2, making

Table 11.1 Water resources in arid areas of Northwest China

Regions	Surface water/10^8 m^3	Ground water/10^8 m^3	Total water/10^8 m^3
North Xinjiang	439.00	43.38	482.38
South Xinjiang	445.00	42.26	487.26
Hexi Corridor	70.00	11.60	81.60
Western Mongolia	15.10	43.10	58.20
Total region	969.10	140.34	1,109.44

it the most scarce in China. In base terms, the Northwest is largely an "irrigated oases", with water resources for oasis agricultural mainly coming from snowmelt and mountain precipitation. Of this, about 93 % is used for agricultural production. The current situation of water resources in arid areas of northwest China is shown in Table 11.1.

11.1 Characteristics of Water Resources in Northwest China

11.1.1 Formation of Water Resources

The arid region of Northwest China, which includes the Tianshan, Kunlun, Altai and Qilian Mountains, is one of the most developed mountainous regions in central Asia and is also an important component of the global water cycle. The Northwest has a unique inland water cycle that maintains a precarious balance. There, the water cycle and water resources have significant characteristics, the main aspects of which are the relatively independent formation and dissipative zones: the mountain areas are the formation zone, and the plains, oases and desert are the dissipative zones. The spatial and temporal distribution of water resources is quite uneven and has been impacted of late by global warming.

Generally, the formation of water resources arises from glaciers, snow, groundwater, lakes and surface runoff. More specifically, water resources are created by high mountain glaciers, seasonal snow melt, mid-mountain precipitation and fractured bedrock in the lower portions of the mountains. In the Tianshan Mountains, runoff is replenished by snowmelt water, which accounts for 30–50 % of water resources in the western part of the mountains, while in the eastern and western Qilian Mountains the amount is somewhat less (10–30 %), and in the northern Kunlun Mountains it is slightly more (40–60 %). In the Altai Mountains, the western Junggar Basin and the eastern portions of the Hexi Corridor, runoff is replenished both by snowmelt water and precipitation. The runoff recharge from seasonal snow in the Altais is up to 50 % and often occurs in the low and middle mountain belts. The precipitation recharge in the eastern part of the Hexi Corridor is larger (upwards of 60 %), while upstream of rivers, the recharge of runoff from groundwater generally accounts for 10–50 % of the total annual runoff, although in some areas that amount is 60–80 %.

11.1.2 Occurrence Form of Water Resources

In the northwest arid region, water resources exist mainly as glacier water, precipitation, runoff, lakes (reservoirs), groundwater, soil and other forms.

11.1.2.1 Precipitation Resources

Precipitation is the fundamental source of all forms of water resources in China's northwest. It not only determines the moisture conditions throughout the region, but also directly affects the runoff, groundwater recharge, glaciers and snow cover distribution. Although precipitation in the Northwest has quite different space allocations, in the main, there is more precipitation in the mountains and less in the plains. The plains areas usually receive less than 100 mm of precipitation annually. Notable exceptions to this rule include the eastern Hexi Corridor, where annual precipitation is around 100–250 mm, and Ruoqiang county (southeastern Tarim Basin), where annual precipitation is less than 10 mm.

Generally speaking, China's precipitation decreases from the sea-fringed southeast to the inland Northwest. Precipitation contour maps reveal that the amount of annual average precipitation in the arid northwest area is more than $5,000 \times 10^8$ m^3, equivalent to an average depth of 175 mm. However, about half of that precipitation falls in the mountains. Moreover, according to estimates from the Xinjiang Meteorological Bureau, only a small portion (around 20 %) of the water vapor in the air actually falls to the ground.

Precipitation in China's northwest also exhibits significant seasonal differences. For instance, the Hexi Corridor and eastern Xinjiang receive 60 % of their annual precipitation in the summer, while northern Xinjiang and the Tarim Basin receive 40 % during the same time frame. In contrast, in the Yili-Tacheng region, most of the precipitation occurs mainly from December to February. Overall, the spatial and temporal distribution of precipitation is extremely imbalanced.

Due to this imbalance, water resources and their consequent use have imposed limitations on further development. The concentration of precipitation in the summer often leads to storms and floods, which has a direct threat on the plains oases. Meanwhile, water shortages in the spring means there is insufficient water for agricultural irrigation and industrial production, or even for personal consumption. Moreover, and perhaps most importantly, the exploitation of water resources has led to ecosystem degradation, species diversity reduction and intensification of the desertification process.

11.1.2.2 Glacier Water Resources

The arid region of northwest China includes the Tianshan, Kunlun, Karakoram, Altun and Qilian Mountains, all of which have extensive glacial coverage. According to statistics, the glacier area in the Northwest measures 32,931 km^2, which is nearly half of the total glaciated area in China. The "Glacier Inventory of China" (1986) estimates the number of glaciers in Xinjiang to be around 18,300 and to cover an area just over 24,721 km^2. Glacier water reserves measure 2,361.12 billion m^3 and account for 46.8 % of the total water reserves in China.

Glaciers in Xinjiang are mainly located in the Tianshan, Kunlun, Karakorum and Altai Mountains. The Kunlun glaciers cover the most ground (12,266.29 km^2) and comprise 42.5 % of the total glaciated area in Northwest, while glaciers in the Tianshans cover 9,235.96 km^2 (32 % of glaciated land in the Northwest) and those in the Altai Mountains cover 296.75 km^2.

Glacier meltwater is the most important water resource in the Northwest. In the Tarim Basin's Aksu, Hotan and Yarkand Rivers, the replenishment of glacial meltwater accounts for 35–65 % of the annual runoff, leading to water shortages in spring and floods in the summer. However, inter-annual variations of water resources are relatively stable.

11.1.2.3 Snow Resources

The vast mountainous regions of China's Northwest provide a favorable space for seasonal snowfall, which in turn supplies a rich source of water for rivers in the arid zone. Seasonal snow is 'stored' on the basin surface during winter and then melts with rising temperatures in the spring to recharge the rivers. Rivers in the Altai Mountains in northern Xinjiang, along with those in the western mountains of Junggar, and the Qilian Mountains south of the Hexi Corridor, are mostly supplied by seasonal snowmelt water. Rivers of this type usually have large flow distribution during the spring, generally accounting for 20–45 % of annual runoff. There are two peak periods annually: one in spring, and the other in summer, although some rivers only have one peak period (spring). Some small rivers, which are mainly supplied by seasonal snowmelt water, show larger inter-annual variability of runoff recharge than glacier-melt rivers because these rivers have no glacial meltwater regulation.

11.1.2.4 Lake Resources

Thousands of lakes dot the Northwest, covering a total area of 17,000 km^2. Of these, more than 400 lakes have an area larger than 1 km^2. These lakes are located in the foothills and upstream of rivers and have a water storage capacity of 300×10^8 m^3 (Zhang and Yang 2001). In addition, numerous salt lakes are located downstream.

Large-scale land development over the past 50 years has led to the interception and diversion of water in the upper and middle rivers. Consequently, lake water shortage has become increasingly evident and some larger lakes have rapidly shrunk or even dried up altogether (Wang and Dou 1998). West Juyanhai Lake, downstream in the Heihe River, dried up in the early 1960s, forming a 350 km^2 barren salt marsh. Similarly, East Juyanhai Lake, a brackish lake, dried up recently due to water shortages.

Xinjiang boasts the most lakes in the Northwest, accounting for 7.1 % of the total lake area in China and covering 5,072 km^2. Of these, 137 lakes are larger than 1 km^2 and 32 are larger than 10 km^2, with an area of 4,828 km^2 (Ma and Wu 2009). Xinjiang-area lakes are categorized according to salinity levels as being either freshwater, brackish, or salt lakes and are mostly located in enclosed or semi-enclosed inland basins. Under the influence of the arid climate, there is a persistent deficit of lake water. Some lakes experience intermittent (or seasonal) shrinking or even dry up completely. For instance, Aibi Lake shrank from 1,070 to 589 km^2 between 1950 and 1972 due to persistent drought, with an average annual reduction of 21.9 km^2. As of 2006, the lake area measured only 550.6 km^2. Manas Lake reduced from 550 km^2 in 1950 to 0 in 1976 due to the disappearance of water in the Manas River. In 2003, Manas Lake reformed a 221 km^2 water area, but later atrophied.

In Xinjiang, Bosten Lake used to be the largest inland freshwater lake, covering an area of about 1,000 km^2. However, its water environment has drastically changed over the past 50 years. As of 2012, the lake water level has dropped to 1,045.05 m, surpassing the critical minimum level. Moreover, the lake's salinity has risen from 1.17 g/L in 2003 to 1.5 g/L in 2012.

The destruction of lakes and the deterioration of water quality in China's northwest are closely related to human activities areas across the zone, with large-scale development of land and water resources being the main reason for the degradation of ecosystems.

11.1.2.5 Shallow Groundwater Resources

Groundwater recharge plays a key role in overall water resources in the Northwest. Basin groundwater recharge and surface water sources are the same two forms and share close hydraulic connection, such that groundwater levels change according to surface runoff and exploitation conditions. During wet periods, the groundwater level is high, whereas during the dry season, groundwater levels are low. In the past 10 years, with the expansion of the oasis and an increase in irrigated areas, exploitation of the groundwater likewise increased and groundwater levels continued to decline. In most regions, the depth of groundwater extraction wells dropped from 100 to 150 m. In some areas, where depths exceed 200 m, serious over-exploitation of groundwater has occurred.

In the Northwest, water circulation mainly occurs between surface water and groundwater, and in the inland basin area, this transformation and cycling is ongoing. For example, in the Hexi Corridor, this conversion process can be repeated 2 or 3 times. Water recycles from the upper piedmont river seepage, overflowing into the alluvial fan front and outcropping into the river. In the water cycle, the amount of water reduces each time until springs gradually disappear. By the second or third cycle, the water is consumed in evaporation or goes underground, thus ending the cycle.

According to the plains hydrogeological survey, by measuring the leakage of the riverbed along with observations of canal seepage, field coefficiency, and comparisons with the water balance control theory, the estimation of precipitation infiltration of the river and canal system in the piedmont region and field penetration can be calculated. In Xinjiang, heavy rains and floods on the piedmont recharge groundwater also have to be considered. According to the statistical calculations of five inland basins in northwest China, shallow plain natural groundwater recharge was 424.63×10^8 m^3, indicating that the 60–86 % of piedmont natural groundwater resources comes from surface runoff transformation.

11.1.3 Basic Characteristics of Water Resources

11.1.3.1 Uneven Spatial and Temporal Distribution

Water shortage and the uneven spatial distribution in the Northwest area greatly reduce the effectiveness of water resources. Table 11.1 shows that the spatial distribution of water resources in the Northwest is mainly concentrated in Xinjiang; furthermore, water resources in Xinjiang account for about 87.4 % of total water resources throughout the arid northwest. Western Inner Mongolia rarely produces runoff, and surface water there accounts for only 1.6 % of water resources. In regional distribution, water and soil resources are poorly matched, which leads to poor resource development, further reducing the effectiveness of water use. Despite its large size, the Northwest has relatively few water resources. For instance, the Hexi Corridor's land area accounts for 68.2 % of Gansu province, but its water resources contribute only 27.3 %.

Inter-annual variation of water resources in the Northwest has an uneven distribution throughout the year, with spring being the most water-scarce period, especially in the Tarim Basin. Due to the lack of large inter-annual regulatory reservoirs, spring droughts become an important feature constraining the production of oasis agriculture.

11.1.3.2 Unique Hydrological Characteristics

Most rivers in the Northwest are inland rivers. The basic topography distribution of the mountain-basin pattern determines obvious inland characteristics of the water resources. Other than for the Irtysh River in northern Xinjiang (which flows into the Arctic Ocean), all other inland rivers in China's arid northwest region originate in the mountains and flow to the plains and basins, forming a heart-shaped water system. Due to the impact of geological structure and natural vertical zonation, each inland river constitutes a separate hydrological unit and complete system that typically flows through landscape units such as hills, piedmonts, oases, plains and deserts, from source to end. Hilly areas are the runoff formation zone, with surface runoff basically representing the total quantity of each river or basin. In contrast, the piedmont oasis plains is the runoff dissipation area. After the surface runoff comes into the piedmont oases, the runoff not only experiences surface-groundwater repeated conversion but also is absorbed by the soil and vegetation through

Table 11.2 Glacier meltwater and their recharge proportions to rivers in the northwest arid area

Inland river system	Glacier area/ hm^2	Percentage of glacier area/%	Runoff/$10^8 m^3$	Glacier melt runoff/$10^8 m^3$	Proportion of runoff recharged by glacier melt/%
The Hexi Corridor	1,334.75	3.77	72.7	9.99	13.8
The Junggar Basin	2,254.10	6.37	125.0	16.89	13.5
The Ili River Basin	2,022.66	5.72	193.0	26.41	13.7
The Tarim River	19,888.81	56.2	347.0	133.42	38.5

From Shi 2000, Glaciers and Related Environments in China

evaporation and transpiration. Then, through industrial and agricultural production usage, the water is consumed and lost to the atmosphere.

Precipitation in the Northwest is limited, with potential evaporation in the plains measuring 2,500–3,000 mm. Under the natural state, and in addition to mountain precipitation (which is greater than evaporation), the inland river basin presents an equivalent means of evaporation and precipitation due to its closeness. Runoff is produced within the watershed basin, while consumption, precipitation and the resulting runoff are utilized by natural inland ecosystems. For example, runoff from a mountain area flows 90 % into the ground when passing by the piedmont Gobi, where it forms groundwater. At the leading edge of the Gobi, 40–50 % of the groundwater overflows and becomes the main source of irrigation water; the other 40–50 % from the underground runoff flows through the lower plains and evaporates.

Mountain runoff in the arid zone is regulated by snowmelt water supply and so remains relatively stable. Meltwater recharge of the Tarim River in southern Xinjiang accounts for about 38.5 %, while snowmelt water from the Qilian Mountains recharges approximately 13.8 % of three inland river system in the Hexi Corridor. The Altai River in northern Xinjiang is mainly supplied by mountain snowmelt water and precipitation. The proportion of different mountain river recharge is relevant to their location and the amount of snow they receive (Table 11.2). As a general rule, rivers dominated by snowmelt water have a stable runoff, and Cv values are usually below 0.3. There are 35 rivers in the Heihe River Basin that originate in the Qilian Mountains, of which 18 are longer than 100 km^2 and 24 have a surface runoff greater than $1,000 \times 10^4$ m^3. In this area, water resources have typical arid inland characteristics, meaning that surface runoff forms mainly in the mountains (by way of precipitation and snowmelt water) and evaporation is high in the plains. There, water formation conditions are poor and have essentially no runoff.

11.1.3.3 Multiple Sources for Water Resources

Water resources in northwest China come mainly from the mountains and are comprised of glacier/snow meltwater in alpine regions, precipitation in the mid-mountain forest areas, and bedrock fissure water in the low mountainous areas. Water recharge shows characteristics of conspicuous vertical zonality. With altitude changes, the natural conditions and precipitation patterns vary, as does the recharge character. Rivers in alpine areas above the snow line are mainly recharged by glacier/snow meltwater, which is an important source of runoff. When the temperature is low, precipitation falls as snow or ice, and part of it is then stored to be developed into a glacier. Glaciers are a special form of water resource in the Northwest and are an important solid water resource as well as an important recharge source for inland river basins. The glacier-melt runoff amounts to approximately 22×10^8 m^3 every year and mainly occurs in the summer (June–August). The river runoff volume is concentrated in these 3 months and generally accounts for 50–60 % of the annual runoff.

Rivers in the middle and low mountains are recharged by seasonal snow melt, precipitation, and also bedrock fissure water (groundwater). The constituents of water resources of some large and medium rivers in the Tianshan Mountains are related to the hydrological regime. These rivers originate from high mountains and flow through the low mountains. The water sources often come from a variety of supply sources, such as glacier/snow melt in the high mountain areas and summer precipitation in the mid-low mountainous areas. The annual distribution of runoff is characterized by a spring flood period followed by a summer flood period, which can extend from May to September.

In this region of China, most rivers originate mainly from mountainous areas that have abundant rainfall, although some rivers are birthed in lower mountainous regions and are formed by precipitation recharge. In addition to a small amount of snowmelt water recharge, precipitation is the main recharge source for small rivers flowing down the southern and northern slopes of the eastern Tianshan Mountains. Rivers in the eastern Qilian Mountains, such as the Shiyang River system and the Heihe River system, are mainly recharged by rainfall, which supplies 60–80 % of the runoff. The pattern of annual runoff distribution of the rivers dominated by rainfall is the same as that of mountainous precipitation, which is characterized by a high proportion during the flood season (runoff during May–September can account for 70–80 %). Water resources in northern Xinjiang are mainly composed of snowmelt and precipitation with less glacial meltwater.

Research has discovered that water resources in the Northwest have significantly changed over the past 50 years. However, the change pattern varies from river to river due to the different recharge proportion of glacier-meltwater. For rivers with a large proportion of glacier-meltwater recharge, the runoff shows notable increases, whereas rivers with little glacier-melt recharge show little increase in runoff. For example, the percentage of glacier-melt runoff is 9.1 % in the Urumqi River in North Xinjiang, with a corresponding runoff increase rate of approximately 3.0 %; however, the glacier-melt runoff is 47 % in the Kumalake River in South Xinjiang, and its corresponding increase rate of runoff is 31.2 %.

11.1.3.4 The High Runoff-Concentration Degree

The uneven distribution in annual streamflows in northwest China is determined by the composition of water resources. Generally, rivers recharged mainly by mountainous snow/glacier-melt water are characterized by a later flood season and a concentration of summer runoff. June–September marks the biggest runoff; during these four consecutive months, 70–85 % of the maximum monthly runoff usually comes in July or August. In contrast, the minimum monthly runoff usually happens in January or February. Rivers recharged by glacial/snow meltwater and precipitation are characterized by a spring flood followed by a summer flood, and have longer flood durations. The maximum monthly runoff occurs in July and the minimum monthly runoff in February.

The runoff-concentration degree and runoff-concentration period reflect the time of a year. The larger the concentration degree, the more concentrated the runoff; otherwise, the more dispersed the runoff. The runoff-concentration period is the time when the largest runoff occurs. The runoff-concentration degree of the rivers is generally 40–50 % in southern China and 60–70 % in northern China. However, in the Altai and Kunlun Mountains, the runoff-concentration degree is over 70 %. The runoff-concentration periods for most rivers are consistent with the highest rainfall. Hence, the runoff-concentration periods occur from mid-July to early August for the rivers in southern China and in September for rivers in northern China. In the Northwest, runoff-concentration periods occur in May for rivers in the western mountainous area of the Junggar Basin, in June for rivers originating from the Altai Mountain, and in July and August for other rivers.

The runoff-concentration degree of inland rivers in the Northwest China is closely related to the recharge pattern. Rivers recharged by groundwater or lake water show relatively uniform distributions of runoff within the year with the runoff-concentration degree below 15 %, while rivers recharged by glacier/snow meltwater show an extremely uneven distribution of runoff within a year, with high runoff-concentration degrees. The Kunlun and Altai mountains boast the highest runoff-concentration degree in China. For example, the runoff-concentration degree is 70 % for the Yurungkash and Karakash Rivers and 65.62 % for the Yarkand River.

Glacier/snow meltwater is an important part of water resources in the arid Northwest. Cv values of rivers recharged by permanent snow and glacier meltwater generally range from 0.19 to 0.25, and the Cv values of rivers recharged by glacier/snow meltwater and precipitation range from 0.10 to 0.20.

Rivers mainly recharged by permanent glacier/snow meltwater are regulated annually by the "mountainous solid reservoir", and the rivers mainly recharged by groundwater are regulated by the "underground reservoir", thus the Cv values are small. Rivers recharged by glacier/snow meltwater and precipitation also have small Cv values. This is because the rivers are recharged by glacier/snow meltwater in the dry years and by precipitation in the wet years. The mutual compensation effect of the alpine meltwater and precipitation in the middle and low mountainous areas for each other leads to the small Cv values of this type of river. Since rivers are recharged by different sources in the different elevation bands (i.e., the alpine regions are recharged mainly by permanent glacier/snow meltwater, and the middle and low mountainous regions are recharged by seasonal snowmelt water and precipitation), the Cv values increase from the river sources to their outlets.

11.1.3.5 Frequent Conversion of Surface Water and Groundwater

Northwest China is characterized by frequent conversion of surface water and groundwater. The river system is mainly composed of inland rivers. Due to the geological structure and vertical variation of natural conditions from the river source to the outlet, the rivers generally flow through two totally different areas (namely, the runoff formation zone and runoff loss areas), so the conversion of surface water and groundwater is intensive. The surface water is recharged by groundwater in the alpine region and the groundwater is recharged by surface water in the piedmont plain by channel leakage and farmland leakage. Data indicate that about 80 % of groundwater in the piedmont plains is recharged by surface water through leakage, while the surface river water comes partly from groundwater through the outflow of springs in Xinjiang and the Gansu Corridor. This kind of surface water transformation of groundwater can be repeated several times.

The Northwest is dominated by inland rivers with closed drainage systems. Affected by westerly flow, this region is characterized by scant water vapor input, intense evaporation, single water cycling pattern, low conversion efficiency of precipitation to runoff, and inactive inner water cycling. Rivers originating from the mountainous areas flow toward the basin, forming a centripetal type of drainage. Mountainous ecosystems with vertical differentiation patterns act as both the water source and water conservation regime. The ecosystem in this region is highly vulnerable and sensitive to water, since runoff recharged by glacier/snow meltwater and precipitation determines the scope and scale of oasis and desert vegetation. Therefore, climate change will directly affect the amount of water and the spatial and temporal distribution by affecting the elements of mountainous water cycle. Thus, climate change influences the fragile ecosystem and landscape patterns and also has a significant impact on economic and social development, both of which are largely based on the agriculture in the arid oasis region.

11.1.3.6 Increasing Instability of Water Resources

Climate change will further increase the uncertainty of water resources in northwest China, resulting in extreme hydrological events such as floods and droughts. Located in middle latitude zone, the arid Northwest China is one of the most ecologically sensitive areas in the world. Since the 1980s, the region has shown a warming trend (0.343 ° C/10a) that is significantly higher than the rate of global warming over the same period, indicating that this area is especially sensitive to global warming. The water system, which mainly relies on the glacier/snow meltwater, is very fragile under the combined effects of climate change and human activities. To some extent, the evaporation and glacier-melt rate increase with temperature, causing variations in glacier/snow meltwater and water volume. With future warming, the temperature is expected to rise more than 1 ° C by 2030. Given the general uncertainty over precipitation amounts, combined with the substantial increase of water use caused by economic development and the further shrinking of glacier/snow cover and lakes, the mountains' water sources and their regulation function are in decline, a situation that will exacerbate runoff instability.

Meanwhile, climate change may also lead to water mutations, which further reduces the stability of the water system and regeneration capacity of water resources in the arid zone and increases the frequency and intensity of extreme climatic/hydrological events. For example, continuous floods occurred in northern Xinjiang in 2006, while severe drought occurred in the south. A few years later, in 2008, Xinjiang experienced the worst drought in 30 years, followed by heavy snow in 2009–2010 (a rarity for the region), which caused avalanches and severe snowmelt flooding. Climate change and the extreme hydrological events it triggers have increased the instability of agricultural production in the oasis region and also increased the risk of safe operation of major projects. Studies indicate that climate change will lead to increased water shortages and instability of water resources in China's northwest.

In conclusion, it is imperative to systematically study the impact of climate change on spatial and temporal distribution of water resources in northwest China and to understand the key processes of the water cycle and the vulnerability mechanism of the water systems. Knowledge about the effects of climate change on water resources can be applied to improve the ability of water systems in order to weather the changes and develop successful coping strategies into the future.

11.2 Problems in Water Resource Development and Utilization

Water shortage in northwest China is the core issue preventing that region's sustainable economic and social development. The problem includes natural water shortages, water shortages due to water-course engineering, and waste-induced water shortages. Proper analysis of the issues surrounding the development and utilization of water resources can promote feasible water resource management and improve water use efficiency.

11.2.1 Resource Water Shortage

Beyond a doubt, water is the most important basic resource for economic and social development in northwest China. Therefore, its appropriate management is crucial for the realization of the nation's grand western development strategy goals. At the same time, water shortage is one of the most prominent issues plaguing the Northwest, where average annual precipitation is about 230 mm and evaporation capacity is 8–10 times that. The area's total water resources measure approximately 197.9 billion m^3, accounting for 5.84% of the nation's total water resources. This is immensely disproportionate to the Northwest's land area in relation to that of the

Table 11.3 Water resources in arid areas of Northwest China

Regions	Surface water/10^8 m^3	Ground water/10^8 m^3	Total water/10^8 m^3
North Xinjiang	439.00	43.38	482.38
South Xinjiang	445.00	42.26	487.26
Hexi Corridor	70.00	11.60	81.60
Western Mongolia	15.10	43.10	58.20
Total region	969.10	140.34	1,109.44

rest of the country (the Northwest covers one-quarter of China's land area). In both the Northwest and the rest of China, 87.35 % of the water resources are surface water and only 12.65 % are groundwater.

Arable land in the Northwest accounts for 5.6 % of the nation's total, while the population in only 2.1 %. The per capita possession of water resources is approximately 80 %, which gives people a false sense that there is no water shortage. However, in measuring the relationship between land area and water resources, we see that the per square kilometer water resource is only 44×10^3 m^3/km^2, making it the lowest in China. In other words, the Northwest is largely an "irrigated oasis", with water resources for oasis agricultural mainly coming from snowmelt and mountain precipitation. Of this, about 93 % is used for agricultural production. The current situation of water resources in northwest China is shown in Table 11.3.

The uneven temporal-spatial distribution of water resources exacerbates the water shortage. For example, about 50 % of Xinjiang's economic and technological strength is concentrated in the Tianshan Slope Economic Zone, which is also the focus of social development in Xinjiang. However, water resources in this region account for only 7.4 % of the area's water resources. The excessive exploitation of groundwater due to severe shortages in some regions (e.g., Urumqi, Shihezi and Changji) has led to varying degrees of funnels. Furthermore, the extremely uneven temporal distribution of water resources results in the formation of spring droughts, summer floods, autumn water shortages and dry winters, especially in southern Xinjiang. The uneven spatial and temporal distribution of water resources also highlights the contradictory needs of the ecology and the economy, with economic and social development clashing with ecological protection.

The differential between water supply and demand first became prominent in the 1950s, when the burgeoning agricultural sector raised the demand for water. Since then, farming in the arid northwest has experienced rapid development, and with the expansion of agricultural land, water use for agricultural purposes has likewise increased. For example, agricultural land in Xinjiang increased from 121×10^4 hm^2 in 1949 to 310×10^4 hm^2 in 1990 and then to 412.46×10^4 hm^2 in 2008. Moreover, water consumption by the primary industries alone is 495.95×10^8 m^3, accounting for about 97.37 % of the area's total water consumption (Xinjiang Statistical Yearbook). Similarly, agricultural land in the Hexi Corridor region increased from 36.3×10^4 hm^2 in 1949 to 60.15×10^4 hm^2 in 1990 and reached 69.71×10^4 hm^2 in 2008 (Wu et al. 2011). Currently, annual grain production in the Hexi Corridor accounts for over 30 % of total grain output in Gansu province, where 70 % of the commodity grain and 90 % of cotton, sugar beet, and more than 40 % of the oil, melons, vegetables and wool are produced.

The extent of land resource exploitation and subsequent socio-economic development is based on the utilization of water resources. Since the 1970s, groundwater has generally been exploited in the arid region, which has gone hand-in-hand with large-scale development and utilization of surface water. By 2000, wells in the Xinjiang region and the Hexi Corridor region numbered more than 6.80×10^4, with corresponding groundwater exploitation pegged at 56.44×10^8 m^3/a and 24.16×10^8 m^3/a, respectively. Consequently, groundwater levels have declined sharply in some areas and the shallow groundwater that maintains the desert vegetation has decreased, causing the degradation or even death of desert vegetation.

In recent years, with the continuous increase in development, water resources in key economic zones have been over-exploited. The water resources development ratio in the Northwest is 46.7 %, which is higher than the national average of 26 %. In the Hexi Corridor and the Junggar and Tarim Basins (which are the most arid plains in China), the development ratio of water resources is over 65 %, far exceeding the 30 % average level of water resources development ratio internationally recognized for arid areas. Regionally, the exploitation rate of water resources in Xinjiang is 58 % and the exploitation rate in the Junggar and Tarim Basins is 65 %. In the Urumqi River Basin, which is the most economically developed and densely populated region in Xinjiang, the surface water volume is 8.51×10^8 m^3, the exploitation volume is nearly 11×10^8 m^3, and the exploitation rate of water resources is around 168.3 %.

In other areas of the Northwest, the exploitation rate of water resources is as follows: Turpan, 147 %; Hexi Corridor, 72.0 %; Shiyang River Basin, 154 %; Heihe River Basin, 95.5 %; and the Shule River Basin, 76.4 %. From these figures, we can see that the excessive exploitation of water clearly exists. It causes dramatic changes of the hydrologic system in the arid zone, such as shortening of the rivers, shrinking or drying up of lakes, and decreasing of regional groundwater levels.

Optimal allocation of water resources should not only meet the needs of economic and social development, but also achieve sustainable use of the resource. The northwest arid area is rich in land, light and heat resources but lacking in water. The growing friction between water supply and demand necessitates a speeding up of the planning and construction of large mountain reservoirs to realize the joint scheduling of control reservoirs and rational utilization of water resources. The aim would be to achieve scientific scheduling among river basins and regions and inter-regional allocation of water resources and adjustments. Coordinating water use between production and ecology, and resolving or mitigating the conflicts arising from the uneven distribution of water resources among regions would improve the coping capacity and adaptability to future climate change. Ultimately, all of these measures would ensure ecological security and sustainable economic and social development. To accomplish these aims, we propose diverting water from outside the Northwest, for example, from the Brahmaputra. With an infusion of water, the vast southern desert area could become an oasis.

11.2.2 Engineering Shortage of Water

Farming in the Northwest mainly depends on irrigation. Most of the key water conservancy projects were built in the 1960s and 1970s and suffer from low standards, poor regulation, imperfect engineering support, and either inordinately low or high canal project seepage rates. Approximately 1/3 of the plains reservoirs are considered dangerous reservoirs by today's standards, and there is a clear lack of controllable mountain regulating reservoirs. Although these issues are gradually being resolved and some of the irrigation projects are being repaired, the water use efficiency and effectiveness of these engineering projects are still low and cannot meet the growing water demands of present and predicted economic and social development.

Currently in the Northwest, about 90 % of the reservoirs are plains reservoirs. Of these, the control capacity is low and it is difficult to protect irrigation water demand. For example, there are 564 reservoirs in Xinjiang, but the total capacity is less than 13.5 billion m^3, and small reservoirs are the mainstay, accounting for about 62 % of the total number of reservoirs. There are few large or medium-sized reservoirs, so it is difficult to fundamentally solve the uneven distribution of water resources in time and space. In spring, during the peak water demand period for a variety of crops, the reservoir storage capacity is not sufficient to fully utilize fallow water, causing drought conditions. Moreover, the plains reservoir storage capacity in summer is limited and thus cannot effectively store the flood waters. Consequently, river erosion is serious and flood disasters frequently occur.

On average, plains reservoirs lose between 20 and 50 % of their stored water through evaporation and leakage. To further complicate matters, plains reservoirs are charged with the anti-regulation task of power plants and play an important role in improving the utilization of water resources. However, alleviating power generation and providing appropriate irrigation are oftentimes conflicting tasks. For this reason alone, it would be prudent to speed up the construction plan of mountain reservoirs and gradually abandon the plains reservoirs altogether.

Farming in the arid zone depends mainly on water channels, but due to low construction standards, inadequate support and aging facilities, the efficiency of water resource use is greatly compromised. In the existing water conservancy channels and channel buildings of some irrigation areas, drainage lines have numerous twists and turns and the seepage rate is low, resulting in water loss, siltation, erosion and frost-related damage. For example, the current rate of channel seepage in Xinjiang is about 32 % and the water utilization rate for agricultural irrigation is only 48.8 %. Furthermore, construction standards are low, and some irrigation water ports are temporary diversion facilities only, so the rate of water diversion is minimal.

In the Northwest China, the rivers often flood in summer. This has become exponentially worse in recent years to do global climate change. As a result, much of the drainage is being washed away by the floods, which affects irrigation as well as the efficiency of power generation. Meanwhile, irrigation projects are essentially incomplete; some irrigation drainage systems cannot provide sufficient water and/or salt drainage, causing the occurrence and development of secondary salinization in the irrigated area. In light of these issues, water conservancy construction and the construction of mountain reservoirs should be speeded up in order to replace at least some (if not all) of the plain reservoirs. This would reduce water loss and increase the available water supply. In particular, the construction of mountain-region controlling and regulating reservoirs should be fast-tracked to resolve issues related to uneven water distribution, water diversion projects, water capacity, alteration of irrigated areas, water supply capacity and water use efficiency. Such reservoirs would also aid in the scientific configuration of water resources, leading to the formation of appropriate storage, generation, transmission and irrigation that would undergird the sustainable economic and social development of the river basin region.

11.2.3 Water Shortage Through Waste

The main problem of water resources in China's northwest arid area is the co-existence of water shortage and water waste. As outlined above, current water resource utilization is imbalanced and low water prices are not fully integrated into market mechanisms. In addition, field irrigation methods are inadequate, resulting in low levels of irrigation, water leaks, flood irrigation and other phenomena. Due to poor production and management and outdated technology, soil drainage water and irrigation fields are still being used in many places so that the amount of water consumed in surface evaporation and seepage is less than 10%. For example, in the Hexi Corridor area, the utilization rate from canal water to field water is only 0.48%; the effective utilization rate from the canal to be absorbed by crops is 0.23%; and the per square aquatic food is only 0.73 kg. In the Altai region in northern Xinjiang, an average 13.4 m^3 of water is used to produce 1 kg food; the gross irrigation quota is generally 18,075 m^3/hm^2; and the comprehensive water efficiency of canals is 0.5% or less. Because the Tarim River mainstream was operating for a long time without engineering controls or oversight, there are 138 unauthorized outlets in the upper and middle reaches, 90% of which have no permanent control mechanisms. This has led to serious wastage of water resources even as water consumption is increasing. Downstream, water deprivation is a very serious problem, resulting in a significant reduction in the amount of water in the river's downstream reaches. In recent years, the situation has improved slightly through the implementation of a renovation project in the Tarim River Basin.

Agriculture is the primary industry in the Northwest and accounts for 93% of the region's total water consumption. Therefore, the focus of water conservation should be on saving water used in agriculture. At present, approximately 96% of the irrigated land is surface-irrigated and waste issues are a crucial factor. Despite these drawbacks, enormous potential for water conservation still exists in agriculture. Following are some strategies for agricultural water conservation: (1) To accelerate the transformation of the traditional extensive irrigated agriculture, promoting water saving agriculture of high quality, high-production and high-efficiency. (2) To accelerate the development of anti-seepage irrigation canal system and the drainage system construction, improving the efficiency of the utilization coefficient of canals, reducing the water quota and improving water utilization efficiency. (3) To vigorously promote the advanced irrigation technology to reduce the invalid irrigation loss of water. Actively implement surface irrigation technology, like trickle furrow irrigation and over membrane irrigation; over-ground drip irrigation such as sprinkler irrigation, drip irrigation; underground irrigation such as subsurface drip irrigation and infiltrating irrigation. (4) To improve the output from unit water and improve the quality of agricultural products and agricultural efficiency. This would lead to a decrease in the proportion of agricultural water use compared to total social water use. The water needed by newly-increased cultivated land at the expense of water for ecological use should be diverted through water-saving potential.

Water pollution and the inefficient use of urban sewage exacerbate the shortage of water resources. The total water supply of cities in the Northwest is about 34.5×10^8 m^3 and characterized by low re-use rates, especially in industries such as mining, electricity generation, petrochemical and nonferrous metals. In addition, chemicals based on energy and raw materials have an extremely high per-unit water consumption that is several or even dozens of times that of advanced levels in developed countries. Compounding this situation is the fact that the compliance rate of industrial water reaching the effluent standard after sewage treatment is less than 45 % and the water recycling rate is less than 1/3 of that in developed countries; even worse, the compliance rate of urban domestic water reaching the effluent standard after sewage treatment is only 15 %. Most of the poor quality water or sewage water is disposed of without any treatment, causing severe water pollution in lakes and rivers. As China's largest inland freshwater lake, Bosten Lake is threatened by pollution from industrial sewage, domestic wastewater and irrigation drainage around the lake. The degree of mineralization of the lake has risen to 1.87 g/L, changing it from a freshwater lake into a brackish one.

The arid northwest is characterized by inland watershed and poor dirt holding capacity. With rapid economic and social development, the urban water supply and wastewater emissions will increase further. To this end, we should vigorously promote a water use strategy for urban and industrial water resources. Specifically, water conservation should be set as a priority and pollution-treatment as a foundation. Water conservation contains elements of both water saving and pollution reduction. The goal in building a water-saving society should be to promote the construction of a sewage treatment plant for urban and industrial sewage water to mitigate water pollution and improve the treatment rate of urban and industrial wastewater and re-use rate.

11.3 Water Supply and Demand

There are 20 prefectures in China's arid northwest, including 106 counties (cities). As of 2010, 27.14 million people lived in this region, accounting for 2 % of the country's total population. Water is an essential resource for supporting

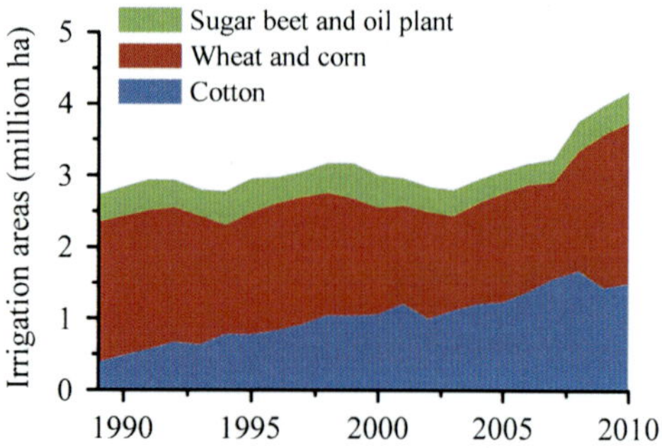

Fig. 11.1 The variations of crops irrigated area in northwestern arid region

socio-economic development and maintaining stability in the fragile ecosystem. However, the area is hampered by a sizeable imbalance between water supply and demand. To resolve this issue, it is crucial to understand how water is utilized in the Northwest in order to help improve the area's water management toward sustainable use. Currently, almost all available water resources are allocated for agriculture (95 %), with industrial and domestic water usage sharing the remaining 5 %. Due to incomplete statistics, this section only investigates temporal and spatial variations of agricultural-use water.

11.3.1 Agricultural Production

Agriculture is mainly carried out in the oases at the edge of the desert and is maintained by river and groundwater. Abundant radiation and heat resources and a relatively stable water supply make the oases agricultural lands an important high-quality cotton production zone; wheat, corn, oil plants and sugar beets are also grown there. These five crops account for more than 80 % of the total crop acreage. By the end of 2010, farmland covered 4.9 million ha, with irrigation reaching 4.682 million ha. Cotton occupied 1.5 million ha, and its acreage and production accounted for 31 and 42.7 % of the country's total cotton crop, respectively.

Water is both an enabler and a major limiting factor in agricultural development. Since 1949, crop acreage in the Northwest has increased year by year, but the area's farmland water-saving projects and water conservancy facilities are outdated and inadequate, which leads to a lower rate of water utilization. According to the "Xinjiang Uygur Autonomous Region Water Resources Bulletin 2010", the annual water usage for irrigation purposes was 48.83 billion m^3, accounting for 91.3 % of total water consumption in Xinjiang. Irrigated areas increased from 2.7 million ha in 1989 to about 4.4 million ha (Fig. 11.1) in 2010. Of these, wheat and corn crops covered 2 million ha, with little change over the past two decades, while beets and oilseed covered 0.4–0.5 million ha and slightly declined in recent years. The fastest growing acreage was cotton, jumping from 0.4 million ha in 1989 to 1.5 million ha in 2008. However, 2010 saw a slight decline in the cotton-sown area to 1.3 million ha, which accounts for 1/3 of the region's total crop.

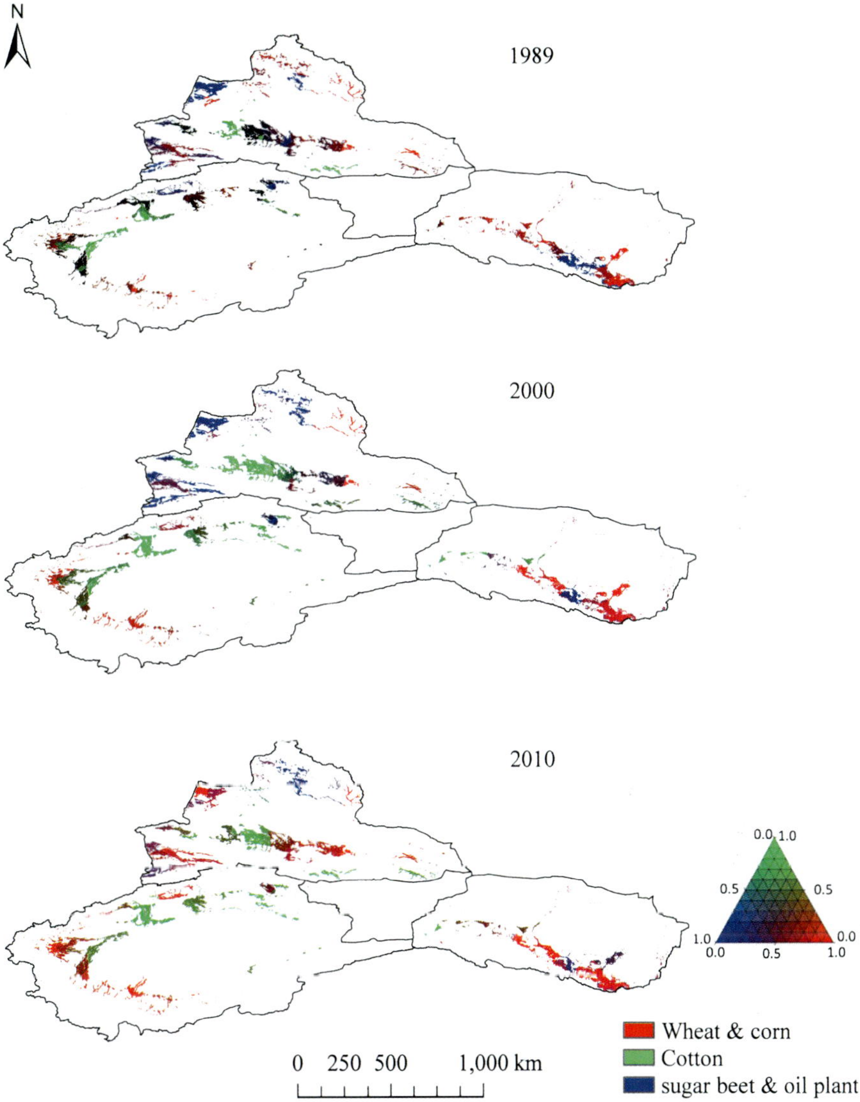

Fig. 11.2 Spatial distribution of crop acreage in the Northwestern Arid Region of China and its change (1989–2010)

Figure 11.2 shows the spatial distribution of the five major crops from 1989 to 2010. In southern Xinjiang, double crops of wheat-corn or single crops of cotton are the major cropping types, while in northern Xinjiang and the Qilian-Hexi region, single crops such as spring wheat or spring corn are mainly grown due to heat constraints. The north and south slopes of the Tianshan Mountains form the transition zone, where sugar beets and oilseed are mainly grown. Figure 11.2 illustrates changes in the structure of the crops in the past 20 years. In 1989, wheat and corn dominated, while cotton was grown mainly in large irrigation districts like Shihezi, Aksu and Kashi, and the acreage at that time accounted for only 10.4% of the total growing area. Oil plants, sugar beets and other cash crops were planted primarily in northern Xinjiang's Ili, Tacheng and Gansu's Minle regions. However, by 2000, with improvements in irrigation facilities and increases in cotton procurement prices, cotton acreage rapidly increased to 1,045 hectares, accounting for 25.6% of the total sown areas. The northern Tianshan Mountains and the oases at the edge of the Tarim Basin saw the most substantial increases, along with Gansu's Dunhuang and Yumen. Changes in oilseed and sugar beet acreages were minor throughout the time period. By the end of 2010, however, cotton was on the decline throughout the Northwest due to water restrictions.

11.3.2 Calculations of Irrigation Water Demand

Based on the crop planting structure and crop coefficient observational data, the agricultural irrigation water demand was estimated by calculating the crop water demand and crop area.

The most commonly used method and the one recommended by the UN Food and Agriculture Organization (FAO) is the Penman-Monteith method. The calculation principle for this method is based on sufficient crop irrigation, with crop water demand determined by weather and crop physiological conditions. In following this method, we calculated the reference crop water demand and the crop coefficient K_c correction to arrive at the crop water demand.

1) The crop irrigation requirement is calculated using the following formula:

$$CWR_i = K_c \cdot ET_0 \tag{11.1}$$

$$IWR_i = S_i \cdot (CWR_i - P_e) / I_c \tag{11.2}$$

where:

CWR_i = crop water requirement for $crop_i$ in the study region
ET_0 = reference evapotranspiration
K_c = crop coefficient
S_i = the acreage of $crop_i$
P_e = effective rainfall, which indicates that the rainfall is transferred to soil moisture
IWR_i = irrigation water requirement of $crop_i$
I_c = irrigation efficiency, by considering the loss in water diversion from the river channel to the field

2) Reference evapotranspiration

Reference evapotranspiration (ET) was calculated using the modified Penman-Monteith equation recommended by FAO (Allen et al. 2005). Reference ET was defined and calculated based on the evapotranspiration of well-watered short grass with a height of 0.12 m, a resistance of 70 s/m, and an albedo of 0.23. This is very similar to the open surface, highly consistent, vigorously growing, ground cover without water shortage (Allen et al. 2005). Reference evapotranspiration is only related to meteorological factors, so it can be used in different regions and has universal significance for inter-comparison. The UN Food and Agriculture Organization (FAO) recommended the Penman-Monteith method to calculate the reference evapotranspiration. The crop water requirement (CWR) was then calculated by multiplying the crop coefficient (K_c) with the reference ET.

The reference ET was calculated using the following equation:

$$ET_0 = \frac{0.408\,\Delta\,(R_n - G) + \gamma \frac{900}{T+273} u_2 (e_s - e_a)}{\Delta + \gamma(1 + 0.34u_2)} \tag{11.3}$$

where ET_0 is the reference evapotranspiration (mm/d); Rn is net radiation at the land surface ($MJ/m^2 \cdot d$); G is the soil heat flux density ($MJ/m^2 \cdot d$); u_2 is 2-meter-height average wind speed in 24 h (m/s); e_s is the saturation vapor pressure (kPa); e_a is actual vapor pressure (kPa); Δ is the slope of the vapor pressure-temperature curve (kPa/°C); and γ is the psychrometric constant (kPa/°C).

3) Crop coefficient during the growing period

Crop coefficient is the ratio between well watered crop evapotranspiration and the reference ET, i.e., ETc/ET_0, commonly written as "K_c". K_c values can generally be obtained through field experiments and reflect the relative water consumption capacity of crops during different growing stages. In general, the crop coefficient varies according to four developing stages: the initial growth period, the rapid development period, mid-maternity, and the mature period. In the initial growth phase, the K_c value is generally very small due to low crop cover, and so most water consumption is from soil evaporation. During the rapid development phase, the leaf area increases rapidly and causes a sharp rise in the K_c value. When crops are fully developed, the K_c value reaches its maximum and remains stable over a period of time. As the crop matures, the K_c value starts to decrease due to foliage senescence. Crop coefficients for typical sub-regions in the Northwest are shown in Table 11.4.

4) Effective rainfall

In agricultural terms, effective rainfall is defined as that portion of rainfall that penetrates the canopy, infiltrates the soil layer, is stored in the crop root zone, and is later used by crops as transpiration. Effective rainfall does not include the interception of the plant canopy, the surface runoff, or the amount of water percolating to the layer below the root zone. Effective rainfall is usually calculated as:

$$P_e = P \times \delta \tag{11.4}$$

Table 11.4 The crop coefficient of the major crops at each growth stage in some sites in northwestern arid area

Sub-region	Crop	Development stage				Sowing (mm/dd)	Harvesting (mm/dd)	Total (days)
		Initial	Developing	Middle	Late			
Yanqi Xinjiang	Spring wheat	0.17	0.17–1.16	1.16	1.16–0.40	3/7	7/5	121
	Spring corn	0.21	0.21–1.20	1.20	1.20–0.35	4/27	9/17	144
	Cotton	0.26	0.26–1.20	1.20	1.20–0.70	4/20	10/16	180
	Oil plant	0.31	0.31–1.15	1.15	1.15–0.35	6/14	9/29	108
Zhangye Gansu	Sugar beet	0.34	0.34–1.21	1.21	0.21–0.70	3/15	9/24	194
	Cotton	0.23	0.23–1.21	1.21	1.21–0.72	4/11	10/11	184
	Spring wheat	0.23	0.23–1.16	1.16	1.16–0.40	3/13	7/17	127
	Spring corn	0.23	0.23–1.20	1.20	1.20–0.35	4/13	9/26	167

Table 11.5 The irrigation efficiency in the study region during past 20 years

Year	1990	1995	2000	2005	2010
Irrigation water efficiency	0.38	0.40	0.42	0.45	0.48

where P is precipitation and δ is an empirically effective utilization coefficient of rainfall, which is adopted as 0.52 in our study region (Xu et al. 2010).

5) Irrigation efficiency

Irrigation efficiency refers to the ratio of the net amount of water used by crops to the total amount of water withdrawn from river channels or reservoirs. It is mainly composed of the canal water utilization coefficient and the field water utilization coefficient. The canal water utilization coefficient is used to measure the degree of utilization of canal water delivery, and the field water utilization coefficient is used to measure the degree of utilization of soil water. Improvements in irrigation infrastructure as well as the development of water-saving technologies have led to significant improvements in irrigation efficiency. According to the "China Water Conservancy Yearbook" and the regional water resources bulletin, the irrigation water utilization coefficient in the northwestern arid region has increased from 0.38 to 0.52 over the past 20 years (Table 11.5). In this study, 226 irrigation water utilization coefficients were used to calculate irrigation requirements.

11.3.3 Spatial and Temporal Variations of Agricultural Irrigation Water Requirements

1) Trends in crop water requirements We calculated the daily ET_0 for each site from 1989 and detected the trends using the Penman-Monteith equation recommended by the FAO. After determining the regional crop growth stages and Kc values, and combining these with the calculation results of ET_0, we obtained the regional crop water requirement (ETc). We discovered some significant differences among the ETc of the five major crops in the Northwest, which were ranked, according to water requirement, as Beet > Cotton > Corn > Wheat > Oil (Table 11.6). The

Table 11.6 Annual average crop water demand and its trend test (1989–2010, Unit: mm/y)

	Cotton	Wheat	Corn	Oil plants	Beet	Annual ET_0	Annual rainfall
Southern MK test	464	353	679	310	749	1126	63
	0.479	−0.085	0.085	0.197	0.31	3.525[a]	0.423
Northern MK test	400	495	614	365	649	1036	176
	2.453[b]	2.397[b]	2.566[b]	2.679[a]	2.876[a]	0.902	0.649
Hexi MK test	519	643	731	376	730	1227	163
	2.34[b]	2.425[b]	2.285[b]	1.241	2.651[a]	2.933[a]	0.931
Average MK test	456	492	670	352	710	1123	131
	1.776	1.805	1.748	2.228[b]	2.228[b]	2.735[a]	0.874

[a] Expressed pass the 0.01 significance test
[b] Expressed pass the 0.05 significance test

average sugar beet crop water requirement (CWR) was 710 mm/y, whereas oilseed had the lowest CWR, at 352 mm. Based on the Mann-Kendall test to ET_C, it was found that the annual water requirement of the five major crops over the past 20 years showed an increasing trend, especially oil plants and sugar beets.

ET_0 is mainly affected by meteorological factors, while Kc values tend to remain relatively stable during a given period. Since 1989, temperatures have increased in the Northwest. The consequent increase in reference evapotranspiration has led to an increase in per-unit irrigation water requirements for the major crops. This trend is more prevalent in southern Xinjiang and the Hexi Corridor than in northern Xinjiang, as the annual reference evapotranspiration in the northern region did not significantly change, likely due to the region's geographical dissimilarity to southern Xinjiang.

The spatial distribution of crop irrigation water requirements for each region showed clear differences. Areas with high crop water requirements were mainly located in Alashan and the east part of southern Xinjiang. The Ili Valley was affected by the westerlies, the northern Altai was bathed in water vapor from the Arctic Ocean, and the marginal areas of the Hexi Corridor were impacted by the monsoon climate. The presence of precipitation increased the relative humidity of farmland and thereby reduced crop evapotranspiration; hence, the crop water requirements in these areas is less than for other arid areas.

In terms of the ETc of each crop, the spatial differences of the ETc of corn were the most obvious, as there is a significant difference among corn crops in northern, southern, and west Xinjiang (Hexi > northern Xinjiang > southern Xinjiang). These differences are mainly due to the spring and summer corn growth characteristics as well as regional differences in growth-period climate. On average, the growth period of spring corn is about 50 days longer than summer corn. In northern, southern and west Xinjiang, the corn growth-period water requirement was 495,353,643 mm. Relative to corn, the ETc spatial difference of other crops is not obvious, although it should be noted that the water requirement for beets is the highest. In the main beet-growing areas of Tacheng (Altay region), the ETc of beets was only about 500 mm. This shows that, to a certain extent, crop ETc spatial differences affect the distribution of crop sowing.

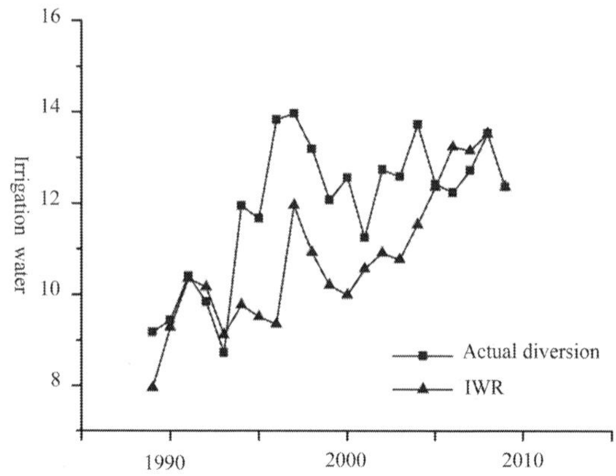

Fig. 11.3 IWR compared with actual diverted in Kuqa County

2) Temporal and spatial variations in irrigation water demand Using the calculation results above for crop water requirements in the Northwest, we considered the effects of crop acreage, irrigation water use coefficient, effective rainfall and other factors in various regions by calculating the irrigation water requirements of different crops according to the formula in 2–2. To verify the calculation results, we selected the water sector agricultural irrigation information from Kuqa County and compared it with the simulated crop irrigation water requirements. Kuqa County is located in the northeast portion of Aksu Prefecture, where agricultural irrigation water is mainly extracted from the Kuqa and Weigan Rivers. Local economic development level is relatively high in Kuqa County, as water conservancy facilities are generally efficient. Hence, crop irrigation requirements are usually easily met.

By comparison, there is a certain degree of error between simulated crop irrigation water requirements and actual water diversion amounts, although both basically show the same trends and the simulation accuracy is good in some years (Fig. 11.3). It should be noted that statistics from the actual amount of water diversion, besides the irrigation water of the five main crops, also included water diversion for the irrigation of fruit trees, forest, pasture, and so on, and thus the statistical results were higher. Overall, the analog value of the crop irrigation water storage capacity in this study was able to reflect the actual situation, and the results of the simulation were able to provide a reference for regional water use.

Figure 11.4 shows the crop spatial distribution of irrigation water in the Northwest in 2010 and how it differs from that in 1989. The coefficient 0.48 was used to calculate the irrigation water demand as being 44.3 billion m^3 in 2010, with northern Xinjiang using 19.8 billion m^3, southern Xinjiang 20.5 billion m^3, and the Hexi Corridor 4.65 billion m^3. This was nearly 13.6 billion m^3 more than the total irrigation water demand in 1989, with northern and southern Xinjiang increasing usage by 6.4 and 7.62 billion m^3, respectively, and the Hexi Corridor decreasing usage by 1.8 billion m^3. The areas requiring more irrigation water are mainly concentrated in Kuitun, Shihezi and Aksu, where agriculture is more developed and cotton acreage accounts for more than half of the total sown area. In contrast, in the Hexi region

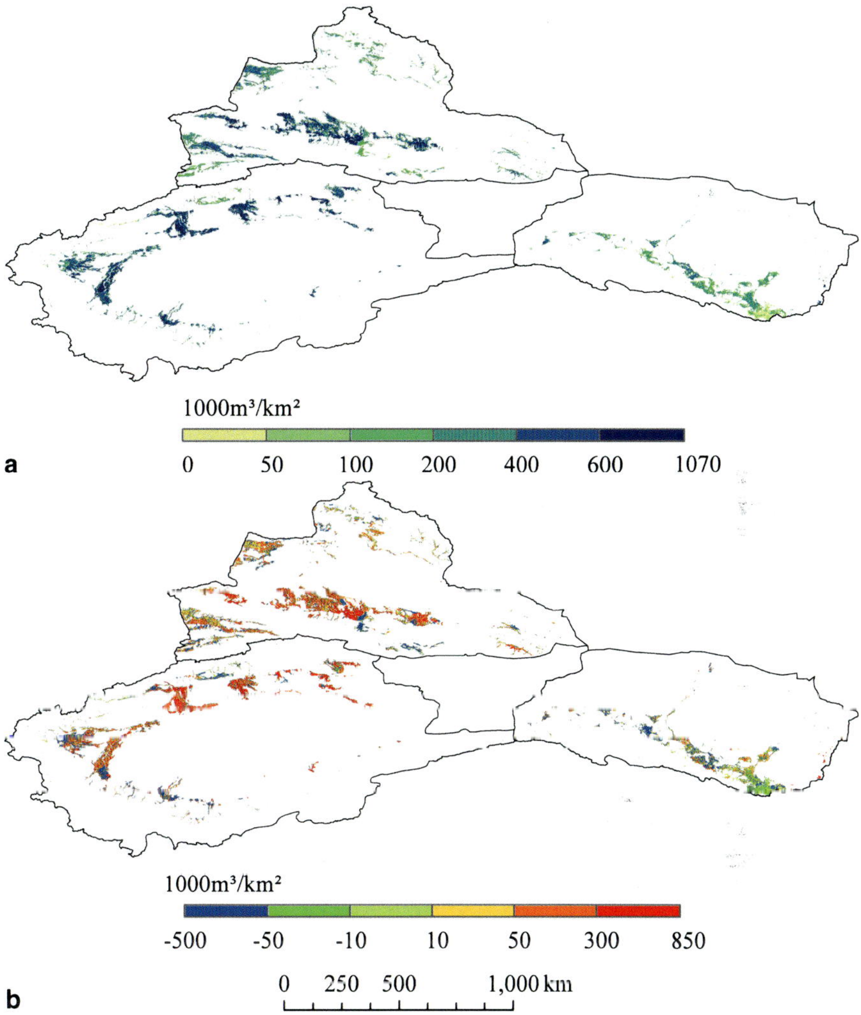

Fig. 11.4 **a** The spatial distribution of irrigation water demand in the arid region of northwestern China in 2010, and **b** Its change compared to 1989

where there are mainly food crops, water requirement per unit is smaller, as is the sown area, and therefore irrigation water demand is lower. Similarly, in the Yili region and east of the Hexi Corridor, irrigation requirements are also relatively low due to a more humid climate that includes sufficient local rainfall.

In recent years, climate change has had an impact on irrigation requirements, showing an overall upward trend. Since 1989, the main crops of wheat, corn and especially cotton have increased their water demand by 10, 28 and 46 mm, respectively. Although local irrigation water usage has attempted to keep pace with the rising demand, it is insufficient to over-ride this phenomenon. Crop water demand

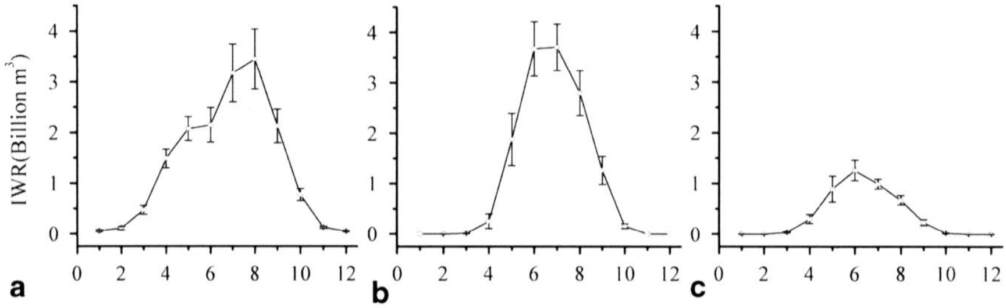

Fig. 11.5 The intra-annual variation of irrigation water demand in different sub-regions **a** South Xinjiang, **b** North Xinjiang, **c** Heixi Corridor in the northwestern arid region (1989–2010)

has seen the most significant increases in the Kashgar region and in the southern and northern oases of the Tianshan Mountains (Fig. 11.4b). This basically coincides with the areas where cotton is grown and reflects the changes in crop farming over the past 20 years.

Despite the boom in cotton cropping in some areas of the Northwest, the planting structure of the Hexi Corridor and northern Altay regions has changed little. Thus, even though the crop water requirement per unit area increased, the regional total irrigation water demand remained relatively stable and even declined in the Hexi region. Further confusing the issue is that large cities such as Urumqi and Shihezi have taken over formerly arable land for urban construction, while desertification and abandonment of some farmland has occurred at the edges of the oases. This has led to a reduction in arable land and a subsequent significant reduction in irrigation water demand in certain parts of the region (Fig. 11.4b).

The multi-year changes in the average irrigation water demand in different regions are shown in Fig. 11.5. Overall, the seasonal distribution of irrigation water requirements in the Northwest was uneven. The peak water demand is mainly concentrated in the summer, and there is a significant difference in the distribution of irrigation water demand in different areas.

As shown in Fig. 11.5a, two peaks occurred in May and August, respectively, and reached maximum levels in August. This is due to the cropping systems being mainly spring wheat and summer corn or growing one-season cotton in this region. In May, winter wheat in southern Xinjiang gradually entered the jointing and heading stage, at which point irrigation water demand jumped to 2 billion m^3/month. Meanwhile, cotton was at the seedling stage. In June, the wheat was harvested and irrigation water demand was reduced, resulting in the first peak seen in this study. After June, water demand in southern Xinjiang was mainly due to cotton and the re-sowing of the summer corn. Cotton reached the boll division and the flowering stage in August. Meanwhile, the re-sown oil plants and summer corn also entered the critical water demand period, and as a result, the irrigation requirement peaked at 3.5 billion m^3/month. After August, the cotton and corn began to enter their mature period and water requirements began to decline. Then, in October, winter wheat began to seed, winter irrigation making winter irrigation water necessary in southern Xinjiang.

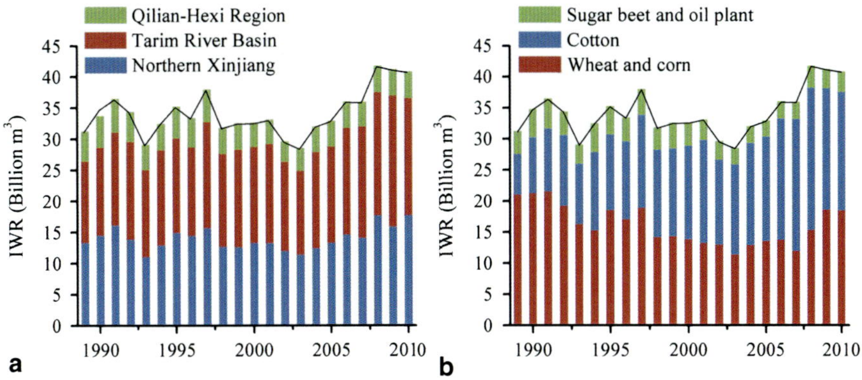

Fig. 11.6 Inter annual variation of irrigation water requirement for different crops and different sub-regions in the northwestern arid region of China

In northern Xinjiang and the Qilian-Hexi region, spring crops such as wheat, corn and cotton are harvested only once a year, and so there is only one peak in water usage (in June and July). At that time, water requirements can be as much as 3.5 billion m^3/month in northern Xinjiang and 1.2 billion m^3/month in the relatively small sowing area of the Qilian-Hexi region. Because of the differences in geographical conditions, the seeding time in northern Xinjiang is earlier than in southern Xinjiang, causing maximum irrigation requirements to occur earlier in the north. Spring wheat, corn and cotton are mainly sown in April, so irrigation water requirements increase rapidly in May and June. The local crop water requirement per unit area in the Hexi region is much larger than in Xinjiang, but the crop acreage is relatively small, so the local total irrigation water requirement is also small. Although the total water demand in the Hexi Corridor is much smaller than that in both southern and northern Xinjiang, severe water shortages still occur here because rivers in the region are fairly short and have little discharge.

To further understand how crop acreage impacts on irrigation water demand in the Northwest, we calculated irrigation requirements in different regions and for different crops based on the calculation of the water demand in the total irrigated area (Fig. 11.6).

The variations in irrigation requirements for different crops in several sub-regions are illustrated in Fig. 11.6. Irrigation requirements have generally risen since 1989 due to reclamation and increases in crop growing areas as well as changes in planting structure. The total irrigation requirement rose 44 % (13.6 billion m^3) over the past 20 years, while the crop growing area increased about 55.6 %, from 35.68×10^5 ha in 1989 to 55.5×10^5 ha in 2010. We can see from this that the increasing range of the total irrigation water demand is less than the increasing range of the sown area. This is the result of water-saving technology implementations and conservation efforts to curb seepage and leakage of irrigation.

The demand for irrigation water has fluctuated within the general increasing trend. In 2003, the irrigation requirement was low, which is consistent with the lower amount of crop being grown in the Northwest that year. However, since 2003,

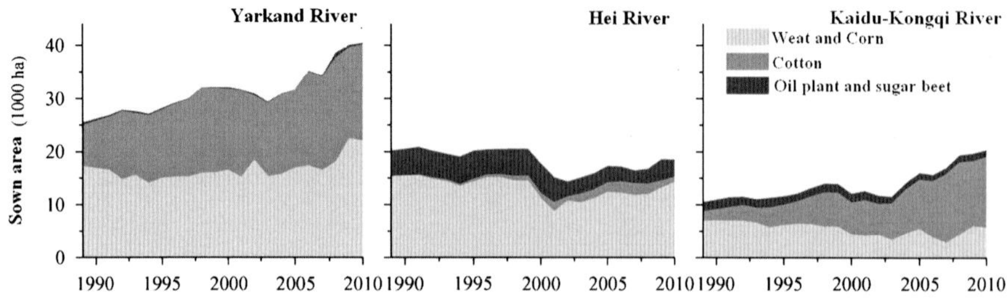

Fig. 11.7 Changes of crop growing area in the three typical basins

crop acreage has expanded rapidly, causing a consequent rapid rise in irrigation water requirement. Cotton is by far the largest water consumer, but oil plants and sugar beets have higher water consumption per unit area, though their smaller acreage has meant that there is almost no change in irrigation requirements during the past 20 years. Looking at the distribution of irrigation water demand, we can see that the Hexi Corridor had no obvious change, while southern Xinjiang had a slightly greater increasing trend than northern Xinjiang mainly because cotton is the major crop in the south. Along with having more growing acreage, cotton has a higher crop water requirement, estimated at around 679 mm/year.

11.3.4 The Supply-Demand Analysis of Water Resources

Given the limited availability of runoff data, we selected the Yarkand, Kaidu-Konqi and Heihe Rivers as typical representative rivers to analyze the water *supply/demand* between available water resources, irrigation water requirements (IWR), and water supply risks. Also analyzed were the water supply and demand situation and variations based on annual average runoff.

1) Variations in irrigated crop-growing areas in the northwestern arid region Among the three typical basins, the Yarkand River Basin has the largest amount of arable land (approx. 4.03 million ha), followed by the Kaidu-Konqi River Basin and the Heihe River Basin, each with approximately 2.00 million ha (Fig. 11.7). Arable land in the Yarkand and Kaidu-Konqi River Basins increased quickly after 2000, while arable land in the Heihe River Basin decreased and then remained relatively stable for the past 10 years. In the Yarkand and Kaidu-Konqi River Basins, the main crops are wheat, corn, and cotton. The acreage of cotton has increased year by year in the Kaidu-Konqi River Basin and is now the region's major crop. The main crops in the Heihe River Basin are wheat and corn, with a small amount of oilseeds and sugar beets, the latter two crops trending downwards.

2) Water requirements and supply risks in typical basins There are significant spatial differences in runoff and irrigation water requirements in the various river basins in the Northwest. Table 11.7 compares results of runoff and irrigation water

Table 11.7 Typical basin runoff characteristics and irrigation water demand (0.1 billion · m^3/y)

Year	Yarkand River region			Heihe River basin			Kaikong River region		
	Q	IWR	IWR/Q (%)	Q	IWR	IWR/Q (%)	Q	IWR	IWR/Q (%)
1989–2000	75.2	35.4	47.1	24.7	22.3	90.3	51.7	14.4	27.9
2001–2010	81.1	36.5	45	27.8	17.4	62.6	62.3	18	28.9
Average	77.4	35.8	46.3	26	20.2	77.7	51.6	15.7	28.2

requirements in different time periods for the three typical river basins. According to the water scarcity index (defined as the ratio of water use to water availability, or WU/Q), when the use to availability ratio is greater than 0.4, the basin is recognized as severely water stressed; when the use to availability ratio is 0.2 < WU/Q < 0.4, the basin is considered moderately water stressed; when the ratio is 0.1 < WU/Q < 0.2, the basin is slightly water stressed; and when the WU/Q < 0.1, the basin is not water stressed at all. However, a ratio of irrigating water requirement to runoff (IWR/R) smaller than 0.4 only occurred in the Kaidu-Konqi River. It had an average value of 28.2 %and could meet the water demand of agriculture, and yet still belonged to a moderate water stressed region. Severe water stress occurred in two other typical river basins. The ratio jumped to 90 % in the 1990s in the Heihe River Basin, but thanks to water saving practices and irrigation technologies, that value decreased in the last 10 years to 62.6 %. Nevertheless, the area is still considered to be under severe water stress due to water withdrawal from irrigation leading to ecological degradation in the downstream regions. The ratio of irrigation water requirement to runoff is over 46 % in the Yarkand River Basin, where severe water stress problem also exist.

Although the total annual runoffs of the three typical basins are greater than their irrigation water requirements, seasonal water shortages persist in some river basins due to inconsistencies in the runoff and agricultural irrigation requirements. In the Northwest, maximum river discharge always occurs in July and August and irrigation water requirements peak at the same time. Therefore, there are no water shortage problems in the flood season, as runoff discharges are much larger than the corresponding irrigation water requirements. However, during April and May, which is a critical time for winter wheat jointing and heading and spring crop seedlings such as spring corn and cotton, irigation water requirements increase significantly. During this period, the rivers are discharged mainly by snowmelt water. Due to still relatively low temperatures in the mountain areas in April and May, the snow/glacier meltwater cannot meet irrigation requirements, and thus water deficits in this period can detrimentally affect production.

Figure 11.8 shows the long-term comparison of irrigation water demand and water availability. Except for the Kaidu-Konqi River, where the discharge is much larger than the demand, the other two rivers experience water deficit problems (e.g., the Yarkand River Basin experiences a water deficit in April and May, and the river discharge is less than the irriagation water requirements in the Heihe River Basin from April to June).

Figure 11.8 also shows a long-term comparison of irrigation water demand and water availability in May, where the IWR is always larger than the discharge in

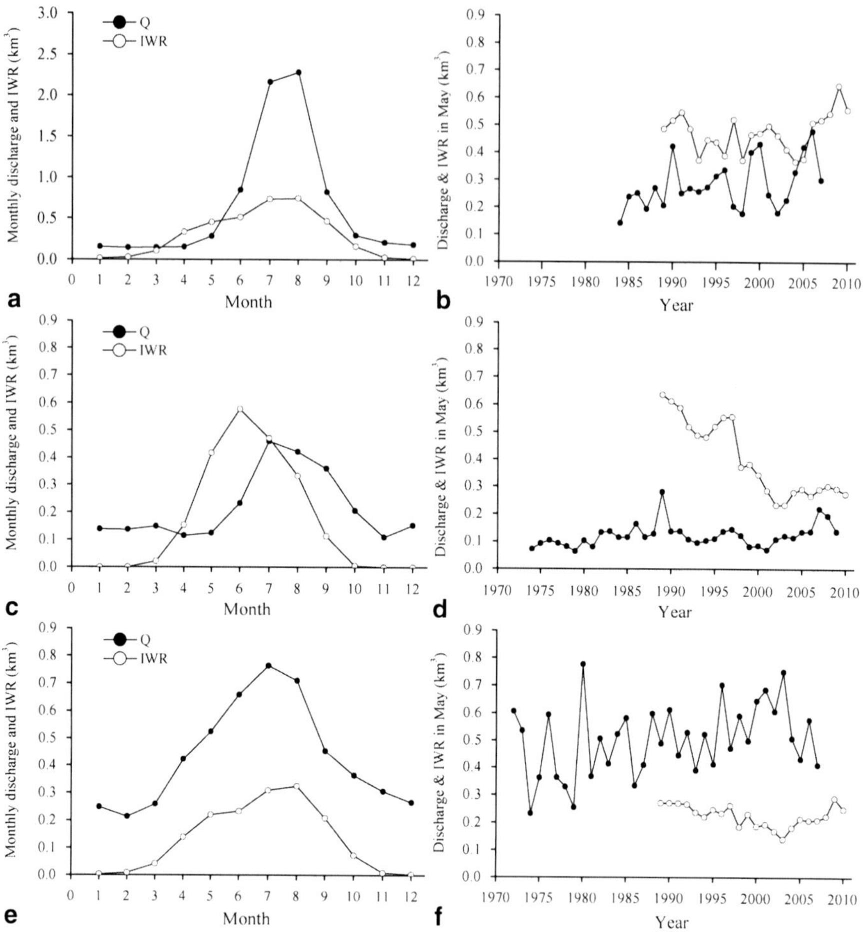

Fig. 11.8 Typical watershed years average monthly runoff, irrigation water demand and the change of the irrigation water supply and demand in May

the Yarkand and Heihe Rivers. Therefore, pumping groundwater or constructing water reservoirs are essential for bridging the deficit. Furthermore, and thanks to water-saving irrigation measures, irrigation water requirements in May are decreasing. For instance, the Kaidu-Konqi River is recharged by plenty of snowmelt, so there is no water shortage in that basin.

3)Evaluation of supply/demand of agricultural water To study the balance of water supply and demand in different basins, we evaluated the irrigation water requirement and water resources based on annual average amounts of available water resources in third-level rivers in the Northwest (Fig. 11.9). In the early 1990s, the water scarcity index of most river basins in the arid zone was already greater

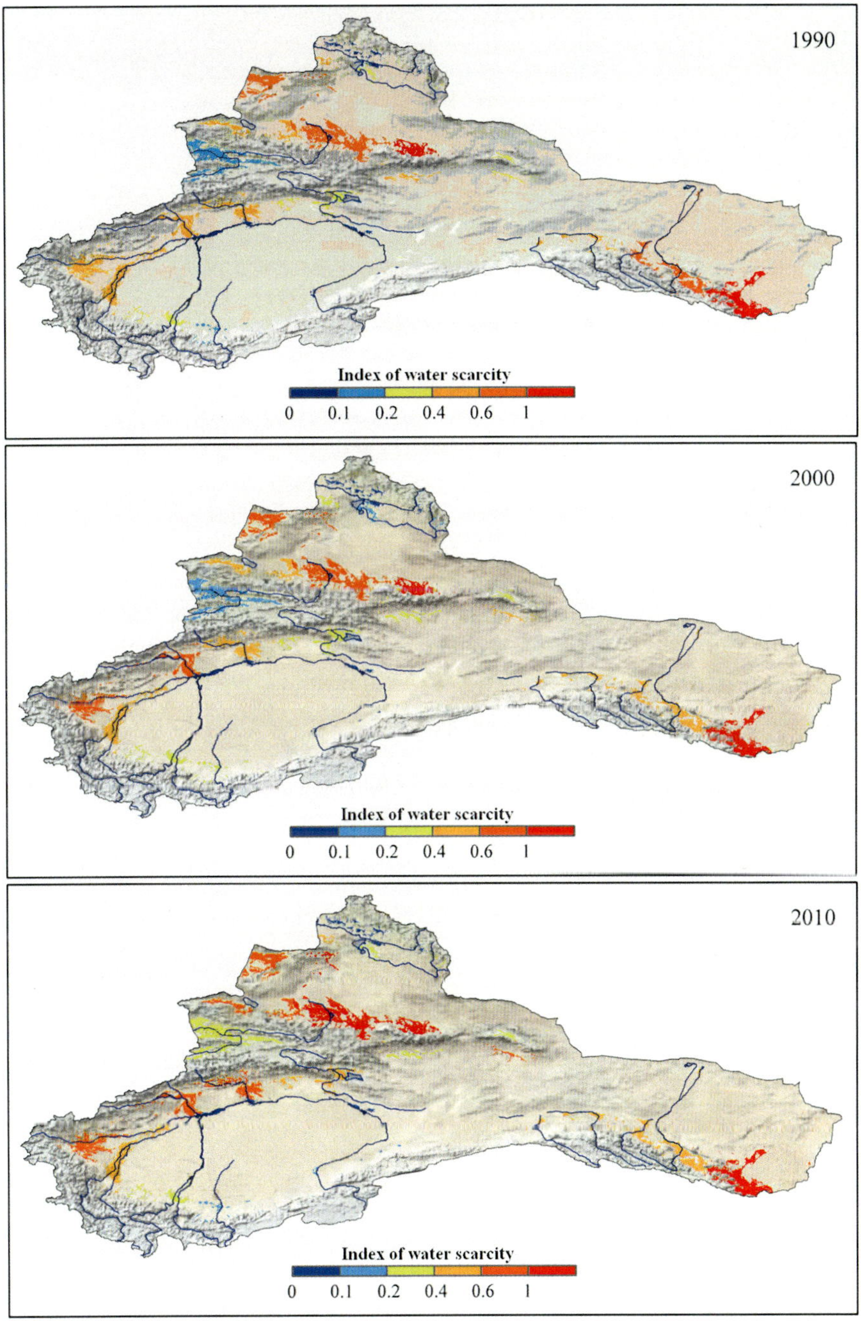

Fig. 11.9 The spatial distribution of water level

than 0.4, indicating that severe water shortage problems existed. The amount of surface water resources in the east Tianshan Mountains and the Shiyang River Basin is relatively small, but irrigation water requirements are generally large. Therefore, these regions are extremely water stressed and an enormous amount of groundwater extraction is required to meet their irrigation needs.

Oases in Xinjiang suffer ongoing water shortages. Regions with less severe water shortages include the northern Altay region and the Ili River Basin. In the Altay, irrigation water requirements are far less than the available surface water due to the absence of arable land. In the Ili, abundant surface water prevents severe water shortages, even with the region's abundant arable land.

Up until 2000, the severity of water stress in northern Xinjiang and the Hexi Corridor region remained relatively unchanged. Meanwhile, the severity of water shortages in the Heihe River Basin was lessened due to the implementation of watershed conservation programs from 1999 onwards. Occasionally, however, severe water shortages still occur, and the severity of water shortages in the Aksu and Kashi River Basins prompt an average water scarity index of around 0.6.

The mismatch between water supply and demand was further aggravated in 2010 when the oasis in the western Tianshan Mountains joined other areas with severe water shortages. From Fukang in the eastern Tianshan Mountains to Kuitun in the mid-west Tianshan Mountains, the water scarcity index rose to over 1.0, and surface water resources could not meet the agricultural water demand, leading to the over-exploitation of groundwater. Moderate water shortages also occur in the Ili River valley. In southern Xinjiang, water shortages are growing increasingly severe and the water scarcity index of the Kuche-Shaya-Xinhe oasis exceeds 0.6. The increase in agricultural water use also diminishes the amount of water allocated for purely ecological purposes (Fig. 11.9).

Summary Northwest China has become an important grain and cotton producing area, with agricultural water consumption accounting for most of the total water use in this region. In the past 20 years, global warming caused an overall increase in crop water requirements, while at the same time crop acreage expansion and changes in the structure of agricultural planting significantly increased irrigation water requirements.

By 2010, irrigation water requirements for the main crops had reached 44.3 km^3, with a net increase of 13.6 km^3 compared to 1989. Cotton is now out-pacing wheat and corn to become not only the principal crop but also the most water-consuming crop in the region. Hence, the regions experiencing the greatest increases in irrigation water requirements are generally those where cotton is grown. Overall, most of the basins in the Northwest are suffering severe water shortage problems. Since 1989, these problems have been exacerbated by rapid increases in irrigation water demand. However, since the end of the twentieth century, water conservation plans such as that in the Heihe River Basin have helped to alleviate at least some of the shortages.

Seasonally, water discharge usually peaks in July/August. In southern Xinjiang, irrigation water requirements have two distinct peaks—May and August—due to differences in cropping systems. In May, irrigation water demand is often greater than surface water runoff, which leads to seasonal water shortage risks in most basins. Thus, hydraulic facilities should be constructed to enhance the seasonal regulation capacity and to ensure an adequate water supply.

11.4 Management Countermeasures of Water Resources

Water resources, irreplaceable and scarce, support fundamental human survival as well as economic and social development. Water management is an orderly intervention activity of development, utilization and protection of water resources which aims to promote the orderly development, efficient allocation and rational utilization of water to ensure that the supply meets the needs of long-term sustainable economic and social development along with the preservation of the natural environment.

11.4.1 Construction of a Water Management System

In the Northwest, water guarantees the sustainable development of the economy and society and also maintains the functioning of the ecosystem. On the surface, the area's water shortages appear to be partly natural and partly artificial. To offset these problems, the development of a scientific management system and operational mechanism that can appropriately allocate water resources need to be established. Through such a system, ongoing issues arising from inefficient water use and natural water shortage can be fundamentally solved. Watershed management problems are currently worsening due to the lack of a unified and effective management system. Meanwhile, with the establishment of the socialist market economic system, the original fee-collecting and management practices are becoming increasingly incompatible with the current economic system and are not suited to water resource shortages and ecological deterioration. With production processes vying for the same scant resources as the natural environment, ecological problems triggered by human economic activity have become more common and also more extreme.

The resource management system currently in place in the Northwest is mostly based on administration. In other words, administration and industry management form the main means of management. The issues existing in this system include highly fragmented management and the overlapping of operations by different departments. With the increased development of economy and society, the water crisis will be further aggravated and a series of new contradictions and problems concerning the development, utilization and protection of water resources will be created. To resolve these issues, a water resource management system that combines valley management and district management should be implemented and watershed management should be strengthened.

11.4.2 Reformation of the Water Resource Management System

The existing watershed management is a combination of basin management and irrigation district management. As such, it involves numerous disadvantages that are mainly political in nature. Therefore, in consideration of severely damaged ecosystems which support economic development in the Northwest, the governing power for watershed management and irrigation management should be split in order to reduce the number of management levels and improve operating efficiency to meet the proper management of water resources and attendant social goals.

Meanwhile, in laws and legal systems related to water resource management, the intensity of water management must be increased and the price management system gradually improved. At the same time, the scientific watershed management and construction of related policies security system need to be accelerated.

11.4.3 Changes in Water Management Property

Water management property changes according to economic and social development. Water is usually formed in the watershed; hence, water management, as a unit of the watershed, should follow unified planning arrangements and establish an authoritative and efficient watershed management system. Only the implementation of such a unified system can ensure the maximum efficient allocation of water resources and achieve a balanced and harmonious development of ecological, economic and social benefits. The management of water resources in time and space will proceed in stages, from watershed management to regional management, and from a single watershed management to inter-basin integrated management. Water management also extends to the integrated eco-hydrological systems management of the natural environment, vegetation, land use and soil erosion.

11.4.4 Unified Management of Surface and Groundwater

The joint scheduling and optimization of multi-source water resources will greatly improve the water use efficiency of surface and groundwater. As these two sources are an indivisible entirety transformed into each other, we must make full use of the characteristics of mutual transformation between the two to achieve sustainable usage. In the arid region, exploitation and overexploitation of groundwater are acute and many areas are on a significant downward trend, which will have a direct and negative impact on ecological processes. Some desert vegetation with shallow roots and weak drought tolerance will die due from the drought. Additionally, the falling groundwater table will accelerate both the ecological degradation and desertification processes, posing a great threat to irrigation based on groundwater.

In the arid inland basin, most rivers have uneven seasonal distribution. Groundwater, which usually serves as the reserve water for agricultural emergency relief, cannot serve as the main irrigation water. The exploitation of groundwater and establishing water sources in the areas with high level groundwater table, not only increases the irrigation water sources and regulates water use in spring, but also reduces the ineffective evaporation, and mitigates soil salinization. Therefore, unified management of surface water and groundwater should be implemented as soon as possible to achieve control objectives of water resources and the sustainable development and management of the economy, society and the ecosystem in the watershed lands.

11.4.5 Broadening Sources and Reducing Use

Despite its fundamentally dry nature, the water saving potential in China's arid northwest is enormous. For instance, the estimated annual evaporation loss from the plains and reservoirs is around 30×10^8 m^3 in the Tarim Basin and up to 10×10^8 m^3 at Bosten Lake. Therefore, building mountain reservoirs and abandoning plains reservoirs to reduce evaporation loss from the plains' large surface areas makes sense. In irrigation management, the joint irrigation of wells and drainage, optimal scheduling, and optimization of the groundwater table are additional means to reduce evaporation losses from phreatic water. There is also huge water-saving potential in agricultural irrigation through the use of advanced water-saving irrigation technology.

Bedrock water provides a wealth of available water resources mainly in the form of groundwater. The karst groundwater system, which includes the exposed karst water system and the cover-buried karst water system, is an important component of regional hydrological systems. Karst groundwater is characterized by fast recharge, short cycles, and good water quality and has gradually become an important source of water for industrial and agricultural production. The rational development and utilization of karst groundwater resources can effectively improve water resource use and, to some extent, ease the agricultural production problems brought on by the uneven distribution of water resources in Xinjiang.

The northwest arid region is characterized by basin-and-range terrain, with rich water resources stored in mountainous regions in the form of glaciers, snow and precipitation. These provide a continuous source of water for basin area oases. The mountains also have abundant groundwater resources. For example, the Tianshan Mountains contain a large area of karst terrain and are also rich in water in the form of glaciers, snow and precipitation. Hence, karst hydrogeology combined with the mountains' unique water conditions indicate the likely presence of natural underground reservoirs and vast untapped groundwater resources. Therefore, it is recommended that exploration, development and utilization of karst groundwater research be carried out as soon as possible.

References

Allen RG, Pereira LS, Smith M, Raes D, Wright JL (2005) FAO-56 dual crop coefficient method for estimating evaporation from soil and application extensions. J Irrigat Drain Eng 131(1): 2–13

Ma L, Wu JL (2009) Climate and hydrological change and environmental response recorded in lacustrine sediment in Xinjiang. Arid Zone Res 26:786–792

Wang SM, Dou HS (1998) Lakes in China. Beijing, Science Press, pp 1–13

Wu X, Zhou H (2011) Influence factor to water transport index in canal system in Xinjiang Irrigation area. Yellow River 09, 93–97 (in Chinese with English abstract)

Xu X, Zhou H, Wang, Z (2010) Study on effective rainfall use efficiency in arid irrigation district. Water Saving Irrigation 12, 44–46 + 50 (in Chinese with English abstract)

Zhang ZK, Yang DY (2001) Lake water resources—environmental problems and its measures in the arid region of Northwest China. J Arid Land Resour Environ 15:7–10

Index

Y. Chen (ed.), *Water Resources Research in Northwest China*,
DOI 10.1007/978-94-017-8017-9, © Springer Science+Business Media Dordrecht 2014

Printed by Publishers' Graphics LLC